MATHEMATICS FOR SCIENCE

MATHEMATICS FOR SCIENCE

PROF. S.M. UPPAL, Ph.D.
Chairman, Department of Mathematics,
Jomo Kenyatta University of Agriculture and Technology,
Nairobi, Kenya

and

H.M. HUMPHREYS, M.Sc.
Lecturer in Mathematics,
Jomo Kenyatta University of Agriculture and Technology,
Nairobi, Kenya

PUBLISHING FOR ONE WORLD

NEW AGE INTERNATIONAL (P) LIMITED, PUBLISHERS
New Delhi • Bangalore • Chennai • Cochin • Guwahati • Hyderabad
Jalandhar • Kolkata • Lucknow • Mumbai • Ranchi
Visit us at **www.newagepublishers.com**

Published by New Age International (P) Ltd., Publishers
First Edition: 1996
Reprint: 2008

Branches:

- 36, Malikarjuna Temple Street, Opp. ICWA, Basavanagudi, **Bangalore**. ✆ (080) 26677815
- 26, Damodaran Street, T. Nagar, **Chennai**. ✆ (044) 24353401
- Hemsen Complex, Mohd. Shah Road, Paltan Bazar, Near Starline Hotel, **Guwahati**. ✆ (0361) 2543669
- No. 105, 1st Floor, Madhiray Kaveri Tower, 3-2-19, Azam Jahi Road, Nimboliadda, **Hyderabad**. ✆ (040) 24652456
- RDB Chambers (Formerly Lotus Cinema) 106A, Ist Floor, S.N. Banerjee Road, **Kolkata**. ✆ (033) 22275247
- 18, Madan Mohan Malviya Marg, **Lucknow**. ✆ (0522) 2209578
- 142C, Victor House, Ground Floor, N.M. Joshi Marg, Lower Parel, **Mumbai**. ✆ (022) 24927869
- 22, Golden House, Daryaganj, **New Delhi**. ✆ (011) 23262370, 23262368

ISBN (10): 81-224-0994-6

ISBN (13): 978-81-224-0994-9

Rs. 200.00

C-08-10-3182

Printed in India at Glorious Printers, Delhi.
Typeset at Anvi Composers, Delhi.

PUBLISHING FOR ONE WORLD
NEW AGE INTERNATIONAL (P) LIMITED, PUBLISHERS
4835/24, Ansari Road, Daryaganj, New Delhi-110002
Visit us at **www.newagepublishers.com**

Preface

During the past thirty years many of the world's older universities have grown, both in terms of their numbers of undergraduate and postgraduate students, and also in the variety of different degree courses offered. Simultaneously many new universities have been born, and have established degree courses in traditional subjects and developed courses in new subjects.

At undergraduate level there is one basic requirement for a student taking a course in Science. Engineering, Medicine or Agriculture, and that is a good understanding of Mathematics. In many of these courses the study of Mathematics at University is compulsory, as is the case in many undergraduate courses in Business Studies, Economics and even the Social Sciences.

This text book "Mathematics for Science" has been primarily prepared with the needs of the undergraduate student who is studying Mathematics in the first year of a university course in Science, Engineering, Agriculture, Medicine or Biological Science, and it is hoped that most of the topics which are required by such a student in the first year of study will be adequately covered. At the same time it is hoped that the material will provide a good foundation for a student who plans to do more specialised work in university Mathematics, especially in the field of Physical Sciences, in the subjects Physics, Chemistry and Mathematics itself. Some or all of the topics may be useful to students who are preparing to join an undergraduate course in university or who are taking various college diploma courses in Agriculture and Engineering. The book is also geared to the needs of the students of Polytechnics and teacher-training colleges where some Mathematics courses form the backbone of the training.

The chapters aim to cover most of the topics which a student will need at an introductory level in Statistics, Probability, Algebra, Trigonometry, Coordinate Geometry and Calculus at university or tertiary level.

While the reader may benefit from studying the chapters in any order it is preferable to study the chapters in the order given. In particular the chapters in Algebra, Trigonometry and Coordinate Geometry should be studied before studying the chapter on Calculus.

Numerous worked examples have been provided with sufficient explanation both to satisfy the non-specialist and to stimulate the curiosity of the specialist.

At the end of each chapter an exercise has been provided in order to give the student adequate practice in each topic covered. Some of the problems in the exercises have been made quite hard in order to challenge the reader, and these have been indicated with an asterisk.

In any case, whatever the level of Mathematics the reader intends to pursue, it is hoped that the reader will gain a good understanding of the topics relevant to the Mathematics required for his/her course of study.

PROF. S.M. UPPAL
H.M. HUMPHREYS

Acknowledgements

It is a great pleasure to acknowledge debts to those who have assisted at different stages in the preparation of this book.

Actually this book owes its origin to a seminar on text book writing given by Mr. A. Machwe of New Age International, New Delhi, India in Mombasa, Kenya in July, 1992 and which one of the authors attended and he started scribbling right there and then! It took us some time to complete the manuscript but Mr. Machwe continued giving expert advice and went on encouraging us. We are extremely grateful to him.

We will like to offer our sincere thanks to Prof. R.W. Michieka, Ph.D., EBS. Vice Chancellor, Jomo Kenyatta University of Agriculture and Technology who constantly encourages his staff to write and publish. He has been a constant source of inspiration to us.

We also desire to express our appreciation to the courtesies extended by New Age International (P) Ltd, Publishers, New Delhi, India for accepting to publish this book and market it at an international level.

Last but not least, we would like to thank Mrs Krishna Uppal for her moral support, forbearance and endless patience over household chores neglected by one of the authors while we have been occupied with this project. Our cups of tea were never empty even when we were working late in the night and deep in discussion!

PROF. S.M. UPPAL
H.M. HUMPHREYS

Contents

1

Statistics

INTRODUCTION

Statistics is the study of collection, arrangement, presentation, summary and analyses of data leading to certain conclusions about the data.

1.1 FREQUENCY DISTRIBUTIONS

1.1.1 Raw Data

In most cases data is collected initially without being organised numerically. Such data is known as raw data. An example of raw data could be the set of heights of 100 students obtained from an alphabetic listing in the university records of a particular university.

1.1.2 Arrays (Statistical)

When raw numerical data has been arranged, either in ascending or descending order of magnitude, the arrangement of the data is said to be a statistical array. For example, the raw data might comprise the marks of twelve students in a test taken from a lecturer's register and might be recorded as :

7, 10, 5, 3, 2, 9, 8, 9, 8, 7, 4, 5

The raw data could then be rearranged in ascending order to form the array:

2, 3, 4, 5, 5, 7, 7, 8, 8, 9, 9, 10.

Another array could be made in descending order:

10, 9, 9, 8, 8, 7, 7, 5, 5, 4, 3, 2.

From the array it can be seen that the highest mark is 10 and the lowest is 2

The difference between the highest and the lowest marks is 10 – 2 = 8, i.e. the RANGE is 8.

For any collection of data the difference between the largest and smallest numbers is known as the RANGE.

1.1.3 Frequency Distributions (Large Amount of Data)

When summarising a large amount of data (say 50 or 100 or even 1000 numbers) it is often helpful to arrange the data in classes (or categories) and then determine the number of individuals in each class, this number being known as the class freqeuncy. From these it is possible to arrange the data by classes together with the corresponding class frequencies in a table known as a *frequency distribution* or *frequency table.*

In a Biological Science practical in a University the students collected and recorded data on the lengths of 60 leaves of a certain species of plant. The raw data were as given below, with length given in centimetres correct to the nearest centimetre.

35	42	49	32	46	37
37	39	26	48	38	39
31	30	48	42	44	40
38	36	35	26	28	33
47	46	41	41	25	37
26	49	33	32	36	37
35	34	44	37	39	38
38	38	33	39	37	31
36	33	44	35	35	30
37	37	44	32	40	44

The students then arranged the data in classes 25—29, 30—34, 35—39, 40—44, 45—49 using a tally. For each measurement in the class a 'match stick' is placed vertically, adjacent to the class, with each fifth 'match stick' crossing the other four. In row 1 the first measurement is 35, so the first 'match stick' is placed by class 35—39. The next measurement is 42, so the second 'match stick' is placed by class 40—44. After recording the measurements in row 1 the 'match sticks' on the tally would appear as:

25—29	
30—34	I
35—39	II
40—44	I
45—49	II

Continuing to record for each measurement in row 2 the 'match sticks' on the tally would appear as

25—29	I
30—34	I
35—39	𝍸 I
40—44	I
45—49	III

Continuing this way the final tally would be :

25—29	𝍸
30—34	𝍸 𝍸 II
35—39	𝍸 𝍸 𝍸 𝍸 𝍸
40—44	𝍸 𝍸 I
45—49	𝍸 II

The table shown below gives the frequency distribution (based on the count of 'match sticks' in the tally) of the lengths (to the nearest centimetre) of the leaves.

Length (cm)	*Number of leaves (Frequency)*
25—29	5
30—34	12
35—39	25
40—44	11
45—49	7

We shall refer to this table quite often in order to calculate or estimate certain summary data (or statistics) later in this section.

The first class (category) comprises lengths from 25 to 29 cm indicated by the symbol 25—29. Five leaves had lengths belonging to this class so the corresponding class frequency is 5.

Data organised and summarised as in the above frequency table are normally called GROUPED DATA. Much of the original detail of the data may be lost as a result of the grouping. However the grouped data have the advantage of showing a clear overall picture and of making evident certain vital relationships.

1.1.4 Class Intervals and Class Limits

A symbol indicating a class such as 25—29 in the above table is called a *class interval*. The end values 25, 29, are called the *class limits* of that particular class. The smaller value (here 25) is known as the *lower class limit* and the larger value (here 29) is known as the *upper class limit*. The two terms, "class" and "class interval" may normally be used interchangeably. In some cases it may be theoretically possible for a class interval to have no upper class limit or no lower class limit indicated. Such a class interval is known as an open class. For instance in age groups of members of a community the class '70 years and over' is an open class interval.

1.1.5 Class Boundaries

In the data given in 1.1.3 the lengths were given to the nearest centimetre. However the lengths recorded in the class interval could theoretically include all lengths from 24.50000000..... to 29.50000000..... cm. For example 25.6, 25.53, 26.27, 24.5000001, all belong to the class, but 29.8, 24.499999, 29.500001 do NOT belong to the class.

The values 24.5 29.5 are called CLASS BOUNDARIES or TRUE CLASS LIMITS. In this case the smaller value (here 24.5) is known as the *lower class boundary* and the larger value (here 29.5) is known as the *upper class boundary*. In practice we may determine the class boundaries by averaging the upper limit of a class interval and the lower limit of the following class interval. In the data given in 1.1.3 the upper class boundary for the class 30—34 can be obtained by averaging the upper limit of the class interval 30—34, which is 34, and the lower limit of the following class interval 35-39, which is 35, to give $\frac{1}{2}$ (34 + 35) = 34.5.

Sometimes it is possible to use class boundaries to symbolise classes, e.g. in the table in 1.1.3 we could have given the data in classes such as 24.5—29.5, 29.5—34.5 etc. In all such cases the person presenting the data should make it clear whether an observation first recorded as 29.5 belongs to the class interval 24.5—29.5 or 29. 5—34.5. The person presenting the data should explain clearly, for instance, that the classes are arranged as say $24.5 \le x < 29.5$, $29.5 \le x < 34.5$ etc.

So 29.499999 goes in the first class

and 29.500000 goes in the second class.

1.1.6 Size or Width of Class Interval

The size or the width of a class interval is the difference between the upper and lower class BOUNDARIES (not necessarily the upper and lower class limits). If all class intervals have equal widths the width is often denoted by the letter 'C'. For the table given in 1.1.3 the width of the class interval C = 29.5—24.5 or 34.5—29.5 (etc.) i.e. C = 5.

A student who claims C = 29—25 = 4 is using the class limits, which in this case are not the same values as the class boundaries, so the answer 'C = 4' is WRONG in this case.

1.1.7 Class Mark/Mid Mark

The class mark or mid mark of any class is the mid point of the class interval. It may be obtained by averaging either the lower and upper class limits or the lower and upper class boundaries. As a result, when we determine measures such as the mean or the standard deviation of grouped data we have to assume that all observations in a given interval coincide with the mid mark e.g. in the table given in 1.1.3 all lengths in the class 25—29 cm are considered to be 27 cm.

1.1.8 General Rules for Forming Frequency Distributions

(1) Determine the largest and smallest values given in the raw data. This enables us to find the range.

(2) Divide the range into a suitable number of class intervals, preferably all of the same size. If this is not practicable then use class intervals of different sizes or, if necessary, open intervals. The best total number of class intervals is normally between 5 and 20 depending on the data which have been given. It is advisable to try to choose class intervals so that the mid marks coincide with the given data, thus reducing the grouping error in subsequent calculations of measures like the mean and standard deviation. As far as possible class boundaries should be chosen so that they do not coincide with observed data.
For most purposes it is advisable to set up between 5 and 10 intervals unless so much detail has to be included as to make a greater number of intervals desirable.

(3) Find the number of observations which fall in each class interval i.e. find the class frequencies. For large amounts of data it is advisable to use a tally as was illustrated in 1.1.3.

1.1.9 Histogram and Frequency Polygon

A histogram is a graph which comprises rectangles with—

(i) bases on the horizontal axis (x-axis) with centres at the mid mark of each class interval and with their base widths proportional to the sizes of the class intervals.

(ii) areas proportional to class frequencies.

If the class intervals are all equal then the heights of the rectangles are proportional to the class frequencies. However in cases where not all the class intervals have equal width the heights of the rectangles MUST be adjusted. This is easily done by naming the vertical axis (y-axis) 'frequency density' and then using the relationship that area (or frequency)

$$= \text{class width} \times \text{frequency density}.$$

A frequency polygon is a line graph of frequency density plotted against class mark (mid mark) and can be obtained by joining the mid-points of the tops of the rectangles in the histogram. For the sake of completion the mid-point of the top of the first rectangle is normally joined to the mid-point of an imagined class interval preceding the first class interval and having the same width as the first class interval, likewise a similar join is made from the mid-point of the last rectangle. For the frequency distribution of 1.1.3 the frequency density is frequency/C = frequency/5, set as

Class boundaries	*Frequency*	*Frequency density*
24.5—29.5	5	1
29.5—34.5	12	2.4
34.5—39.5	25	5
39.5—44.5	11	2.2
44.5—49.5	7	1.4

So the histogram and frequency polygon take the form shown below. Fig. 1.1.9(a)

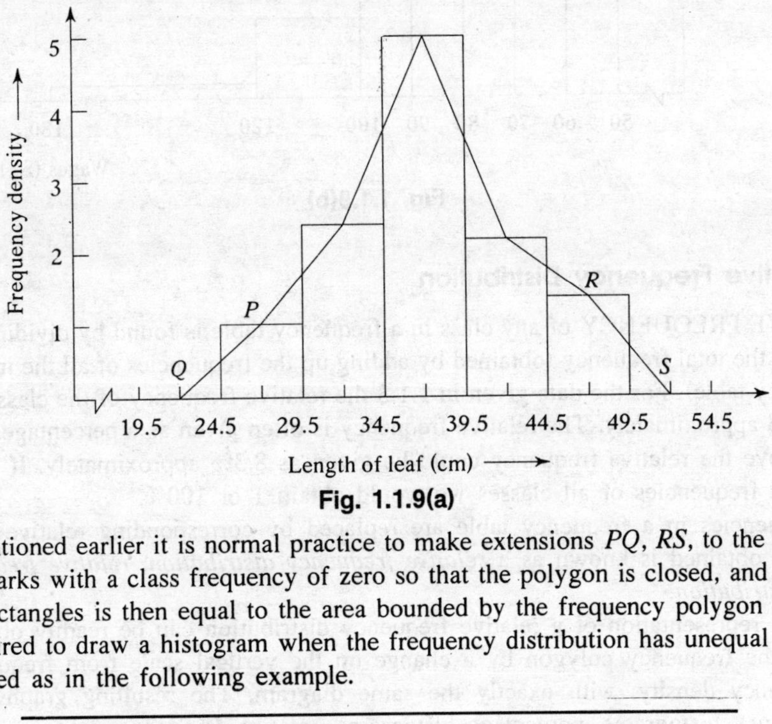

Fig. 1.1.9(a)

As was mentioned earlier it is normal practice to make extensions *PQ*, *RS*, to the next lower and higher class marks with a class frequency of zero so that the polygon is closed, and the sum of the areas of the rectangles is then equal to the area bounded by the frequency polygon and the *x*-axis. If we are required to draw a histogram when the frequency distribution has unequal class intervals, we may proceed as in the following example.

Wages £x per month	*Frequency*	*C*	*Frequency density*
$50 \le x < 60$	8	10	0.8
$60 \le x < 70$	10	10	1.0
$70 \le x < 80$	16	10	1.6
$80 \le x < 90$	15	10	1.5
$90 \le x < 100$	10	10	1.0
$100 \le x < 120$	8	20	0.4
$120 \le x$	3	60	0.05

(assume upper boundary 180)

See Fig. 1.1.9 (b) below.

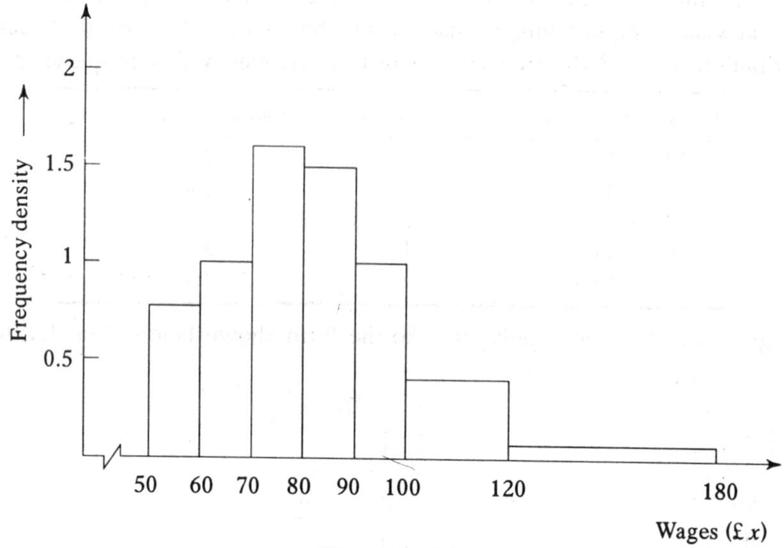

Fig. 1.1.9(b)

1.1.10 Relative Frequency Distribution

The RELATIVE FREQUENCY of any class in a frequency table is found by dividing the frequency of the class by the total frequency (obtained by adding up the frequencies of all the individual classes in the frequency table). For the data given in 1.1.3 the relative frequency of the class 25—29 is 5/60 = 1/12 = 0.083 approximately. The relative frequency is often given as a percentage. In the example mentioned above the relative frequency could be given as 8.3% approximately. If we were to add up the relative frequencies of all classes we would obtain 1 or 100%.

When frequencies in a frequency table are replaced by corresponding relative frequencies the resulting table obtained is known as a *relative frequency distribution, relative frequency table*, or *percentage distribution.*

A graphical representation of a relative frequency distribution can be readily obtained from the histogram or the frequency polygon by a change on the vertical scale from frequency density to relative frequency density, with exactly the same diagram. The resulting graphs are known as relative frequency histograms, percentage histograms, relative frequency polygons etc.

1.1.11 Cumulative Frequency and Ogives

In a frequency table the total frequency of all values less than the upper class bound ary of a given class interval is called the cumulative frequency up to and including the class interval, e.g. in the data of 1.1.3 the cumulative frequency up to and including the class interval 35—39 is 5 + 12 + 25 = 42, indicating that 42 leaves have lengths less than 39.5 cm.

A table may be drawn up to represent cumulative frequencies based on a given frequency table. Such a table is known as a *cumulative frequency distribution, cumulative frequency table* or *cumulative distribution*. The table is shown below for the data of 1.1.3. It is strongly advised that you include a column for the upper class boundary so that you will then be able to estimate values of median, quartiles and percentiles of a distribution by a correct method (whether the method is that of interpolation or that using the cumulative frequency curve).

Length (cm)	*Upper class boundary (ucb)*	*Frequency*	*Cumulative frequency (cf)*
25—29	29.5	5	5
30—34	34.5	12	17
35—39	39.5	25	42
40—44	44.5	11	53
45—49	49.5	7	60

A graph which plots the cumulative frequency less than any upper class boundary is called a *cumulative frequency curve*, often drawn as a curve through the plotted points, but sometimes drawn as a polygon with straight line segments joining the points. The curve is also known as an *ogive*. Below is the ogive for the data of 1.1.3. for which we calculated the cumulative frequencies above. See Fig. 1.1.11(a).

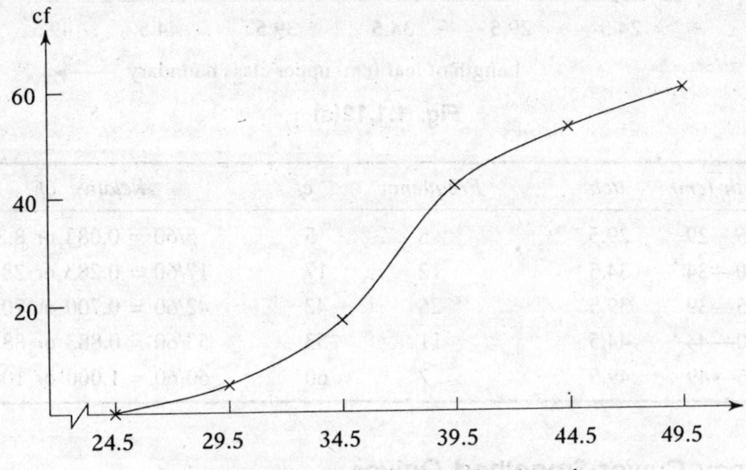

Fig. 1.1.11(a)

1.1.12 Relative Cumulative Frequency Distributions Percentage Ogives

The relative cumulative frequency or percentage cumulative frequency can be found by dividing the cumulative frequency (up to and including a given class interval) by the total frequency. In the distribution of 1.1.11 for which we had calculated the cumulative frequencies the relative cumulative frequency of lengths less than 39.5 cm is 42/60 = 0.70 or 70%, indicating that 70% of the leaves have lengths less than 39.5 cm.

When relative cumulative frequencies are used in place of cumulative frequencies the resulting tables are called *relative cumulative frequency distributions* or *percentage cumulative distributions*, and the corresponding graphs for these tables are known as *relative cumulative frequency polygons* (if adjacent points are joined by straight line segments) or *percentage ogives* (if adjacent points are joined to form a curve).

The relative cumulative frequency distribution and percentage ogive for the data of 1.1.3 are given below in Fig. 1.1.12(a).

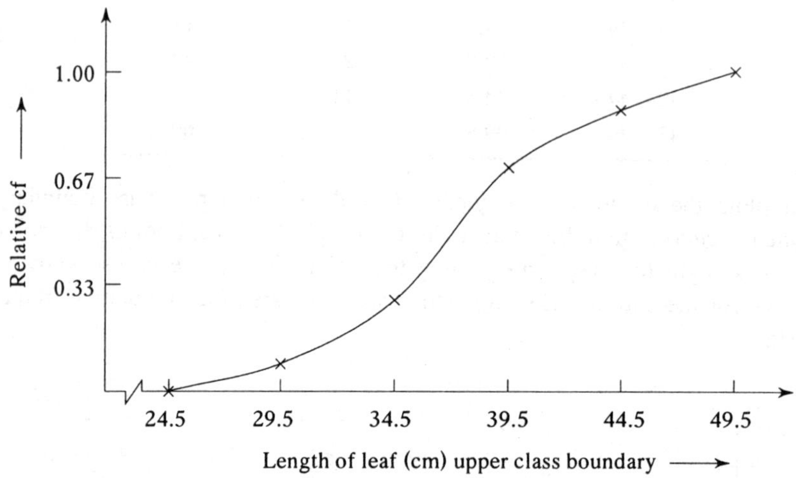

Fig. 1.1.12(a)

Length (cm)	*ucb*	*Frequency*	*cf*	*Relative cf*
25—29	29.5	5	5	5/60 = 0.083 or 8.3%
30—34	34.5	12	17	17/60 = 0.283 or 28.3%
35—39	39.5	25	42	42/60 = 0.700 or 70.0%
40—44	44.5	11	53	53/60 = 0.883 or 88.3%
45—49	49.5	7	60	60/60 = 1.000 or 100.0%

1.1.13 Frequency Curves-Smoothed Ogives

We normally think of collected data as belonging to a sample drawn from a large population. We define population and sample below.

Population

A collection of data under discussion—normally assigned to quantitative data such as lengths, volumes, weights, number of goals, number of prizes etc. The population may be finite e.g. the population of lions in a particular game park, or infinite (perhaps in certain cases tending to be infinite) e.g. the number of all babies born in this world since creation.

Sample

A sample is any subset of a population. Many observations can normally be made in a population. It is therefore possible in theory to choose class intervals in continuous data (such as length or volumes) to be very small and still have sizeable numbers falling within every class. In such cases it would be possible to imagine a frequency polygon or relative frequency polygon for a large population to consist of a large number of broken line segments which could closely approximate to curves known as *frequency curves* or *relative frequency curves* respectively. Such theoretical curves could be approximated by smoothing frequency polygons or relative frequency polygons, the approximation becoming better as the sample size becomes larger. For this reason a frequency curve is sometimes called a smoothed frequency polygon.

In the same way smoothed ogives—normally drawn as curves rather than polygons since it is easy to do this—can be drawn, so that a cumulative frequency polygon approximates to a cumulative frequency curve. Even for small samples it is customary practice to drawn an ogive rather than a cumulative frequency polygon when we wish to give a graphical representation of cumulative frequency.

1.1.14 Types of Frequency Curve

The frequency polygons for most frequency distributions can be found to approximate to one of the curves shown below in Figs. 1.1.14 (a) to (h).

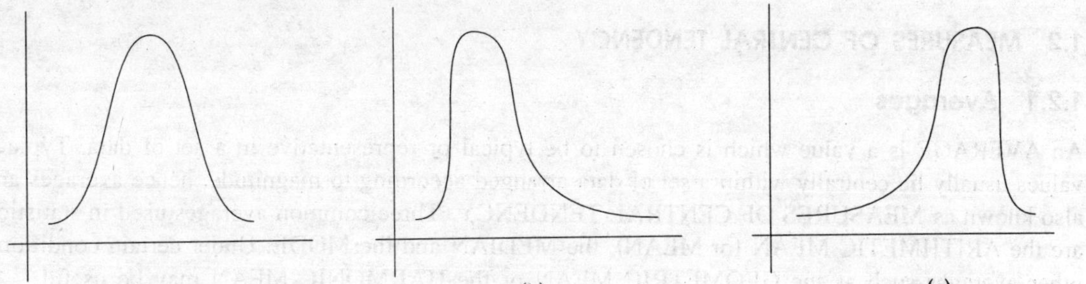

(a) Symmetrical or Bell shaped (e.g. Normal Curve)

(b) Skewed to the Right (Positive Skewness)

(c) Skewed to the left (Negative Skewness)

Fig. 1.1.14 (a) (b) and (c)

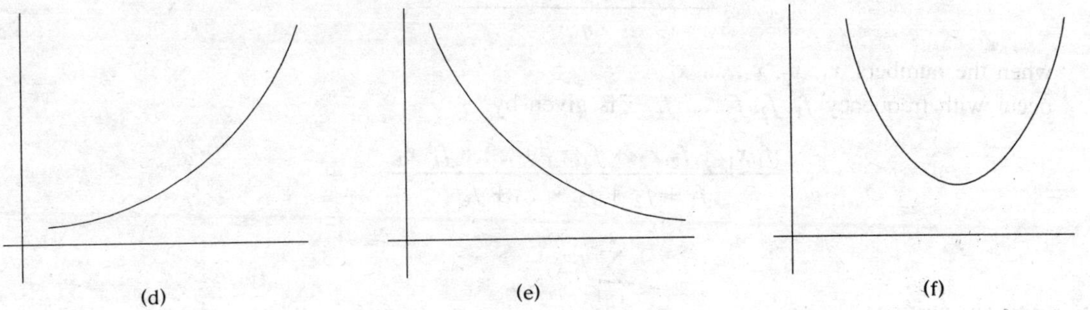

(d) J-shaped Figure

(e) Reverse J-shaped

(f) U-shaped

Fig. 1.1.14(d) (e) and (f)

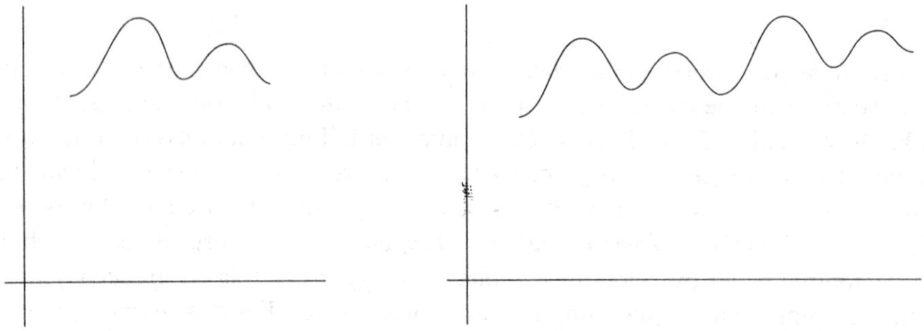

(g)
Bimodal

(h)
Multimodal (in case of three or more local modes—for definition of mode see 1. 2. 4)

Fig. 1.1.14(g) and (h)

When a curve (or distribution) is skewed the tail of the curve (or distribution) to one side of the mode is longer than that to the other. When the longer tail occurs to the right the curve is said to be *skewed to the right* or *positively skewed.* When the larger tail occurs to the left the curve is said to be *skewed to the left* or *negatively skewed.*

1.2 MEASURES OF CENTRAL TENDENCY

1.2.1 Averages

An AVERAGE is a value which is chosen to be typical or representative in a set of data. Typical values usually lie centrally within a set of data arranged according to magnitude, hence averages are also known as MEASURES OF CENTRAL TENDENCY. Three common averages used in statistics are the ARITHMETIC MEAN (or MEAN), the MEDIAN and the MODE. Under certain conditions other averages such as the GEOMETRIC MEAN or the HARMONIC MEAN may be useful.

1.2.2 Arithmetic Mean

For a set of n numbers $x_1, x_2, x_3, \ldots. x_n$ the mean (denoted often by $\bar{x}$ or m) is defined as

$$\frac{x_1 + x_2 + x_3 + \ldots. + x_n}{n}$$

when the numbers $x_1, x_2, x_3, \ldots. x_k$
occur with frequency $f_1, f_2, f_3, \ldots. f_k$ $\bar{x}$ is given by

$$\frac{f_1 x_1 + f_2 x_2 + f_3 x_3 + \ldots. + f_k x_k}{f_1 + f_2 + f_3 + \ldots. + f_k}$$

Commonly written

$$\frac{\sum_{i=1}^{k} f_i x_i}{\sum_{i=1}^{k} f_i}$$

or
$$\frac{\sum fx}{\sum f} \text{ or } \frac{\sum fx}{N}$$

where $N = \Sigma f$ or $\sum_{i=1}^{k} f_i$ is the total frequency.

In many cases it is possible to leave out the subscript (in this case i) provided this practice does not lead to ambiguity.

In order to ease the calculation process we often use the method of CODING, whereby we make a guess at the mean, say $x = a$ (a is the *assumed mean*)

In a frequency distribution each value $x_i = a + d_i$ say, and d_i is called the deviation of x_i from a.

$$\bar{x} = \frac{\sum_{i=1}^{k} f_i x_i}{N} = \frac{\sum_{i=1}^{k} f_i (a + d_i)}{N}$$

$$= \frac{\sum_{i=1}^{k} (f_i a + f_i d_i)}{N} = \frac{a\sum_{i=1}^{k} f_i + \sum_{i=1}^{k} f_i d_i}{N}$$

$$= \frac{a\sum_{i=1}^{k} f_i}{N} + \frac{\sum_{i=1}^{k} f_i d_i}{N}$$

$$= \frac{aN}{N} + \frac{\sum_{i=1}^{k} f_i d_i}{N} \quad \left(as\ N = \sum_{i=1}^{k} f_i\right)$$

$$= a + \frac{\sum_{i=1}^{k} f_i d_i}{N}$$

or $a + \dfrac{\sum fd}{N}$ if the subscript is omitted.

In other words $\bar{x}$ is equal to the assumed mean (a) plus the mean deviation from the assumed mean.

In grouped data all values which fall in a given class interval are treated as if they were coincident with the class mark or mid mark of that interval. The coding formula for $\bar{x}$ may now be further adjusted in cases where all class intervals have a common width, say C. In such cases we should take a (the assumed mean) to be one of the mid marks in the frequency table, and each d_i can be written as $d_i = Cu_i$ (all mid marks will lie at a multiple of the class interval above or below the assumed mean).

From the previous formula

$$\bar{x} = a + \frac{\sum_{i=1}^{k} f_i\, d_i}{N}$$

$$= a + \frac{\sum_{i=1}^{k} f_i\, Cu_i}{N}$$

$$= a + \frac{C \sum_{i=1}^{k} f_i\, u_i}{N}$$

$$= a + \frac{C \sum fu}{N}$$

or

$$a + C\,\bar{u}$$

The formula $\bar{x} = a + C\bar{u}$ is an application of the coding method for calculating the mean, the standard deviation, and several other measures which we may require to find for a particular frequency distribution. It is a very efficient method but it should only be applied when class interval sizes are equal. In coding the values of the variable x are transformed into the variable u, via the formula $x = a + Cu$, as will be illustrated in calculating the mean length for the data given in 1.1.3.

Mean determined without coding

Length (cm)	*Frequency f*	*Mid Mark (x)*	*fx*
25—29	5	27	135
30—34	12	32	384
35—39	25	37	925
40—44	11	42	462
45—49	7	47	329
	$\Sigma f = 60$		$\Sigma fx = 2235$

$\bar{x} = 2235/60$

$= 37.25$ cm

N.B.: (1) As this is a frequency distribution do not simply add the entries in the column labelled x. Add entries in column labelled fx.

(2) This method can become cumbersome with large values for f or x.

Mean determined with coding (assumed mean $a = 37$) and width of class interval $c = 5$

Length (cm)	*Frequency f*	*Mid Mark x*	*y = x — 37*	*u = y/5*	*fu*
25—29	5	27	– 10	– 2	– 10
30—34	12	32	– 5	– 1	– 12/ – 22
35—39	25	37	0	0	0
40—44	11	42	5	1	11
45—49	7	47	10	2	14/25
	$\Sigma f = 60$				$\Sigma fu = 3$

$$\bar{x} = 37 + \frac{5 \times 3}{60}$$

$$= 37.25 \text{ cm}$$

N.B: (1) It is helpful to add the negative and positive terms in Σfu separately—in this case these came to –22 and 25.

(2) If all class intervals have an equal width you will discover that when you subtract the assumed mean from each class mid mark the answer is a multiple of the class width. If this does not happen check your working!

1.2.3 Median

The MEDIAN of a set of numbers when they are *arranged in order of magnitude* in the array is the middle value or the arithmetic mean of the two middle values.

Example 1.2.3(1)

What is the median of the numbers

7, 12, 11, 14, 16, 16, 13, 12, 17?

First arrange in ascending order as

7, 11, 12, 12, 13, 14, 16, 16, 17

↑

Median

There are nine numbers, and the median is the fifth number, which means that the median is 13.

In any set of $(2n + 1)$ numbers arranged in order of magnitude the median is the $(n + 1)$th number. This is possible provided there is an odd total of numbers to be considered.

Example 1.2.3 (2)

What is the median of the numbers

18, 11, 12, 19, 21, 17, 12, 19 ?

First arrange in ascending order as

11, 12, 12, 17, 18, 19, 19, 21

↑

Median = 1/2 (17 + 18) = 17.5

In this case there are 8 numbers and so there is no 'middle' number.

We find the median by taking the average of the fourth and fifth numbers, i.e. 1/2(17 + 18) = 17.5.

In any set of $2n$ numbers, the number having been arranged in order of magnitude, the median is the average of the nth and $(n + 1)$th numbers.

For grouped data we normally treat the data as continuous even if the original data were discrete when we estimate the median. This will also apply later when we estimate quartiles. Before we proceed it is necessary to define what is a discrete variable and what is a continuous variable.

Discrete Variable

A variable which can only take one of a finite set of values (or in some cases a countably infinite set) is called DISCRETE

e.g. the number of goals scored by a hockey team must be 0 or 1 or 2 or 3 or .....

In this case we might find from our records that the most goals scored by the team in any match was 10.

The number of goals would then be limited to the set (0, 1, 2, 3, 4, 5, 6, 7, 8, 9, 10).

However we might argue that if the opposition was very weak the team might be able to score as many goals as it liked, perhaps far more than 10.

We could then represent the possible number of goals by the set (0, 1, 2, 3, 4, ...) where the dots indicate 'and so on' for any whole number of goals. In this case the values come from a countably infinite set in the sense that we could go on counting as far as we like to say 100 goals, 1000 goals, 10^6 goals, ... infinity.

We cannot record that a hockey team scored 3.58 goals. In any match we know that the number of goals scored by the team must be a whole number.

Continuous Variables

A variable which can assume any value between two given values is called CONTINUOUS

e.g. the lengths of leaves measured in the data collected in 1.1.3. We saw earlier that it could be possible for a leaf whose length lay in the class 24.5—29.5 cm to have a length which was any value between 24.5 and 29.5 cm.

In general measurements normally give rise to continuous data whereas enumeration (or counting) will normally give rise to discrete data.

We might be required to draw a graph, say a histogram or a cumulative frequency curve for the number of goals 'x'. We would know initially that x can only take the values 0, 1, 2, 3, ... say as given in the frequency table below.

Number of goals (x)	*Frequency (f)*
0	2
1	6
2	1
3	1
	Total Frequency 10

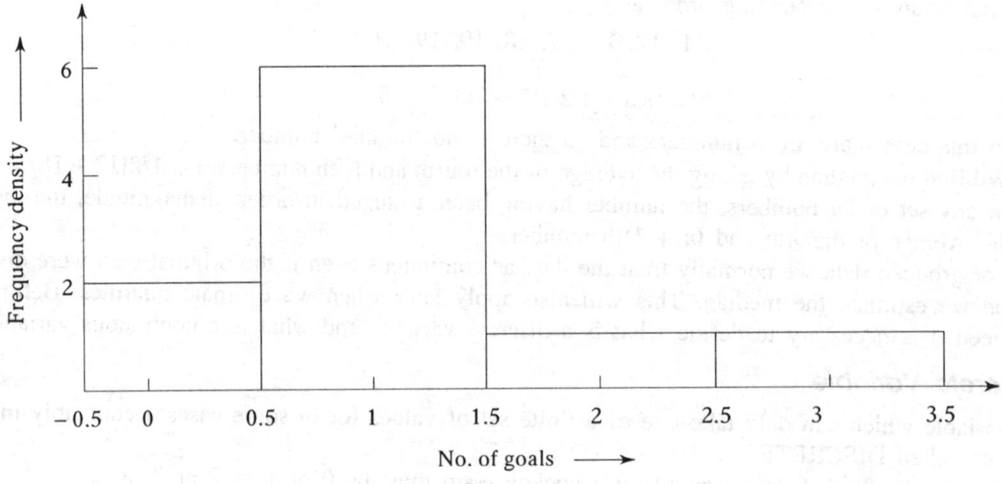

Fig. 1.2.3(a)

Looking at the histogram (Fig. 1.2.3(a)) it would seem to be logical to take each score (0, 1, 2, 3,) at the mid-point of the base of each rectangle, with a class width of 1 for each class. We would then assume that for $x = 1$ these would be a lower class boundary of 0.5 and an upper class boundary of 1.5, even though we have never seen a hockey team scoring 1.5 goals!

This idea can also be used in drawing the cumulative frequency curve. The table and corresponding cumulative frequency curve are shown below in Fig. 1.2.3(b).

Upper class boundary	*Cumulative Frequency*
0.5	2
1.5	8
2.5	9
3.5	10

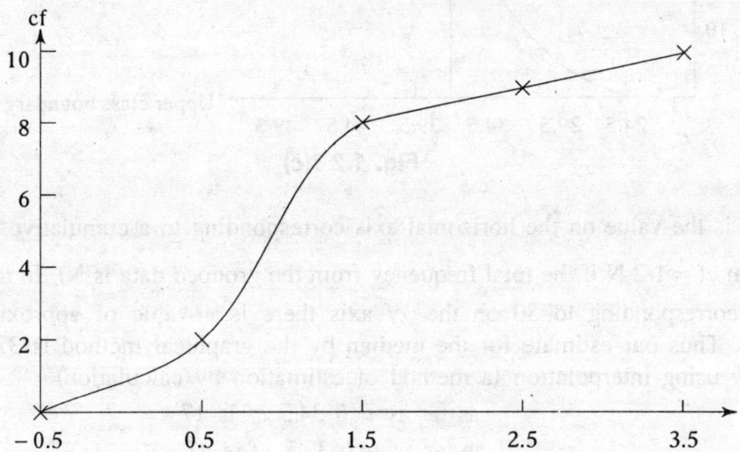

Fig. 1.2.3(b)

Geometrically the median is the value of x corresponding to a vertical line which divides a histogram with two parts having equal areas. In the case of the frequency table for the number of goals scored by the hockey team given above it is easy to see that the vertical line $x = 1$ divides the histogram into two parts each of area 5.

However it is normally easier in cases of a frequency table (whether discrete or continuous data) to find an estimate for the median using either a graphical method (reading from the cumulative frequency curve) or by interpolation (a calculation method). Re-call for the data for 1.1.3.

Length (cm)	*ucb*	*Frequency*	*cf*
25—29	29.5	5	5
30—34	34.5	12	17
35—39	39.5	25	42
40—44	44.5	11	53
45—49	49.5	7	60

Cumulative Frequency Curve for Lengths of Leaves see Fig. 1.2.3(c).

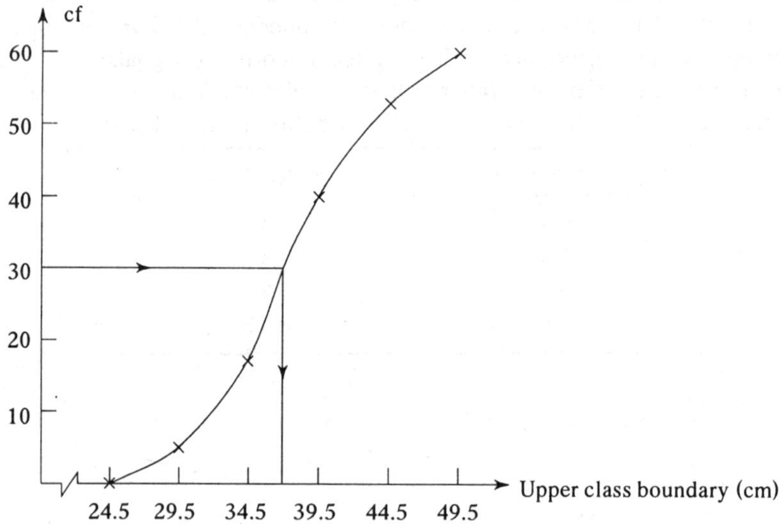

Fig. 1.2.3(c)

The median is the value on the horizontal axis corresponding to a cumulative frequency of $\frac{1}{2}$ N (always when at cf = 1/2 N if the total frequency from the grouped data is N). In this case $\frac{1}{2}$ N = 30. On the curve corresponding to 30 on the *cf* axis there is a value of approximately 37 on the horizontal axis. Thus our estimate for the median by the graphical method is 37 cm.

Alternatively using interpolation (a method of estimation by calculation)

as far as *ucb* 34.5 *cf* is 17

as far as *ucb* 39.5 *cf* is 42

so the median, corresponding to *cf* 30, lies in the class 34.5—39.5. The class 34.5—39.5 has a frequency of 25.

We want to identify the value of x in this class which corresponds to *cf* = 30.

At the boundary 34.5 *cf* is 17. To reach *cf* 30 we require 30 – 17 = 13. If we assure that the 25 leaves in the class 34.5—39.5 cm are evenly spaced along the interval it seems reasonable to locate the median at the point which is 13/25 along the interval from its lcb 34.5

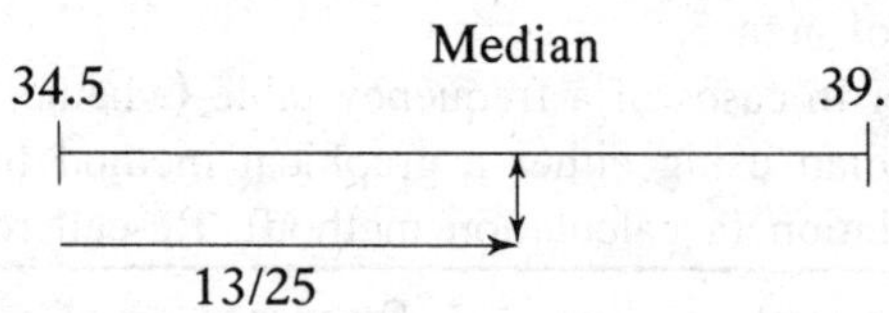

The interval has also width of (39.5—34.5) cm (or 5 cm)

Our estimate of the median is therefore

$$34.5 + 13/25\ (39.5 - 34.5)$$

$$= (34.5 + 13/25 \times 5)\ \text{cm}$$

$$= (34.5 + 2.6)\ \text{cm} = 37.1\ \text{cm}.$$

If we now estimate the median for the frequency table of the number of goals scored by the hockey team mentioned earlier in this section the relevant data are

Number of goals	*u.c.b.*	*Frequency*	*Cf*
0	0.5	2	2
1	1.5	6	8
2	2.5	1	9
3	3.5	1	10

The median lies at the value corresponding to cf 5

as far as u.c.b 0.5 cf is 2

as far as u.c.b 1.5 cf is 8

The class 0.5—1.5 has frequency = 6
We want to identify the value of x in this class which corresponds to cf = 5
At the boundary 0.5 cf is 2
To reach cf 5 we require 5 – 2 = 3.
Assuming that the 6 goals in the class 0.5—1.5 are evenly spaced along the interval it seems reasonable to locate the median at the point which is 3/6 along the interval from its l.*c.b.* 0.5.

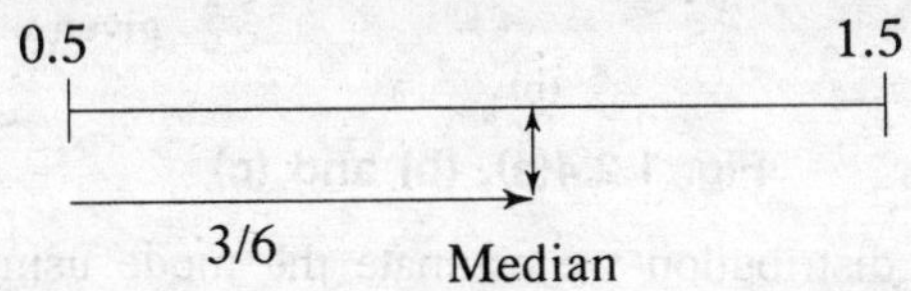

The interval has a class width of 1.5—0.5 = 1 goal
Our estimate of the median is therefore

$$0.5 + 3/6\,(1.5 - 0.5)$$

$$= 0.5 + 0.5$$

$$= 1 \text{ goal.}$$

1.2.4 Mode

The Mode of any set of numbers is that value which occurs with the greatest frequency, i.e. the most common value. The mode may not exist and even if one does exist it may not be unique.

Example 1.2.4 (1) Find the mode for the data

12, 12, 15, 17, 19, 19, 19, 20, 21, 22, 28

Clearly as 19 has a frequency of 3, the others all have a frequency of 1 or 2, then the mode = 19.

Example 1.2.4 (2) Find the mode for the data

28, 30, 27, 33, 44, 17

First arrange in order :

17, 27, 28, 30, 33, 44

Since all the values have a frequency of 1 there is no mode.

Example 1.2.4 (3) Find the mode of the data

105, 104, 102, 107, 109, 103, 104, 107, 107, 104, 105

First arrange in order:

102, 103, 104, 104, 104, 105, 105, 107, 107, 107, 109.

Here these are Two modes i.e. 104 and 107 (both with frequency 3)

For grouped data if we fit a frequency curve the mode will be the value or values of x corresponding to the maximum point [see Figs. 1.2.4(a), (b) (c].

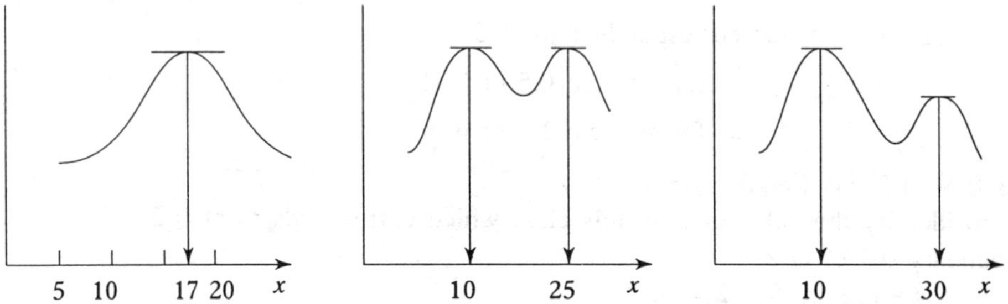

Mode: $x = 17$

(a)

Modes: $x = 10, x = 25$

(b)

Mode: $x = 10$
($x = 30$ gives a local maximum but not the maximum in the given set of values)

(c)

Fig. 1.2.4(a), (b) and (c)

In practice for a frequency distribution we estimate the mode using either the histogram or a formula (calculation method). For the data of 1.1.3 the histogram is given below in Fig. 1.2.4 (d).

Histogram for distribution of lengths of leaves.

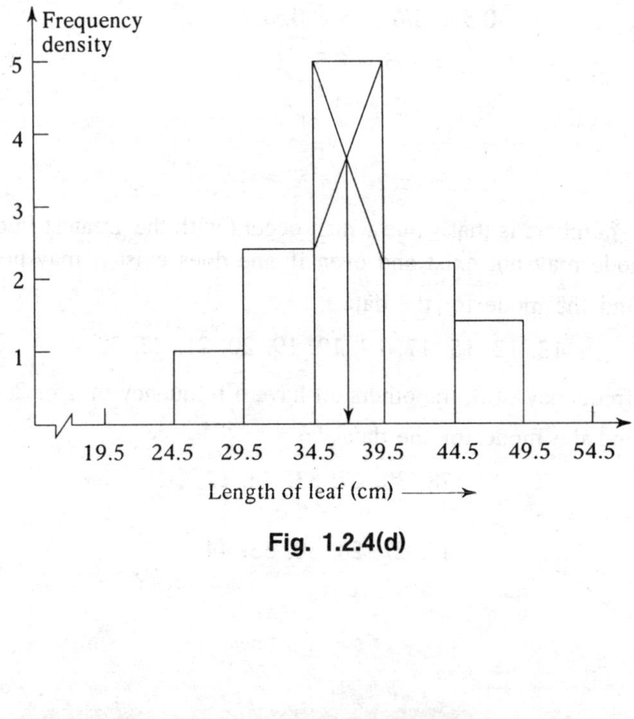

Fig. 1.2.4(d)

The modal class is 34.5—39.5 with area 5 units.

We can estimate the mode using the histogram by joining the top left-hand corner of the modal rectangle to the top left-hand corner of the rectangle which follows the modal rectangle, and then by joining the top right-hand corner of the modal rectangle to the top right-hand corner of the rectangle which precedes the modal rectangle. The mode is read from the value on the x-axis corresponding to the point where the two joins interest.

From the graph the mode is approximately 37 cm.

If we wish to estimate the mode by calculation we may proceed as follows (see the histogram)

$$\text{The mode} = 34.5 + l_1$$

where $$\frac{l_1}{l_2} = \frac{13}{14} \quad \text{(corresponding to the frequencies)} \quad (1)$$

and $$l_1 + l_2 = 5 \quad (2)$$

so l_1 can be found by substituting l_2 from (1) into (2) to give $l_1 + 14l_1/13 = 5$

or $$l_1 = \frac{13}{27} \times 5 \simeq 2.4$$

Hence the mode is 34.5 + 2.4 or 36.9 cm approximately.

We were able to use this method based an frequency because all classes in the given data had a uniform width, and we had therefore used the fact that frequencies were proportional to frequency densities. Try re-working the above calculation using $\frac{l_1}{l_2} = \frac{2.6}{2.8}$ (based on frequency densities of 2.4, 5 and 2.2 for the classes 29.5—34.5, 34.5—39.5, 39.5—44.5 respectively).

We may formulate in general, for classes with UNIFORM width C, for the mode as follows:

$$\text{Mode} = L + \left(\frac{\Delta_1}{\Delta_1 + \Delta_2}\right) C$$

where L is the $l\ c\ b$ of the modal class,

Δ_1 is the difference between the frequency of the modal class and the frequency of the class preceding the modal class, Δ_2 is the difference between the frequency of the modal class and the frequency of the class following the modal class.

This formula can be checked using similar triangles as illustrated on the diagram of the histogram. Modification of the formula is needed when the classes may not all have a uniform width. In this case we may take the mode to be given by

$$\text{Mode} = L + \left(\frac{d_1}{d_1 + d_2}\right) C_M$$

where L is the $l\ c\ b$ of the modal class, d_1 is the difference between the frequency density of the modal class and the frequency density of the class preceding the modal class, d_2 is the difference between the frequency density of the modal class and the frequency density of the class following the modal class,

C_M is the class width of the modal class.

1.2.5 Empirical Relation between Mean, Median and Mode

For a unimodal curve which is moderately skewed we have an empirical relation which is APPROXIMATELY true from observation [see diagrams Figs. 1.2.5 (a) (b)] Mean-Mode $\simeq$ 3(Mean-Median)

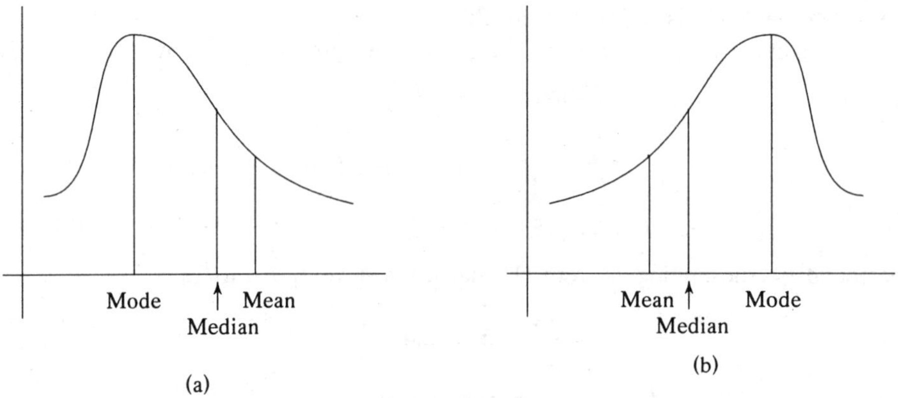

Fig. 1.2.5(a) and (b)

For the data of 1.1.3 for which we estimated the mean in 1.2.2, the median in 1.2.3, and the mode in 1.2.4, the results are:

$$\text{Mean-Mode} = 37.25 - 36.9 = 0.35 \text{ cm}$$

$$3\ (\text{Mean-Median}) = 3(37.25—37.10)$$

$$= 3 \times 0.15 = 0.45 \text{ cm}$$

This does not quite fit the relation, which is the reason why it is regarded as an approximation and not an exact relation.

The diagrams above illustrate the relative positions of the three averages for skewed curves. For symmetric curves (distributions) the three averages have the same value.

1.3 MEASURES OF DISPERSION

1.3.1 Same Average but Different Variability

Consider the percentage marks of two students who were each assessed on five tests (all marks out of 100)

Test	*Student A*	*Student B*
1	15	35
2	35	45
3	55	55
4	75	65
5	95	75

Both students have the same mean score of 55. However, it is clear from the scores that B is more consistent than A, i.e. his marks have less 'scatter' or 'dispersion' or 'variabilities'.

There are a number of measures such as the range, the semi-interquartile range, the 10-90 percentile range, the mean absolute deviation and the standard deviation, which can be used to measure the amount of variability.

1.3.2 Range

The range is the difference between the largest and smallest numbers in a set.

Example 1.3.2(1) For the data given on the two students in section 1.3.1 for A the range = 95 –15 = 80.

For B the range = 75 – 35 = 40.

1.3.3 Semi-Interquartile Range (Quartile Deviation)

The semi-interquartile range (or Quartile Deviation) is given by $\frac{1}{2}$ (Upper Quartile-Lower Quartile) often written as $\frac{1}{2}(Q_3 - Q_1)$

Q_3 (the Upper Quartile) is the score just exceeded only by the top 25% of any distribution.

Q_1 (the Lower Quartile) is the score which just exceeds the score of the bottom 25% of any distribution.

For arrays the data should normally be set in ascending (or descending) order before finding the quartiles. The following examples will illustrate the method.

Example 1.3.3(1) Find the semi-interquartile range for the data 5, 9, 12, 10, 7, 13, 14

Rearranging in ascending order we have

5	7	9	10	12	13	12
	↑		↑		↑	
	Q_1		Median		Q_3	

The median is 10.

In this case there are three values below the median score, so the lower quartile is the middle value i.e. $Q_1 = 7$

Similarly, there are three values above the median score. The upper quartile is the middle value of these i.e. $Q_3 = 13$.

In this case the semi-interquartile range is $\frac{1}{2}(Q_3 - Q_1) = \frac{1}{2}(13 - 7) = 3$

Example 1.3.3(2) Determine the semi-interquartile range for the marks of students A and B as given in Section 1.3.1

A:	15	35	55	75	95
	↑		↑		↑
	Q_1		Median		Q_3
	$= \frac{1}{2}(15 + 35)$				$= \frac{1}{2}(75 + 95)$
	= 25				= 85

In this case with five scores the median is 55. However there are two scores below the median. The lower quartile is the average of these hence $Q_1 = 1/2\ (15 + 35) = 25$. Similarly Q_3 can be found to be 85.

For A the semi-interquartile range is $\frac{1}{2}(Q_3 - Q_1) = \frac{1}{2}(85 - 25) = 30$.

A similar calculation can be made for B's scores.

B:	35		45		55		65		75
		↑			↑			↑	
		Q_1			Median			Q_3	
		$= \frac{1}{2}(35 + 45)$						$= \frac{1}{2}(65 + 75)$	
		$= 40$						$= 70$	

For B the semi-interquartile range $= \frac{1}{2}(70 - 40) = 15$.

Example 1.3.3(3) Determine the median and the semi-inter quartile range for the data 7.5, 10, 5, 3, 2, 9, 8, 7,4, 5, 8.5, 9.5

Rearranging in ascending order we get

2	3	4		5	5		7	7.5	8	8.5		9	9.5	10
			↑			↑					↑			
			Q_1			Median					Q_3			
			$= \frac{1}{2}(4 + 5)$			1/2 (7 + 7.5)					$= \frac{1}{2}(8.5 + 9)$			
			$= 4.5$			$= 7.25$					$= 8.75$			

The median is 7.25.

In this case there are 6 values below the median value of 7.25 (there are 12 values altogether). The lower quartile is the average of the third and fourth values i.e. $Q_1 = \frac{1}{2}(4 + 5) = 4.5$ Similarly $Q_3 = 1/2(8.5 + 9) = 8.75$

The semi-interquartile range is $\frac{1}{2}(8.75 - 4.5) = 2.13$ approximately.

For grouped discrete or continuous distributions the lower quartile Q_1 is the measure which has relative cumulative frequency of 0.25 (or 25%) and the upper quartile Q_3 is the measure which has relative cumulative frequency of 0.75 (or 75%).

These are sometimes called the 25th and 75th percentiles, in that exactly 25% of the measures are less than Q_1, and exactly 25% of the measures are greater than Q_3' (or exactly 75% of the measures are less than Q_3).

Similarly to the methods available for estimating the median, the quartiles (and also other percentiles such as the tenth and ninetieth percentiles) may be estimated using either the cumulative frequency curve or interpolation. See Fig. 1.3.3(a) below.

By interpolation we may use the cumulative frequency table for the data in section 1.1.3

u.c.b.	*cf*
29.5	5
34.5	17
39.5	42
44.5	53
49.5	60

We shall look at the calculation of Q_1 in some detail, but that for the other percentiles will be summarised.

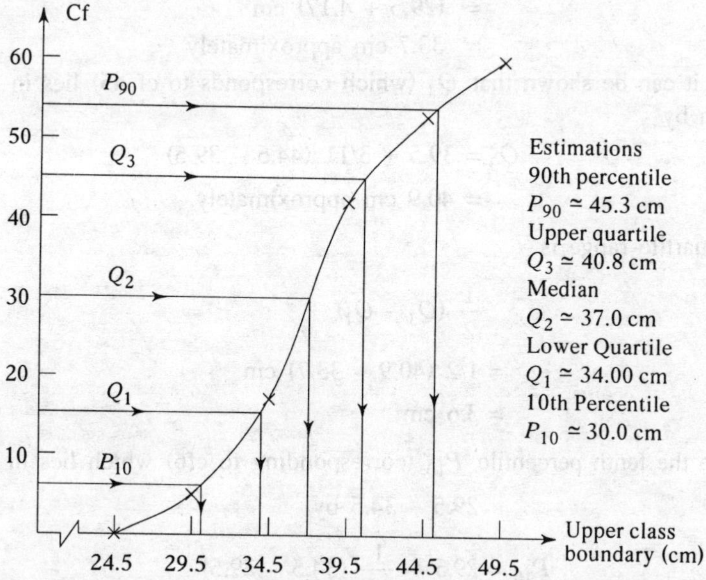

Fig. 1.3.3(a)

Q_1 has a cumulative frequency 15 (total frequency is 60 so 25% of 60 = 15).

As far as u.c.b 29.5 cf is 5

As far as u.c.b 34.5 cf is 17

so the lower quartile Q_1, corresponding to cf 15 lies in the class 29.5 – 34.5. The class 29.5 – 34.5 has a frequency of 12.

We want to identify the value (Q_1) in this class which corresponds to cf = 15.

At the lower boundary 29.5 cf is 5. To reach cf 15 we require 15 – 5 = 10. If we assume that the 12 leaves in the class 29.5—34.5 cm are evenly spaced along the interval it seems reasonable to locate Q_1 at the point which is 10/12 along the interval from its lcb. 29.5. See Fig. 1.3.3(b).

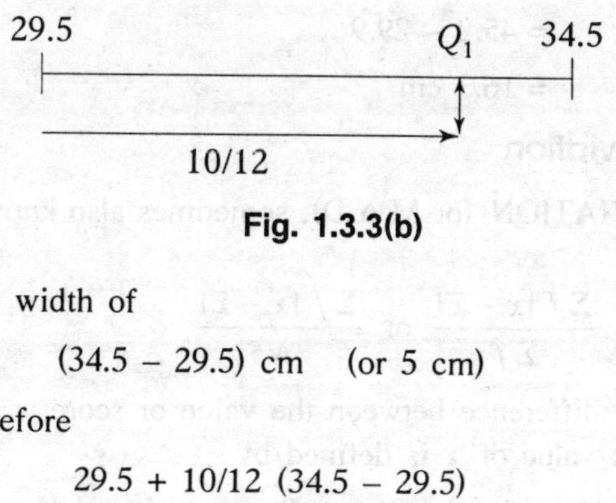

Fig. 1.3.3(b)

The interval has a class width of

$$(34.5 - 29.5) \text{ cm} \quad (\text{or } 5 \text{ cm})$$

Our estimate of Q_1 is therefore

$$29.5 + 10/12\ (34.5 - 29.5)$$
$$= (29.5 + 10/12 \times 5) \text{ cm}$$

$$= (29.5 + 4.17) \text{ cm}$$
$$= 33.7 \text{ cm approximately}$$

In a similar way it can be shown that Q_3 (which corresponds to cf 45) lies in the interval 39.5—44.5 and is given by

$$Q_3 = 39.5 + 3/11\ (44.5 - 39.5)$$
$$= 40.9 \text{ cm approximately.}$$

The semi-inter quartile range is

$$\frac{1}{2}(Q_3 - Q_1)$$
$$= 1/2\ (40.9 - 33.7) \text{ cm}$$
$$= 3.6 \text{ cm}$$

We may estimate the tenth percentile P_{10} (corresponding to cf6) which lies in the interval 29.5 – 34.5 by

$$P_{10} = 29.5 + \frac{1}{12}(34.5 - 29.5)$$
$$= 29.9 \text{ cm}$$

We may estimate the ninetieth percentile P_{90} (corresponding to cf 54) which lies in the interval 44.5 – 49.5 by

$$P_{90} = 44.5 + \frac{1}{7}(49.5 - 44.5)$$
$$= 45.2 \text{ cm.}$$

Another measure of dispersion (or variability) which can be used is the 10—90 Percentile Range (i.e. $P_{90} - P_{10}$)

In this case the 10—90 Percentile Range

$$P_{90} - P_{10}$$
$$= 45.2 - 29.9$$
$$= 16.3 \text{ cm.}$$

1.3.4 Mean Absolute Deviation

The MEAN ABSOLUTE DEVIATION (or M.A.D), sometimes also known as the Mean Deviation, is defined by

$$\frac{\Sigma f\,|x - \bar{x}|}{\Sigma f} \quad \text{or} \quad \frac{\Sigma f\,|x - \bar{x}|}{N}$$

where $|x - \bar{x}|$ is the absolute difference between the value or score x and its mean.

The function $|x|$ on absolute value of x is defined by

$$|x| = x \text{ if } x \geq 0$$
$$|x| = -x \text{ if } x < 0$$

e.g. $|4| = 4$, $|0| = 0$, $|-5| = 5$,

$|-3.2| = 3.2$

For the data in 1.1.3 we would need to calculate the mean $\bar{x}$ (done earlier in 1.2.2) before we determine the M.A.D.

Length (cm)	*Mid Point* x	*Frequency* f	$\lvert x - \bar{x}\rvert = \lvert x - 37.25\rvert$	$f\,\lvert x - 37.25\rvert$
24.5—29.5	27	5	10.25	51.25
29.5—34.5	32	12	5.25	63.00
34.5—39.5	37	25	0.25	6.25
39.5—44.5	42	11	4.75	52.25
44.5—49.5	47	7	9.75	68.25
		$\Sigma f = 60$		$\Sigma f \lvert x - \bar{x}\rvert = 241.00$

M.A.D. $= \Sigma f\,|x - \bar{x}|/N$

$= 241/60 = 4.02$ cm

Readers who may be wondering how the values for $|x - \bar{x}|$ were derived can look at the method given below for the entry corresponding to $x = 27$

We require the absolute value of

$27 - 37.25$ i.e. the absolute value of

-10.25 i.e. $|-10.25| = 10.25$.

It will be noticed that computationally this measure can bring difficulties in that the value of $|x - \bar{x}|$ will not very often be a whole number. It is possible to carry out the calculation using coding, and the problem is repeated using an assumed mean of 37 and scaling down by the class width (in this case 5)

Length (cm)	*Mid-point* x	*Frequency* f	$\lvert y\rvert$ $\lvert x - 37\rvert$	$\lvert u\rvert$ $= \lvert x - 37\rvert/5$	$f\,\lvert u\rvert$
24.5—29.5	27	5	10	2	10
29.5—34.5	32	12	5	1	12
34.5—39.5	37	25	0	0	0
39.5—44.5	42	11	5	1	11
44.5—49.5	47	7	10	2	14
		$\Sigma f = 60$			$\Sigma f\,\lvert u\rvert = 47$

On scaling up by scale factor 5 (the class width) $5(\Sigma f\,|u|) = 235$. However this gives an error of 6 compared will the value $\Sigma f\,|x - \bar{x}| = 241$ obtained earlier

In the course of coding, counting in the scale factor of 5, in the calculation using coding, for class 24.5—29.5 the contribution to $\Sigma f\,|x - \bar{x}|$ became $5 \times 5 \times 2 = 50$ instead of $5 \times 10.25 = 51.25$ an error of 1.25 or 5×0.25 (based on frequency 5)

In the class 29.5—34.5 the contribution using coding was $5 \times 12 \times 1 = 60$, instead of $12 \times 5.75 = 63.00$, an error of 3 or 12×0.25 (based on frequency 12)

In the class 34.5—39.5 the contribution using coding was 0, instead of $25 \times 0.25 = 6.25$, an error of 6.25 or 25×0.25 (based on frequency 25) a correction of $(5 + 12 + 25) \times 0.25$ has to be made, as we used 37 as assumed mean instead of 37.25, the true mean, thus reducing each absolute difference by 0.25, as shown below:

$$|27 - 37| = 10 \text{ but } |27 - 37.25| = 10.25$$
$$|32 - 37| = 5 \text{ but } |32 - 37.25| = 5.25$$
$$|37 - 37| = 0 \text{ but } |37 - 37.25| = 0.25$$

However our overall error so far is $1.25 + 3 + 6.25 = 10.50$, which is more than the difference of 6 between the true $\Sigma f\,|x - \bar{x}| = 241$ and our coded sum $= 235$.

We have yet to consider the error in the remaining classes.

In the class 39.5—44.5 the contribution by the coded method gave $5 \times 11 \times 1 = 55$ compared with the value 52.25 in the first method, an error of 5.75 or 11×0.25 too much. Similarly in the class 44.5—49.5 the contribution by the coded method gave an error of $5 \times 7 \times 2 - 68.25 = 3.75 = 7 \times 0.25$ too much

However if we adjust our coded total by an amount

$(5 + 12 + 25) \times 0.25$ (for classes where the deviation was 0.25 too little)

$-(11 + 7) \times 0.25$ (for classes where the deviation was 0.25 too much)

we have $(42 - 18) \times 0.25 = 24 \times 0.25 = 6$

which exactly recovers the error when we used 37 (assumed mean) instead of 37.25 (time mean) hence the M.A.D. has the coded method

$$= \frac{\Sigma f\,|x - a|}{N} + \text{correction}$$

where a is the assumed mean,
and in this case it is

$$5 \times 47/60 + \{(5 + 12 + 25) \times 0.25 - (11 + 7) \times 0.25)\}/60$$
$$= 241/60$$
$$= 4.02 \text{ cm}$$

If there is a difference d between the time mean $\bar{x}$ and the assumed mean a then in calculating the correction for the M.A.D. we need to add fd for each class where the coding method makes the contribution too little, and to subtract fd for each class where the coding method makes the contribution too much.

1.3.5 Variance and Standard Deviation

All the previously mentioned measures of dispersion (range, semi-interquartile range, 10—90 percentile range, mean absolute deviation) are useful. However the most useful measures for further statistical analysis of a set of data are the VARIANCE and the STANDARD DEVIATION.

The variance of grouped data is defined as

$$\frac{\Sigma f\,(x - \bar{x})^2}{N}$$

where each class has mid-point x with corresponding frequency f, $\Sigma f = N$, and $\bar{x} = \Sigma fx / N$ (the mean)

A corresponding definition of variance for ungrouped data is $\Sigma (x - \bar{x})^2/N$ where each x is assumed to have a frequency 1.

The standard deviation is the square root at the variance. In symbols,

$$\text{standard deviation} = \sqrt{\left\{\frac{\Sigma f (x - \bar{x})^2}{N}\right\}}$$

for grouped data

or $$\sqrt{\left\{\frac{\Sigma (x - \bar{x})^2}{N}\right\}} \text{ for ungrouped data.}$$

In calculating the variance we may use coding, without involving a correction at the end of the calculation, as happens in the calculation of the M.A.D. by coding.

We may code for the formula for variance using $x = a + Cu$, $\bar{x} = a + C\bar{u}$ in the notation of section 1.2.2.

$$\text{Variance} = \frac{\sum_{i=1}^{k} f_i (x_i - \bar{x})^2}{N}$$

$$= \frac{\sum_{i=1}^{k} f_i (a + Cu_i - a - C\bar{u})^2}{N}$$

$$= \frac{\sum_{i=1}^{k} f_i (Cu_i - C\bar{u})^2}{N}$$

$$= \frac{\sum_{i=1}^{k} f_i \{C (u_i - \bar{u})\}^2}{N}$$

$$= \frac{\sum_{i=1}^{k} f_i C^2 (u_i - \bar{u})^2}{N}$$

$$= \frac{C^2 \sum_{i=1}^{k} f_i (u_i - \bar{u})^2}{N}$$

$$= C^2 \left\{\frac{\sum_{i=1}^{k} f_i (u_i^2 - 2u_i \bar{u} + \bar{u}^2)}{N}\right\}$$

$$= C^2 \left\{ \frac{\sum_{i=1}^{k} f_i u_i^2}{N} - \frac{2\bar{u} \sum_{i=1}^{k} f_i u_i}{N} + \frac{\bar{u}^2 \sum_{i=1}^{k} f_i}{N} \right\}$$

$$= C^2 \left\{ \frac{\sum_{i=1}^{k} f_i u_i^2}{N} - 2\bar{u} \cdot \bar{u} + \bar{u}^2 \frac{N}{N} \right\}$$

$$\text{Variance} = C^2 \left\{ \frac{\sum_{i=1}^{k} f_i u_i^2}{N} - \bar{u}^2 \right\}$$

So the standard deviation (the square root of the variance) can be given by

$$C \sqrt{\left\{ \frac{\sum_{i=1}^{k} f_i u_i^2}{N} - \bar{u}^2 \right\}}$$

If we take $C = 1$ and $a = 0$ then

$$x = 0 + 1\ u \text{ or } x = u$$

$$\bar{x} = 0 + 1\ \bar{u} \quad \text{or } \bar{x} = \bar{u}$$

Substituting $C = 1$, $a = 0$, $u = x$, $\bar{u} = \bar{x}$ in the formula we have just derived for variance gives

$$\text{variance} = \frac{\sum_{i=1}^{k} f_i x_i^2}{N} - \bar{x}^2$$

This last formula is useful when the original values of x are easy to work with e.g. $x = 0$, $x = 1$, $x = 2$, $x = 3$, ...
but it is *not* recommended in cases when x is say 2011.5, 2013.5, 2015.5, ... where a coding method should be used.

For any data it is essential to calculate the mean before calculating the variance. We shall now calculate the variance and the standard deviation for the data of 1.1.3 using each of the three formulae.

Example 1.3.5(1) Calculate the variance and the standard deviation of the distribution given below using the formula variance

$$= \frac{\Sigma f\,(x - \bar{x})^2}{N}$$

Length (cm)	Frequency
25—29	5
30—34	12
35—39	25
40—44	11
45—49	7

Mid-point x	Frequency f	fx	$x - \bar{x}$	$(x - \bar{x})^2$	$f(x - \bar{x})^2$
27	5	135	– 10.25	105.0625	525.3125
32	12	384	– 5.25	27.5625	330.7500
37	25	925	– 0.25	0.0625	1.5625
42	11	462	4.75	22.5625	248.1875
47	7	329	9.75	95.0625	665.4375
	$\Sigma f = 60$	$\Sigma fx = 2235$			$\Sigma f(x - \bar{x})^2 = 1771.25$

$$\text{Mean } \bar{x} = 2235/60$$
$$= 37.25 \text{ cm}$$
$$\text{Variance} = 1771.25/60$$
$$= 29.52 \text{ cm}^2$$
$$\text{Standard Deviation} = \sqrt{29.52}$$
$$= 5.43 \text{ cm}$$

If you use a calculator keep as many figures as the calculator will allow, as rounding error can accumulate in your addition and lead to a large error in the calculation.

As can be seen the calculation can become laborious, especially if there are many class intervals. If students use this method there are three common mistakes:

(i) Adding the column for $(x - \bar{x})^2$ [instead of $f(x - \bar{x})^2$] and then dividing the value obtained by the total frequency.

(ii) Squaring the numbers recorded for the frequency f, instead of those for $(x - \bar{x})$, in the calculation of the numbers in the column for $f(x - \bar{x})^2$.

(iii) Squaring both the numbers recorded for the frequency f and those recorded for $(x - \bar{x})$, in the calculation of the number in the column for $f(x - \bar{x})^2$.

In the following example the benefits of using the coding method to make the calculation easier will become apparent.

Example 1.3.5(2) Calculate the variance and the standard deviation of the distribution below using the formula

$$\text{variance} = C^2 \left\{\Sigma f u^2 / N - \bar{u}^2\right\}$$

Length (cm)	Frequency
25—29	5
30—34	12
35—39	25
40—44	11
45—49	7

Mid-Point x	Frequency f	$u = (x - 37)/5$	fu	fu^2
27	5	-2	-10	20
32	12	-1	$-12/-22$	12
37	25	0	0	0
42	11	1	11	11
47	7	2	14/25	28
	$\Sigma f = 60$		$\Sigma fu = 3$	$\Sigma fu^2 = 71$

$\bar{u} = \Sigma fu/N = 3/60 = 0.05$

Mean $\bar{x} = 37 + 5 \times 3/60 = 37.25$ cm

$$\text{Variance} = C^2\left\{\Sigma fu^2/N - \bar{u}^2\right\}$$
$$= 5^2\ \{71/60 - 0.05^2\}$$
$$= 25 \times 1.180\ 8$$
$$= 29.52\ \text{cm}^2$$

$$\text{Standard Deviation} = \sqrt{29.52}\ \text{cm}$$
$$= 5.43\ \text{cm}$$

N.B: The easiest method of computing entries for the column fu^2 is by multiplying together entries for u and fu in the previous columns.

Example 1.3.5(3) Calculate the variance and the standard deviation of the distribution below using the formula

$$\text{Variance} = \Sigma fx^2/N - \bar{x}^2$$

Length (cm)	Frequency
25—29	5
30—34	12
35—39	25
40—44	1
45—49	7

Mid-point x	Frequency f	fx	fx^2
27	5	135	3645
32	12	384	12284
37	25	925	34225
42	11	462	19404
47	7	329	15463
	$\Sigma f = 60$	$\Sigma fx = 2235$	$\Sigma fx^2 = 85025$

Mean $\bar{x} = \Sigma fx/N = 2235/60 = 37.25$

Variance $= \Sigma fx^2/N - \bar{x}^2$

$= 85025/60 - 37.25^2$
$= 29.52\ cm^2$

Standard Deviation $= \sqrt{29.52}$
$= 5.43$ cm

Accuracy with a calculator is enhanced, as seen earlier, by keeping all the figures in your working until the last step.

N.B: The easiest method of computing entries for the column fx^2 is by multiplying together entries for x and fx in previous columns.

EXERCISE I

1. Find the mean, the median and the mode for the following set of discrete data:

147, 151, 152, 151, 146, 148, 156, 154, 155, 157, 148, 148

2. The histogram below [Fig. 1.4.1(a)] shows the grades *A-E* obtained by 100 students in a test where the marks were integers between 0 and 99 inclusive.
Copy and complete the table below and hence compute the mean mark giving your answer correct to three significant figures.

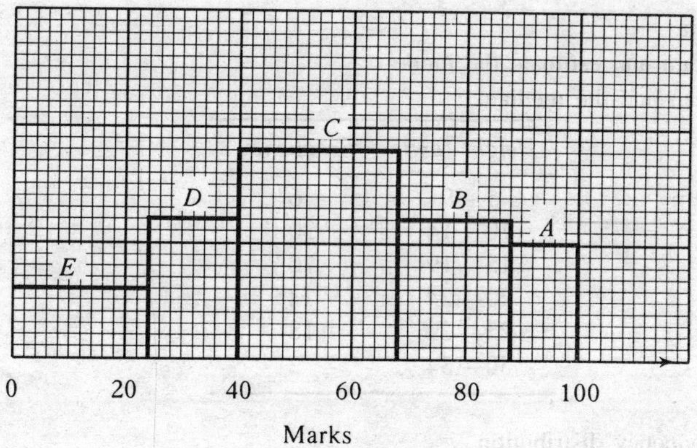

Fig. 1.4.1(a)

Grade	*Mark*	*No. of Students*
A	88—99	10
B		
C		
D	24—39	
E		

3. The frequency table below gives the scores obtained by 50 students in a statistics quiz:

Score	0—4	5—9	10—14	15—19
Frequency	1	2	8	11

Score	20—24	25—29	30—34	35—39
Frequency	14	8	4	2

(i) Draw a cumulative frequency graph to represent this data.
(ii) Hence estimate the quartile deviation of the scores

4. The ages of 60 women who visited 'youthful hair saloon' in the first week of January were distributed as follows.

Age	*Frequency*
10—14	1
15—19	3
20—24	8
25—29	10
30—34	18
35—39	14
40—44	4
45—49	2

Calculate to two decimal places
(a) the median age,
(b) the quartile deviation of the ages.

5. For the following data
(a) draw the histogram and estimate the mode,
(b) calculate the mean and the median.

Height (cm)	*Frequency*
5—9	6
10—14	30
15—19	70
20—24	74
25—29	18
30—34	2

6. For the following frequency distribution
(a) state the class width,
(b) state the lower and upper class boundaries of the class 45—49
(c) state the modal class,
(d) calculate the mean,
(e) calculate the standard deviation,
(f) calculate the median and quartiles.

Height (cm)	*Number of Plants*
40—44	6
45—49	10
50—54	25
55—59	11
60—64	8

7. For the following frequency distribution
   (a) state the mid point of the class 40—44,
   (b) state the upper and lower class boundaries of the class 55—59,
   (c) calculate the mean of the distribution,
   (d) calculate the standard deviation of the distribution,
   (e) calculate the median and quartiles of the distribution.

Volume (cm^3)	*Frequency*
40—44	8
45—49	20
50—54	45
55—49	25
60—64	2

8. For a set of data let M denote the mean absolute deviation from the mean and S the standard deviation. It is known that an estimate of S can be given by

$$S \simeq \frac{5}{4} M.$$

Calculate M for the data below and use the relation above to estimate S.

x	30—34	35—39	40—44	45—49	50—54	55—59
f	5	10	18	16	9	2

9. The masses of 100 students in 'Keep Fit University' are shown in the following frequency distribution given below. For this frequency distribution:
   (a) state the size of each class interval,
   (b) state the upper and lower class boundaries for the class 63—67,
   (c) state the modal class,
   (d) calculate the mean and the standard deviation
   (e) calculate the minimum acceptable mass if the heaviest 10% of the students are accepted for special weight training.

Mass (kg)	*Frequency*
53—57	2
58—62	12
63—67	12
68—72	25
73—77	27
78—82	10
83—87	9
88—92	3

10. The marks of 200 students have been recorded in a grouped frequency table as shown below. The table has been extended for the purpose of calculating the variance of the distribution of the marks. Copy and complete the table and calculate the standard deviation of the distribution.
    Calculate also the value of the semi-interquartile range of this distribution. Find the percentage error in using the semi-interquartile range as an approximation for the standard deviation.

Mark (X)	*Frequency (f)*	*d*	d^2	*fd*	fd^2
1—10	7	− 5	25	− 35	175
11—20	9	− 4	16	− 36	144
21—30	12	− 3	9	− 36	108
31—40	21	− 2	4	− 42	84
41—50	39	− 1	1	− 39	39
51—60	44	0	0	0	0
61—70	29	1	1	29	29
71—80	18	2	4	36	72
81—90	14	3	9	42	126
91—100	7	4	16	28	112
	$\Sigma f = 200$			$\Sigma fd =$	$\Sigma fd^2 =$

11. 105 pages of a manuscript are checked and the number of typing errors on each page are recorded as shown in the table below.

Number of errors	0	1	2	3	4	5
Frequency	15	30	28	18	9	5

Construct a cumulative frequency table and use it to calculate the median and the quartile deviation. Check your answer graphically.

12. Construct a cumulative frequency table and use it to estimate the median and quartile deviation of the following data. Check your answers graphically.

Mass (kg)	*Frequency*
50	4
52	7
54	12
56	11
58	9
60	7

13. Copy and complete the following table and hence calculate the mean and standard deviation for the distribution shown below.

Score x	*Frequency f*	*z*	*fz*	fz^2
50	2	−3	−6	18
55	12	−2	−24	48
60	12			
65	25			
70	27			
75	10			
80	9			
85	3			
			$\Sigma fz =$	$\Sigma fz^2 =$

14. Calculate the mean and the standard deviation for the following marks:

Mark	*Number of Students*
90—99	3
80—89	32
70—79	43
60—69	21
50—59	1

15. The electrical resistances in ohms of 150 carbon rods tested at the same temperature are given below.

Resistance (ohms)	310	311	312	313	314	315
Frequency	1	2	6	24	29	34

Resistance (ohms)	316	317	318	319	320
Frequency	27	18	5	3	1

Calculate the mean and the standard deviation.

16. The table below shows the distribution of the points scored by 1000 beautiful young girls in their pre-university examination prior to their entering their first year at the 'University of Utopia' in the country of Buranea.

Points	*Frequency*
16—19	100
14—15	300
12—13	500
10—11	100

If the dowries payable are as shown below determine the mean and standard deviation of the dowries in Buranian dollars.

Points	*Dowry (in thousands Buranian dollars)*
16—19	200
14—15	100
12—13	80
10—11	50

17. The examination marks for 200 students were distributed as follows:

Marks	*Frequency*
0—9	4
10—19	6
20—29	15
30—39	27
40—49	39
50—59	45
60—69	25
70—79	20
80—89	14
90—99	5

(a) Construct a cumulative frequency table and use it to estimate the median mark.

(b) It was decided that 60% of the students be awarded a pass and that the best 15% of the students be awarded a distinction. Determine the minimum mark for which a student will qualify for (i) a pass (ii) a distinction.

18. The masses (in grams) of 50 small fruits are as shown in the following table.

Mass (grams)	*Number of Fruits*
50—59	4
60—69	5
70—79	9
80—89	14
90—99	7
100—109	6
110—119	5

(a) Draw a histogram to represent the data and use it to estimate the mode of the distribution. Check by calculation.

(b) Calculate also the mean and the median of the distribution.

19. A small branch of a flowering plant had forty leaves. The length of the leaves were measured. The histogram given below represents the information gathered. The region A represents the number of leaves whose length is more than zero centimetres but less than or equal to three centimetres.

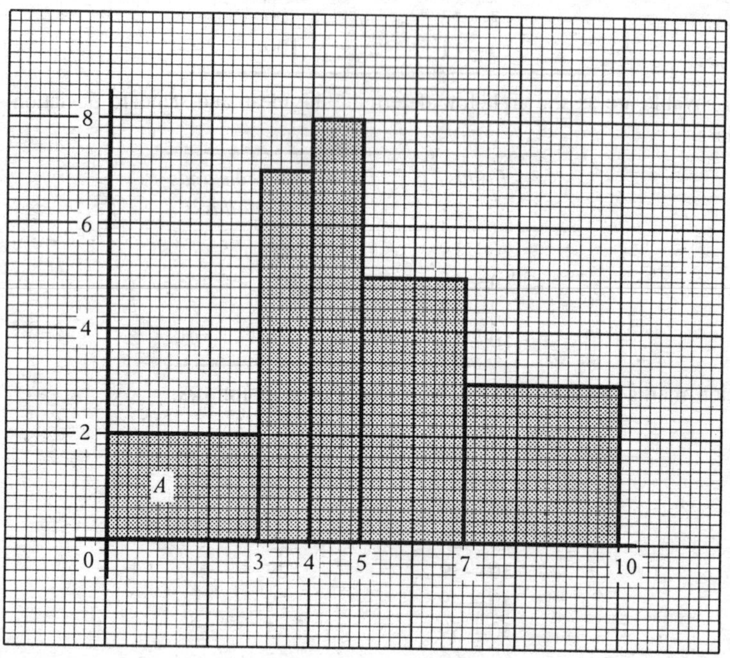

Determine a cumulative frequency table from this information and use it to estimate the median length.

***20.** (a) 50 students from TA University take a mathematics examination in which the highest possible mark is 100.

The cumulative frequency curve for the marks of these 50 students is shown below.

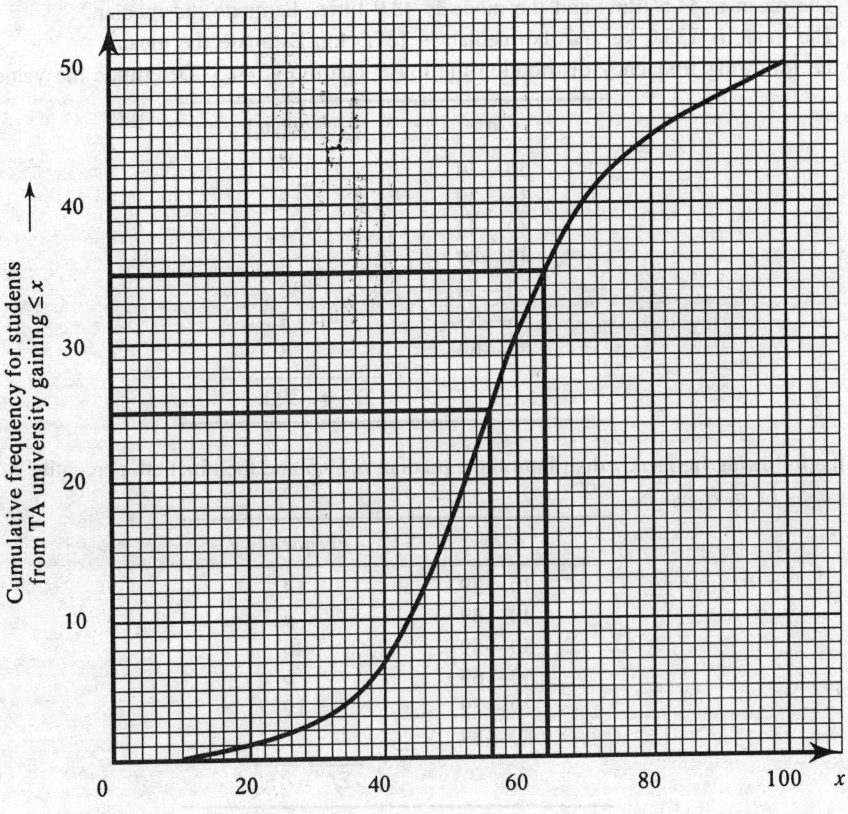

(i) What is the median mark for this group of students?

(ii) If 80% of the students pass how many students fail?

(iii) What is the pass mark?

(b) (i) 80 students from HP university take the same examination and their median mark is 65. Their upper quartile mark is 80 and the inter quartile range is 25.

What is Q_1, the lower quartile mark for the students from HP university?

(ii) The top 5 students from the group in HP university score more than 90 marks and the lowest 5 marks are in the interval $25 < x \le 45$.

Using this and all the information in (b), (i) copy and complete the cumulative frequency table shown below for the marks of the students from HP university, replacing the letter Q_1 in the table by your answer to (b) (i).

Mark (x)	≤ 25	≤ 45	$\le Q_1$	≤ 65	≤ 80	≤ 90	≤ 100
Cumulative Frequency							

21. For certain data it was found that several summary statistics were not readily available. Use the empirical relation between the mean, the median and the mode to estimate the required statistic in each case
    (a) The mean is 35.3 cm, the median is 33.9 cm. Estimate the mode.
    (b) The mean is 44.5 litres and the mode is 43.9 litres. Estimate the median.
    (c) The mode is 135.0 kg and the median is 137.5 kg. Estimate the mean.
22. For the following frequency distribution the lower quartile is 44.5. Determine the values of a and b.

Class	*Frequency f*
30—34	7
35—39	12
40—44	a
45—49	b
50—54	38
55—59	15
60—64	8
	$\Sigma f = 200$

23. For the following frequency distribution the median is 59.5 and the ninetieth percentile is 79.5. Determine the values of a, b and c.

Class	*Frequency f*
30—39	14
40—49	a
50—59	40
60—69	b
70—79	19
80—89	c
	$\Sigma f = 150$

*24. For the following frequency distribution the modal class is 140—159 and the upper quartile is in the class 160—179.
Determine the ranges of possible values for a and b, and write down the upper quartile when a has its maximum value.

Class	*Frequency f*
100—119	30
120—139	136
140—159	a
160—179	b
180—199	25
	$\Sigma f = 400$

2

Probability

INTRODUCTION

Probability is the branch of Mathematics which deals with the likelihood of an event occurring. It provides us with a forecast of what is likely to happen in the future? It tries to answer questions like: will it rain tomorrow? Will student Kamau be late for his lecture next Friday? What will be a good estimate for the median mark in the next CAT taken by a particular class of students?

For example, if in a class of 45 students there are 9 students having a box hair style, then the proportion of students who have a box hair style is 9/45 or 0.2. This ratio tells us that the chance that a student chosen at random will have a box hair style is 1/5 or 0.2.

2.1 DEFINITIONS OF PROBABILITY

2.1.1 Experimental Probability

Example 2.1.1(1)
Go to the side of a busy road and, for each of the first hundred cars you observe passing, record the units digit of its registration number.

If you have recorded the digit '6' on twelve of the hundred cars then for you as the observer the probability of seeing a car whose registration number has unit digit '6' will be 12/100.
We may write this as

$$P(6) = 12/100 \text{ (or } 6/50 \text{ or } 3/25)$$

or

$$0.12$$

In the same way you can find the probability of seeing a car whose registration number has units digit 0, 1, 2, etc. by reference to the number of occasions out of 100 when you observed a car having that particular units digit in its registration number.

Suppose you found the recordings to be

Units digit n	0	1	2	3	4	5	6	7	8	9
Number of Occasions it was recorded	8	11	13	7	8	9	12	10	11	11
Probability $P(n)$	0.08	0.11	0.13	0.07	0.08	0.09	0.12	0.10	0.11	0.11

Since the digit '0' has appeared on 8 occasions out of 100 we may say that the probability of seeing a car with units digit '0' in its registration number is 8/100

or $$P(0) = 8/100 = 0.08$$

In the same way we may determine $P(n)$ when $n = 5$, by using the fact that there were 9 occasions when a car was recorded as having units digit '5' in its registration number in that

$$P(5) = 9/100 = 0.09$$

What is the probability of seeing a car with units digit
(a) '8' (b) '4'
in its registration number?
What is the probability of seeing a car with
(c) an even registration number?
(d) an odd registration number?
From the recordings you should be able to conclude that the required probabilities are
(a) 0.11 (b) 0.08 (c) 0.52 (d) 0.48

Example 2.1.1(2)

Toss a coin 20 times and at each toss record the outcome as heads (H) or tails (T), according to the uppermost face of the coin on landing. Your table of records might look like this (but not exactly)

Toss Number	1	2	3	4	5	6	7	8	9	10
Uppermost Face	H	H	T	T	H	T	T	H	H	H
Toss Number	11	12	13	14	15	16	17	18	19	20
Uppermost Face	T	H	H	T	T	H	T	H	H	T

From your own observations what is the probability of the coin landing on (a) heads (b) tails? To answer this question you would use

(a) $$\frac{\text{Number of times heads turned up}}{20}$$

(b) $$\frac{\text{Number of times tails turned up}}{20}$$

Try this again, but this time record the outcomes for 50 tosses, then 100 tosses, in each case recording the answers for the probability of the coin landing heads and the probability of the coin landing tails.

Using a large number of tosses it appears that we can find an answer for the probability that a head turns up on one toss of a coin in an experimental way.

Based on the experiment, if H = head turns up, T = tail turns up,

$$P(H) = \frac{\text{Number of times tails turns up}}{\text{Total Number of tosses}}$$

$$= \frac{\text{Frequency of Heads}}{N}$$

This can be written as $P(H) = \dfrac{f(H)}{N}$

Similarly $\quad P(T) = \dfrac{\text{Number of times a tail turns up}}{\text{Total Number of tosses}}$

$$= \frac{\text{Frequency of Tails}}{N} = \frac{f(T)}{N}$$

You probably have an intuitive idea of what the actual values of $P(H)$, $P(T)$ should be, and that it is likely that we ought to find experimental probabilities which will become closer to these values if we increase the total number of tosses.

We write 'N approaches infinity' or $N \to \infty$, to indicate that we may make N as large as we like (100, 1000, even a million or more, if we had time) and write

$$P(H) = \lim_{N \to \infty} \frac{f(N)}{N}$$

This is the experimental definition of the probability of obtaining a head in one toss of a coin.

In general if we are interested in finding the probability of an outcome A we may conduct an experiment of N trials to find out how many times the event A occurs, say A occurs $f(A)$ times.

We can then define the experimental probability of event A as

$$P(A) = \lim_{N \to \infty} \frac{f(A)}{N}$$

In practice we may have to accept a relatively small finite value for N, and on the basis of the experiment for that value of N, take as our experimental probability

$$P(A) = \frac{f(A)}{N}$$

where $f(A)$ = frequency of event A or number of times A occurs.

It will be clear if you compare your results for the above mentioned experiments with another student that it is unlikely that the experimental probabilities you both obtained for any particular event are the same. In this sense we cannot give a unique correct answer for probabilities, based on experiment. However if a sufficient number of trials is conducted we may expect to arrive at a reasonable answer. As a rule of thumb at least twenty trials should be made for any experiment.

2.1.2 Theoretical Probability

In some cases it is possible to deduce probability without use of an experiment. Such probability is known as Theoretical or a Priori probability.

In some cases it may not always be possible to carry out a sufficient number of trials to find a reasonable answer $\left(\dfrac{f(A)}{N}\right)$ to determine the probability of an event A experimentally. In these cases we may look for a theoretical way of determining a probability.

We use the word mutually exclusive in relation to two or more events, where the occurrence of any one event prevents the occurrence of any of the others. For example in tossing a coin there are two possibilities: either a head or a tail turns up. If a head turns up on one toss of a coin then a tail cannot turn up on the same toss. Similarly if a tail turns up on toss of a coin then a head cannot turn up on the same toss. For one toss of a coin the events 'head turns up' and 'tail turns up' are mutually exclusive outcomes.

Example 2.1.2(1)

You visit the animal orphanage in Nairobi Game Park one day. In the orphanage there are 3 lions, 4 cheetahs and 5 zebras—no other animals. If you choose an animal at random to be brought to the boundary fence what is the probability that the animal is

(a) a Lion (b) a Zebra (c) a Lion or a Zebra (d) a Lion or a Cheetah or a Zebra (e) a Tiger?

Youre reasoning probably follows the idea set out below:

(a) there are 12 animals to choose from and of these 12 animals there are 3 Lions

so $$P(\text{Lion}) = 3/12 = 1/4$$

(b) On a similar argument

$$P(\text{Zebra}) = 5/12$$

(c) Of the 12 animals there are 3 Lions, 5 Zebras and there is no animal which can be both a lion and a zebra (these are mutually exclusive events), so any one of eight animals (3 lions, 5 zebras) is acceptable and we say

$$P(\text{Lion or Zebra}) = \frac{3+5}{12} = \frac{8}{12} = \frac{2}{3}$$

(d) Of the 12 animals all are acceptable

so $$P(\text{Lion or Cheetah or Zebra}) = \frac{12}{12} = 1$$

(e) Of the 12 animals there is no Tiger

so $$P(\text{Tiger }) = \frac{0}{12} = 0$$

By studying the last example you will have noted that we have been looking for the number of outcomes which are favourable to an event happening, and also the total number of possible outcomes, assuming all the outcomes are equally likely.

In general if E is the (universal) set of all possible outcomes and A is the set representing the event of interest, we may define the probability of event A as

$$P(A) = \frac{n(A)}{n(E)}$$

$$= \frac{\text{(Number of favourable outcomes for event } A)}{\text{(Number of possible outcomes)}}$$

This is the definition of Theoretical Probability.

2.1.3 Impossibility and Certainty

Look back at example 2.1.2(1). It should be clear from the answers to all the parts that for any event A the value of $P(A)$ must lie between 0 and 1 i.e. $0 \le P(A) \le 1$

$P(A) = 0$ means that A is an impossible event, for example what is the probability that you are older than your grandmother?

$P(A) = 1$ means that A is a certain event we are sure that it will happen, for example if a lecturer calls a student from his class what is the probability that the student is less than 200 years old?

In some cases the sum of probabilities of mutually exclusive events may be 1, e.g. in tossing a coin once

$$P(H) + P(T) = 1$$

e.g. in throwing a die once, $P(1)$ = probability that the face showing '1' lands uppermost, with similar definitions for the other probabilities,

$$P(1) + P(2) + P(3) + P(4) + P(5) + P(6) = 1$$

The six outcomes of throwing a die are mutually exclusive.

The symmetry of a coin implies that Heads and Tails are equally likely and since $P(H) + P(T) = 1$ then

$$P(H) = P(T) = 1/2$$

On a similar argument for a fair die each of the scores (1 to 6) is equally likely, so

$$P(1) = P(2) = P(3) = P(4) = P(5) = P(6) = 1/6$$

Our intuitive idea about equally likely events led us to such examples of theoretical probability.

Many events do not have theoretical probabilities. An example is the throwing of a drawing pin, where P(point up) and P(point down) are not equal. These events are not equally likely, since the pin is lighter nearer the point. The only method available for determining P(point up) and P(point down) for one throw of a drawing pin is the experimental method.

In general if there are n outcomes in one trial and all the outcomes are equally likely then each outcome has a probability of $1/n$.

Example 2.1.3(1)

Determine the probability of an odd number in one throw of a die.

Out of 6 equally likely outcomes (1, 2, ..., 6) there are three odd numbers (1, 3, 5) so

$$P(\text{odd number}) = 3/6 = 1/2$$

Example 2.1.3(2)

A box contains 6 identical red, 4 identical blue and 5 identical yellow beads. If a bead is drawn at random, what is the probability that it is (a) blue (b) yellow (c) red or blue

(a) $$P(\text{blue}) = \frac{\text{Number of ways of choosing a blue bead}}{\text{Total number of ways}} = \frac{4}{15}$$

(b) $$P(\text{yellow}) = \frac{\text{Number of ways of choosing a yellow bead}}{\text{Total number of ways}} = \frac{5}{15} \text{ or } \frac{1}{3}$$

(c) $$P(\text{red or blue}) = \frac{\text{Number of ways of choosing a red or blue bead}}{\text{Total number of ways}}$$

$$= \frac{10}{15} \text{ or } \frac{2}{3}.$$

2.2 SUMS AND PRODUCTS OF PROBABILITIES

2.2.1 Possibility Spaces

Suppose we are interested in the probability of throwing a 2 or a 3 with one throw of a single die.

All possible outcomes are 1, 2, 3, 4, 5, 6 which make up the Possibility Space. We are interested in throwing 2 or 3, which are two members of the possibility space. P(2 or 3) = 2/6 or 1/3.

Example 2.2.1(1)

What is the probability of obtaining

(a) two heads when two coins are tossed?

(b) three tails when three coins are tossed?

(a) tabulate the outcome for the two tosses.

The possibility space is HH, HT, TH, TT

so $$P(HH) = 1/4$$

(b) For three tosses there are eight outcomes.

The possibility space is *HHH, HHT, HTH, THH, HTT, THT, TTH, TTT*

$$P(3 \text{ tails}) = P(TTT) = 1/8$$

N.B.: When one coin is tossed the number of outcomes in the possibility space is 2 or 2^1; when two coins are tossed the number of outcomes in the possibility space is 2^2; when three coins are tossed the number of outcomes in the possibility space is 2^3; when n coins are tossed the number of outcomes in the possibility space is 2^n.

Example 2.2.1(2)

Two dice are thrown. What is the probability that the sum of the numbers on the two uppermost faces will be (a) 10 (b) less than 5?

The solution for a problem involving two dice can often be readily obtained after a table is drawn up, showing all possible outcomes of the two throws. [Fig. 2.2.1(a)].

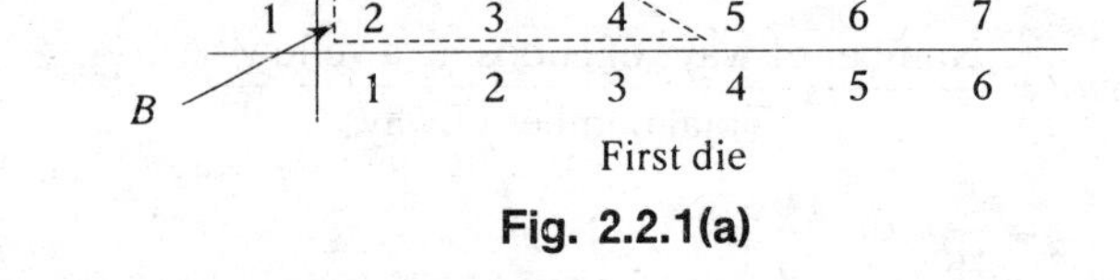

Fig. 2.2.1(a)

A represents 'total = 10'

$$P(A) = 3/36 = 1/12$$

since there are altogether 36 outcomes, of which 3 outcomes give a total of 10.

B represents 'total < 5'

$$P(B) = 6/36 = 1/6$$

If *B* represents the event 'total < 5' the event not-*B* (often written as ~*B* or *B*′) is 'total = 5 or total > 5'.

For the event *B* we know that *B* or not-*B* must happen, and these events are mutually exclusive

$$P(B) + P(\sim B) = 1$$

hence $$P(\sim B) = 1 - 1/6 = 5/6$$

In general we may say that for any event *X*, there is an associated event not-*X* (written ~*X* or *X*′) which is the event '*X* does Not happen'.

Since *X* and ~*X* are mutually exclusive and one of them must happen we conclude that

$$P(X) + P(\sim X) = 1$$

Other examples of this principle are

(i) On one toss of a coin, if *X* denotes 'heads' then ~*X* denotes 'tails'

$$P(X) = 1/2, \ P(\sim X) = 1/2$$

(ii) On one throw of a die, if *X* denotes the event 'a six appears on the uppermost face' then ~*X* denotes the event 'any other number apart from six appears on the uppermost face'

$$P(X) = 1/6, \ P(\sim X) = 5/6$$

Example 2.2.1(3)

Two dice are thrown. What is the probability that the difference of the numbers shown on the two uppermost faces will equal (a) 2 (b) 3 or more (c) less than 3.

Draw up a table of possible outcomes, but this time, instead of adding the scores, take the difference of the scores, ignoring whether the higher scores may have occurred on the first or the second die. See Fig. 2.2.1(b).

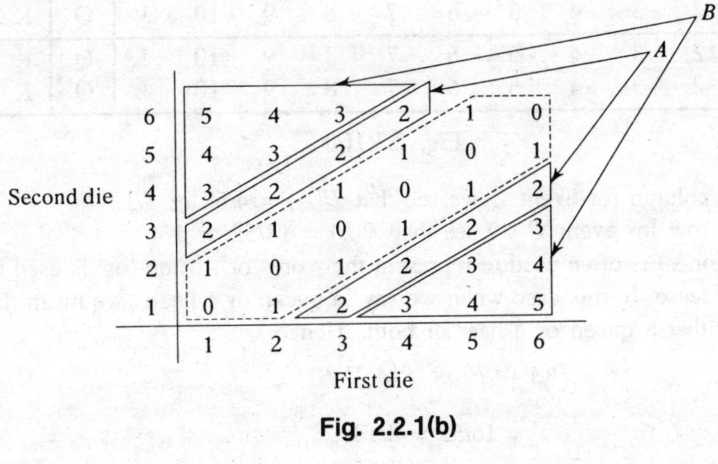

Fig. 2.2.1(b)

e.g. If the score on the first die is 6, the score on the second die is 1, then the difference in the scores is 5.

A represents the event 'difference = 2'

B represents the event 'difference is 3 or 4 or 5'

(a) $$P(A) = 8/36 = 2/9$$

(b) $$P(B) = 12/36 = 1/3$$

(c) $$P(\sim B) = 1 - P(B) = 1 - 12/36$$
$$= 24/36 = 2/3$$

(This may be checked by counting all the outcomes where the difference is 0, 1 or 2)

2.3 COMPOUND EVENTS

2.3.1 'And', 'Or'

Sometimes we may be interested in the occurrence of two or more events simultaneously or in the probability of an event which obeys more than one condition. The example below will illustrate these ideas.

Example 2.3.1(1)

A card is drawn from an ordinary pack of 52 cards. Find the probability of each of the events:

(a) drawing a queen
(b) drawing a heart
(c) drawing a queen or a heart
(d) drawing the queen of hearts

Let *A* be the event of drawing a queen and *B* be the event of drawing a heart.

The table below shows the possibility space [Fig. 2.3.1(a)]

												A ↓		
Diamond	ace	2	3	4	5	6	7	8	9	10	J	Q	K	
Hearts	ace	2	3	4	5	6	7	8	9	10	J	Q	K	← B
Clubs	ace	2	3	4	5	6	7	8	9	10	J	Q	K	
Spades	ace	2	3	4	5	6	7	8	9	10	J	Q	K	

Fig. 2.3.1(a)

(a) Looking at the column for event *A* we see that $P(A) = 4/52$ or $1/13$
(b) Looking at the row for event *B* we see that $P(B) = 13/52$ or $1/4$
(c) The symbol $\cup$(or V) is often used to represent the word 'or', when 'or' is used in mathematics in its inclusive sense. In this case when we say 'a queen or a heart' we mean that the card we want may be either a queen or a hear or both. Hence

$$P(A \cup B) \text{ or } P(A \vee B)$$

$$= 16/52 = 4/13$$

(This is not the same as $P(A) + P(B) = 4/52 + 13/52 = 17/52$—if you look carefully at the possibility space you will notice that if we add $P(A)$ and $P(B)$ we will have counted one card, the queen of hearts twice.)

(d) The symbol $\cap$(or $\wedge$) is often used in mathematics to mean 'and', when we write $A \cap B$ (or$A \wedge B$) we mean that the card is both a queen and a heart, i.e. in this case the queen of hearts.

Looking at the possibility space we see that there is only one card which is both a queen and a heart, so we conclude that

$$P(A \cap B) \text{ or } P(A \wedge B) = 1/52.$$

In the last example we saw that we could make compound events out of two different events using the conjunctions 'or' and 'and'. In general if X and Y are any two events we use the notation $P(X \cup Y)$ or $P(X \vee Y)$ to mean the probability that event X happens, or event Y happens, or both events happen.

We use the notation $P(X \cap Y)$ or $P(X \wedge Y)$ to mean the probability that event X happens and event Y happens.

Example 2.3.1(2)

In a university science class there are 100 students. Of these students 40 take Physics, 75 take Biology, and 20 take both Physics and Biology. Determine the probabilities that a student chosen at random takes Physics or Biology.

It will be useful to represent the situation, using A as the event 'the student takes Physics' and B as the event 'the student takes Biology', in a diagram known as Venn diagram. Such a diagram comprises a rectangle inside which circles are drawn to represent the events A and B, where the intersection of the two circles represents the event '$A \cap B$'. See Fig. 2.3.1(b). From the information there are 20 students who take both subjects, so there are 20 students who take Physics but not Biology, and 55 students who take Biology but not Physics.

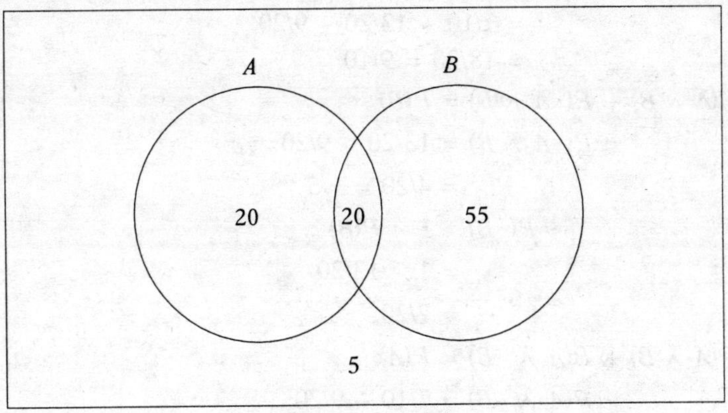

Fig. 2.3.1 (b)

In order to arrive at a total of 100 students there must therefore be 5 students who study neither subject.

From the diagram

$$P(A \vee B) = \frac{20 + 20 + 55}{100} = 95/100, \text{ or } \; 0.95$$

It may be noted from looking at the diagram that $P(A) = 40/100$. $P(B) = 75/100$, $P(A \wedge B) = 20/100$ and that we could have arrived at the result using

$$P(A \vee B) = P(A) + P(B) - P(A \wedge B)$$

i.e.

$$P(A \vee B) = 40/100 + 75/100 - 20/100$$
$$= 95/100$$

This example illustrates a general result for the probability of $X \vee Y$, where X and Y are any two events, in that

$$P(X \vee Y) = P(X) + P(Y) - P(X \wedge Y)$$

In the case where X and Y are mutually exclusive $P(X \wedge Y) = 0$ and

$$P(X \vee Y) = P(X) + P(Y)$$

Also for any events X and Y, if event X happens then X must happen with either Y or $\sim Y$.

i.e.

$$P(X \wedge Y) + P(X \wedge \sim Y) = P(X)$$

since the events $X \wedge Y$, $X \wedge \sim Y$, are mutually exclusive.

Example 2.3.1(3)

For two events A and B, $P(\sim A) = 3/10$ and $P(B) = 13/20$. If $P(A \wedge B) = 9/20$, find (a) $P(A \vee B)$ (b) $P(\sim A \wedge B)$ (c) $P(A \vee \sim B)$

Solution:

(a) $P(\sim A) = 3/10$ so $P(A) = 1 - 3/10 = 7/10$

Since we are given $P(A \wedge B) = 9/20$, $P(B) = 13/20$

Using

$$P(A \vee B) = P(A) + P(B) - P(A \wedge B)$$
$$= 7/10 + 13/20 - 9/20$$
$$= 18/20 = 9/10$$

For (b) using $P(A \wedge B) + P(\sim A \wedge B) = P(B)$

then

$$P(\sim A \wedge B) = 13/20 - 9/20$$
$$= 4/20 = 1/5$$

For (c) using

$$P(\sim B) = 1 - P(B)$$
$$= 1 - 13/20$$
$$= 7/20$$

and $P(A \wedge B) + P(A \wedge \sim B) = P(A)$

giving

$$P(A \wedge \sim B) = 7/10 - 9/20$$
$$= 5/20$$

Then

$$P(A \vee \sim B) = P(A) + P(\sim B) - P(A \wedge \sim B)$$
$$= 7/10 + 7/20 - 5/20$$
$$= 16/20$$
$$= 4/5.$$

The probabilities may also be represented in the venn diagram shown below in Fig. 2.3.1(c).

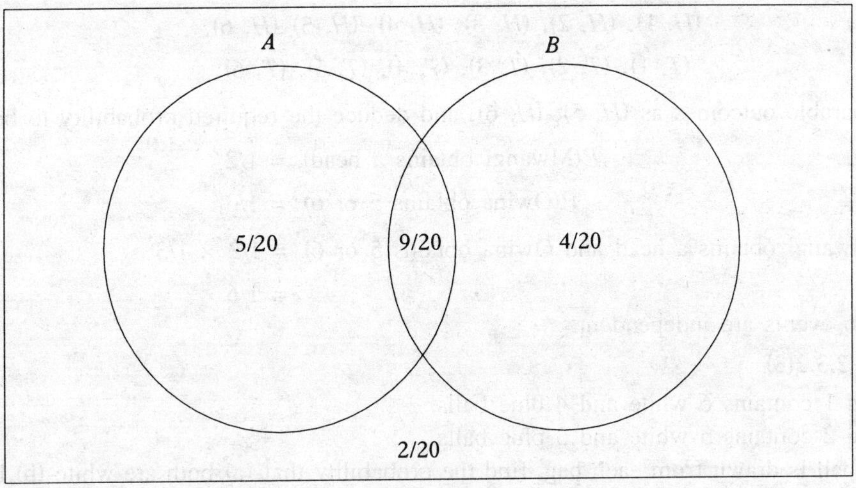

Fig. 2.3.1(c)

2.3.2 Independent Events

Two events X and Y are said to be independent if the occurrence of one does not affect the occurrence of the other, i.e. the occurrence of X does not affect the occurrence of Y, and the occurrence of Y does not affect the occurrence of X. An example involving the use of conditional probabilities will be shown in Section 2.3.4 in order to help us to arrive at a definition in terms of conditional probabilities but in the meantime we shall look at several examples where the independence of the two events will be obvious.

Example 2.3.2(1)

A fair coin is thrown twice. What is the probability that on each throw a head appears?

Solution:

We could list the possibility space $\{HH, HT, TH, TT\}$ and conclude that $P(HH) = 1/4$. However it may also be observed that

$$P(\text{Head on 1st throw}) = 1/2$$

$$P(\text{Head on 2nd throw}) = 1/2$$

and that P(Head on 1st throw and Head on 2nd throw) = P(Head on 1st throw) $\times$ P(Head on 2nd throw)

$$= 1/2 \times 1/2 = 1/4$$

It is obvious that the outcome of the first throw does not affect the outcome of the second throw, and it appears that the outcomes of the two throws are independent and that we may multiply the probabilities of the two individual events, head on first throw and head on second throw, in order to arrive at the probability of obtaining a head on each of the throws.

Example 2.3.2(2)

Mwangi tosses a coin and Owina throws a die. What is the probability that Mwangi obtains a head on one toss of the coin and Owina obtains 5 or 6 on one throw of the die?

solution:

We could list the possibility space

$$(H, 1), (H, 2), (H, 3), (H, 4), (H, 5)\ (H, 6),$$
$$(T, 1), (T, 2), (T, 3), (T, 4), (T, 5), (T, 6);$$

with favourable outcomes as $(H, 5)$, $(H, 6)$, and deduce the required probability to be $2/12 = 1/6$.

or

$$P(\text{Mwangi obtains a head}) = 1/2$$
$$P(\text{Owina obtains 5 or 6}) = 1/3$$

so $P(\text{Mwangi obtains a head and Owina obtains 5 or 6}) = 1/2 \times 1/3$

$$= 1/6$$

as the two events are independent.

Example 2.3.2(3)

Bag 1 contains 6 white and 4 blue balls.
Bag 2 contains 5 white and 3 blue balls.

If one ball is drawn from each bag, find the probability that (a) both are white (b) both are blue (c) one is blue and one is white.

Solution:

In this case the colour of ball from one bag is not affected by the colour of ball from the other bag, i.e. the events we are looking at in (a), (b), (c) are independent.

(a) Two white balls occur when a white ball is taken from each bag. If we let

W_1 means a white ball is drawn from bag 1,

W_2 man a white ball is drawn from bag 2,

then

$$P(W_1 \wedge W_2) = P(W_1) \times P(W_2)$$
$$= 6/10 \times 5/8 = 3/8$$

(b) Let B_1 mean a blue ball is drawn from bag 1,
B_2 mean a blue ball is drawn from bag 2,

then

$$P(B_1 \wedge B_2) = P(B_1) \times P(B_2)$$
$$= 4/10 \times 3/8 = 3/20$$

(c) We may obtain one ball of each colour by either drawing a blue ball from bag 1 and then a white ball from bag 2, or by drawing a white ball from bag 1 and then a blue ball from bag 2.

These events $B_1 \wedge W_2$, $W_1 \wedge B_2$, are mutually exclusive, as it is not possible to draw a ball that is coloured both blue and white from bag 1 at the same draw.

$$P(B_1 \wedge W_2) = P(B_1) \times P(W_2)$$
$$= 4/10 \times 5/8$$
$$P(W_1 \wedge B_2) = P(W_1) \times P(B_2)$$
$$= 6/10 \times 3/8$$

hence $P(\text{one of each colour})$

$= P((B_1 \wedge W_2) \vee (W_1 \wedge B_2))$

$= P(B_1 \wedge W_2) + P(W_1 \wedge B_2)$ (The events are mutually exclusive)

$= 4/10 \times 5/8 + 6/10 \times 3/8$

$= 38/80$

$= 19/40.$

2.3.3 Mutually Exclusive Events

As we have seen earlier, two events X and Y are said to be mutually exclusive if the occurrence of one prevents the occurrence of the other. In such cases X and Y cannot happen together, and so by implication X and Y are mutually exclusive whenever $P(X \wedge Y) = 0$, with the result that in such cases

$$P(X \vee Y) = P(X) + P(Y).$$

Example 2.3.3(1)

In a single draw from a pack of 52 playing cards A is the event of drawing an ace, B is the event of drawing a card marked '10', and C is the event of drawing any card marked between 2 and 5 inclusive.

Find (a) $P(A \vee B)$ (b) P(C)

Solution:

(a) Clearly A and B are mutually exclusive. So $P(A \vee B) = P$(drawing either an ace or a ten)

$= 4/52 + 4/52$

$= 8/52$

$= 2/13$

(b) Event C can be thought of as 'either a two or a three or a four or a five' and all four events are mutually exclusive. The probability of any one of these events is 4/52.

so $P(C) = 4/52 + 4/52 + 4/52 + 4/52 = 16/52 = 4/13.$

Example 2.3.3(2)

A fair die is thrown once.

Events X, Y, Z, T, V are defined as follows.

X = the number on the uppermost face is even

Y = the number on the uppermost face is odd

Z = the number on the uppermost face is prime

T = the number on the uppermost face is less than 5

U = the number on the uppermost face is more than 4

Find $P(X \vee Y)$, $P(X \vee Z)$, $P(Y \vee Z)$, $P(T \vee V)$. Name any pairs of events which are mutually exclusive.

Solution:

From possibility space of (1, 2, 3, 4, 5, 6)

$$X = (2, 4, 6),\ Y = (1, 3, 5),$$
$$Z = (2, 3, 5),\ T = (1, 2, 3, 4),\ U = (5, 6)$$
$$P(X \vee Y) = P(X) + P(Y) - P(X \wedge Y) = 3/6 + 3/6 - 0$$
$$= 3/6 + 3/6 = 1$$
$$P(X \vee Z) = P(X) + P(Z) - P(X \wedge Z)$$
$$= 3/6 + 3/6 - 1/6 = 5/6$$
$$P(Y \vee Z) = P(Y) + P(Z) - P(Y \wedge Z)$$
$$= 3/6 + 3/6 - 2/6$$
$$= 4/6 = 2/3$$
$$P(T \vee U) = P(T) + P(U) - P(T \wedge U)$$
$$= 4/6 + 2/6 - 0$$
$$= 1$$

the pairs (X, Y), (T, U) are mutually exclusive.

2.3.4 Conditional Probability and Independence

For any two events A and B there are four possibilities:

(1) Both events happen, i.e. event $A \wedge B$ happens;

(2) A happens but B does not, i.e. event $A \wedge \sim B$ happens;

(3) B happens but A does not, i.e. event $\sim A \wedge B$ happens;

(4) Neither event happens, i.e. $\sim A \wedge \sim B$ happens.

We now define $P(B \mid A)$ as the conditional probability that B happens, given that A has already happened. We shall now investigate how this probability can be related to probabilities which involve events A, B or both events.

Example 2.3.4(1)

A number is chosen at random from the numbers 1, 2, 3, ... 30. Find the probability that it is (a) divisible by 3 (b) divisible by 4 (c) divisible by both 3 and 4 (d) divisible by 4, given that it is divisible by 3.

Solution:

Let A = {numbers divisible by 3}

and B = {numbers divisible by 4}

based on E = {1, 2, 3, ..., 30}

so A = {3, 6, 9, 12, 15, 18, 21, 24, 27, 30}

and B = {4, 8, 12, 16, 20, 24, 28}

(a) $P(A) = 10/30 = 1/3$

(b) $P(B) = 7/30$

(c) $P(A \wedge B) = 2/30 = 1/15$

as there are only two numbers (12 and 24) which are divisible by both 3 and 4.

(d) We wish to find the probability that event B will happen, given that we already know that event A has happened. We wish to know the probability of event B, based on the probability space for event A which is {3, 6, 9, 12, 15, 18, 21, 24, 27, 30}

The possibility space for A has ten numbers of which only two numbers (12 and 24) are divisible by 4.

Hence $$P(B \mid A) = 2/10 \text{ or } 1/5$$

However looking at the answers for parts (a) and (c) we observe that

$$P(A) = 1/3 = 10/30$$

$$P(B \wedge A) = P(A \wedge B) = 1/15 = 2/30$$

$$\frac{P(A \wedge B)}{P(A)} = \frac{2/30}{10/30} = 1/5$$

This leads to the definition of $P(B \mid A)$, using probabilities of events A, B or both, as:

$$P(B \mid A) = \frac{P(B \wedge A)}{P(A)}$$

N.B: $$P(B \mid A) \neq P(A \mid B)$$

$P(A \mid B)$ means the probability of event A, given that event B has happened. In this case we require the probability that a number is divisible by 3, given that it is divisible by 4.

Using the possibility space for B i.e. {4, 8, 12, 16, 20, 24, 28} we observe that there are 7 numbers of which 2 numbers (12 and 24) are divisible by 3, leading to

$$P(A \mid B) = 2/7$$

Using the definition $$P(A \mid B) = \frac{P(A \wedge B)}{P(B)}$$

$$= \frac{2/30}{7/30} = 2/7$$

This is not the same probability as

$P(A \mid B) = 1/5$ obtained in (d).

Example 2.3.4(2)

Two dice, coloured red and green respectively, are thrown in a game. The player has to score a total of 9 or more in order to stay in the game. Find the probability that

(a) the player scores a total of 9 or more

(b) the player scores a total of 9 or more given that the score on the green die is 3,

(c) the player scores a total of 9 or more, given that the score on the green die is 6.

Solution:

Let event A be : total is 9 or more

event B be : the score on the green die is 3,

event C be : the score on the the green die is 6.

We need to find $P(A)$, $P(A \mid B)$, and $P(A \mid C)$. The possibility space for the scores on the two dice is shown below. [Fig. 2.3.4(a)]

Red die						
6	7	8	9	10	11	12
5	6	7	8	9	10	11
4	5	6	7	8	9	10
3	4	5	6	7	8	9
2	3	4	5	6	7	8
1	2	3	4	5	6	7
	1	2	3	4	5	6
	Green die					

(B: column for green die 3; C: column for green die 6; A: region of totals 9 or more)

Fig. 2.3.4(1)

(a) There are 36 possible outcomes of which 10 outcomes give a score of 9 or more on the two dice,

so $$P(A) = 10/36 = 5/18$$

(b) Using possibility space, the possibility space for event B is {(3, 1), (3, 2), (3, 3), (3, 4), (3, 5), (3, 6)} (taking the score on the green die first). Of these outcomes only one outcome (3, 6) gives a total of 9 or more hence

$$P(A \mid B) = 1/6$$

Using the probability definition

$$P(A \mid B) = P(A \wedge B)/P(B)$$

In the possibility space for all outcomes of the two throws $P(A \wedge B) = 1/36$ (i.e. we need a score of 3 recorded on the green die and a total score of 9 or more)

$$P(B) = 6/36$$

Hence $$P(A \mid B) = 1/36 \;/\; 6/36 = 1/6$$

(c) Using possibility space, the possibility space for event C is {(6, 1), (6, 2), (6, 3), (6, 4), (6, 5), (6, 6)}. Of these outcomes four outcomes {(6, 3), (6, 4), (6, 5), (6, 6)} give a total score 9 or more hence

$$P(A \mid C) = 4/6 = 2/3$$

Using the probability definition

$$P(A \mid C) = P(A \wedge C)/P(C) = 4/36 \;/\; 6/36 = 4/6 = 2/3$$

It will be noticed that if the score on the green die is known to be 6, there is a greater probability of success in reaching the required total of 9 or more.

Example 2.3.4(3)

For two events A and B, $P(\sim A) = 1/3$, $P(\sim B) = 1/2$ and $P(A \vee B) = 3/4$. Find $P(A \mid B)$.

Solution:

$$P(\sim A) = 1/3 \text{ so } P(A) = 2/3$$
$$P(\sim B) = 1/2 \text{ so } P(B) = 1/2$$

Using $$P(A \vee B) = P(A) + P(B) - P(A \wedge B)$$
$$3/4 = 1/2 + 2/3 - P(A \wedge B)$$

so $$P(A \wedge B) = 1/2 + 2/3 - 3/4$$
$$= 5/12$$

so $$P(A \mid B) = \frac{P(A \wedge B)}{P(B)} = \frac{5/12}{1/2}$$
$$= 5/6$$

In this example we had to use the definition given by the probabilities as it was not possible to list the possibility spaces.

We are now going to look at the meaning of independent events, using conditional probabilities. In some cases two events, which do not appear at first sight to be independent, may be shown to be independent.

Example 2.3.4(4)

A set of cards, identical in size and shape, is numbered from 1 to 15 inclusive. A card is drawn at random from the pack and its number n is noted.

A is the event : n is prime
B is the event : $n \leq 10$

Find (a) $P(A)$ (b) $P(B)$ (c) $P(A \mid B)$ (d) $P(A \mid \sim B)$.

Record your observations about the answers for (a), (c) and (d).

Solution:

Setting out the possibility space and indicating all the numbers associated with event A by ringing them we have [Fig. 2.3.4(b)]

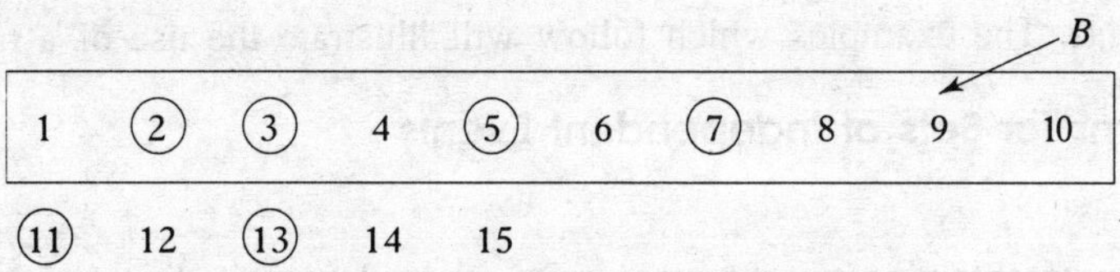

Fig. 2.3.4(b)

So (a) $P(A) = 6/15 = 2/5$

(b) $P(B) = 10/15 = 2/3$

(c) $P(A \wedge B) = 4/15$

so $$P(A \mid B) = P(A \wedge B)/P(B)$$
$$= 4/15/2/3$$
$$= 12/30$$

$$= 2/5$$

(d) $P(A \wedge \sim B) = 2/15$

$$P(A \mid \sim B) = P(A \wedge \sim B)/P(\sim B)$$
$$= 2/15/5/15$$
$$= 2/5$$

Looking at the answers we see that $P(A) = P(A \mid B) = P(A \mid \sim B)$ in other words the probabilities of drawing a prime number remains the same, whether the draw is made from numbers lying between 1 and 10 inclusive, or from numbers lying between 11 and 15 inclusive.

This last example illustrates the definition of independent events which we shall use from now onwards:

In general two events X and Y are said to be independent whenever $P(X) = P(X \mid Y) = P(X \mid \sim Y)$.

Note (1) In practice it will be sufficient to show that two of the three probabilities, $P(X)$, $P(X \mid Y)$, $P(X \mid \sim Y)$ are equal;

(2) When X and Y are independent

then $$P(X) = P(X \mid Y)$$

so $$P(X) = \frac{P(X \wedge Y)}{P(Y)}$$

so $$P(X \wedge Y) = P(X)\ P(Y)$$

(3) When X and Y are not independent

then $$P(X \mid Y) = \frac{P(X \wedge Y)}{P(Y)}$$

so that $$P(X \wedge Y) = P(Y)\ P(X \mid Y)$$

2.4 TREE DIAGRAMS

A tree diagram is a useful aid in solving problems where probabilities will be involving two or more sets of compound events. The examples which follow will illustrate the use of a tree diagram.

2.4.1 Tree Diagrams for Sets of Independent Events

Example 2.4.1(1)

What is the probability of obtaining two or more sixes when three fair dice are thrown?

Solution:

For the first throw set up two branches representing S(Six), and ~S(other score), and mark on each branch the respective probabilities as in Fig. 2.4.1(a).

For the second throw add two more branches to each branch already drawn, and then mark on each branch the respective probabilities [see Fig. 2.4.1(b)].

The probabilities on the new branches are still 1/6 for a six, and 5/6 for another score, because the outcomes on the second throw are independent of those on the first throw. For the third throw we

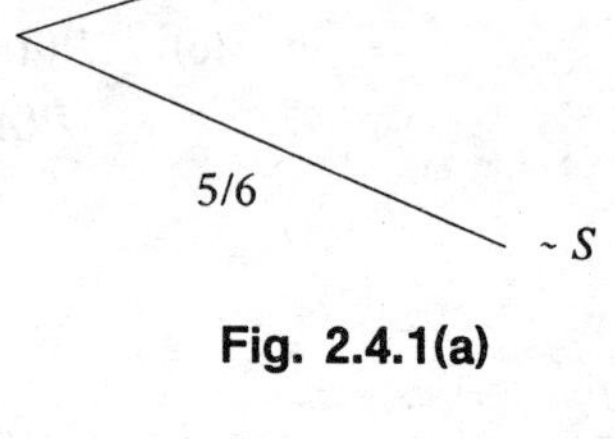

Fig. 2.4.1(a)

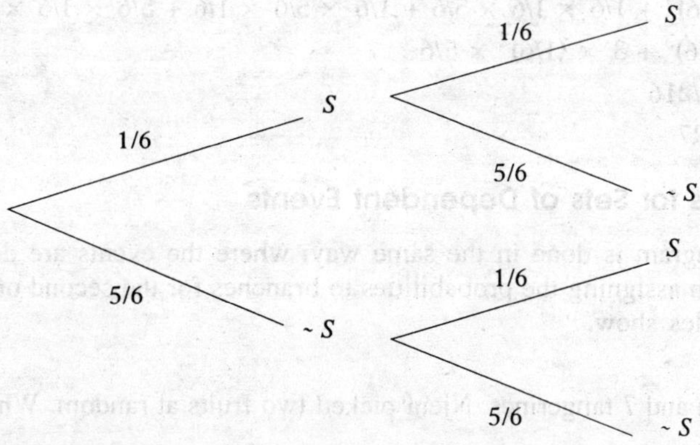

Fig. 2.4.1(b)

add two more branches to each branch already drawn for the second throw. The final diagram has eight branches added to the previous diagram [see Fig. 2.4.1(c)]

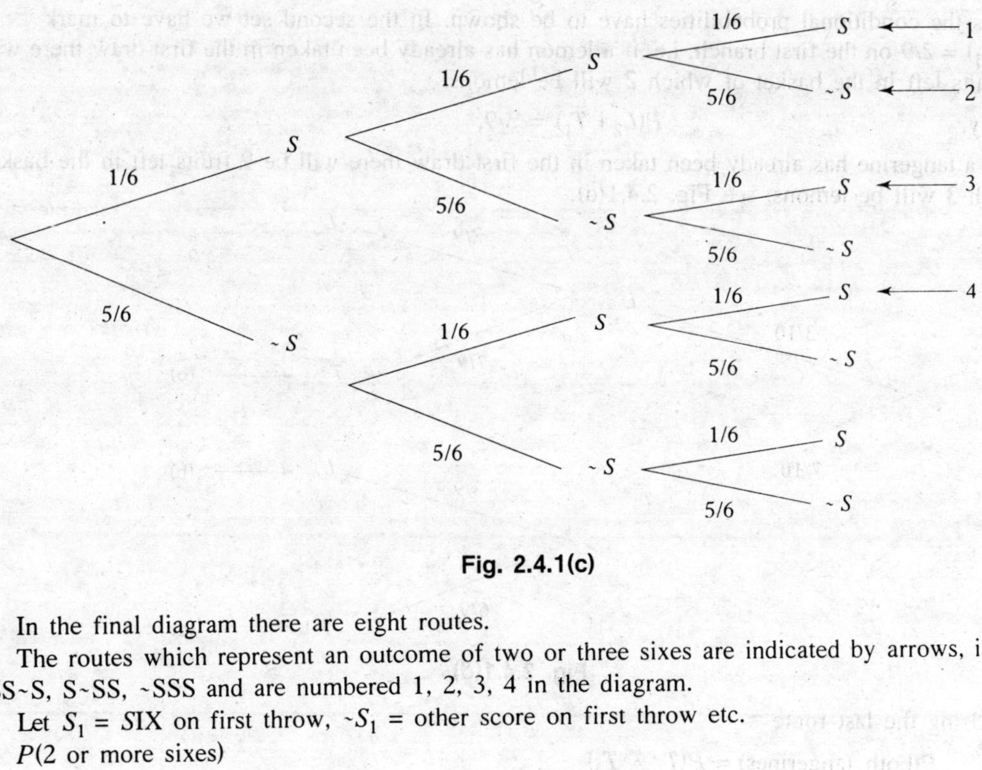

Fig. 2.4.1(c)

In the final diagram there are eight routes.

The routes which represent an outcome of two or three sixes are indicated by arrows, i.e. SSS, SS~S, S~SS, ~SSS and are numbered 1, 2, 3, 4 in the diagram.

Let S_1 = *S*IX on first throw, $\sim S_1$ = other score on first throw etc.

P(2 or more sixes)

$$= P(S_1 \wedge S_2 \wedge S_3) + P(S_1 \wedge S_2 \wedge \sim S_3)$$
$$+ P(S_1 \wedge \sim S_2 \wedge S_3) + P(\sim S_1 \wedge S_2 \wedge S_3)$$

$= (1/6)^3 + 1/6 \times 1/6 \times 5/6 + 1/6 \times 5/6 \times 1/6 + 5/6 \times 1/6 \times 1/6$
$= (1/6)^3 + 3 \times (1/6)^2 \times 5/6$
$= 16/216$
$= 2/27$

2.4.2 Tree Diagrams for Sets of Dependent Events

Building up the tree diagram is done in the same way, where the events are dependent, but some modification is needed in assigning the probabilities to branches for the second or subsequent events, as the following examples show.

Example 2.4.2(1)

A basket holds 3 lemons and 7 tangerines. Njeni picked two fruits at random. What is the probability that

(a) both were tangerines?
(b) there was one lemon and there was one tangerine?
(c) there was at least one tangerine?

Solution:

Using L for lemon T for tangerine, the tree diagram can be drawn. However in the second set of branches the conditional probabilities have to be shown. In the second set we have to mark $P(L_2 \mid L_1) = 2/9$ on the first branch, i.e. if a lemon has already been taken in the first draw there will be 9 fruits left in the basket of which 2 will be lemons.

Similarly, $$P(L_2 \mid T_1) = 3/9,$$

since if a tangerine has already been taken in the first draw there will be 9 fruits left in the basket of which 3 will be lemons, see Fig. 2.4.1(d).

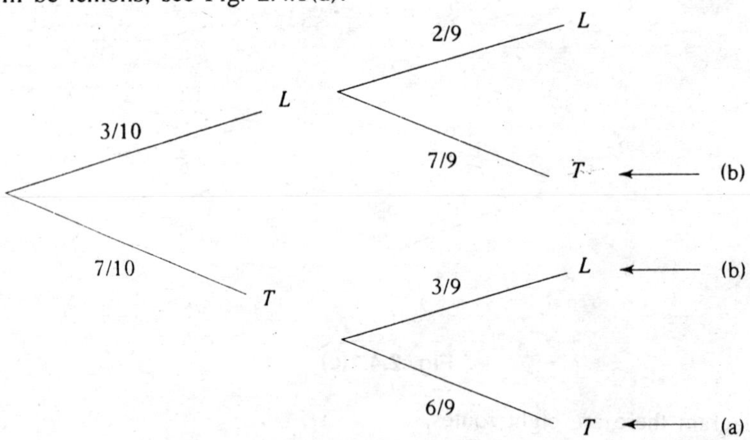

Fig. 2.4.1(d)

(a) Using the last route

$$P(\text{both tangerines}) = P(T_1 \wedge T_2)$$
$$= P(T_1)\, P(T_2 \mid T_1)$$
$$= 7/10 \times 6/9$$
$$= 7/15$$

(b) Using the routes marked (b)

$$\begin{aligned}&P(\text{one lemon and one tangerine})\\&= P(L_1 \wedge T_2) + P(T_1 \wedge L_2)\\&= P(L_1)\, P(T_2 \mid L_1) + P(T_1)\, P(L_2 \mid T_1)\\&= 3/10 \times 7/9 + 7/10 \times 3/9\\&= 7/15\end{aligned}$$

(c) The quickest method of answering this part is to use

$$\begin{aligned}P(\text{at least one tangerine}) &= 1 - P(\text{no tangerine})\\&= 1 - P(L_1 \wedge L_2)\end{aligned}$$

Using the first route we get

$$\begin{aligned}&= 1 - 3/10 \times 2/9\\&= 1 - 1/15\\&= 14/15\end{aligned}$$

The alternative is to add together the probabilities for the routes marked (a) and (b)

i.e. $$7/10 \times 6/9 + 3/10 \times 7/9 + 7/10 \times 3/9$$

to arrive at the same answer of 14/15.

Example 2.4.2(2)

A box holds 8 red and 4 blue beads. Three beads are taken from the box and not replaced. What is the probability that

(a) all three are red?

(b) there are at least two blue beads?

Solution:

We set up the tree diagram with three sets of branches. In the second and third sets attention must be paid to the number of beads of each colour which remain after a previous draw [as in Fig. 2.4.2(a)].

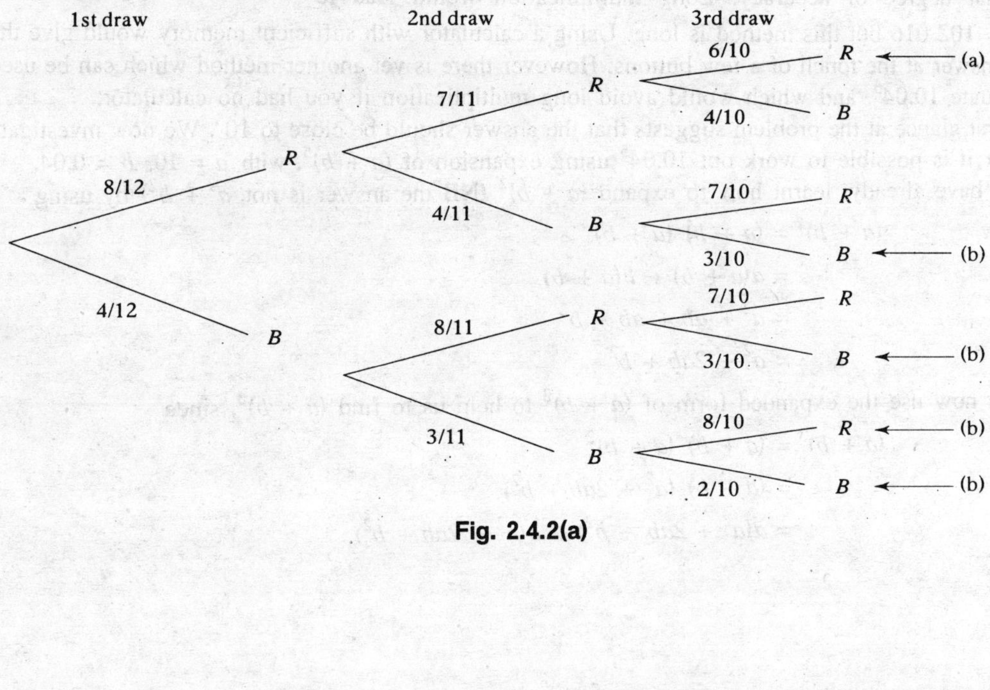

Fig. 2.4.2(a)

(a) Using the arrow marked (a)

$$P(\text{all three are red}) = P(R_1 \wedge R_2 \wedge R_3)$$
$$= 8/12 \times 7/11 \times 6/10$$
$$= 14/55$$

(b) Using the arrows marked (b)

P(at least two blue beads)

$$= P(R_1 \wedge B_2 \wedge B_3) + P(B_1 \wedge R_2 \wedge B_3)$$
$$+ P(B_1 \wedge B_2 \wedge R_3) + P(B_1 \wedge B_2 \wedge B_3)$$
$$= 8/12 \times 4/11 \times 3/10 + 4/12 \times 8/11 \times 3/10$$
$$+ 4/12 \times 3/11 \times 8/10 + 4/12 \times 3/11 \times 2/10$$
$$= 39/165 \text{ or } 13/55.$$

2.5 BINOMIAL DISTRIBUTION

2.5.1 Binomial Theorem

We shall look at one way of establishing the result for the binomial theorem (we shall look at another way in chapter 3 section 3.5.7). The theorem has applications in probability.

There are several methods available for evaluating 10.04^5. If no degree of accuracy was asked for you would probably work out the problem using logarithms and give an answer of 1.02×10^5 correct to three significant figures.

No	log
10.04	1.0017
	× 5
1.02×10^5	5.0085

Suppose the degree of accuracy required was the nearest whole number. Tables of four figure logarithms would not be sufficient to give that degree of accuracy. Long multiplication would lead to answer 102 016 but this method is long. Using a calculator with sufficient memory would give the same answer at the touch of a few buttons. However there is yet another method which can be used to evaluate 10.04^5, and which would avoid long multiplication if you had no calculator.

A first glance at the problem suggests that the answer should be close to 10^5. We now investigate whether it is possible to work out 10.04^5 using expansion of $(a + b)^5$, with $a = 10$, $b = 0.04$.

You have already, learnt how to expand $(a + b)^2$ (**NB** the answer is not $a^2 + b^2$) by using

$$(a + b)^2 = (a + b)(a + b)$$
$$= a(a + b) + b(a + b)$$
$$= a^2 + ab + ab + b^2$$
$$= a^2 + 2ab + b^2$$

We can now use the expanded form of $(a + b)^2$ to help us to find $(a + b)^3$, since

$$(a + b)^3 = (a + b)(a + b)^2$$
$$= (a + b)(a^2 + 2ab + b^2)$$
$$= a(a^2 + 2ab + b^2) + b(a^2 + 2ab + b^2)$$

$$= a^3 + 2a^2\,b + ab^2 + a^2b + 2ab^2 + b^3$$
$$= a^3 + 3a^2b + 3ab^2 + b^3$$

In a similar way we can work out $(a + b)^4$ using the expansion of $(a + b)^3$ as $(a + b)^4 = (a + b)\,(a + b)^3$

$$= (a + b)(a^3 + 3a^2\,b + 3ab^2 + b^3)$$
$$= a(a^3 + 3a^2\,b + 3ab^2 + b^3) + b(a^3 + 3a^2\,b + 3a\,b^2 + b^3)$$
$$= a^4 + 3a^3\,b + 3a^2\,b^2 + ab^3 + a^3\,b + 3a^2\,b^2 + 3ab^3 + b^4$$
$$= a^4 + 4a^3\,b + 6a^2\,b^2 + 4ab^3 + b^4$$

In the same way you may try for yourself to find the expansion of $(a + b)^5$ and $(a + b)^6$. The answers are

$$(a + b)^5 = a^5 + 5a^4\,b + 10a^3\,b^2 + 10a^2\,b^3 + 5ab^4 + b^5$$
$$(a + b)^6 = a^6 + 6a^5\,b + 15a^4\,b^2 + 20a^3\,b^3 + 15a^2\,b^4 + 6ab^5 + b^6$$

Is it possible to deduce the coefficients for each of the terms in the expansion of $(a + b)^7$ and higher powers without first multiplying out the brackets?

Arrange the coefficients by rows in a triangular array, starting with

$$(a + b)^0 = 1$$
$$(a + b)^1 = a + b,$$

and see if a pattern emerges [see Fig. 2.5.1(a)].

Power	Coefficients
0	1
1	1 1
2	1 2 1
3	1 3 3 1
4	1 4 6 4 1
5	1 5 10 10 5 1
6	1 6 15 20 15 6 1

Fig. 2.5.1(a)

Can the coefficients for any row be deduced from those in the previous row? The coefficients in the expansion of $(a + b)^5$ are known from the row 1 5 10 10 5 1. If we did not know the expansion for $(a + b)^6$ could the coefficients in the row 1 5 10 10 5 1 be used to help us to find them?

Observing the pattern we notice that there is a '1' at the beginning and end of each row. The coefficient for each of the remaining terms can be found by adding the two coefficients adjacent to it in the row above, e.g.

for $(a + b)^5$ 1 5 10 10 5 1

for $(a + b)^6$ 1 (6) (15) (20) (15) (6) 1

When '1' has been placed at the beginning and end of the row, the other coefficients may be obtained using 1 + 5 = 6, 5 + 10 = 15, 10 + 10 = 20, 10 + 5 = 15, 5 + 1 = 6.

This procedure may then be extended in order to find the coefficients of the terms in the expansions of $(a + b)^7$, $(a + b)^8$ etc. The resulting array of coefficients is known as Pascal's triangle, named after the famous French mathematician Blaise Pascal (1623—1662) [see Fig. 2.5.1(b)]

Power	Coefficients
0	1
1	1 1
2	1 2 1
3	1 3 3 1
4	1 4 6 4 1
5	1 5 10 10 5 1
6	1 6 15 20 15 6 1
7	1 7 21 35 35 21 7 1
8	1 8 28 56 70 56 28 8 1
9	1 9 36 84 126 126 84 36 9 1
10	1 10 45 120 210 252 210 120 45 10 1

Fig. 2.5.1(b)

The coefficients are always symmetrical about the central vertical line of the triangle, e.g. in the row for the expansion of $(a + b)^7$ the entries read 1 7 21 35 35 21 7 1, and so the coefficients of $a^5 b^2$ and of $a^2 b^5$ are both equal to 21.

In order to find the coefficients of $(a + b)^n$ where n is a positive integer, look for the row beginning 1 n (where n is the second coefficient), e.g. for the expansion of $(a + b)^9$ we look for the row which begins 1 9 ... i.e. 1 9 36 84 126 126 84 36 9 1. From Pascal's triangle it is then possible to write down the expansion of

$$\begin{aligned}(a + b)^9 &= a^9 + 9a^8 b + 36a^7 b^2 \\ &\quad + 84a^6 b^3 + 126a^5 b^4 + 126\, a^4 b^5 \\ &\quad + 84a^3 b^6 + 36a^2 b^7 + 9ab^8 + b^9\end{aligned}$$

In the examples which follow we shall make use of Pascal's triangle to find particular binomial expansions or individual terms of an expansion.

Example 2.5.1 (1)

Write down the expansion of $(5x - y)^5$

Solution:

Here $a = 5x,\ b = -y$

Powers	$(5x)^5$	$(5x)^4 (-y)$	$(5x)^3(-y)^2$	$(5x)^2 (-y)^3$	$(5x)(-y)^4$	$(-y)^5$
Row of Pascal's triangle	1	5	10	10	5	1

so $(5x - y)^5 = (5x)^5 + 5(5x)^4(-y) + 10(5x)^3(-y)^2$

$+ 10(5x)^2(-y)^3 + 5(5x)(-y)^4 + (-y)^5$

$= 3125x^5 - 3125\,x^4y + 1250\,x^3y^2$

$- 250\,x^2y^3 + 25xy^4 - y^5$

Example 2.5.1(2)

Write down and simplify the expansion

$$(\sqrt{6} + \sqrt{2})^4 - (\sqrt{6} - \sqrt{2})^4$$

Solution:

In the first expression $(\sqrt{6} + \sqrt{2})^4$

$$a = \sqrt{6},\ b = \sqrt{2}$$

Powers	$(\sqrt{6})^4$	$(\sqrt{6})^3\sqrt{2}$	$(\sqrt{6})^2(\sqrt{2})^2$	$\sqrt{6}(\sqrt{2})^3$	$(\sqrt{2})^4$
Rows of Pascal's triangle	1	4	6	4	1

Expansion is $(\sqrt{6})^4 + 4(\sqrt{6})^3\sqrt{2} + 6(\sqrt{6})^2(\sqrt{2})^2 + 4\sqrt{6}(\sqrt{2})^3 + (\sqrt{2})^4$

In the second expression $(\sqrt{6} - \sqrt{2})^4$,

$$a = \sqrt{6},\ b = -\sqrt{2}$$

Powers	$(\sqrt{6})^4$	$(\sqrt{6})^3(-\sqrt{2})$	$(\sqrt{6})^2(-\sqrt{2})^2$	$\sqrt{6}(-\sqrt{2})^3$	$(-\sqrt{2})^4$
Rows of Pascal's triangle	1	4	6	4	1

Expansion is

$$(\sqrt{6})^4 + 4(\sqrt{6})^3(-\sqrt{2}) + 6(\sqrt{6})^2(-\sqrt{2})^2 + 4(\sqrt{6})(-\sqrt{2})^3 + (-\sqrt{2})^4$$

so $(\sqrt{6} + \sqrt{2})^4 - (\sqrt{6} - \sqrt{2})^4$

$= (\sqrt{6})^4 + 4(\sqrt{6})^3(\sqrt{2}) + 6(\sqrt{6})^2(\sqrt{2})^2 + 4(\sqrt{6})(\sqrt{2})^3 + (\sqrt{2})^4$

$- \{(\sqrt{6})^4 - 4(\sqrt{6})^3(\sqrt{2}) + 6(\sqrt{6})^2(\sqrt{2})^2 - 4(\sqrt{6})(\sqrt{2})^3 + (\sqrt{2})^4\}$

$= (\sqrt{6})^4 + 4(\sqrt{6})^3(\sqrt{2}) + 6(\sqrt{6})^2(\sqrt{2})^2 + 4(\sqrt{6})(\sqrt{2})^3 + (\sqrt{2})^4$

$- (\sqrt{6})^4 + 4(\sqrt{6})^3(\sqrt{2}) - 6(\sqrt{6})^2(\sqrt{2})^2 + 4(\sqrt{6})(\sqrt{2})^3 - (\sqrt{2})^4$

$= 2 \times 4(\sqrt{6})^3\sqrt{2} + 2 \times 4(\sqrt{6})(\sqrt{2})^3$

$= 8 \times 6\sqrt{6}\sqrt{2} + 8\sqrt{6} \times 2\sqrt{2}$

$= 64\sqrt{6}\sqrt{2}$

$= 64\sqrt{2}$

$= 64 \times 2\sqrt{3}$

$= 128\sqrt{3}$

Example 2.5.1(3)

Find (a) the coefficient of x^4,

(b) the sixth term, in the expansion of $(2 + 3x)^8$

Solution:

(a) In the expansion $a = 2$, $b = 3x$. Since the total coefficient (Not just the binomial coefficient from the row in Pascal's triangle) is required we need to find the binomial coefficient of the term in $2^4(3x)^4$. The relevant row of binomial coefficients in Pascal's triangle is 1, 8, 28, 56, 70, 56, 28, 8, 1, so the term is $70 \times 2^4 \times (3x)^4$

$$= 70 \times 2^4 \times 3^4\, x^4$$

and the required coefficient is

$$70 \times 2^4 \times 3^4 = 90\ 720$$

(b) As $a = 2$, $b = 3x$ in the expansion of $(2 + 3x)^8$ the first term is 2^8, i.e. it does not contain x, and the second term is $8 \times 2^7\,(3x)$, i.e. it contains x^1. The sixth term contains x^5 and it will be

$$56 \times 2^3 \times (3x)^5 = 108\ 864\ x^5$$

Example 2.5.1(4)

Evaluate 10.04^5 correct to the nearest whole number using the binomial theorem.

Solution:

We saw earlier that the answer is approximately 10^5.

If we expand $(10 + x)^5$ we get

$$10^5 + 5 \times 10^4 x + 10 \times 10^3 x^2 + 10 \times 10^2 x^3 + 5 \times 10\, x^4 + x^5$$

Substitute $x = 0.04$

Evaluating successive terms we may add as follows:

$$\begin{aligned} 10^5 &= 100\ 000 \\ 5 \times 10^4 \times 0.04 &= 2\ 000 \\ 10 \times 10^3 \times 0.04^2 &= 16 \\ 10 \times 10^2 \times 0.04^3 &= 0.064 \\ \hline & \quad 102\ 016.064 \end{aligned}$$

The terms are becoming smaller and the fourth term is less than 0.1, so we may state that $10.04^5 = 102\ 016$ to the nearest whole number.

Example 2.5.1(5)

Expand and simplify the terms in ascending powers of x up to and including the term in x^3 in the expansion of $(4 - x)^6$, and hence evaluate 3.995^6 correct to two decimal places.

Solution:

$$\begin{aligned} (4 - x)^6 &= 4^6 + 4 \times 4^5(-x) + 6 \times 4^4(-x)^2 + 15 \times 4^3(-x)^3 + \ldots \\ &= 4096 - 4096\, x + 1536x^2 - 960\, x^3 + \ldots \end{aligned}$$

In order to evaluate 3.995^6 put

$$x = 4 - 3.995 = 0.005$$

$$3.995^6 = 4096 - 4096 \times 0.005 + 1536 \times 0.005^2 - 960 \times 0.005^3 + \ldots$$

Separating positive and negative terms

$$\begin{aligned} 4096 &= 4096 \\ 1536 \times 0.005^2 &= 0.0384 \\ &\overline{4096.0384} \\ 4096 \times 0.005 &= 20.48 \\ 960 \times 0.005^3 &= 0.00012 \\ &\overline{20.48012} \\ 3.995^6 &= 4096.0384 \\ &\ -20.48012 \\ &\overline{4075.55828} \end{aligned}$$

so $3.995^6 = 4075.56$ to two decimal places.

Example 2.5.1(6)

Expand $(3 - 2x + x^2)^6$ in ascending powers of x up to and including the term in x^3.

Solution:

Here the expression we wish to expand contains three terms, but we require an expansion in ascending powers of x. Taking $a = 3$, $b = -2x + x^2 = x(-2 + x)$, it is possible to write the expansion of $(3 - 2x + x^2)^6$ as $3^6 + 6 \times 3^5 x(-2 + x) + 15 \times 3^4 x^2(-2 + x^2)^2 + 20 \times 3^3 x^3(-2 + x)^3 + ...$

Other terms may be omitted as they contain powers of x which are higher than x^3.

Then expand the powers of $(-2 + x)$ as far as is required to give all the terms up to x^3 in the original expansion.

Hence

$$\begin{aligned} (3 - 2x + x^2)^6 &= 729 - 2916x + 1458x^2 \\ &+ 1215\ x^2(4 - 4x + x^2) + 540\ x^3(-8 + ...) + ... \\ &= 729 - 2916\ x + 1458\ x^2 \\ &+ 4860\ x^2 - 4860\ x^3 - 4320\ x^3 + ... \quad \text{(taking all terms in powers of } x \text{ up to } x^3) \\ &= 729 - 2916x + 6318x^2 - 9180\ x^3 + ... \end{aligned}$$

2.5.2 Binomial Coefficients

The coefficients for the expansion of $(a + b)^n$ for any positive integer n can be worked out by building up successive rows of Pascal's triangle. This procedure may take a long time for large n, say $n = 12$ or $n = 16$. A table of binomial coefficients is provided in many books of mathematical tables up to $n = 20$, and such a table is provided below. In chapter III on algebra, where further work on the binomial theorem is covered, we shall discuss an alternative way of working out binomial coefficients which will be applicable for higher values of n, and the coefficients may also be evaluated using an electronic calculator.

Table 2.5.2(a) Binomial coefficients $\binom{n}{i}$

n \ i	0	1	2	3	4	5	6	7	8	9	10
1	1	1									
2	1	2	1								
3	1	3	3	1							
4	1	4	6	4	1						
5	1	5	10	10	5	1					
6	1	6	15	20	15	6	1				
7	1	7	21	35	35	21	7	1			
8	1	8	28	56	70	56	28	8	1		
9	1	9	36	84	126	84	36	9	1		
10	1	10	45	120	210	252	210	120	45	10	1
11	1	11	55	165	330	462	462	330	165	55	11
12	1	12	66	220	495	792	924	792	495	220	66
13	1	13	78	286	715	1287	1716	1716	1287	715	286
14	1	14	91	364	1001	2002	3003	3432	3003	2002	1001
15	1	15	105	455	1365	3003	5005	6435	6435	5005	3003
16	1	16	120	560	1820	4368	8008	11440	12870	11440	8008
17	1	17	136	680	2380	6188	12376	19448	24310	24310	19448
18	1	18	153	816	3060	8568	18564	31824	43758	48620	43758
19	1	19	171	969	3876	11628	27132	50388	75582	92378	92378
20	1	20	190	1140	4845	15504	38760	77520	125970	167960	184756

The following examples illustrate how we may use the table.

Example 2.5.2(1)

Write down the expansion of $(1 + x)^{15}$ in ascending powers of x up to and including the term in x^3. find also the coefficients of x^6 and x^{12}.

Solution:

In the table it is possible to determine the coefficient of $a^{n-i} b^i$ in the expansion of $(a + b)^n$ for any given values of n, i up to $n = 20$. In the table the value of n is given in the first left hand column. Find the row which starts with $n = 15$. The entries which follow in the row are 1 below $i = 0$, 15 below $i = 1$, 105 below $i = 2$, and 455 below $i = 3$.

Hence the coefficient of $a^{15} b^0$ is 1

the coefficient of $a^{14} b^1$ is 15

the coefficient of $a^{13} b^2$ is 105

the coefficient of $a^{12} b^3$ is 455

The value of i is the appropriate power of b in the expansion of $(a + b)^n$. Since we have $a = 1$, $b = x$ in this case, we may write

$$(1 + x)^{15} = 1 + 15x + 105\,x^2 + 455\,x^3 + \ldots$$

In order to find the coefficient of x^6 read the entry for the row for $n = 15$ under the heading i = 6. In this case the entry is 5005. The coefficient of x^6 in the expansion is therefore 5005.

For the coefficient of x^{12} you will observe that there is no heading $i = 12$. In general the symmetry of the binomial coefficients in Pascal's triangle means that the binomial coefficient of $a^{n-i} b^i$ is the same as that for $a^i b^{n-i}$. In this case the binomial coefficient for $1^{15-12} x^{12}$ will be the same as that for $1^{12} x^{15-12}$ i.e. that for $1^{12} x^3$. Read the entry for $n = 15$, $i = 3$. The entry is 455. So the coefficient of x^{12} is that for x^3, i.e. 455.

Example 2.5.2(2)

Write down the first three terms in descending powers of x in the expansion of $(3 - x)^{17}$

Solution:

We need the coefficients of $(-x)^{17}$, $3^1(-x)^{16}$ and $3^2(-x)^{15}$, taking $n = 17$, $a = 3$, $b = -x$. Using the row in the table of binomial coefficients which begins with $n = 17$ it is not possible to read the coefficients directly from the table. However the coefficient of b^{17} is the coefficient of a^{17} which is 1. The coefficient of $a\,b^{16}$ is the coefficient of a^{16} *b which is 17. The coefficient of* $a^2 b^{15}$ is the coefficient of $a^{15}\,b^2$ which is 136.

The required terms are $(-x)^{17} + 17 \times 3(-x)^{16} + 136 \times 3^2 (-x)^{15}$

$$= -x^{17} + 51x^{16} - 1224\,x^{15}$$

Example 2.5.2(3)

Find the constant term in the expansion of $(4x^2 - 1/4x)^{15}$

Solution:

We know that in the expansion of $(a + b)^{15}$ the general term contains $a^{15-i}\,b^i$.

Here $a = 4x^2\, b = -1/4x.$

The general term contains

$$(4x^2)^{15-i}\,(-1/4x)^i$$

$$= 4^{15-i}\quad x^{30-2i}\,\frac{(-1)^i}{4^i\,x^i}$$

$$= 4^{15-2i}\,x^{30-3i}\,(-1)^i$$

We require the constant term which is the term in x^0, so the power of x which is $30 - 3i$ mus be zero

i.e. $$30 - 3i = 0$$

or $$i = 10$$

From the table of binomial coefficients with $n = 15$, $i = 10$ the binomial coefficient is 3003.

The constant term is therefore

$$3003 \times 4^{15-2\times 10}\,x^0\,(-1)^{10}$$

$$= 3003 \times 4^{-5}$$

or $$\frac{3003}{1024}.$$

2.5.3 Binomial Theorem Applied to Probability

A student who had done no revision for a test had to answer the questions, which were multiple choice, by guessing each answer.

Find the probability for the number of correct answers for

(i) one question attempted,
(ii) two questions attempted,
(iii) three questions attempted,
(iv) four questions attempted,

if in each question there were three choices.

(i) Clearly for one question there is a probability of 2/3 that he fails to guess the correct answer (F), and a probability of 1/3 that he succeeds in guessing the correct answer (S). The tree diagram for one question is shown below [Fig. 2.5.3(a)].

(ii) If he attempts two questions the outcomes would be fails in both questions, fails in the first and succeeds in the second, succeeds in the first and fails in the second, succeeds in both. The events may be thought of as FF, FS, SF, SS and are illustrated in the tree diagram below [Fig. 2.5.3(b)].

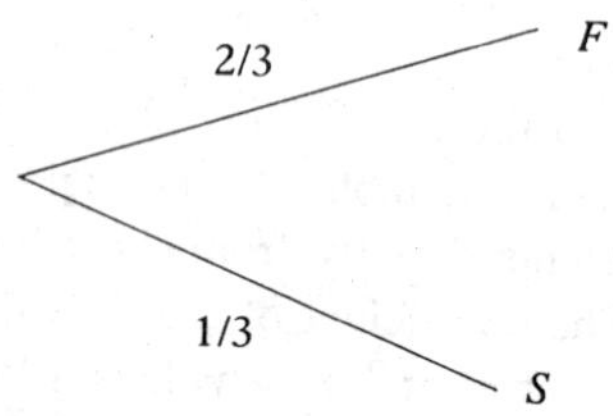

Fig. 2.5.3(a)

From the tree diagram, for two questions

P(neither question correct) = $P(0)$ = P(FF) = $2/3 \times 2/3 = 4/9$ [route (a)]

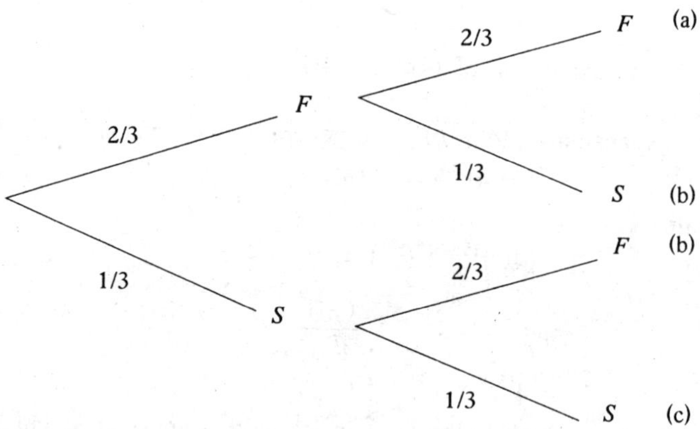

Fig. 2.5.3(b)

P(exactly one question correct) = $P(1) = P(FS) + P(SF)$

$= 2/3 \times 1/3 + 1/3 \times 2/3 = 4/9$ [two routes marked (b)]

P(both questions correct) = $P(2)$ = P(SS) = $1/3 \times 1/3 = 1/9$ [route (c)]

(iii) If he attempts three questions the outcomes would be FFF, FFS, FSF, SFF, FSS, SFS, SSF, SSS, as shown in the tree diagram below [Fig. 2.5.3(c)]

P(no question correct) = $P(0)$

= P(FFF) = $(2/3)^3 = 8/27$ [route (a)]

P(exactly one question correct) = $P(1)$

= P(FFS) + P(FSF) + P(SFF)

$= 2/3 \times 2/3 \times 1/3 + 2/3 \times 1/3 \times 2/3 + 1/3 \times 2/3 \times 2/3 = 3 \times (2/3)^2 \times 1/3$

[routes (b)] (three routes)

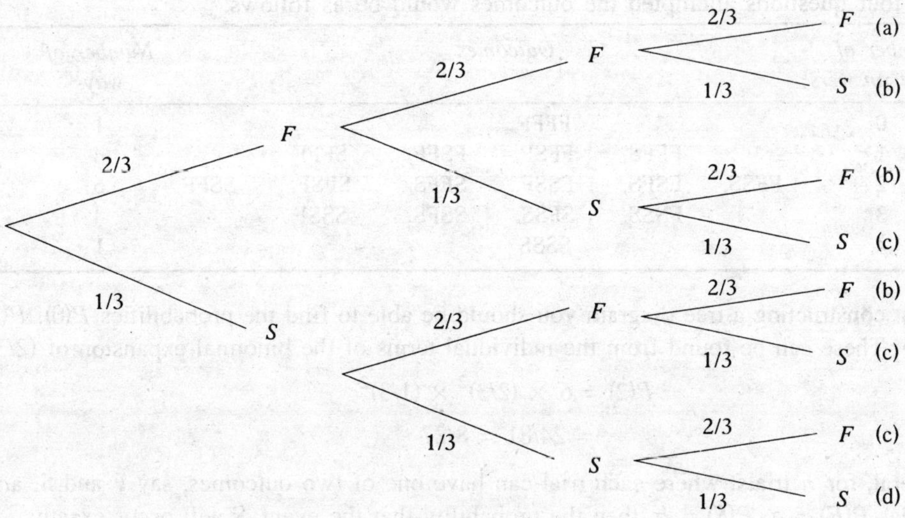

Fig. 2.5.3 (c)

i.e. $$P(1) = 12/27 \text{ or } 4/9$$

P(exactly two questions correct) $= P(2)$

$$= P(FSS) + P(SFS) + P(SSF)$$

$$= 2/3 \times 1/3 \times 1/3 + 1/3 \times 2/3 \times 1/3 + 1/3 \times 1/3 \times 2/3$$

$$= 3 \times 2/3 \times (1/3)^2$$

i.e. $P(2) = 6/27$ or $2/9$ [route (c)] (three routes)

P(all three questions correct) $= P(3)$

$$= P(SSS) = (1/3)^3 = (1/27) \text{ [route (d)]}$$

Looking at the probabilities it may be observed that they could have been obtained from individual terms in the binomial expansion of $(2/3 + 1/3)^3$

e.g. $$P(2) = 3 \times 2/3 \times (1/3)^2$$

For two questions attempted could the probabilities be obtained from the terms in the expansion of $(2/3 + 1/3)^2$? Check the results obtained in (ii).

In forming the tree diagram the number of routes for the different probabilities follow the pattern of the rows of Pascal's triangle, omitting the '1' at the top of the triangle.

Regarding each question as a 'trial' we have

Number of trials				
1		1	1	
2	1	2	1	
3	1	3	3	1

(iv) For four questions attempted the outcomes would be as follows:

Number of correct answers			*Outcomes*				*Number of ways*
0			FFFF				1
1		FFFS,	FFSF,	FSFF,	SFFF		4
2	FFSS,	FSFS,	FSSF,	SFFS,	SFSF,	SSFF	6
3		FSSS,	SFSS,	SSFS,	SSSF		4
4			SSSS				1

Without constructing a tree diagram you should be able to find the probabilities $P(0)$, $P(1)$, $P(2)$, $P(3)$, $P(4)$. These can be found from the individual terms of the binomial expansion of $(2/3 + 1/3)^4$

e.g.

$$P(2) = 6 \times (2/3)^2 \times (1/3)^2$$
$$= 24/81 = 8/27$$

In general, for n trials, where each trial can have one of two outcomes, say F and S, and where in each trial $P(F) = a$, $P(S) = b$, then the probability that the event S will occur exactly i times in n trials is given by the term in b^i in the binomial expansion of $(a + b)^n$.

Note 1 Since F and S are the only two outcomes in a single trial these events are mutually exclusive and exhaustive. Hence $a + b = 1$ in each case. For n trials the sum of all the probabilities will be $(a + b)^n = 1$.

Note 2 Care must be taken to ensure that the probabilities for F and S remain constant from trial to trial before the terms of the binomial expansion are used to find the required probabilities.

Example 2.5.3(1)

A fair coin is tossed 8 times. Find the probability of obtaining (a) exactly 4 heads (b) at least 6 heads (c) at least 1 head.

Solution:

As the coin is fair we assume that for any trial $P(\text{head}) = P(\text{tail}) = 1/2$.

For 8 tosses we may use the terms from the binomial expansion of

$$\left(\frac{1}{2}+\frac{1}{2}\right)^8 \quad \text{i.e. } a = b = 1/2$$

(a)

$$P(4 \text{ heads}) = P(4) = 70(1/2)^{8-4}(1/2)^4$$
$$= 70(1/2)^8$$
$$= 35/128$$

(b)

$$P(\text{at least 6 heads}) = P(6 \text{ heads or } 7 \text{ heads or } 8 \text{ heads})$$
$$= P(6) + P(7) + P(8)$$
$$= 28\,(1/2)^{8-6}(1/2)^6 + 8(1/2)^{8-7}(1/2)^7 + (1/2)^8$$
$$= 37/256$$

(c)

$$P(\text{at least one head}) = 1 - P(\text{no head})$$
$$= 1 - (1/2)^8$$
$$= 255/256$$

Example 2.5.3(2)

From past records Sonny Surekick has an 80% success rate in kicking penalties in football. In his next twelve penalty shots what is the probability that (a) he succeeds in exactly ten shots (b) he succeeds in at least ten shots? Answer to two decimal places.

Solution:

We may use the expansion of $(0.2 + 0.8)^{12}$ regarding the next twelve shots as twelve trials, with the probability of success $b = 0.8$ in each trial.

(a) P(exactly ten successes)

$$= P(10)$$
$$= 66 \times 0.2^2 \times 0.8^{10}$$
$$= 0.28 \text{ (two decimal places)}$$

(b) P(at least ten successes)

$$= P(\text{ten or eleven or twelve successes})$$
$$= P(10) + P(11) + P(12)$$
$$= 66 \times (0.2)^2 \times (0.8)^{10} + 12 \times (0.2) \times (0.8)^{11} + (0.8)^{12}$$
$$= 0.56 \text{ (two decimal places)}$$

In cases where a sample is taken from a large batch it is possible to use the binomial distribution as it is called, where terms of the binomial expansion may be used to determine the probabilities for a given number of trials. The next two examples illustrate the idea.

Example 2.5.3(3)

A factory produces a large output of light bulbs every day. If a sample of 10 bulbs is taken at the end of the day, what is the probability that

(a) exactly two bulbs are faulty?

(b) at least three bulbs are faulty, if 5% of the day's production are faulty? Answer to three decimal places.

Solution:

Use for one trial $P(\text{good}) = 0.95$
$P(\text{faulty}) = 0.05$

where we are interested in the number of faulty bulbs.

Required probabilities may be derived from the binomial expansion of $(0.95 + 0.05)^{10}$ when a sample of 10 bulbs is taken.

(a) $P(\text{exactly two faulty}) = 45 \times (0.95)^8\,(0.05)^2$

$$= 0.075$$

(b) P(at least three faulty)

$$= 1 - [P(\text{none}) + P(\text{one}) + P(\text{two})]$$
$$= 1 - [0.95^{10} + 10 \times 0.95^9 \times 0.05 + 45 \times 0.95^8 \times 0.05^2]$$
$$= 0.012$$

Example 2.4.3(4)

In the University of Utopia there is a class of 200 students of whom 40% come from Kisheka Province. If four students are chosen for a trip, what is the probability that exactly two of them come from Kisheka Province?

Solution:

First done using a tree diagram [Fig. 2.5.3(d)]

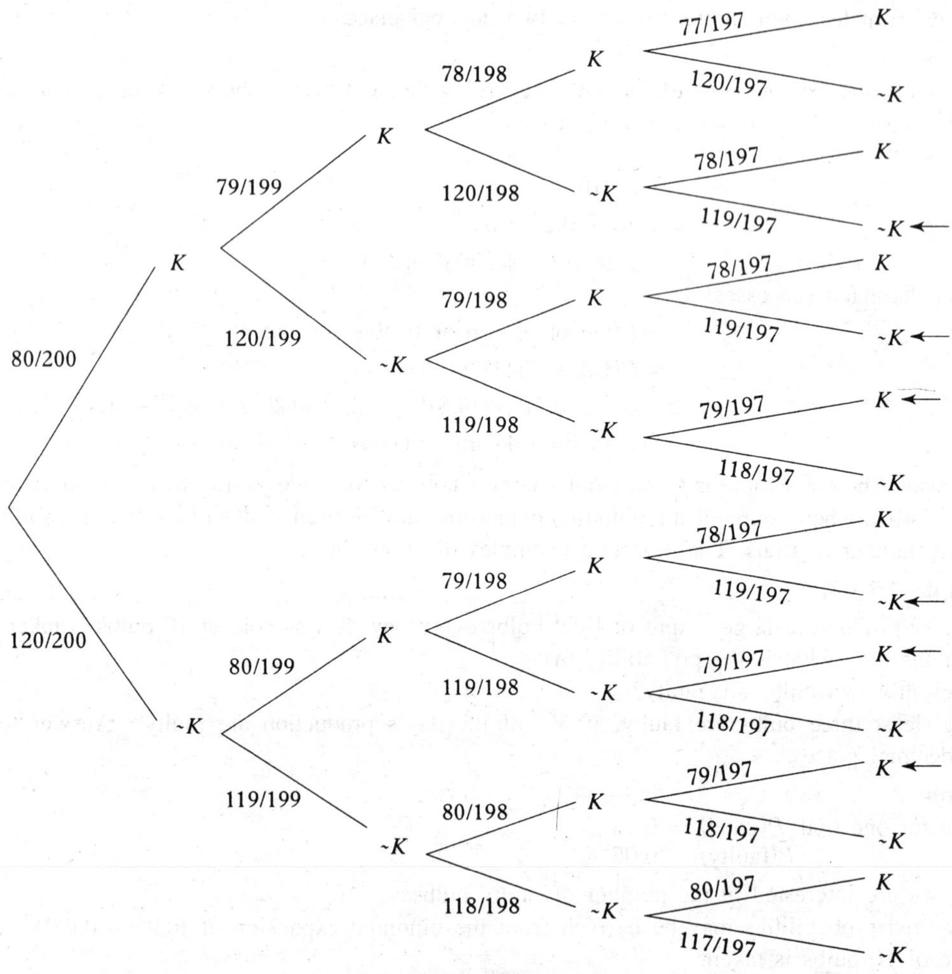

Fig. 2.5.3(d)

Using the routes indicated by the arrow

$$P(2\text{K and } 2 \sim K)$$

$$= \frac{80}{200} \times \frac{79}{199} \times \frac{120}{198} \times \frac{119}{197} + \frac{80}{200} \times \frac{120}{199} \times \frac{79}{198} \times \frac{119}{197}$$

$$+ \frac{80}{200} \times \frac{120}{199} \times \frac{119}{198} \times \frac{79}{197} + \frac{120}{200} \times \frac{80}{199} \times \frac{79}{198} \times \frac{119}{197}$$

$$+ \frac{120}{200} \times \frac{80}{199} \times \frac{119}{198} \times \frac{79}{197} + \frac{120}{200} \times \frac{119}{199} \times \frac{80}{198} \times \frac{79}{197}$$

$$\text{or} \quad = \frac{6 \times 80 \times 79 \times 120 \times 119}{200 \times 199 \times 198 \times 97} = 0.3488 \text{ (to four decimal places)}$$

Secondly using the binomial expansion of $(3/5 + 2/5)^4$ and choosing the term in $(3/5)^2$ $(2/5)^2$ (since we require two students from Kisheka Province and two students who are not from that Province)

We have $6(2/5)^2$ $(3/5)^2$ = 0.3456 (four decimal places)

Both methods give an answer of 0.35 to two decimal places. The error in using the shorter method of the binomial distribution is 0.0032 and as a percentage this is $\dfrac{0.0032 \times 100}{0.3488} = 0.9\%$.

So no'harm' has been done in using the binomial expression as a shorter but more 'approximate' method, and in all cases of sampling from a large batch we shall normally use the appropriate terms of the binomial expansion to find the required probabilities.

EXERCISE 2

1. Two fair six faced dice are thrown. What is the probability that the sum of the numbers on the two uppermost faces will be
   (a) 7
   (b) at least 7?
2. Two fair dice are each in the form of a regular tetrahedron, with the faces numbered 1, 2, 3, 4 on each die. Each die is thrown once, and the score of the face which lands on the floor is taken. What is the probability that the sum of the scores will be
   (a) 4 (b) 6 (c) at least 6 (d) an odd number?
3. There are one hundred identical cards, numbered from 1 to 100 inclusive, and the cards are thoroughly mixed and placed in a large bag. One card is drawn out and its number is noted.
   What is the probability that the number will be
   (a) less than 50
   (b) a multiple of 3
   (c) a multiple of 5
   (d) a multiple of 7
   (e) a multiple of both 3 and 5
   (f) a multiple of both 3 and 7
   (g) a multiple of both 5 and 7
   (h) a multiple of 3 or a multiple of 5
   (i) a multiple of 3 or a multiple of 7
   (j) a multiple of 5 or a multiple of 7
   (k) a multiple of 3 and of 5 and of 7?
4. Two fair six faced dice are thrown and the final score X is obtained by multiplying the two numbers shown on the topmost faces of the dice.
   Determine each of the following probabilities.
   (a) $P(X \text{ is prime})$ (b)$P(X = 12)$ (c) $P(X > 16)$
5. Two fair six faced dice are thrown and the events A, B are defined as follows:
   A : the score on the topmost face of the first die is 5
   B : the sum of the scores on the topmost faces of both dice is at least 9

Determine each of the following probabilities

(a) $P(A)$ (b) $P(B)$

(c) $P(A \wedge B)$ (d) $P(A \vee B)$

(e) $P(A \mid B)$ (f) $P(B \mid A)$

(g) $P(B \mid \sim A)$

6. Bag A holds 2 red balls and 3 blue balls. Bag B also holds 2 red balls and 3 blue balls
   (a) A student picks one ball from each bag. What is the probability that the balls are the same colour?
   (b) The balls are now taken from both bags and all of them are placed in a third bag C. The student now draws two balls without replacement from bag C. What is the probability that the balls are the same colour?
   (c) Explain the difference in your answers for parts (a) and (b) of the question.
7. A bag contains 4 white beads and 1 black bead. The experiment of drawing a bead from the bag is repeated three times.
   (a) Find the probability of drawing three white beads from the bag.
      (i) if the bead is replaced after each draw,
      (ii) if the bead is not replaced after each draw.
      (b) In the case of (ii), what is the probability that the third bead drawn will be black?
8. A bag holds 6 blue balls, 4 red balls and 2 green balls. Three balls are drawn at random from the bag without replacement. What is the probability that
   (a) all three balls will be of different colours
   (b) all three balls are of the same colour?
9. In a University class 3 students have registration numbers with the initial digits 40, 22 have registration numbers with the initial digits 41, and 6 have registration numbers with the initial digits 42. If three students out of the class of 31 are called for a trip, what is the probability that all three have registration numbers with the same initial digits? (answer to three decimal places)
10. In a school a house captain has 12 students available for games of whom 6 are good athletes. If he selects three students for training at random, what is the probability that
    (a) all three students are good athletes
    (b) at least one student is a good athlete?
11. Three athletes Kaman, Mpalirwa and Shah have probabilities 1/2, 1/3 and 1/6 respectively, of qualifying for the finals of the high jump.
    If their performances are independent of each other find the probability that
    (a) all three qualify,
    (b) none qualifies,
    (c) exactly two of them will qualify.
12. In a large batch of hens in a market the ratio of black : brown : white is 3:2:1. A farmer buys three hens at random.
    Determine the probability that
    (a) all three are black,
    (b) all three are of different colours.

*13. In an examination there are three equations. Past records studied by one of the students show that for that student each of questions one and two is equally likely to be easy, moderately hard, or very hard, but that question three has a probability of $\frac{2}{3}$ of being moderately hard and a probability of $\frac{1}{3}$ of being very hard.
For the next time this particular student attempts this particular examination, find the probability that
(a) all three questions are very hard,
(b) exactly two of the questions are moderately hard.

14. Expand $(5 + x)^4$ and simplify your answer. Use your expansion to evaluate 5.02^4 correct to two decimal places.
15. Use the binomial theorem to expand and simplify $(10 + x)^6$ in ascending powers of x up to and including the term in x^3. Use these terms in your expansion to evaluate 10.02^6 correct to the nearest whole number.
16. Use the binomial theorem to expand $(10 - x)^8$ in ascending powers of x up to and including the term in x^3, and simplify your answer.
Use the terms in your expansion to evaluate 9.98^8 correct to the nearest whole number.
17. A fair coin is tossed six times. What is the probability that
(a) there will be exactly 4 heads
(b) there will be at least 4 heads?
18. A fair six faced die is thrown four times. What is the probability that
(a) there will be exactly two sixes
(b) there will be at least two sixes?
19. In the army past records show that 60% of the soldiers will take standard jacket size for their uniform.
(a) In a group of 5 soldiers chosen at random, what is the probability that at least two soldiers take standard jacket size?
(b) In two groups of 5 soldiers, what is the probability that in each group there will be at least two soldiers taking standard jacket size? (Answer to two significant figures)
20. According to an agriculturalist' genetic theory a Friesian bull and Ayrshire cow have a constant probability 0.8 of producing a Fresian calf. In a small farm the bull is Friesian and all the cows are Ayrshire. Eight calves are born during one month.
Determine to two decimal places the probability of producing
(a) exactly six Friesian calves,
(b) more than five Friesian calves.
21. There are 12 athletes in the high jump, each of whom has a probability of 1/3 of qualifying for the finals. Determine the probability that
(a) exactly six athletes will qualify,
(b) between two and four (inclusive) athletes will qualify.
Answer to three decimal places.
22. For a certain couple the probability that the couple has a boy child is $\frac{1}{5}$.
If the couple plans to have 5 children, find the probability that
(a) there will be exactly 2 boys,
(b) there will be less than 2 boys.
Answer to three decimal places.
23. In a large estate the ratio of brown goats to black goats is 2 : 1. The farmer selects 9 goats for sale. Determine to three decimal places the probability that
(a) he will sell exactly six brown goats
(b) he will sell more than six brown goats.
24. A couple plans to have six children. Assuming that the chances of their having a boy or a girl are equally likely, find the probability that they have
(a) at least 4 girls
(b) at most 1 boy.

*25. A patient has a chance of 0.6 of finding the doctor in his surgery on a random visit. On the next six visits what is the probability
(a) that the patient finds the doctor is in on exactly four visits
(b) that the patient finds the doctor is in on at least four visits?
Answer to three decimal places.

3

Algebra

INTRODUCTION

At school when you first studied algebra you were probably told that algebra was generalised arithmetic, in which a number of everyday problems, which you had previously solved using given numbers (values) for the units you were counting or measuring, were written as a general formula with letters (symbols) representing the numbers. However, algebra has broadened its horizons in the last hundred years, and can now be defined as the study of the properties of operations between two or more elements of a given set. We may consider the set of all two-dimensional vectors, or the set of all matrices of a given order, rather than the set of numbers, and investigate whether certain laws which are true for addition or multiplication of numbers are also true for operations on elements in these other sets. For example, the commutative law $a + b = b + a$, $a\,b = b\,a$, is true for addition and multiplication of numbers, but for the set of all square matrices of order 2, we know that for any two such matrices A, B, it is true that $A + B = B + A$ in all cases, but it is not true in all cases that $AB = BA$.

3.1 RATIONAL AND IRRATIONAL NUMBERS

3.1.1 Sets of Numbers

From previous work in Mathematics you will be familiar with the different kinds of numbers you have met in various problems on the subject. These numbers can be grouped together and classified in different sets of numbers such as:

(i) the set of Natural Numbers (Counting Numbers)

$$N = \{1, 2, 3, 4, ...\}$$

(ii) the set of Whole Numbers

$$W = \{0, 1, 2, 3, 4, ...\}$$

(iii) the set of Integers

$$Z = \{....., -3, -2, -1, 0, 1, 2, 3, ...\}$$

where the dots indicate that we can extend the numbers as far as we please to the right in the cases of N and W, and as far as we please to the left and to the right in the case of Z.

We also have met numbers in

(iv) the set Q of rational numbers, i.e. all numbers of the form $\frac{p}{q}$ where p and q are integers.

Examples of Rational Numbers are $\frac{3}{5}, -\frac{7}{3}, \frac{10}{4}$, 2 and – 3. The latter two numbers can be written as $\frac{2}{1}$ and $\frac{-3}{1}$ respectively. When all the common factors of p and q have been divided out of the numerator and the denominator then we say that the number $\frac{p}{q}$ is in its lowest form. For example the rational number $\frac{36}{27}$ is *not* in its lowest form but if we divide each of 36 and 27 by the common factor 9 we obtain $\frac{4}{3}$, which is now in its lowest form.

It is possible to represent rational numbers as points on the number line, if we mark an origin O to represent the number zero.

Positive numbers are on the right of the origin (e.g. 1, 2, 7/3) and negative numbers to the left of the origin (e.g. – 1, – 2, – 2/5) [see Fig. 3.1.1(a)]

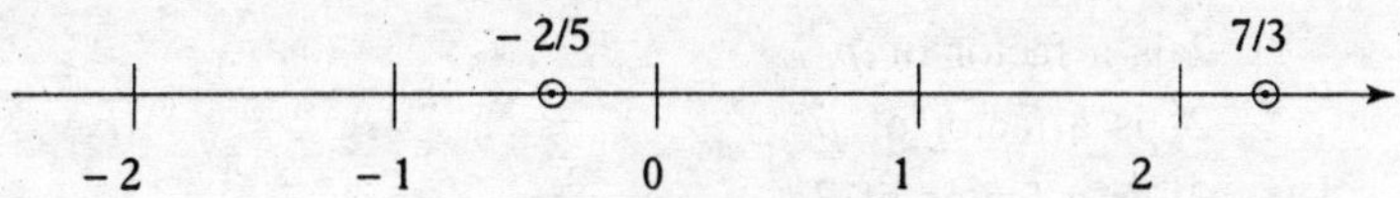

Fig. 3.1.1(a)

It might be imagined that the number line can be filled up completely with rational numbers but this is not true. It is possible by a geometrical construction as shown in Fig. 3.1.1(b) to represent the number $\sqrt{2}$ on the number line.

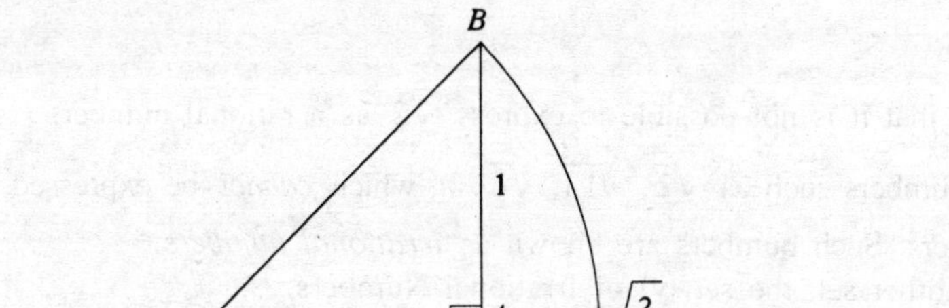

Fig. 3.1.1(b)

In Fig. 3.1.1(b) OA is 1 unit; A is representing 1 on the number line. If AB is of length 1 and at right angles to OA then by Pythagoras theorem OB is of length $\sqrt{2}$.

Setting the compass point at O and drawing an arc, centre O, radius OB, to cut the number line in C will make OC of length $\sqrt{2}$, and C will represent $\sqrt{2}$ on the number line. In practice we may estimate $\sqrt{2}$ to be approximately 1.4 by measurement (however $1.4^2 = 1.96$) but it is not possible to represent $\sqrt{2}$ in the form $\frac{p}{q}$ where p and q are integers. A proof by contradiction has been given for interest but readers who find the proof difficult on a first reading may omit further study of it in the meantime and return to it later.

To prove that $\sqrt{2}$ is not rational

Suppose $\sqrt{2}$ is a rational and can be expressed in the form $\frac{p}{q}$ where p and q have no common factors, i.e. we have written $\sqrt{2}$ as $\frac{p}{q}$ in its lowest form.

On squaring $$2 = \frac{p^2}{q^2}$$

so $$2q^2 = p^2$$

Now since p and q are integers then $2q^2$ and p^2 are integers and they are equal and therefore 2 must be a factor of p^2.

If 2 is a factor of p^2(i.e. of $p \times p$)

then 2 is a factor of p
then 2^2 is a factor of p^2
then 2^2 is a factor of $2q^2$
then 2 is a factor of q^2
then 2 is a factor of q

This means that 2 is a factor of p and also of q. This contradicts the initial assumption that $\sqrt{2} = \frac{p}{q}$ in lowest form.

Hence we deduce that it is not possible to express $\sqrt{2}$ as a rational number.

There are many numbers such as $\sqrt{2}$, $\sqrt{13}$, $\sqrt[3]{4}$, π, which *cannot* be expressed in the form $\frac{p}{q}$ where p, q are integers. Such numbers are known as *irrational numbers*.

Thus, we have another set, the set (v) of Irrational Numbers.

The number line is completely covered by rational and irrational numbers, and we name the set formed by all the rational and irrational numbers as (vi) the set of Real Numbers.

3.1.2 Surds

A surd is a number of the form $\sqrt[n]{(p/q)}$ where p, q and n are integers but which cannot be expressed in the form $\frac{a}{b}$ where a, b are integers. Thus surds are a special type of irrational number.

For example $\sqrt{5}$, $\sqrt[4]{8}$, $\sqrt{(5/2)}$ are surds but $\sqrt[4]{16} = 2$, and $\sqrt[3]{(125/27)} = 5/3$, are rational numbers.

In many cases we may simplify expressions containing surds by looking for factors which are exact powers (e.g. exact squares or cubes) of the number whose root we require.

3.1.3 Factorisation Under the Root Sign

In finding say $\sqrt[3]{216}$ by factors it is possible to write $216 = 2^3 \times 3^3$ and deduce that $\sqrt[3]{216}$ is $2 \times 3 = 6$. However $\sqrt{216}$ is irrational but can be re-written in the form $p\sqrt{q}$ where p, q are integers by observing that

$$\sqrt{216} = \sqrt{(2^3 \times 3^3)}$$

$$= \sqrt{(2^2 \times 3^2 \times 2 \times 3)} = \sqrt{2^2} \times \sqrt{3^2} \times \sqrt{(2 \times 3)}$$

$$= 2 \times 3 \sqrt{(2 \times 3)} = 6\sqrt{6}$$

Check by squaring $(6\sqrt{6})^2$

$$= 6 \sqrt{6} \times 6 \sqrt{6} = 6 \times 6 \times \sqrt{6} \times \sqrt{6}$$

$$= 36 \times 6 = 216$$

Example 3.1.3(1)

Express $\sqrt{80}$ in the form $p\sqrt{q}$ where p, q are integers and q contains no factors which are squares of integers.

Solution:

$$\sqrt{80} = \sqrt{(2^4 \times 5)}$$

$$= 2^2\sqrt{5} = 4\sqrt{5}.$$

Example 3.1.3(2)

Express $\sqrt[4]{48}$ in the form $p\sqrt[4]{q}$ where p, q are integers and q contains no factors which are exact fourth powers of integers.

Solution:

$$\sqrt[4]{48} = \sqrt[4]{(2^4 \times 3)}$$

$$= 2\sqrt[4]{3}.$$

Example 3.1.3(3)

Express $\sqrt[3]{\left(\frac{81}{625}\right)}$ in the form $p\sqrt[3]{q}$ where p, q are rational numbers and q contains no factors which are exact cubes of rational numbers.

Solution:

$$\sqrt[3]{\left(\frac{81}{625}\right)} = \sqrt[3]{\left(\frac{3^4}{5^4}\right)}$$

$$\sqrt[3]{\left(\frac{3^3 \times 3}{5^3 \times 5}\right)} = \frac{3}{5}\sqrt[3]{\left(\frac{3}{5}\right)}.$$

3.1.4 Expressing a Given Surd in the Form $\sqrt[n]{(p/q)}$ (where *n, p, q* are integers)

Sometimes it may be necessary to re-write a surd where it will be expressed finally as a root of a number, as in the following example.

Example 3.1.4(1)

Express $6\sqrt{5}$ in the form $\sqrt{p}$ where p is an integer.

Solution:

$$6\sqrt{5} = \sqrt{(6^2 \times 5)}$$

$$= \sqrt{180}.$$

Example 3.1.4(2)

Express $7\sqrt[4]{3}$ in the form $\sqrt[4]{p}$ where p is an integer.

Solution:

$$7\sqrt[4]{3} = \sqrt[4]{(7^4 \times 3)}$$

$$= \sqrt[4]{7203}.$$

Example 3.1.4(3)

Express $\frac{8}{3}\sqrt[3]{\left(\frac{5}{9}\right)}$ in the form $\sqrt[3]{\left(\frac{p}{q}\right)}$ where p, q are integers.

Solution:

$$\frac{8}{3}\sqrt[3]{\left(\frac{5}{9}\right)} = \sqrt[3]{\left(\frac{8^3 \times 5}{3^3 \times 9}\right)}$$

$$= \sqrt[3]{\left(\frac{2560}{243}\right)}.$$

3.1.5 Addition and Multiplication of Surds

We may in some cases simplify expressions containing surds using the four basic arithmetical operations. Addition of like terms may be applied for surds, e.g.

$$6\sqrt{3} + 4\sqrt{3} = (6 + 4)\sqrt{3} \text{ (using the distributive law)}$$
$$= 10\sqrt{3},$$

and $5\sqrt{3} + \sqrt{27}$ (which might not appear to be possible to simplify at first sight) can be re-written

$$5\sqrt{3} + \sqrt{(3^2 \times 3)} = 5\sqrt{3} + 3\sqrt{3} = 8\sqrt{3}.$$

It is possible to check these calculations approximately by use of an electronic calculator if the reader wishes to do this. However in some cases it may not be possible to do further simplification of an addition of two surds.
Please avoid the mistake of stating that

$$\sqrt{2} + \sqrt{3} = \sqrt{5}.$$

A check using the electronic calculator should show you that $\sqrt{2} + \sqrt{3}$

$$\simeq 1.4142136 + 1.7320508 = 3.1462644$$

whereas $\sqrt{5} \simeq 2.236068$

so $\sqrt{2} + \sqrt{3} \neq \sqrt{5}$

In multiplication, if we look first at a calculation where the square roots are integers, we can see for example that

$$\sqrt{4} \times \sqrt{81} = 2 \times 9 = 18$$

but $\sqrt{(4 \times 81)} = \sqrt{324} = 18$

We shall therefore assume that it will be possible to multiply surds of the *same order* in this way. Suppose we wish to simplify $\sqrt{2} \times \sqrt{7}$. The answer is $\sqrt{(2 \times 7)} = \sqrt{14}$, as the surds are of the same order (both square roots).

This can be calculated approximately using an electronic calculator as

$$\sqrt{2} \times \sqrt{7} \simeq 1.4142136 \times 2.6457513$$
$$\simeq 3.7416575$$

and $3.7416575^2 \simeq 14.000001$

On the electronic calculator

$$\sqrt{14} \simeq 3.7416574.$$

Example 3.1.5(1)
Simplify $(5\sqrt{2} + 3\sqrt{3})(4\sqrt{2} - 5\sqrt{3})$
Solution: Using binomial multiplication

$$(5\sqrt{2} + 3\sqrt{3})(4\sqrt{2} - 5\sqrt{3})$$
$$= 20\sqrt{2}\sqrt{2} + 12\sqrt{3}\sqrt{2} - 25\sqrt{2}\sqrt{3} - 15\sqrt{3}\sqrt{3}$$

$$= 20 \times 2 + 12\sqrt{6} - 25\sqrt{6} - 15 \times 3$$

$$= -5 - 13\sqrt{6}.$$

Example 3.1.5(2)

Simplify $\sqrt{\left(\frac{85}{21}\right)} \times \sqrt{\left(\frac{63}{136}\right)}$

Solution: The surds are of the same order so $\sqrt{\left(\frac{85}{21}\right)} \times \sqrt{\left(\frac{63}{136}\right)} = \sqrt{\left(\frac{85 \times 63}{21 \times 136}\right)}$. Now cancel out any common factors to give

$$\sqrt{\left(\frac{5 \times 3}{8}\right)} = \sqrt{\left(\frac{15}{8}\right)} \text{ or } \frac{1}{2}\sqrt{\left(\frac{15}{2}\right)}$$

After studying Section 3.1.6 you will be able to show that the answer may also be written as $\frac{1}{4}\sqrt{30}$.

Example 3.1.5(3)

Simplify $\dfrac{\sqrt[3]{18} \times \sqrt{48}}{\sqrt[4]{216}}$.

Solution: Here the surds are of different orders. We have a square root, a cube root and a fourth root. However if all the surds can be expressed as twelfth roots (since the LCM of 2, 3 and 4 is 12) it will be possible to simplify the expression

$$\frac{\sqrt[3]{18} \times \sqrt{48}}{\sqrt[4]{216}} = \frac{\sqrt[12]{18^4} \times \sqrt[12]{48^6}}{\sqrt[12]{216^3}}$$

$$= \sqrt[12]{\left(\frac{18 \times 18 \times 18 \times 18 \times 48 \times 48 \times 48 \times 48 \times 48 \times 48}{216 \times 216 \times 216}\right)}$$

$$= \sqrt[12]{(18 \times 4 \times 4 \times 4 \times 48 \times 48 \times 48)} \quad \text{(on cancelling)}$$

$$= \sqrt[12]{(2 \times 3^2 \times 2^2 \times 2^2 \times 2^2 \times 2^4 \times 3 \times 2^4 \times 3 \times 2^4 \times 3)}$$

$$= \sqrt[12]{(2^{19} \times 3^5)}$$

$$= \sqrt[12]{(2^{12} \times 2^7 \times 3^5)}$$

$$= 2\sqrt[12]{(2^7 \times 3^5)}$$

$$= 2\sqrt[12]{31104}.$$

3.1.6 Rationalisation of Denominator

For some calculations it is easier to express a surd with a rational denominator. Some problems may be set in which no tables or calculator will be allowed in the calculations, such as in several of the examples which follow.

Example 3.1.6(1)

Given that $\sqrt{7} = 2.6457513$ correct to seven decimal places. Calculate $1/\sqrt{7}$ correct to six decimal places without using tables or calculator.

Solution: Most readers would take a long time to work out 1/2.6457513 using long division.

However $\dfrac{1}{\sqrt{7}} = \dfrac{\sqrt{7}}{\sqrt{7}\sqrt{7}}$ (multiplying numerator and denominator by $\sqrt{7}$)

so
$$\frac{1}{\sqrt{7}} = \frac{\sqrt{7}}{7} \simeq \frac{2.6457513}{7}$$
$$\simeq 0.3779644$$
$$\simeq 0.377964 \text{ to six decimal places.}$$

Example 3.1.6(2)

Simplify $\dfrac{\sqrt{17}-\sqrt{11}}{\sqrt{17}+\sqrt{11}}$ by expressing with a rational denominator.

Solution: At first sight it might be tempting to solve the problem by multiplying numerator and denominator by $\sqrt{17}+\sqrt{11}$, however the denominator would then become $(\sqrt{17}+\sqrt{11})$ $(\sqrt{17}+\sqrt{11})$

$$= 17 + \sqrt{187} + \sqrt{187} + 11$$
$$= 28 + 2\sqrt{187} \quad \text{and would } \textit{not} \text{ be rational.}$$

However you may have observed that

$$(\sqrt{17}+\sqrt{11})\,(\sqrt{17}-\sqrt{11})$$
$$= 17 + \sqrt{187} - \sqrt{187} - 11$$
$$= 17 - 11 = 6$$

and 6 is rational.

So we multiply numerator and denominator by $\sqrt{17} - \sqrt{11}$ to get

$$\frac{\sqrt{17}-\sqrt{11}}{\sqrt{17}+\sqrt{11}} = \frac{(\sqrt{17}-\sqrt{11})}{(\sqrt{17}+\sqrt{11})}\cdot\frac{(\sqrt{17}-\sqrt{11})}{(\sqrt{17}-\sqrt{11})}$$

$$= \frac{17 - \sqrt{187} - \sqrt{187} + 11}{17 - 11}$$

$$= \frac{28 - 2\sqrt{187}}{6}$$

$$= \frac{2(14 - \sqrt{187})}{6}$$

$$= \frac{14 - \sqrt{187}}{3}$$

N.B : $(\sqrt{5} - \sqrt{2})(\sqrt{5} + \sqrt{2}) = 5 - \sqrt{10} + \sqrt{10} - 2$

$= 3$ which is rational

$$(\sqrt{7} + 2\sqrt{3})(\sqrt{7} - 2\sqrt{3}) = 7 + 2\sqrt{21} - 2\sqrt{21} - 4 \times 3$$

$$= -5$$

which is rational

and $(\sqrt{a} + \sqrt{b})(\sqrt{a} - \sqrt{b})$

$$= a + \sqrt{(ab)} - \sqrt{(ab)} - b$$

$$= a - b$$

which is rational, if a and b are integers.

Hence if a denominator is a fraction of the form $\sqrt{a} + \sqrt{b}$ (with a, b integers) multiplication of the numerator and the denominator by $\sqrt{a} - \sqrt{b}$ will make the denominator become $a - b$. This procedure is known as rationalising the denominator, and a pair of surds of form $\sqrt{a} + \sqrt{b}$, $\sqrt{a} - \sqrt{b}$ are known as *conjugate surds.*

Example 3.1.6(3)

Simplify $\frac{2\sqrt{5} - 3}{3\sqrt{5} - 2}$ by expressing with a rational denominator.

Solution:

$$\frac{2\sqrt{5} - 3}{3\sqrt{5} - 2} = \frac{(2\sqrt{5} - 3)(3\sqrt{5} + 2)}{(3\sqrt{5} - 2)(3\sqrt{5} + 2)}$$

$$= \frac{6 \times 5 - 9\sqrt{5} + 4\sqrt{5} - 6}{9 \times 5 - 4}$$

$$= \frac{24 - 5\sqrt{5}}{41}.$$

Example 3.1.6(4)

Given that $\sqrt{35} = 5.9160798$ correct to seven decimal places, evaluate $\dfrac{\sqrt{7}-\sqrt{5}}{\sqrt{7}+\sqrt{5}}$ correct to six decimal places without use of tables or calculator.

Solution: Rationalising the denominator

We have
$$\frac{\sqrt{7}-\sqrt{5}}{\sqrt{7}+\sqrt{5}} = \frac{(\sqrt{7}-\sqrt{5})(\sqrt{7}-\sqrt{5})}{(\sqrt{7}+\sqrt{5})(\sqrt{7}-\sqrt{5})}$$
$$= \frac{7-\sqrt{35}-\sqrt{35}+5}{7-5}$$
$$= \frac{12-2\sqrt{35}}{2}$$
$$= \frac{2(6-\sqrt{35})}{2}$$
$$= 6-\sqrt{35}$$
$$\simeq 6 - 5.9160798$$
$$\simeq 0.0839202$$
$$\simeq 0.083920 \quad \text{to six decimal places.}$$

Example 3.1.6(5)

Simplify without tables or calculator

$$3\cos^2 45° \cos 42° + \tan^2 60° \sin 48° - 9\cos 60° \cos 42°.$$

Solution: If necessary see Sections 4.4.2 and 4.5.2 in Chapter 4. Using the exact values for the trigonometric ratios of the angles of 45° and 60° we have

$$3\cos^2 45° \cos 42° + \tan^2 60° \sin 48° - 9\cos 60° \cos 42°$$
$$= 3\left(\frac{1}{\sqrt{2}}\right)^2 \cos 42° + (\sqrt{3})^2 \sin 48° - 9 \times \frac{1}{2}\cos 42°$$
$$= 3 \times \frac{1}{2}\cos 42° + 3\sin 48° - 9 \times \frac{1}{2}\cos 42°$$
$$= 3\sin 48° + \frac{3}{2}\cos 42° - \frac{9}{2}\cos 42°$$
$$= 3\cos 42° + \frac{3}{2}\cos 42° - \frac{9}{2}\cos 42° \quad (\text{as } \sin 48° = \cos 42°)$$

$$= 3\cos 42^\circ + \frac{(3-9)}{2}\cos 42^\circ$$

$$= 3\cos 42^\circ - \frac{6}{2}\cos 42^\circ$$

$$= 3\cos 42^\circ - 3\cos 42^\circ$$

$$= 0.$$

Example 3.1.6(6) Simplify without tables or calculator

$$\frac{\sin^2 315^\circ\,(1-\tan^2 210^\circ)}{(1+\cos 120^\circ)\,(1+\tan^2 330^\circ)}$$

Solution: As in the previous example use the exact values for the trigonometric ratios of angles of 30°, 45°, 60° (see Sections 4.4.2 and 4.5.2 in Chapter 4 if necessary)

$$\frac{\sin^2 315^\circ\,(1-\tan^2 210^\circ)}{(1+\cos 120^\circ)\,(1+\tan^2 330^\circ)}$$

$$= \frac{\left((-\sin 45^\circ)^2\,(1-\tan 30^\circ)^2\right)}{\left((1-\cos 60^\circ)\,(1+(-\tan 30^\circ)^2\right)}$$

$$= \frac{\left(-\frac{1}{\sqrt{2}}\right)^2\left(1-\left(\frac{1}{\sqrt{3}}\right)^2\right)}{\left(1-\frac{1}{2}\right)\left(1+\left(\frac{-1}{\sqrt{3}}\right)^2\right)}$$

$$= \frac{\frac{1}{2}\left(1-\frac{1}{3}\right)}{\frac{1}{2}\left(1+\frac{1}{3}\right)}$$

$$\frac{\frac{2}{3}}{\frac{4}{3}} = \frac{2}{3}\times\frac{3}{4} = \frac{1}{2}.$$

3.2 LOGARITHMS

Introduction

From your knowledge of index notation the statement $2^3 = 8$ has been derived from $2 \times 2 \times 2 = 8$.

In this statement 3 is the power to which 2 is raised to give 8.

An alternative way of writing the statement is to say that 3 is the *Logarithm*, to base 2, of 8.

We normally write this as

$$3 = \log_2 8$$

where the word logarithm has been abbreviated to log, and the base we are using (here 2) is written as a subscript.

Thus the statements

$$2^3 = 8$$

and $$3 = \log_2 8,$$

are equivalent.
Similarly the statement

$3^2 = 9$ in index form may be written as

$\log_3 9 = 2$ in what we shall call 'log' form.

Similarly the statements $5^{-4} = 1/625$ and $\log_5 (1/625) = -4$ are equivalent.
Similarly the statements

$$(1/4)^{-3} = 64$$

and $$\log_{1/4} 64 = -3$$

are equivalent.

The base we choose for a logarithm can be any *positive* real number (why can we not use a negative real number? try finding out what power of – 2 is required to give + 8). The logarithms which you have used regularly either on your calculator or in tables have been what we call *common logarithms*, and these are base 10.

We normally omit mentioning the base if we are using base 10, and write simply:

$$\log_{10} 3 = 0.4771$$

as $\log 3 = 0.4771$

Other bases should be stated, but later in the chapter on calculus you will meet another base which is very often used in advanced work in Mathematics. This is base *e*, where the number *e* is approximately 2.718. In most sets of mathematical tables you will find one or two pages giving you the logarithms of numbers, with base 10, and one or two pages giving you the logarithms of numbers, with base *e*. The latter are often described as 'Natural Logarithms' or 'Napierian Logarithms' (named in honour of John Napier who produced the first set of logarithm tables in Edinburgh, Scotland, UK, in 1614).

When we write 'the logarithm to base *e* of 2' there are two acceptable ways of doing this. These are:

$$\log_e 2 \text{ or } \ln 2$$

Those who have an electronic calculator will notice that there are two buttons labelled 'log' and 'ln'. The first of these is the button which will give you the logarithm to base ten of the number which you have input on your calculator, and the second of these will give you the logarithm, to base *e*, of the number which you have input on your calculator.

In general if

$$y = x^Z \qquad (\text{ where } x \text{ is a positive real number})$$

then this is equivalent to saying that

$$\log_x y = Z.$$

3.2.1 Laws of Logarithms

You should already be familiar with the three laws of indices related to powers of the same base x

i.e.
$$x^m x^n = x^{m+n} \qquad \text{(I)}$$

$$x^m \cdot \frac{1}{x^n} = x^{m-n} \qquad \text{(II)}$$

$$(x^m)^p = x^{mp} \qquad \text{(III)}$$

From these three laws it is possible to derive the three laws of logarithms which you have used extensively in earlier numerical work, i.e. when multiplying two numbers you add their logarithms, when dividing two numbers you subtract their logarithms, and when taking the power of a number you multiply the logarithm of the number by that power.

Law I of Logarithms

For any base $x > 0$,

$$\log_x (ab) = \log_x a + \log_x b$$

Proof Let $\log_x a = m$

and $\log_x b = n$

Taking these two statements to equivalent index form

$$a = x^m$$

and $b = x^n$

and so $ab = x^m x^n$

$= x^{m+n}$ by law I of indices

The equivalent statement in 'log' form is then

$$\log_x (ab) = m + n$$

i.e. $\log_x (ab) = \log_x a + \log_x b.$

Law II of Logarithms

For any base $x > 0$,

$$\log_x (a/b) = \log_x a - \log_x b$$

The method of proof is very similar to that for Law I and is left as an exercise to the reader.

Law III of Logarithms

For any base $x > 0$,

$$\log_x a^p = p \log_x a$$

Proof Let $\log_x a = m$

then in index form $a = x^m$ and so $a^p = (x^m)^p$

$= x^{mp}$ by Law III of indices

Taking $a^p = x^{mp}$ back to 'log' form with base x

$$\log_x a^p = mp$$

$$= pm$$

i.e. $\log_x a^p = p \log_x a.$

3.2.2 Change of Base for Logarithms

Most sets of tables and electronic calculators will only provide logarithms to base 10 and to base e.

In practice it may be necessary to find logarithms of numbers to bases other than 10 and e. The question of change of base can be dealt with either by the approach used in 3.2.1 (where we move from index to 'log' form or 'log' to index form as appropriate) or by deriving a general formula for change of base.

Example 3.2.2(1)

Calculate to three decimal places the value of $\log_{12} 2$.

Solution:

Let $\log_{12} 2 = x$

Change to index form $2 = 12^x$

Take logarithms of base 10

$$\log 2 = x \log 12$$

Hence $x = \dfrac{\log 2}{\log 12} = \dfrac{0.3010}{1.0792}$

$= 0.2789$

i.e. $\log_{12} 2 = 0.279$ (to three decimal places)

It is possible to prove a general formula for obtaining the logarithm of a number to base u, say, assuming that the logarithms to another base v are available. The proof of this formula is given below and follows in general terms the same principle as in example 3.2.2(1).

Proof

We require to find the logarithm of a number t, to base u say, assuming that the logarithms of all numbers to base v are available.

Let $\log_u t = x$

Change to index form $t = u^x$

Take logarithms to base v,

$$\log_v t = x \log_v u$$

so $$x = \frac{\log_v t}{\log_v u}$$

i.e. $$\log_u t = \frac{\log_v t}{\log_v u}$$

Since logarithms of all numbers to base v are available, the result follows.

A useful result which follows is that the logarithm, to a base, of a given number, is the reciprocal of the logarithm, to that given number as base, of the original base,

i.e. $$\log_u v = \frac{1}{\log_v u}$$

This result may be obtained from

$$\log_u t = \frac{\log_v t}{\log_v u}$$

by writing $t = v$ in the above mentioned formula, so that

$$\log_u v = \frac{\log_v v}{\log_v u}$$

But since $\log_v v = 1$ then

$$\log_v v = \frac{1}{\log_v u}.$$

3.2.3 Exponential and Logarithmic Equations

The word *exponent* may often be used instead of *index*, and expressions in which the variable or unknown is in the index, such as 3^x, 10^{7x}, are called *Exponential Functions.*

We are now going to look at methods of solving certain types of equations in which the unknown is in the index. Such equations are known as exponential equations. In other equations we may find the unknown expressed in logarithmic form, and such equations are known as logarithmic equations. Often the method of solution will involve changing from index to 'log' form or from 'log' to index form, as the next few examples show.

Example 3.2.3(1)

Solve the equation $5^x = 14$ correct to three significant figures.

Solution: The equation can be re-written in 'log' form and then x can be found using Law III of logarithms.

$$5^x = 14$$

i.e. $$\log 5^x = \log 14$$

i.e. $$x \log 5 = \log 14$$

i.e. $$x = \frac{\log 14}{\log 5}$$

$$= \frac{1.1461}{0.6990}$$

$$= 1.64 \quad \text{(three significant figures)}$$

Example 3.2.3(2)

Solve the equation $3^x.7^{2x+1} = 37$ correct to three significant figures.

Solution: The approach is the same as in the previous example, using both Laws I and Law III of logarithms

$$3^x \cdot 7^{2x+1} = 37$$

i.e. $x \log 3 + (2x + 1) \log 7 = \log 37$

i.e. $x \log 3 + 2x \log 7 + \log 7 = \log 37$

i.e. $x (\log 3 + 2 \log 7) = \log 37 - \log 7$

i.e. $x = \dfrac{\log 37 - \log 7}{\log 3 + 2 \log 7}$

$$= \frac{1.5682 - 0.8451}{0.4771 + 2 \times 0.8451}$$

$$= \frac{0.7231}{2.1673}$$

$$= 0.3336$$

$$= 0.334 \quad \text{(to three significant figures)}$$

Example 3.2.3(3)

Solve the simultaneous equations

$$xy = 160 \qquad (1)$$

$$\log x - 3 \log y = 1 \qquad (2),$$

$$\text{for } x > 0,\ y > 0$$

Solution: In (2) we can re-write the equation as $\log (x/y^3) = 1$. Using Laws II and III of logarithms.

Change to index form $x/y^3 = 10^1$

i.e. $x = 10y^3$

Substituting in (1) for x

$$10y^3 \cdot y = 160$$

i.e. $y^4 = 16$

i.e. $y = 2$ as $y > 0$

Substituting in (1) for y,

$$2x = 160$$

i.e. $x = 80$

so $x = 80,\ y = 2$

Example 3.2.3(4)

Solve the equation

$$\log_2 (x + 2) + \log_2(x - 2) = 5$$

$$\text{for } x > 0$$

Solution: Using Law I of logarithms

$$\log_2 \{(x + 2)(x - 2)\} = 5$$

Change to index form $\quad (x + 2)(x - 2) = 2^5$

i.e $\quad x^2 - 4 = 32$

i.e. $\quad x^2 = 36$

i.e. $\quad x = 6$ as $x > 0$

Example 3.2.3(5)

Solve the equation

$$\log_x 2 + 48 \log_2 x = 14$$

Giving your answers to three significant figures.

Solution: The main difficulty on the left hand side is that the logarithms have been expressed to different bases. However if we set both logarithms in base x, using the rule for change of base (See 3.2.2 of this section) the equation becomes

$$\log_x 2 + \frac{48}{\log_x 2} = 14$$

write $\quad \log_x 2 = y$

then $\quad y + \dfrac{48}{y} = 14$

i.e. $\quad y^2 + 48 = 14y$

i.e. $\quad y^2 - 14y + 48 = 0$

i.e. $\quad (y - 6)(y - 8) = 0$

(the method of factorising a quadratic expression will be reviewed in 3.3 of this chapter)

i.e. $\quad y = 6$ or $y = 8$

i.e. $\quad \log_x 2 = 6$ or $\log_x 2 = 8$

if $\quad \log_x 2 = 6$

then $\quad 2 = x^6$

i.e. $\quad \log 2 = 6 \log x$

i.e. $\quad \log x = \dfrac{1}{6} \log 2$

$$= \frac{1}{6} \times 0.301$$

$$= 0.05017$$

i.e. $$x = 10^{0.05017}$$

i.e. $$x = 1.12 \text{ (to three significant figures)}$$

The reader is advised to use a similar method, starting with $\log_x 2 = 8$, to show that a second solution is

$$x = 1.09 \text{ (to three significant figures)}$$

Hence $$x = 1.12 \text{ or } 1.09.$$

Example 3.2.3(6)

Determine the value of x without tables or electronic calculator if

$$x = \frac{\log 75 + \log 9 + \log 5}{\log 5 + \log 45}$$

Solution: Using law I of logarithms

$$x = \frac{\log(75 \times 9 \times 5)}{\log(5 \times 45)}$$

$$= \frac{\log(5^2 \times 3 \times 3^2 \times 5)}{\log(5 \times 3^2 \times 5)}$$

(expressing the numbers whose logarithms are to be taken as products of prime factors)

i.e. $$x = \frac{\log(3^3 \times 5^3)}{\log(3^2 \times 5^2)}$$

Beware: x is *not* found by cancelling out the powers at this stage (i.e. x is *not* either 15 or log 15) However using Law III of logarithms

$$x = \frac{3 \log 3 + 3 \log 5}{2 \log 3 + 2 \log 5}$$

i.e. $$x = \frac{3 (\log 3 + \log 5)}{2 (\log 3 + \log 5)}$$

$$= \frac{3 \log 15}{2 \log 15}$$

$$= \frac{3}{2}$$ (On cancelling out the common factor log 15 on numerator and denominator)

Example 3.2.3(7)

Solve the equation

$$4^x - 15(2^x) + 56 = 0$$

giving the solutions where necessary to three significant figures.

Solution: Great care is needed here. This equation cannot be solved in the same way as those of examples 3.2.3(1) and (2). On the left hand side the terms have to be added and subtracted and *not* multiplied, so it is not possible to start the method of solution by writing any logarithmic form.

However the reader will notice that

$$4^x = (2^2)^x$$
$$= 2^{2x} = (2^x)^2$$

and if we write $y = 2^x$ the equation becomes

$$y^2 - 15y + 56 = 0$$

This equation can be factorised to give

$$(y - 8)(y - 7) = 0$$

i.e. $y = 8$ or $y = 7$

i.e. $2^x = 8$ or $2^x = 7$

If $2^x = 8$ then $2^x = 2^3$

so $x = 3$

If $2^x = 7$

then $x \log 2 = \log 7$

leading to $x = 2.81$ to three significant figures (The reader should check this)

Hence $x = 3$ or 2.81.

3.3 QUADRATIC EQUATIONS AND FUNCTIONS

Introduction

An expression of the form $ax^2 + bx + c$ in which x is a variable (i.e. x can take up different values on the set of real numbers) and a, b, c are known constants, is called a quadratic expression. The highest power of x is 2, and so the expression is of degree 2. We shall be studying various methods of solving quadratic equations and then studying the properties of quadratic functions and their graphs.

3.3.1 Quadratic Equations

The material in this part of the section may be familiar to the reader, and in this case the reader may omit 3.3.1 and proceed to 3.3.2.

Quadratic equations may be solved by any one of four different methods, i.e. by factors (where possible), by completing the square, by formula, or by graph.

(a) **Solution by Factors**

The easiest types of equation are those where the coefficient of x^2 is 1.

Example 3.3.1(1)

Solve the equation $x^2 - 5x + 6 = 0$.

Solution: The constant is + 6, so if the left hand side can be expressed as a product of two factors of the form $x + k$, $x + l$, the product + 6 itself must be the product of two positive numbers or the product of two negative numbers, that is the numbers k and l are either both positive or both negative.

If we consider the factors $(x + k)(x + l)$ it will therefore be seen that the connecting signs in the brackets which separate the two factors will be the same (either both positive or both negative)

Since the middle term in the left hand side is $- 5x$ this can only be possible if these connecting signs are both negative. Now look for two different numbers whose product is 6 and whose sum is 5. These are 2 and 3.

Since the connecting signs of the terms in the brackets are both negative factors are $(x - 2)$ $(x - 3)$.

Hence in our equation

$$(x - 2)(x - 3) = 0$$

i.e. one of the factors *must* be zero, so $x - 2 = 0$ or $x - 3 = 0$

i.e. $x = 2$ or $x = 3$.

Example 3.3.1(2)

Solve the equation

$$x^2 + 7x + 12 = 0$$

As the constant is + 12 the connecting signs of the terms in the bracket (when the factors are formed) will be the *same*. Since the middle term on the left hand side is $+ 7x$ this can only be possible if both connecting signs are *positive*.

Now look for two different numbers whose product is 12 and whose sum is 7. These are 3 and 4. Since the connecting signs in the brackets are both positive the factors are $(x + 3)(x + 4)$. Hence in our equation

$$(x + 3)(x + 4) = 0$$

i.e. $x = -3$ or $x = -4$.

Example 3.3.1(3)

Solve the equation

$$x^2 + 9x - 22 = 0$$

The constant is $- 22$, which can only be expressed as a product of two numbers when one of the numbers is positive and the other number is negative. This means that in forming factors of the form $(x + k)(x + l)$ the connecting signs in the brackets which separate the two factors will be *different*.

The middle term is + $9x$ which means that it has been formed by adding two terms in x of different signs of which the 'greater' numerically is positive.

We now require to find two numbers whose product is 22 and whose difference is 9 (since the connecting signs are different). These are 11 and 2 of which 11 is numerically greater. Hence the factors are $(x + 11)(x - 2)$

i.e. $$(x + 11)(x - 2) = 0$$

i.e. $$x = -11 \text{ or } 2$$

Example 3.3.1(4)

Solve the equation

$$x^2 - 8x - 20 = 0$$

As the constant is $- 20$, the connecting signs in the brackets which separate the two factors will be *different*.

The middle term is $- 8x$, which has been formed by adding together two terms in x of different signs of which the 'greater' numerically is negative.

We now require to find two numbers whose product is 20 and whose difference is 8. These are 10 and 2 of which 10 is numerically greater. Hence the factors are $(x - 10)(x + 2)$ since the 'greater' numerically is negative.

i.e. $$(x - 10)(x + 2) = 0$$

i.e. $$x = 10 \text{ or } - 2$$

The next three examples will extend the same idea in cases where the coefficient of x^2 is not equal to 1.

Example 3.3.1(5)

Solve the equation

$$6x^2 - 13x + 6 = 0$$

In this case the expanded form of the left hand side has come from factors of the form $(gx + k)\cdot(hx + l)$ and the expansion

$$ghx^2 + (kh + gl)x + kl$$

means that, equating the coefficients of x^2, x and the constant term,

$$gh = 6$$
$$kh + gl = - 13$$
$$kl = 6.$$

We have to find two numbers (say g, h) whose product is 6 and two numbers (say k, l) whose product is 6.

As before the constant term 6 is *positive* and so the connecting signs in the brackets which separate the two factors will be the *same*.

However the middle term $- 13x$ tells us that the connecting *signs* will both be *negative*

Considering $$gh = 6,$$
$$kh + gl = - 13$$
$$kl = 6$$

We can see that it is possible to set up two columns $\begin{pmatrix} g \\ h \end{pmatrix}, \begin{pmatrix} k \\ l \end{pmatrix}$ in the form $\begin{matrix} g & k \\ h & l \end{matrix}$ whereby the product of the numbers in the first column can give 6, the product of the numbers in the second column can give 6, and the cross-products (ignoring sign for the moment) give a sum of 13.

By trial and error we see that the arrangement $\begin{matrix} 3 & 2 \\ 2 & 3 \end{matrix}$ (crossed) will give us the correct coefficients for the terms in each bracket, as $3 \times 3 + 2 \times 2 = 13$. The coefficients along the first row (3 2) give us the coefficients of the first factor and the coefficients of the second row (2 3) give us the coefficients of the second factor. We already know that both the connecting signs will be negative, hence the factors will be $(3x - 2)(2x - 3)$. The reader should check by expanding the brackets that these are indeed the factors of the given left hand side of the equation.

The reader should check that other arrangements which look as if they could be correct, cannot work, e.g. $\begin{matrix} 6 & 1 \\ 1 & 6 \end{matrix}$ (crossed) would appear to be a possible arrangement but the cross-products, 6×6 and 1×1, will add to 37 and *not* to 13.

From the factors $(3x - 2)(2x - 3)$ we have

$$(3x - 2)(2x - 3) = 0$$

i.e. $$3x - 2 = 0 \quad \text{or} \quad 2x - 3 = 0$$

i.e. $$3x = 2 \quad \text{or} \quad 2x = 3$$

i.e. $$x = 2/3 \text{ or } 3/2.$$

Example 3.3.1(6)

Solve the equation

$$12x^2 + 19x + 4 = 0$$

The constant term + 4 tells us that the connecting signs in the brackets which separate the two factors will be the *same*. The middle term + 19x tells us that the connecting signs will both be *positive*.

By trial and error we find that the arrangement $\begin{matrix} 4 & 1 \\ 3 & 4 \end{matrix}$ (crossed) gives product of factors in first column = 12, product of factors in second column = 4, and the cross-products $4 \times 4 + 1 \times 3 = 19$.

Hence the factors are $(4x + 1)(3x + 4)$ and the reader can check this by expanding the brackets.

So $$(4x + 1)(3x + 4) = 0$$

i.e. $$4x + 1 = 0 \quad \text{or} \quad 3x + 4 = 0$$

i.e. $$4x = -1 \quad \text{or} \quad 3x = -4$$

i.e. $$x = -1/4 \quad \text{or} \quad x = -4/3.$$

Example 3.3.1(7)

Solve the equation

$$35x^2 - x - 12 = 0$$

The constant term -12 tells us that the connecting signs in the brackets which separate the two factors will be *different*. The middle term $-x$ tells us that the 'greater' numerically of the numbers in the cross-products will be negative.

By trial and error we find that the arrangement $\begin{smallmatrix}7 & & 3\\ & \times & \\ 5 & & 4\end{smallmatrix}$ gives product of factors in first column $= 35$, product of factors in second column $= 12$, and the difference in the cross-products is $28 - 15 = 13 \neq 1$.

However the arrangement $\begin{smallmatrix}7 & & 4\\ & \times & \\ 5 & & 3\end{smallmatrix}$ gives product of factors in first column $= 35$, product of factors in second column $= 12$, and the difference in the cross-products $21 - 20 = 1$.

Bearing in mind that the 'greater' numerically of these cross-prodcuts is negative, then the arrangement of the factors $(7x + 4)(5x - 3)$ will ensure that the middle term $-x$ will be obtained from $+4 \times 5x$ and $-3 \times 7x$.

Check that the expansion of $(7x - 4)(5x + 3)$ would *not* give $35x^2 - x - 12$. But $35x^2 + x - 12$.

If you have doubts about your factors, always check by expanding to see whether you arrive back at the expression you were given.

Proceeding with the correct factors we have

$$(7x + 4)(5x - 3) = 0$$

i.e. $$7x + 4 = 0 \text{ or } 5x - 3 = 0$$

i.e. $$7x = -4 \text{ or } 5x = 3$$

i.e. $$x = -4/7 \text{ or } x = 3/5.$$

(b) **Solution by Completing the Square**

Example 3.3.1(8)

Solve the equation $x^2 = 29$. In this case it is clear that we can proceed to take the square root of each side.

Solution: In obtaining the full solution we need to take both the positive and the negative square root,

so $$x = \pm\sqrt{29}$$

i.e. $$x = \pm 5.39 \quad \text{to two decimal places.}$$

Example 3.3.1(9)

Solve the equation $(x + 2)^2 = 49$.

Solution: The left hand side of this equation is already written as a 'square', and, although this equation may also be solved by the method of factors, we shall proceed to use the method shown in the previous example by taking the square root of each side

so $$x + 2 = \pm\sqrt{49}$$

i.e. $$x + 2 = \pm 7$$

when $$x + 2 = +7 \text{ then } x = 5$$

when $x + 2 = -7$ then $x = -9$

i.e. $x = 5$ or -9

Example 3.3.1(10)

Solve the equation $(x + 3)^2 = 7$.

Solution: Even if you expand the left hand side you will discover that there are *no* rational factors available for solving this equation. Using the method of the previous two examples

$$x + 3 = \pm \sqrt{7}$$

i.e. $$x = -3 \pm \sqrt{7}$$

If we require the solutions (or roots) to two decimal places of accuracy we now both at the two answers separately

i.e. when $x = -3 + \sqrt{7}$

then $x = -3 + 2.6458$

$= -0.3542$

$x = -0.35$ to two dicimal places

when $x = -3 - \sqrt{7}$

Then $x = -3 - 2.6458$

$= -5.6458$

i.e. $x = -5.65$ to two decimal places

So $x = -0.35$ or -5.65 correct to two decimal places.

Example 3.3.1(11)

Solve the equation $(x + 0.4)^2 = 0.21$, and give the roots of the equation correct to two places of decimals.

Solution: Proceeding as in the previous three examples

$$x + 0.4 = \pm \sqrt{0.21}$$

i.e. $$x = -0.4 \pm \sqrt{0.21}$$

When $x = -0.4 + \sqrt{0.21}$

then $x = -0.4 + 0.4583$

$= 0.0583$

i.e. $x = 0.06$ to two decimal places.

When $x = -0.4 - \sqrt{0.21}$

then $\qquad x = -0.4 - 0.4583$

$\qquad = -0.8583$

i.e. $\qquad x = -0.86$ $\qquad$ to two decimal places.

So $\qquad x = 0.06$ or -0.86 $\qquad$ correct to two decimal places.

In the last four examples the equation has been given in the form $(x + h)^2 = l$ and it has been easy to solve the equation by taking the square root of each side. However if the equation is given in the form $ax^2 + bx + c = 0$ where a, b, c are constants, initial steps are necessary in order to write the equation in the required form $(x + h)^2 = l$.

Consider the expansion of

$$(x - 4)^2$$

This is

$$x^2 - 8x + 16$$

Note that the coefficient of x is

$$-8 = 2 \times (-4)$$

and that the constant term is

$$16 = (-4)^2$$

i.e. the constant term is the square of half the coefficient of x.

The reader should try expanding $(x + 3)^2$, $\left(x + \frac{1}{2}\right)^2$, $(x + 0.3)^2$, and check in each case that the constant term is the square of half the coefficient of x.

Now consider the expression

$$x^2 - 8x.$$

What has to be added to $x^2 - 8x$ to make the expression become a perfect square?

Earlier we showed that $x^2 - 8x + 16$ was the expansion of $(x - 4)^2$. So we add the constant $+ 16$, which we saw earlier was the square of half the coefficient of x. In a similar way what has to be added to the expression $x^2 + \frac{1}{3}x$ to make the expression become a perfect square?

Here the coefficient of x is 1/3 and half the coefficient of x is 1/6, so we add the square of half the coefficient of x i.e. $(1/6)^2$. So the expression becomes

$$x^2 + \frac{1}{3}x + (1/6)^2 \text{ or } x^2 + \frac{1}{3}x + 1/36$$

$$= (x + 1/6)^2$$

The reader should now find out what has to be added to each of the expression $x^2 - \frac{2}{3}x$, $x^2 + \frac{1}{8}x$, $x^2 - \frac{1}{7}x$, to make a perfect square, and should check that when the appropriate constant has been added the resulting expression will become $(x - 1/3)^2$, $(x + 1/16)^2$ and $\left(x - \frac{1}{14}\right)^2$ respectively.

Example 3.3.1(12)

Solve the equation $x^2 - 8x + 11 = 0$ by completing the square.

Solution: First transpose the given constant to the right hand side

so $\qquad x^2 - 8x = -11$

Now add the square of half the coefficient of x, here $\left(\frac{1}{2} \times -8\right)^2 = 16$, to *both sides* of the equation. (Many students forget to add this number to the right hand side).

i.e. $$x^2 - 8x + (-4)^2 = -11 + 16$$

The left hand side is now $$(x-4)^2$$

so $$(x-4)^2 = 5$$

i.e. $$x - 4 = \pm\sqrt{5}$$ (Remember to take both positive and the negative square root of the right hand side)

i.e. $$x = 4 \pm \sqrt{5}$$

If it is required to write the roots to two decimal places then take the roots for x separately

When $$x = 4 + \sqrt{5}$$

then $$x = 4 + 2.2361$$
$$= 6.2361$$

i.e. $$x = 6.24$$ to two decimal places

When $$x = 4 - \sqrt{5}$$

then $$x = 4 - 2.2361$$
$$= 1.7639$$

i.e. $$x = 1.76$$ to two decimal places.

so $$x = 6.24 \text{ or } 1.76$$ to two decimal places.

Example 3.3.1(13)

Determine the roots of the equation $4x^2 + 4x - 1 = 0$ and give the roots correct to two decimal places.

Solution: The reader will notice that in this example the coefficient of x^2 is *not* equal to 1. In such a case our very first step *must* be to divide every term in the equation by the coefficient of x^2 in order to ensure that in the subsequent steps we have an equation where the coefficient of x^2 is 1.

So the equation becomes

$$x^2 + x - \frac{1}{4} = 0$$ on dividing every term by the coefficient of x^2 (in this case 4).

Now transpose the constant term to the right hand side

$$x^2 + x = 1/4$$

Now add the square of half the coefficient of x, here $\left(\frac{1}{2} \times 1\right)^2 = \frac{1}{4}$ to both sides.

i.e. $$x^2 + x + \left(\frac{1}{2}\right)^2 = \frac{1}{4} + 1/4$$

i.e. $$\left(x+\frac{1}{2}\right)^2=\frac{1}{2}$$

i.e. $$x+\frac{1}{2}=\pm\sqrt{\left(\frac{1}{2}\right)}$$

i.e. $$x=-\frac{1}{2}\pm\sqrt{\left(\frac{1}{2}\right)}$$

Taking the roots separately

when $$x=-\frac{1}{2}+\sqrt{\left(\frac{1}{2}\right)}$$

then
$$\begin{aligned}x&=-\frac{1}{2}+0.7071\\&=-0.5+0.7071\\&=0.2071\\&=0.21\qquad\text{to two decimal places}\end{aligned}$$

when $$x=-\frac{1}{2}-\sqrt{\left(\frac{1}{2}\right)}$$

then
$$\begin{aligned}x&=-\frac{1}{2}-0.7071\\&=-0.5-0.7071\\&=-1.2071\\&=-1.21\qquad\text{to two decimal places}\end{aligned}$$

so $$x=0.21\text{ or }-1.21\quad\text{to two decimal places.}$$

Te reader may wonder whether we should not be writing $\pm\left(x+\frac{1}{2}\right)=\pm\sqrt{\left(\frac{1}{2}\right)}$ i.e. writing the positive and negative square roots of both sides, rather than only the right hand side, at the point where we take the square root of both sides. In fact it would appear that we may be having four roots, rather than two roots.

The reader should check this by showing that the solutions of the four resulting equations

$$x+\frac{1}{2}=+\sqrt{\left(\frac{1}{2}\right)},\ x+\frac{1}{2}=-\sqrt{\left(\frac{1}{2}\right)},\ -\left(x+\frac{1}{2}\right)=+\sqrt{\left(\frac{1}{2}\right)},\ -\left(x+\frac{1}{2}\right)=-\sqrt{\left(\frac{1}{2}\right)},$$

will still only give two roots.

For interest the reader may also check, in a similar manner, some of the earlier examples.

(c) **Solution by Quadratic Formula**

This method is the most complete method of all, since it will always give the roots of a quadratic equation to any desired accuracy within the bounds of an electronic calculator or a computer. However it is a very cumbersome method in cases where the quadratic equation has factors that can be found easily. The derivation of the solution by the quadratic formula follows the same steps as are used in obtaining the solution by completing the square. The derivation of the formula is given below.

For the equation $ax^2 + bx + c = 0$ where a, b, c are real numbers and $a \neq 0$, then $x^2 + \frac{bx}{a} + \frac{c}{a} = 0$

i.e. $$x^2 + \frac{b}{a}x = -\frac{c}{a}$$

i.e. $$x^2 + \frac{bx}{a} + \left(\frac{b}{2a}\right)^2 = \left(\frac{b}{2a}\right)^2 - \frac{c}{a}$$

i.e. $$\left(x + \frac{b}{2a}\right)^2 = \frac{b^2}{4a^2} - \frac{4ac}{4a^2}$$

i.e. $$\left(x + \frac{b}{2a}\right)^2 = \frac{b^2 - 4ac}{4a^2}$$

i.e. $$x + \frac{b}{2a} = \pm\sqrt{\left\{\frac{(b^2 - 4ac)}{(4a^2)}\right\}}$$

i.e. $$x + \frac{b}{2a} = \frac{\pm\sqrt{(b^2 - 4ac)}}{2a}$$

i.e. $$x = \frac{-b}{2a} + \frac{\pm\sqrt{(b^2 - 4ac)}}{2a}$$

i.e. $$x = \frac{-b \pm \sqrt{(b^2 - 4ac)}}{2a}$$

An observant reader will have noticed that the *nature* of the roots (types of roots) depends on the quantity $b^2 - 4ac$, underneath the square root sign. The types of root will be different, depending on whether $b^2 - 4ac$ is positive, zero or negative. For this reason the quantity $b^2 - 4ac$ is often called the *discriminant*.

When $b^2 - 4ac > 0$ there will be two different real roots for x so the roots are said to be *real* and *distinct*.

When $b^2 - 4ac = 0$ there will be two real roots for x which are both the same number, so the roots are said to be *real* and *equal*.

When $b^2 - 4ac < 0$ it is not possible to find any real roots for x, and then the roots are said to be *complex*. The study of complex numbers is beyond the scope of this book.

Example 3.3.1(14)

Solve the equation

$$3x^2 - 8x - 2 = 0$$

by the quadratic formula, giving the roots correct to two decimal places.

Solution:

When $3x^2 - 8x - 2 = 0$

then $a = 3,\ b = -8,\ c = -2$

Substituting in the formula

i.e. $$x = \frac{-b \pm \sqrt{(b^2 - 4ac)}}{2a}$$

$$x = \frac{-8 \pm \sqrt{\{(-8)^2 - 4 \times 3 \times (-2)\}}}{2 \times 3}$$

$$= \frac{8 \pm \sqrt{\{64 + 24\}}}{6}$$

$$= \frac{8 \pm \sqrt{88}}{6}$$

$$= \frac{8 \pm 9.3808}{6}$$

When $x = \dfrac{8 + 9.3808}{6} = \dfrac{17.3808}{6} = 2.8968$

then $x = 2.90$ (to two decimal places)

when $x = \dfrac{8 - 9.3808}{6} = \dfrac{-1.3808}{6} = -0.2301$

then $x = -0.23$ (to two decimal places)

So $x = 2.90$ or -0.23 both roots correct to two decimal places

Most mistakes that occur when using the formula result from algebraic errors. Great care needs to be taken to ensure that the correct sign is used when numbers have to be added or multiplied, especially when one or more of the coefficients in the given equation may be negative.

Example 3.3.1(15)

Determine the nature of the roots of each of the equations

(i) $5x^2 - 12x + 6 = 0$

(ii) $196\ x^2 + 476\ x + 289 = 0$

(iii) $6x^2 + 2x + 7 = 0$

Solution:

(i) For the equation $5x^2 - 12x + 6 = 0$

$a = 5,\ b = -12,\ c = 6,$

and the discriminant

$$b^2 - 4ac = (-12)^2 - 4 \times 5 \times 6$$
$$= 144 - 120$$
$$= 24 > 0$$

So the roots are real and distinct.

(ii) For the equation $196x^2 + 476x + 289 = 0$

$$a = 196,\ b = 476,\ c = 289$$

and the discriminant

$$b^2 - 4ac = 476^2 - 4 \times 196 \times 289$$
$$= 226576 - 226576$$
$$= 0$$

So the roots are real and equal.

(iii) For the equation $6x^2 + 2x + 7 = 0$

$$a = 6,\ b = 2,\ c = 7,$$

and the discriminant

$$b^2 - 4ac = 2^2 - 4 \times 6 \times 7$$
$$= 4 - 168$$
$$= -164 < 0$$

So the roots are complex.

N.B. We have not been asked to solve the equations in this question.

Example 3.3.1(16)

Find the possible values of c if the equation

$$x^2 + (c + 2)x + c^2 = 0 \qquad \text{has equal roots.}$$

Solution: The discriminant is zero when the roots are equal so

$$(c + 2)^2 - 4c^2 = 0$$

i.e. $$c^2 + 4c + 4 - 4c^2 = 0$$

i.e. $$0 = 3c^2 - 4c - 4$$

i.e. $$0 = (3c + 2)(c - 2)$$

i.e. $$c = -2/3 \text{ or } 2$$

(d) **Solution by Graph**

It is possible to obtain solutions to a quadratic equation by drawing suitable graphs, although in such cases the expected accuracy will rarely be more than two significant figures. In some cases thought will need to be given regarding the most suitable graphs to draw.

Example 3.3.1(17)

Draw the graph of $y = x^2$ for $-3 \le x \le 3$.

By adding suitable lines to your graph

(i) estimate $\sqrt{7}$

(ii) estimate the roots of the equation $x^2 - x - 1 = 0$

(iii) estimate the roots of the equation $4x^2 + x - 1 = 0$.

Solution: The graphs are shown in Figure 3.3. A table of values is drawn up for the graph of $y = x^2$.

x	– 3	– 2.5	– 2	– 1.5	– 1	– 0.5	0	0.5	1	1.5	2	2.5	3
$y = x^2$	9	6.25	4	2.25	1	0.25	0	0.25	1	2.25	4	6.25	9

Plot these points on the graph paper and join the points to form a smooth curve {Fig. 3.3.1(a)}.

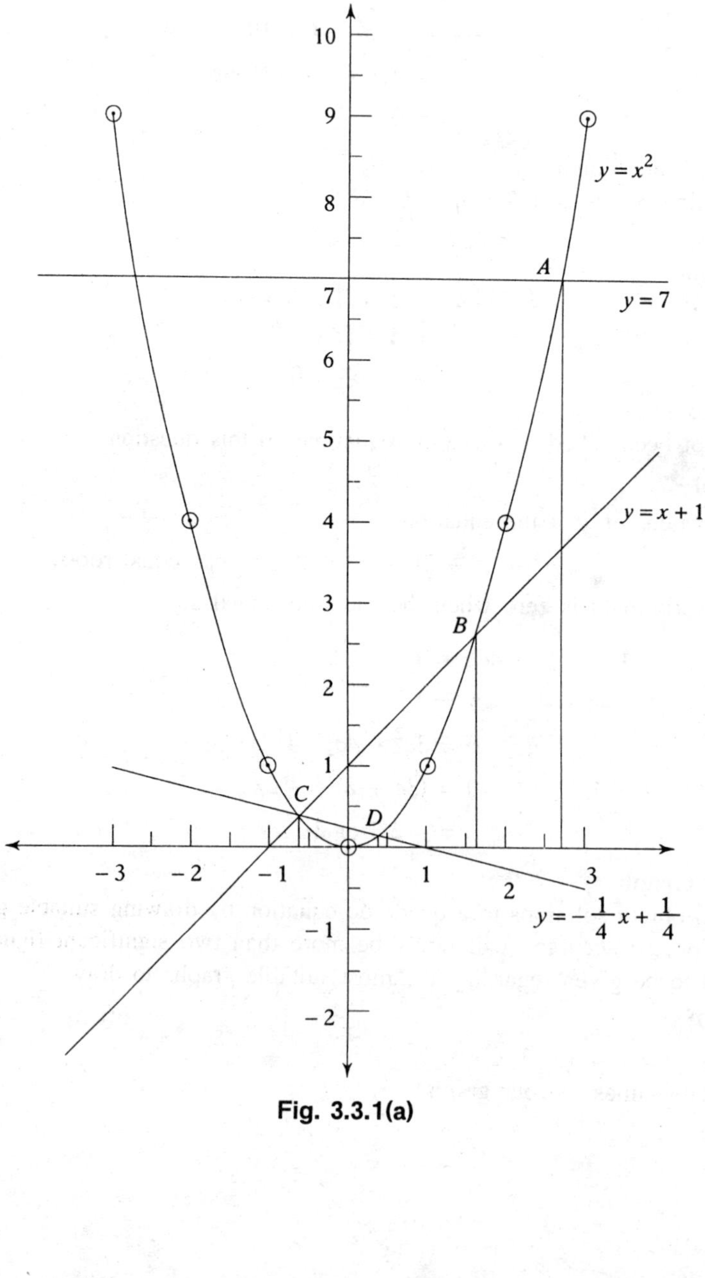

Fig. 3.3.1(a)

(i) In order to find $\sqrt{7}$, draw the line $y = 7$ and read the positive x-coordinate of the point A where the line $y = 7$ meets the graph of $y = x^2$. The x-coordinate is approximately 2.65

so $$\sqrt{7} \simeq 2.65$$

(ii) Given the equation $x^2 - x - 1 = 0$ we rearrange to make x^2 the subject, as we already have the graph of $y = x^2$.

As the equation is $x^2 = x + 1$. then $x^2 = x + 1$.

So the roots we require can be read by the x-coordinates at the intersection of the graphs of $y = x^2$ and $y = x + 1$. The latter graph is a straight line which passes through the points $(-1, 0)$ and $(0, 1)$.

The two graphs meet at B and C and the x-coordinates of these points are approximately 1.6 and 0.6.

Hence the roots of the equation

$$x^2 - x - 1 = 0 \text{ are } x = 1.6 \text{ or } x = -0.6.$$

(iii) Rearrange the equation $4x^2 + x - 1 = 0$ to make x^2 the subject so that $x^2 = -\frac{1}{4}x + \frac{1}{4}$. Draw the graph of $y = -\frac{1}{4}x + \frac{1}{4}$.

The x-coordinates of the points C, D, where the graphs of $y = x^2$ and $y = -\frac{1}{4}x + \frac{1}{4}$ intersect will give the roots we require.

These are $x = 0.4$, $x = -0.6$ approximately.

Example 3.3.1(18)

Draw the graph of $y = 2x^2 - 12x + 19$ for $1 \leq x \leq 5$

By adding suitable lines to your graph.

(i) solve the equation $x^2 - 6x + 6 = 0$

(ii) solve the equation $4x^2 - 25x + 28 = 0$.

Solution: Draw up a table of values for $y = 2x^2 - 12x + 19$ as shown below.

x	1	1.5	2	2.5	3	3.5	4	4.5	5
x^2	1	2.25	4	6.25	9	12.25	16	20.25	25
$2x^2$	2	4.5	8	12.5	18	24.5	32	40.5	50
$-12x$	−12	−18	−24	−30	−36	−42	−48	−54	−60
19	19	19	19	19	19	19	19	19	19
$y = 2x^2 - 12x + 19$	9	5.5	3	1.5	1	1.5	3	5.5	9

Plot these points and then join the points to form a smooth curve. See the graph in Fig. 3.3.1(b)

(i) We need to reset the equation $x^2 - 6x + 6 = 0$ so that the left hand side reads $2x^2 - 12x + 19$ and the right hand side is an expression of two terms, a term in x and a constant.

Observe that our given equation has the coefficient of x^2 equal to 1. In order to obtain an equation containing $2x^2$ we multiply our given equation by 2 to get

$$2x^2 - 12x + 12 = 0$$

The left hand side now contains two of the terms we want, i.e. $2x^2$ and $-12x$. In order to get $2x^2 - 12x + 19$ in the left hand side we add 7 to each side.

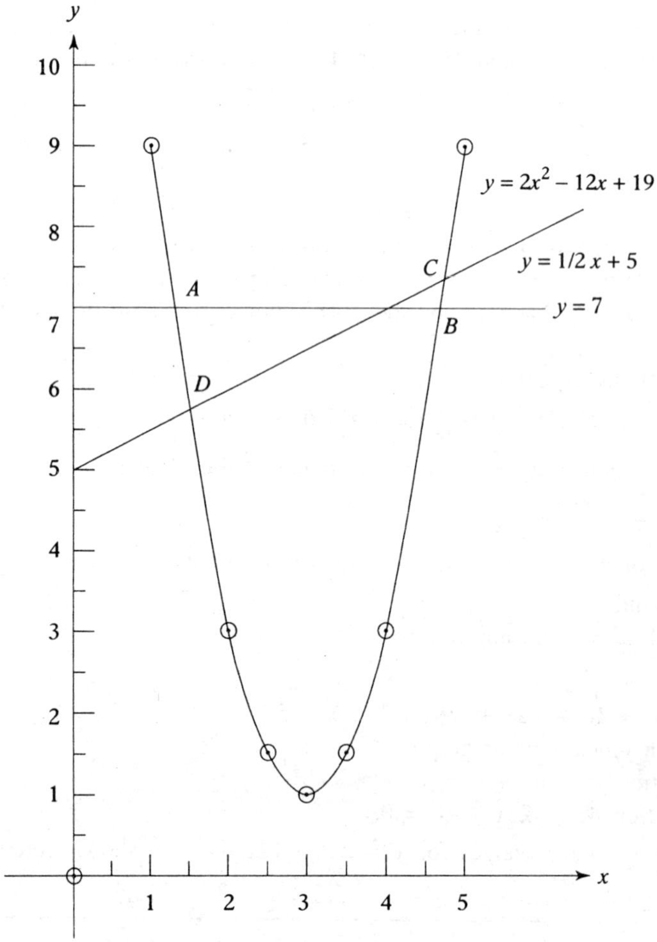

Fig. 3.3.1(b)

This gives $2x^2 - 12x + 19 = 7$. In this case there is no term in x on the right hand side (it is Ox), and we have to add the line $y = 7$ to our graph. Now read the x-coordinates of the points A and B where the graphs of $y = 2x^2 - 12x + 19$ and $y = 7$ meet. These are approximately 1.3 and 4.7.

So the roots are $x = 1.3$ or $x = 4.7$ approximately.

(ii) As in (i) we need to re-set the given equation $4x^2 - 25x + 28 = 0$ so that the left hand side reads $2x^2 - 12x + 19$.

First divide the terms of the given equation by 2 to make the coefficient of x^2 equal to 2, thus $2x^2 - 12.5x + 14 = 0$.

By adding $0.5x + 5$ to each side we get $2x^2 - 12x + 19 = 0.5x + 5$.

Add the line $y = 0.5x + 5$

or $$y = \frac{1}{2}x + 5 \text{ to the graph.}$$

Now read the x-coordinates of the points C and D where the graphs of $y = 2x^2 - 12x + 19$ and $y = \frac{1}{2}x + 5$ meet. These are approximately 1.45 and 4.8.

So the roots are $x = 1.45$ or $x = 4.8$ approximately.

3.3.2 Quadratic Functions and Inequalities

It can be tedious drawing up a table of values of x and y every time we would like to see a graph of a quadratic function of x. It is possible to obtain a good picture of such a graph, without necessarily plotting points, by making a *sketch*. A good sketch should include the important features such as the shape of the curve, the points where the curve meets the x– and y–axes (if these can be found quickly), and also the coordinates of the turning points. If the turning point of a quadratic function is a minimum the curve will be shaped thus ∪ like a valley. If the turning point of a quadratic function is a maximum the curve will be shaped thus ∩ like a mountain. A clear sketch can also be useful in helping us to solve quadratic inequalities, as the next two examples will illustrate.

Example 3.3.2(1)

Sketch the graph of $y = x^2 - 5x + 6$ and use your sketch to solve the inequality $x^2 - 5x + 6 < 0$.

Solution: The curve meets the y-axis when $x = 0$, i.e. when $y = 6$.

So the curve passes through (0, 6). The cuve meeets the x-axis when $y = 0$.

i.e. when $x^2 - 5x + 6 = 0$

i.e. $(x - 2)(x - 3) = 0$

i.e. $x = 2$ or $x = 3$.

So the curve passes through (2, 0) and (3, 0).

The turning point can be determined in two ways for a quadratic function.

(i) By completing the square

$$y = x^2 - 5x + (-5/2)^2 + 6 - (-5/2)^2$$

(We have added the square of half the coefficient of x to make the first three terms into a perfect square but we then have to subtract the same amount to maintain the same function)

i.e. $y = (x - 5/2)^2 + 6 - 25/4$

i.e. $y = (x - 5/2)^2 - 1/4$

The bracketed expression $(x - 5/2)^2$, being a perfect square must always be positive or zero. Its least value is zero when $x = 5/2$. Since the other expression on the right hand side is – 1/4, a constant, then y has its minimum value when $x = 5/2$ and then $y = -1/4$.

So the curve has a minimum turning point at (5/2, – 1/4).

The sketch can now be made on plain paper. See Fig. 3.3.2(a). The line $x = 2\frac{1}{2}$ is the axis of symmetry for the curve.

To solve the inequalities $x^2 - 5x + 6 < 0$ note that the curve meets the x-axis where $x = 2$ and where $x = 3$.

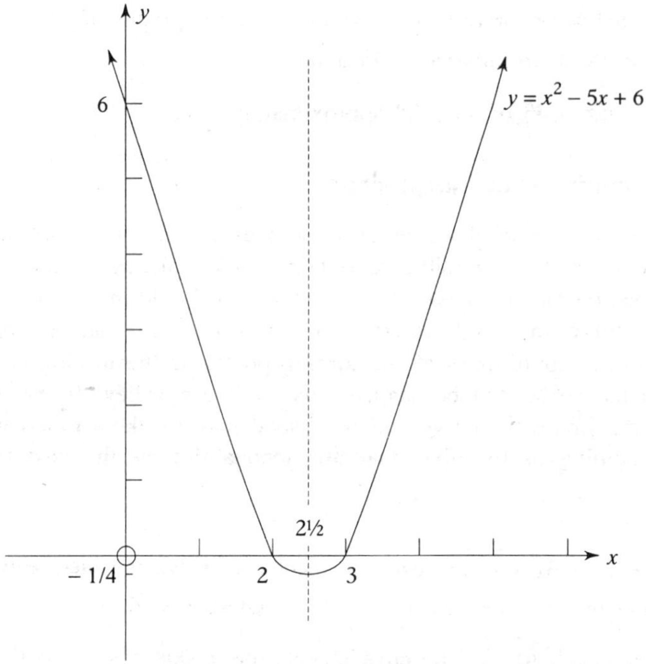

Fig. 3.3.2(a)

The curve is *below* the *x*-axis when $2 < x < 3$ [see Fig. 3.3.2(a)] i.e. the function $y = x^2 - 5x + 6$ is negative (i.e. less than zero) when $2 < x < 3$. For all other values of x, $y \geq 0$.

So the solution of the inequality

$$x^2 - 5x + 6 < 0$$

is $$\{x : 2 < x < 3\}$$

which means the set of all values of x for which $2 < x < 3$. See Fig. 3.3.2(b)(Heavy line)

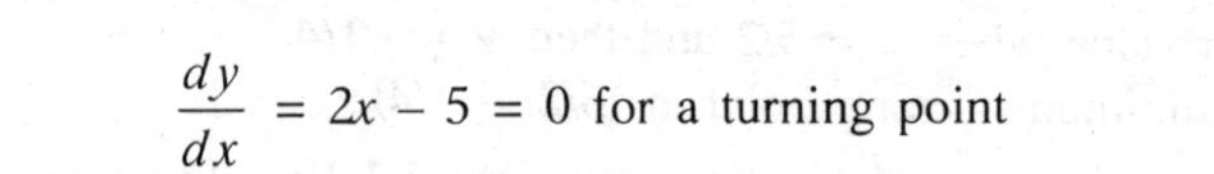

Fig. 3.3.2(b)

(ii) The turning point may also be determined by use of the calculus (see also chapter 6 Section 6.2.4).

$$\text{If } y = x^2 - 5x + 6$$

then the gradient $$\frac{dy}{dx} = 2x - 5 = 0 \text{ for a turning point}$$

i.e. $$x = 5/2 \text{ and by substitution}$$

$$y = (5/2)^2 - 5 \times 5/2 + 6$$

$$= -1/4$$

Testing the sign of the derivative before and after $x = 5/2$ gives

x	5/2 −	5/2	5/2 +
$\frac{dy}{dx} = 2x - 5$	−	0	+
y	↘	→	↗

The sign test shows that y is decreasing before $x = 5/2$ and increasing after $x = 5/2$, hence (5/2, – 1/4) is a minimum turning point. The sketch and the solution of the inequality $x^2 - 5x + 6 < 0$ then follow as before.

Example 3.3.2(2)

Sketch the graph of $y = -3x^2 + 5x + 2$ and use your sketch to solve each of the inequalities

(i) $-3x^2 + 5x + 2 > 0$

(ii) $-3x^2 + 5x + 2 \geq 0$

(iii) $-3x^2 + 5x + 2 < 0$

Solution: The curve meets the y-axis when $x = 0$ (i.e. when $y = 2$).

So the curve passes through (0, 2). The curve meets the x-axis when $y = 0$

i.e. when $-3x^2 + 5x + 2 = 0$

i.e. when $3x^2 - 5x - 2 = 0$

i.e. $(3x + 1)(x - 2) = 0$

i.e. $x = -1/3$ or 2

So the curve passes through (– 1/3, 0) and (2, 0).

We shall find the turning point for the curve $y = -3x^2 + 5x + 2$ by completing the square. This is a fairly demanding piece of algebra. If the algebra is too difficult the reader may leave this method in the meantime and read it later, while concentrating on the alternative method given later, using calculus.

The goal is to re-write y in the form $k - a(x + h)^2$

where $-a$ is the coefficient of x^2, and k, a, h are constants.

In this case the coefficient of x^2 is – 3.

From $$y = -3x^2 + 5x + 2.$$

First fit the terms in x^2 and x under a "common" factor of – 3 (even if the resulting expression may look untidy to the reader)

$$y = -3(x^2 - 5/3x) + 2$$

Secondly *add* and then subtract the square of half the coefficient of x *inside* the bracket, so as to ensure that the final expression on the right hand side remains the same.

$$y = -3[x^2 - 5/3x + (-5/6)^2 - (-5/6)^2] + 2$$

i.e. $$y = -3\left[(x - 5/6)^2 - \frac{25}{36}\right] + 2$$

On forming a perfect square inside the main bracket.

Thirdly expand the *outer* bracket then simplify

$$y = -3(x - 5/6)^2 + \frac{25}{12} + 2$$

i.e. $$y = \frac{49}{12} - 3\left(x - \frac{5}{6}\right)^2$$

The first expression on the right hand side is constant and the second is -3 times a perfect square containing x.

In this case y has a *maximum* when the expression containing the perfect square is zero, i.e. when $x = \frac{5}{6}$. So when $x = \frac{5}{6}$, y has its maximum turning value of $\frac{49}{12}$.

i.e. the maximum turning point is $\left(\frac{5}{6}, \frac{49}{12}\right)$.

The sketch is now shown in Fig. 3.3.2(c).

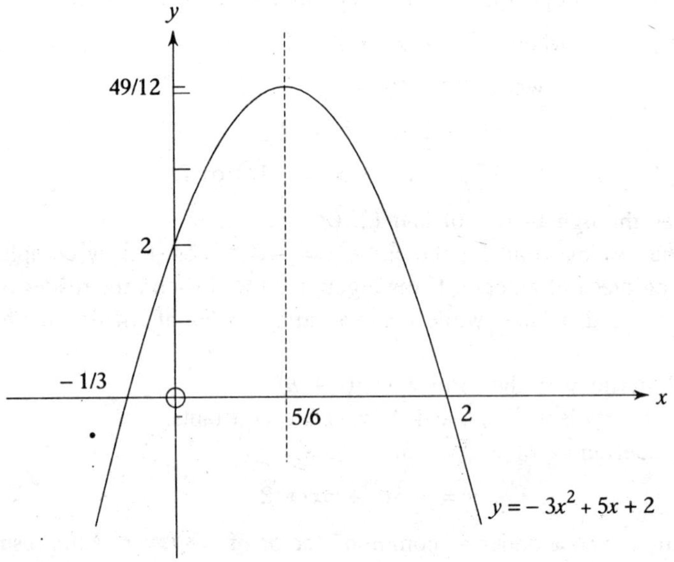

Fig. 3.3.2(c)

The line $x = \frac{5}{6}$ is the axis of symmetry for the curve.

Note: (i) For a graph of a quadratic function, if the turning point has x-coordinate equal to k, then the line $x = k$ is the axis of symmetry for the curve.

(ii) If the coefficient of x^2 is positive in the equation given for the quadratic function, the graph will be shaped like a valley and the turning point will be a minimum, but if the coefficient of x^2 is negative in the equation given for the quadratic function then the graph will be shaped like a mountain and the turning point will be a maximum.

We now use the sketch to solve the given inequalities (i), (ii) and (iii).

In (i) $$-3x^2 + 5x + 2 > 0$$

The curve meets the x-axis

where $x = -1/3$ and where $x = 2$.

Referring back to Fig. 3.3.2(c) the curve is *above* the x-axis when $-1/3 < x < 2$, i.e. the function $y = -3x^2 + 5x + 2$ is positive (i.e. is greater than zero) when $-1/3 < x < 2$.

For all other values of x, $y \leq 0$.

So the solution of the inequality $-3x^2 + 5x + 2 > 0$ is $\{x : -1/3 < x < 2\}$. [See Fig. 3.3.2(d) Heavy line)]

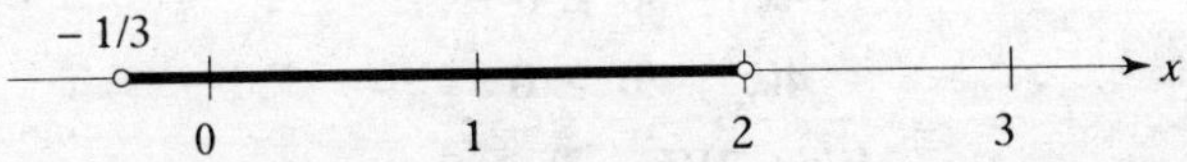

Fig. 3.3.2(d)

Note that in this Fig. 3.3.2(d) and also in Fig. 3.3.2(b) the boundary points have been shown in clear circles indicating that they are *not* in the solution.

For instance in Fig. 3.3.2(d) we can allow x to be 1.999999, as 1.999999 is less than 2. But when $x = 2$ the expression $-3x^2 + 5x + 2$ will be zero, and therefore $x = 2$ is *not* in the solution of the inequality $-3x^2 + 5x + 2 > 0$.

In (ii) $$-3x^2 + 5x + 2 \geq 0.$$

In this case all the points on the x-axis in the solution for (i) are included.

However this time the inequality includes points for which $-3x^2 + 5x + 2 = 0$. The solution consists of all points in Fig. 3.3.2(c) for which the curve is *above* the x-axis and in which the curve meets the x-axis. Hence the solution is $\{x : -1/3 \leq x \leq 2\}$ and is illustrated in Fig. 3.3.2(e).

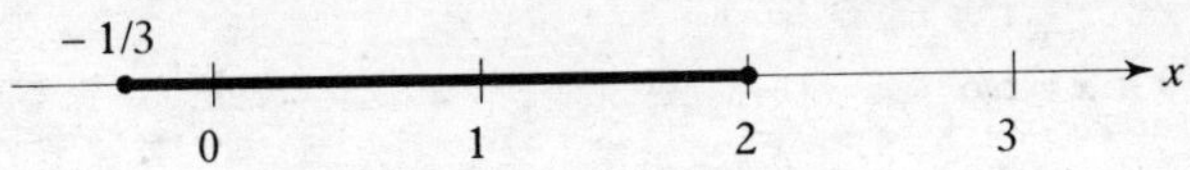

Fig. 3.3.2(e)

In this case the points $x = -\frac{1}{3}$, $x = 2$, which are the boundary points, have been included in the solution, so they have been shown in filled-in circles.

In (iii) $$-3x^2 + 5x + 2 < 0.$$

The curve is below the x-axis when $x < -1/3$ or when $x > 2$, referring back to Fig. 3.3.2(c). The solution of the inequality $-3x^2 + 5x + 2 < 0$ is $\{x : x < -1/3 \text{ or } x > 2\}$ and is illustrated in Fig. 3.3.2(f).

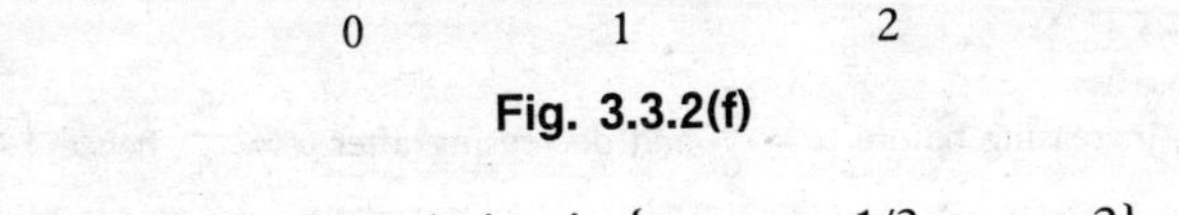

Fig. 3.3.2(f)

Note that the correct way to write the solution is $\{x : x < -1/3 \text{ or } x > 2\}$ as the solution comprises the union of two disjoint (non-intersecting) sets of numbers on the real number line. A common mistake is to write $\{x : x < -1/3 \text{ and } x > 2\}$ which is nonsense, since there is *no* real number which can satisfy both the conditions $x < -1/3$ *and* $x > 2$ simultaneously. For instance $x = 3$ satisfies $x > 2$ but it is obviously false that $3 < -1/3$.

Example 3.3.2(3)

Determine the values of k for which the equation $x^2 + 2(k + 1)x + 2(k + 5) = 0$ has real roots.

Solution: For real roots the discriminant

$$4(k + 1)^2 - 4 \times 2\,(k + 5) \geq 0$$

i.e. $$4k^2 + 8k + 4 - 8k - 40 \geq 0$$

i.e. $$4k^2 - 36 \geq 0$$

i.e. $$4(k^2 - 9) \geq 0$$

i.e. $$4(k + 3)(k - 3) \geq 0$$

A sketch of the curve of the equation $y = (k + 3)(k - 3)$ shows that $y \geq 0$ when $k \geq 3$ or when$k \leq -3$ [see Fig. 3.3.2(g)]

Hence the solution set is

$$\{k : k \geq 3 \text{ or } k \leq -3\}$$

and this is the required set of values of k.

---

*Example 3.3.2(2)

For readers who prefer to use the calculus in order to determine the coordinates of the turning point, and also the nature of the turning point, the method is shown below.

If $$y = -3x^2 + 5x + 2$$

$$\frac{dy}{dx} = -6x + 5 = 0 \text{ if } x = 5/6$$

and $$y = -3\left(\frac{5}{6}\right)^2 + 5 \times \frac{5}{6} + 2$$

which simplifies to $\frac{49}{12}$

Testing the sign of $\frac{dy}{dx}$ before $x = 5/6$, at $x = 5/6$, and after $x = 5/6$ we have

x	$\frac{5}{6}^-$	$\frac{5}{6}$	$\frac{5}{6}^+$
$\frac{dy}{dx} = -6x + 5$	+	0	–
y	↗	→	↘

The sign test shows that y is increasing before $x = \frac{5}{6}$ and decreasing after $x = \frac{5}{6}$, hence $\left(\frac{5}{6}, \frac{49}{12}\right)$ is a maximum turning point.

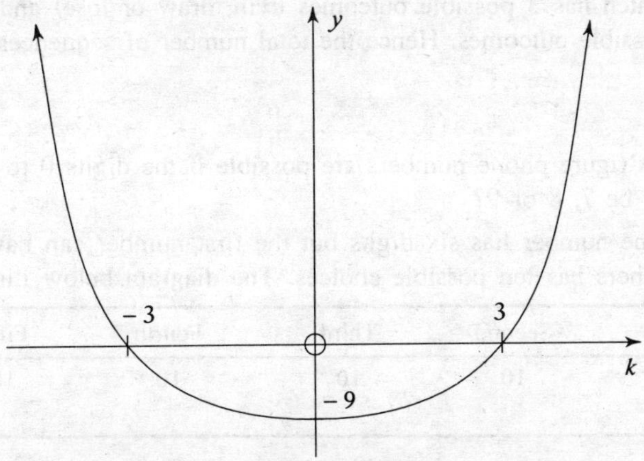

Fig. 3.3.2(g)

3.4 PERMUTATIONS AND COMBINATIONS

Introduction

This topic deals with ways of finding the number of different arrangements or selections which are possible from a set of given objects. This is often a difficult topic, as there are often no clear rules or steps to follow, unlike say in solving a quadratic equation, and the method of solution often depends on the ability of the reader to think the problem out using any suitable method.

3.4.1 Permutations

If a number (k) of independent choices have to be made, with n_1 possibilities for the first choice, n_2 for the second choice, etc., then the total number of choices is

$$n_1 \times n_2 \times .... \times n_k.$$

One particular important application of this principle is the calculation of the number of ways n different objects can be arranged in a specified order.

We may proceed as follows: the first object may be chosen in n ways; now there are only $(n - 1)$ objects remaining, so the second object can be chosen in $(n - 1)$ ways; when the first and second objects have been chosen there are $(n - 2)$ ways of choosing the third; and so on until only one choice is remaining for the last object.

So the total number of possible arrangements is

$$n\ (n - 1)(n - 2)\ .\ .\ .\ 3\ .\ 2\ .\ 1$$

This expression is normally abbreviated to $n!$ (read as n factorial). Any arrangement of different objects in a specified order is called a *Permutation* of the objects so the number of permutations of n different objects is $n!$

Example 3.4.1(1)

In hockey a match can be won, drawn or lost. If a team plays five matches how many different sequences of results are possible?

Solution: The first match has 3 possible outcomes (win, draw or lose) and similarly each of the other matches has 3 possible outcomes. Hence the total number of sequences of results is $3 \times 3 \times 3 \times 3 \times 3 = 3^5 = 243$.

Example 3.4.1(2)

How many different six-figure phone numbers are possible if the digits 0 to 9 are allowed, except that the first digit must be 7, 8 or 9?

Solution: The telephone number has six digits but the first number can have only three possible choices. Each of the others has ten possible choices. The diagram below illustrates this.

Digit:	First	Second	Third	Fourth	Fifth	Sixth
Number of choices	3	10	10	10	10	10

The total number of possibilities is

$$3 \times 10^5 \text{ or } 300\,000$$

Example 3.4.1(3)

A car registration in the country of Burania consists of three letters, followed by three digits. The first letter is always A, and the other two letters should not include A, I or O and must be different. The first digit must not be zero, but otherwise any digits may be used in any of the three places. How many different registrations are there?

Solution: Setting out the possibilities we have

Letters			Digits		
First	Second	Third	First	Second	Third
1	23	22	9	10	10

So altogether there are $1 \times 23 \times 22 \times 9 \times 10 \times 10 = 455\,400$ different registrations.

Example 3.4.1(4)

In how many ways can eleven books be arranged on a shelf?

Solution: This is the same as asking how many permuations of 11 objects there are, hence the answer is

$$11! = 39\,916\,800$$

Example 3.4.1(5)

In how many ways can 10 people sit at a round table?

Solution: On a round table we assume that the position of the people relative to the table is of no consequence, unless a special place of honour (head of the table) is specified. This means that if all the occupants were to move one place to the left, or one place to the right, the new arrangement would still be regarded as the same arrangement, as each occupant would be finding the other occupants in the same places relative to him/her.

In this problem we fix one person. The other nine can then be arranged in 9! ways or 362 880 ways.

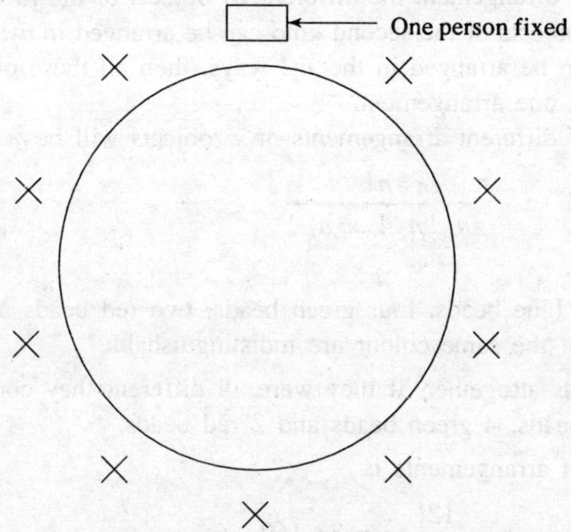

Fig. 3.4.1(a)

Example 3.4.1(6)

In how many ways can the letters of the word TITTER be arranged in a row?

Solution: There are six letters, and if these were all different the letters could be arranged in 6! ways.

However let us label the three T's as T_1, T_2, T_3 and look at all the arrangements when the letters appear as in the original word TITTER.

The arrangements are

$$\begin{array}{cccccc} T_1 & I & T_2 & T_3 & E & R \\ T_1 & I & T_3 & T_2 & E & R \\ T_2 & I & T_1 & T_3 & E & R \\ T_2 & I & T_3 & T_1 & E & R \\ T_3 & I & T_1 & T_2 & E & R \\ T_3 & I & T_2 & T_1 & E & R \end{array}$$

This means that in this arrangement, and indeed in every other arrangement, the T's can be arranged in 3! = 6 ways *without* altering the appearance of the letters in the arrangement.

So the number of different arrangements is

$$\frac{6!}{3!} = 120$$

since the three T's are the same.

This principle may be extended to finding the number of ways of arranging n objects in a row, if there are n_1 alike of one kind, n_2 alike of a second kind, ... n_k alike of kth kind, where $n_1 + n_2 + ... n_k < n$. Since in any one arrangement the different n_1 objects of the first kind can be arranged in $n_1!$ ways, the different n_2 objects of the second kind can be arranged in $n_2!$ ways, ..., the different n_k objects of the kth kind can be arranged in the $n_k!$ ways, then all these objects can be arranged in $n_1!\ n_2!\ ...\ n_k!$ ways within one arrangement.

Hence the total number of different arrangements of n objects will be

$$\frac{n!}{n_1!\,n_2!\,.....\,n_k!}.$$

Example 3.4.1(7)

In how many ways can five blue beads, four green beads, two red beads and one white bead be arranged in a row if beads of the same colour are indistinguishable?

Solution: There are 12 beads altogether. If they were all different they could be arranged in 12! ways. But there are 5 blue beads, 4 green beads and 2 red beads.

So the number of different arrangements is

$$\frac{12!}{5!\,4!\,2!} = 83\ 160$$

Example 3.4.1(8)

How many even numbers greater than 4000 can be formed from the digits 1, 4, 6, 8, if each digit may only be used once?

Solution: Since the numbers are greater than 4000 the thousands place can only be filled in 3 ways (using 4, 6 or 8). Since the number has to be even then the units place can now only be filled in 2 ways, since we have already filled the thousands place. For instance, suppose we put the digit 8 in the thousands place, then we can only use 4 or 6 to fill the units place.

Once the thousands and units places have been filled, any of the remaining two digits can be used to fill the hundreds place, and the remaining one digit to fill the tens place.

The problem may be set out as below (arrows indicating the order in which we consider the different places):

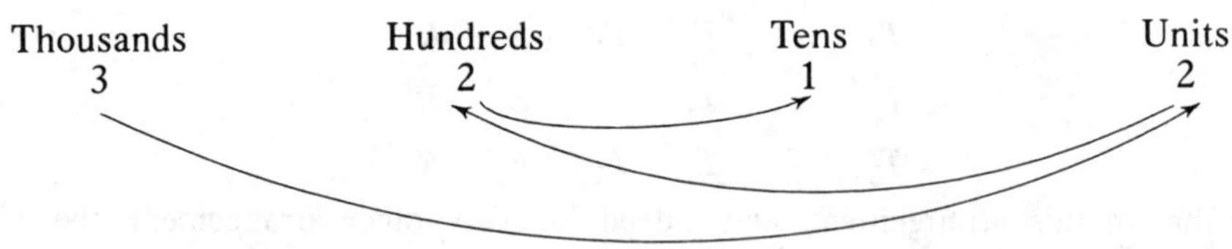

So we can form $3 \times 2 \times 1 \times 2 = 12$ numbers.

Example 3.4.1(9)

How many odd numbers greater than 70 000 can be formed using the digits 0, 1, 4, 7, 8, 9

(a) without repetition

(b) if repetitions are allowed?

Solution: (a) The required numbers may have five or six digits.

Consider the five digit numbers. The ten thousands place can only have 7, 8 or 9 as the required number is over 70 000. The number of choices for the units digit will depend on whether 7 or 9 has already been used in the ten thousands place. Set out two different possibilities.

7 or 9 in ten thousands place means only 2 choices will be available to fill the units place.

TTh	Th	H	T	U
2	4	3	2	2

i.e. $2 \times 4 \times 3 \times 2 \times 2 = 96$ ways

8 in ten thousands place means 3 choices will be available to fill the units place.

TTh	Th	H	T	U
1	4	3	2	3

i.e. $1 \times 4 \times 3 \times 2 \times 3 = 72$ ways

Consider the six digit numbers.

The zero cannot appear in the hundred thousands place, and once again the units place can only have three choices from 1, 7 and 9.

Once the units place has been fixed then there will remain only four choices for the hundred thousands place.

HTh	TTh	Th	H	T	U
4	4	3	2	1	3

i.e. $4 \times 4 \times 3 \times 2 \times 1 \times 3 = 288$ ways

Altogether if there are no repetitions there will be $96 + 72 + 288 = 456$ possible numbers.

(b) Consider the five digit numbers. Any of the digits can fill the thousands, hundreds and tens places, but only three (7, 8, 9) can fill the ten thousands places, and only three (1, 7, 9) can fill the units place, thus

TTh	Th	H	T	U
3	6	6	6	3

i.e. $3 \times 6^3 \times 3 = 1944$ ways.

Consider the six digit numbers. Any of the digits (apart from zero) can be chosen in the hundred thousands place, but only three (1, 7, 9) can fill the units place. Otherwise the other places have no restriction, thus

HTh	TTh	Th	H	T	U
5	6	6	6	6	3

i.e. $5 \times 6^4 \times 3 = 19\ 440$

Altogether 1944 + 19440 = 21384 possible numbers.

Example 3.4.1 (10)

A man dines at the same hotel for three consecutive days and the menu each day is one of any four types of goat dish, or two types of chicken dish, or one type of vegetarian dish.

In how many ways can he arrange his lunches over the three days if he does not have a goat dish two days running nor repeat any dish?

Solution: Clearly the number of arrangements will depend on which days, if any, he takes a goat dish. List the possibilities based on whether he takes a goat dish on the first day (which implies he cannot take a goat dish on the second day), or on whether he takes a goat dish on the second day (which implies he cannot take a goat dish on the first or third days)

Suppose he takes a goat dish on the first day, then the possible arrangements would be:

First day (Goat dish)	Second day (Other dish)	Third day (Any remianing dish)
4	3	5

i.e. $4 \times 3 \times 5 = 60$ ways

Suppose he takes a goat dish on the second day, then the possible arrangements would be:

First day (Other dish)	Second day (Goat dish)	Third day (Any one of the two remaining others)
3	4	2

i.e. $3 \times 4 \times 2 = 24$ ways

Suppose he does not take a goat dish on either of the first or second days. The possible arrangements would be:

First day (Other dish)	Second day (Any one of the two remaining others)	Third day (Any goat dish or or the single remaining other dish)
3	2	5

i.e. $3 \times 2 \times 5 = 30$ ways

Altogether there will be 60 + 24 + 30 = 114 ways of arranging his lunches.

Example 3.4.1(11)

In how many ways can 3 objects be taken from 8 objects and arranged in a row?

The answer is fairly simple, as the first object can be chosen in 8 ways, the second in 7 ways, and the third in 6 ways, i.e. $8 \times 7 \times 6$ ways i.e. 280 ways.

3.4.2 Combinations

Let us look at the last example again. When we worked it out we used straight multiplication of $8 \times 7 \times 6$. However that calculation can be re-written as $\dfrac{8 \times 7 \times 6 \times 5 \times 4 \times 3 \times 2 \times 1}{5 \times 4 \times 3 \times 2 \times 1}$ i.e. $\dfrac{8!}{5!}$

Suppose now the problem is *not* one of arranging, but one of *selecting* any three objects from the eight objects. How can we determine the number of selections?

Suppose the selections included one selection of three objects A, B, C. These three objects A, B, C can arrange themselves as follows:

$ABC, ACB, BAC, BCA, CAB, CBA$, i.e. there are 3! arrangements. However that means that if we had used our previous answer of $\frac{8!}{5!}$ we would have counted each selection 3! times over.

So if we want to know the number of selections of 3 objects from 8 objects, when the order of arranging the objects does not matter, the answer is $\frac{8!}{5!} \div 3! = \frac{8!}{5!\,3!}$.

In general if we wish to know the number of ways of selecting r objects from n available objects the same type of argument can be used.

If we are arranging the r objects this can be done in

$$n(n-1)(n-2)\ \dots\ (n-r+1) \text{ ways}$$

or

$$\frac{n(n-1)(n-2)\ \dots\ (n-r+1)(n-r)\ \dots\ \times 3 \times 2 \times 1}{(n-r)\ \dots\ \times 3 \times 2 \times 1}$$

i.e.

$$\frac{n!}{(n-r)!} \text{ ways}$$

But any one selection of r objects can be arranged in $r!$ ways, so the total number of selections is given by $\frac{n!}{(n-r)!} \div r!$ ways

or

$$\frac{n!}{(n-r)!} \times \frac{1}{r!} = \frac{n!}{(n-r)!\,r!} \text{ ways.}$$

This expression, which gives the number of ways of selecting r objects from n available objects, (or of combining r objects from n available objects), is known as n combination r or abbreviated to $n\,C_r$.

This can also be written ${}^nC_r$ or $\binom{n}{r}$.

Now we shall look at a few examples where selections (rather than arrangements) will be the object of our interest.

Example 3.4.2(1)

Evaluate ${}_{10}C_4$

Solution:

$${}_{10}C_4 = \frac{10!}{6!\,4!} = \frac{10 \times 9 \times 8 \times 7}{4 \times 3 \times 2 \times 1} \text{ (on cancelling 6!)}$$

i.e. ${}_{10}C_4 = 210$ on cancelling the remaining factors.

The reader may feel that such a problem can be solved more quickly using a calculator or a table of binomial coefficients.

Example 3.4.2(2)

Evaluate ${}_{200}C_3$

Solution: The problem cannot be performed directly with a supplied function on the calculator and is out of range of most tables of binomial coefficients.

However
$$ {}_{200}C_3 = \frac{200!}{197!\,3!}$$

$$= \frac{200 \times 199 \times 198}{3!} = 1313\ 400$$

using the multiplication and division operations on a calculator.

Example 3.4.2(3)

In how many ways can a customer at the super market select 3 different types of soda from 30 available types, and 10 different packets of biscuits from 12 different available packets?

Solution: Any selection of sodas can be combined with any selection of biscuits so there are

$${}_{30}C_3 \times {}_{12}C_{10} \text{ ways}$$

i.e.
$$\frac{30!}{27!\,3!} \times \frac{12!}{2!\,10!} = 4060 \times 66$$

$$= 267\ 960 \text{ ways.}$$

Example 3.4.2(4)

In how many ways can a class of 20 children be split into two groups of 8 and 12 respectively if there are two twins in the class who must not be separated?

Solution: Once the group of 8 has been selected then the remaining 12 children will automatically comprise the other group.

For the selection of those to join the group of 8 we have two cases

(i) the twins are included,

(ii) the twins are excluded.

If the twins are included we have to select 6 other children (from 18) i.e. this can be done in ${}_{18}C_6$ ways.

If the twins are excluded we have to select 8 children, but still from 18 children, i.e. this can be done in ${}_{18}C_8$ ways.

The total number of ways will be

$${}_{18}C_6 + {}_{18}C_8$$

$$= \frac{18!}{12!\,6!} + \frac{18!}{10!\,8!}$$

$$18\ 564 + 43\ 758$$

$$= 62\ 322 \text{ ways.}$$

Example 3.4.2(5)

Five books are to be selected from twenty books, of which eight are paperback and twelve are hardback. How many selections are possible if at least one paperback book has to be included?

Solution: It is possible to work out the number of selections by considering the different cases where 1, 2, 3, 4, 5 paperback books are included. However it is quicker to find out the number of selections when all the books chosen are hardback, then subtract that answer from the total number of available selections.

If there is no restrictions these are

$$_{20}C_5 = \frac{20!}{15!\,5!} = 15\ 504 \text{ ways}$$

If only hardback books are selected this can be done in

$$_{12}C_5 = \frac{12!}{7!\,5!} = 792 \text{ ways}$$

So there are $15\ 504 - 792$

$= 14712$ ways of selecting at least one paperback book.

Example 3.4.2(6)

Nine players are available to play for a table tennis team of four players. In how many ways can the team be selected if two of the players are brothers and must either both be included or both be excluded, and if two other players have recently quarrelled and should not both play in the team?

Solution: Let the brothers be A and B, and the two players who have quarrelled be X and Y.

There are six possibilities to be considered.

(i) A, B both play and X, Y are both excluded.

This leaves 5 players available for 2 remaining places. The number of selections = $_5C_2$ = 10 ways.

(ii) A, B both play and X also plays. This leaves 5 players available for just 1 place. The number of selections = $_5C_1$ = 5 ways.

(iii) A, B both play and Y also plays. This is the same situation as (ii) i.e. there are $_5C_1$ = 5 ways.

(iv) Neither A nor B plays, but X plays. This leaves 5 players available for 3 places. The number of selections = $_5C_3$ = 10 ways.

(v) Neither A nor B plays, but Y plays. This is the same situation as (iv) i.e. there are $_5C_3$ = 10 ways.

(vi) A, B, X, Y are all excluded. This leaves 5 players available for 4 places. The number of selections = $_5C_4$ = 5 ways.

Altogether there are 10 + 5 + 5 + 10 + 10 + 5 = 45 ways of selecting the team.

3.5 SEQUENCES AND SERIES

3.5.1 Sequences

Consider the sets of numbers

(a) $501, -603, 1111, -2048, 3112, \ldots$

(b) $1.4, -1.7, 2.5, -3.7, -3.6, 4.7, \ldots$

If you were asked to give the next number in the set you would be likely to say that you had no idea, as the numbers appear to be in a random arrangement, without any 'rule' or 'pattern' which would give you a clue as to the identity of the next number.
On the other hand consider the sets of numbers

(c) $5, 10, 15, 20, \ldots$

(d) $1, -3, 9, -27, \ldots$

(e) $27, 64, 125, 216, \ldots$

If you were asked to give the next number in the set you would be able to do so. There is an obvious rule for giving the next number and as many subsequent numbers as we would like to find.

In (c) we could continue by giving multiples of 5, i.e. the next few numbers would be 25, 30, 35, 40, ...

In (d) you should have spotted that the numbers are all powers of -3, i.e. we multiply the previous number by -3 to obtain the next numbers, so the next few numbers would be $81, -243, 729, -2187, \ldots$

In (e) we appear to be using the rule that the numbers are cubes of consecutive integers in ascending order, starting with 3^3, hence the next few numbers would be 343, 512, 729, 1000, ...

Sets of numbers such as (c), (d), (e), where there is a definite rule which gives us the next number in the set, are called *Sequences* and any member (or number) in the set is called a *Term* of the sequence. For ease of reference we often use the term u_n to denote the nth term of a sequence. Thus in (e) $u_1 = 3^3$, $u_2 = 4^3$, $u_3 = 5^3$, $u_4 = 6^3$, and so $u_n = (n + 2)^3$. Why?

As an exercise try to identify u_1, u_2, u_3, u_4, in each of the sequences (c) and (d) and then find a formula for u_n.

We may define a sequence as a set of terms in a defined order with a rule for obtaining each of the terms.

3.5.2 Series

When the terms of a sequence are added, e.g. $5 + 10 + 15 + 20 + \ldots$ a series is formed.

A series which stops after a finite number of terms is called a *Finite Series.*

$$5 + 10 + 15 + 20 + 25 + 30 + 35 + 40$$

is a finite series of eight terms.

If a series doe not stop but continues indefinitely then such a series is called an *Infinite Series.*

$$1 + \frac{1}{4} + \frac{1}{16} + \frac{1}{64} + \frac{1}{256} + \frac{1}{1024} + \ldots.$$

is an infinite series.

Consider the series $5 + 10 + 15 + 20 + 25 + 30 + 35 + 40$.

Each term is a multiple of 5 and we can write the series in the form

$$5 \times 1 + 5 \times 2 + 5 \times 3 + 5 \times 4 + 5 \times 5 + 5 \times 6 + 5 \times 7 + 5 \times 8$$

All the terms of the series are of the form $5 \times r$ or $5r$, so that $5r$ is a general term.

We can then define the series as the sum of terms of the form $5r$ where r can take all integer values in order from 1 to 8 inclusive.

3.5.3 Sigma Notation

If we use the symbol Σ(pronounced *Sigma*, which is the symbol for the Greek capital S in the Greek alphabet) this symbol can be taken to mean 'the sum of terms such as'. Using the sigma notation we can define the series $5 \times 1 + 5 \times 2 + ... + 5 \times 8$ in short form as $\Sigma 5r$, r taking all integer values from 1 to 8 inclusive, or briefly,

$$\sum_{r=1}^{8} 5r$$

If we put the lowest and highest value that r can take, below and above the sigma symbol respectively, this will indicate that r also takes each of the integer values between the extreme values.

The series $3^3 + 4^3 + ... + 10^3$ could be written as

$$\sum_{r=3}^{10} r^3$$

or as

$$\sum_{r=1}^{8} (r + 2)^3$$

In the first expression $\sum_{r=3}^{10} r^3$ means the sum of all terms of the form r^3 where r takes all integer values from 3 to 10 inclusive.

In the second expression $\sum_{r=1}^{8} (r + 2)^3$ means the sum of all terms of the form $(r + 2)^3$ where r takes all integer values from 1 to 8 inclusive.

Try to write the same series, $3^3 + 4^3 + ... + 10^3$ in sigma notation using an appropriate form, where r takes all integer values from 0 to 7 inclusive.

N.B: (i) When you write out a finite series you should always write out the last term even if some intermediate terms are omitted e.g. $3^3 + 4^3 + ... + 10^3$, or $5 + 10 + 15 + ... + 40$.

The series
$$1 + \frac{1}{4} + \frac{1}{16} + \frac{1}{64} + \frac{1}{256} + ...$$

can also be written using the sigma notation. When the dots continue after the last mentioned term this is the indication that the series is infinite (i.e. the series has no last term).

Is it possible to find the general term of the series $1 + \frac{1}{4} + \frac{1}{16} + \frac{1}{64} + \frac{1}{256} + ...$?

You will notice that each term of the series is a power of $\frac{1}{4}$, so the general term can be written as $\left(\frac{1}{4}\right)^r$. However the first term is $1 = \left(\frac{1}{4}\right)^0$, so if we use the sigma notation the first value that r takes is zero. As there is no last term for this series we say there is no upper limit for the value of r. In such cases we say that r can be as large as we like (1000, 1 million, 10^{20}, 10^{100}, etc.), or that r approaches infinity or $r \to \infty$. We can abbreviate this by writing the series as

$$\sum_{r=0}^{\infty}\left(\frac{1}{4}\right)^r$$

N.B: (ii) When any series is written using the sigma notation it is advisable to check that the first few values of r will give the correct first few terms of the series.

When we write a series using the sigma notation, apart from the benefit of an abbreviated form, we are also able to select any term of a series without writing down all the previous terms.

Example 3.5.3(1)

Determine the first four terms and the tenth term of the series $\sum_{r=3}^{120} (r^2 - 3r + 6)$

Solution: The first term (u_1) is the value of $r^2 - 3r + 6$ when $r = 3$

i.e. $u_1 = 3^2 - 3 \times 3 + 6 = 6$

The second term (u_2) is the value of $r^2 - 3r + 6$ when $r = 4$.

i.e. $u_2 = 4^2 - 3 \times 4 + 6 = 10$

Similarly $u_3 = 5^2 - 3 \times 5 + 6 = 16$

$u_4 = 6^2 - 3 \times 6 + 6 = 24$

The tenth term (u_{10}) is the value of $r^2 - 3r + 6$ when $r = 10 + 2 = 12$ (since the series was started with the value $r = 3$, *Not* the value $r = 1$)

so $u_{10} = 12^2 - 3 \times 12 + 6 = 114.$

3.5.4 Arithmetic Progression

Consider the sequence 5, 9, 13, 17, ..., 101. We see that successive terms differ by a constant number 4.

Here $u_1 = 5$

$u_2 = 9 = 5 + 4$

$u_3 = 13 = 9 + 4 = 5 + 4 + 4 = 5 + 2 \times 4$

$u_4 = 17 = 13 + 4 = 5 + 2 \times 4 + 4 = 5 + 3 \times 4$

Similarly $u_5 = 21 = 5 + 4 \times 4$

In general $u_r = 5 + (r - 1)4.$

This particular sequence is one example of an arithmetic progression.

An *arithmetic progression* (A.P.) is a sequence in which any term differs from the previous term by a constant number. This constant number, which may be positive or negative is known as the *common difference* (C.D.).

Example 3.5.4(1)

Determine the first six terms of the A.P. whose first term is 18 and whose common difference is – 5.

Solution:

$$u_1 = 18$$
$$u_2 = 18 + (-5) = 13$$
$$u_3 = u_2 + (-5) = 13 - 5$$
$$= 8$$

or use $$u_3 = 18 + 2(-5) = 8$$

Similarly check for yourself that

$$u_4 = 3,\ u_5 = -2,\ u_6 = -7.$$

Example 3.5.4(2)

Determine the fourth, twelfth and nth terms of the A.P. whose first term is 19 and whose common difference is 6.

Solution:

$$u_1 = 19$$
$$u_2 = 19 + 6 = 25$$
$$u_3 = 19 + 2 \times 6 = 31$$

so $$u_4 = 19 + 3 \times 6 = 37$$

Similarly $$u_{12} = 19 + 11 \times 6 = 85$$

and $$u_n = 19 + (n - 1)6$$

or $$19 + 6(n - 1)$$

In general if an A.P. has its first term a and its common difference d the first four terms are

$$u_1 = a$$
$$u_2 = a + d$$
$$u_3 = a + 2d$$
$$u_4 = a + 3d$$

and its nth term $u_n = a + (n - 1)d$.

Hence such an A.P. with n terms can be written as $a,\ a + d,\ a + 2d,\ a + 3d,\ ...,\ a + (n - 1)d$.

The next question we might ask is how we may be able to find the sum of a given number of terms in an A.P. Where the first term, the common difference, and the last term (or the number of terms required to be added) are known, we could simply write down all the terms and add them by the normal means of addition. However it would be quicker if it was possible to find a formula for the sum of an A.P. The famous mathematician Euler was in Primary School when he discovered a quick method for adding all the whole numbers from 1 to 100 inclusive. It was on a day near the end of the school term and his teacher had to do a lot of work in marking end-of-term examinations and entering the marks on the record sheets and students' reports. The teacher decided that a good way to keep the students occupied would be to tell all the students to add up all the whole numbers from 1 to 100 inclusive, expecting that this exercise would take up a considerable amount of time. The teacher was amazed when young Euler gave the correct answer within two minutes.

This is how young Euler solved the problem.

Write the sum required as $1 + 2 + 3 + ... + 98 + 99 + 100$. Write the sum required with the terms in the reverse order as $100 + 99 + 98 + ... + 3 + 2 + 1$

Add the terms vertically $101 + 101 + 101 + ... + 101 + 101 + 101$

There are 100 terms and so $2 \times$ (the sum required) must be $100 \times 101 = 10100$

So the sum required is $10100 \div 2 = 5050$.

It is possible to use a similar method to Euler's when finding the sum of any A.P. where the first term, the common difference, and the number of terms to be added, are all known. This method is known as the method of first principles, as is illustrated in the next example.

Example 3.5.4(3)

Find using first principles the sum of the series

$$12 + 17 + 22 + 27 + ... + 102.$$

Solution: We know the first term is 12 and the common difference is 5 but we do not know how many terms have to be added.

However if we let 102 be u_n (the nth term) then

$$102 = 12 + (n - 1)5$$

i.e. $$102 = 12 + 5n - 5$$

i.e. $$95 = 5n$$

i.e. $$19 = n$$

so there are 19 terms to be added.

Write the sum to n terms as S_n then taking the terms in the order given,

$$S_n = 12 + 17 + 22 + ... + 92 + 97 + 102$$

Taking the terms in reverse order $S_n = 102 + 97 + 92 + ... + 22 + 17 + 12$

so $$2S_n = 114 + 114 + 114 + ... + 114 + 114 + 114$$

$$= 19 \times 114 \text{ as there are 19 terms}$$

so the required sum $$S_n = \frac{19 \times 114}{2}$$

$$= 1083$$

If the same process is applied to a general A.P., whose first term is a and whose common difference is d, where the sum is required to n terms, it is possible to establish a formula for this sum, which may normally be quoted and used directly unless a formal proof using first principles is asked for.

Proof

Let the last term of the A.P. be l, where $l = a + (n - 1)d$.
Writing the terms in the order given

$$S_n = a + a + d + a + 2d + \dots\ l - 2d + l - d + l$$

In reverse order $\quad S_n = l + l - d + l - 2d + \dots + a + 2d + a + d + a$

Adding $\quad 2S_n = a + l + a + l + a + l + \dots\ a + l + a + l + a + l$

As there are n terms to be added

$$2S_n = n\ (a + l)$$

i.e. $\quad S_n = \frac{n}{2}(a + l)$

Also the last term $l = a + (n - 1)d$, so the sum S_n can also be written as $S_n = \frac{n}{2}[a + a + (n - 1)d]$

or $\quad S_n = \frac{n}{2}[2a + (n - 1)d]$

Either formula may be used to find the sum to n terms of an A.P.

Example 3.5.4(4)

Find the sum of the following series.

(a) an A.P. of 10 terms whose first term is 7 and whose last term is 10

(b) $\sum_{r=1}^{16}\left(5 - \frac{r}{4}\right)$

Solution: (a) Since we know the first term, the last term and the number of terms, i.e. we know $a = 7$, $l = 10$, and $n = 10$

Using $\quad S_n = \frac{n}{2}(a + l)$

we require $\quad S_{10} = \frac{10}{2}(7 + 10) = 85$

(b) $\quad \sum_{r=1}^{16}\left(5 - \frac{r}{4}\right) = 5 - 1/4 + 5 - 2/4 + \dots + 5 - 16/4$

This is an A.P. with 16 terms where $a = 4\frac{3}{4}$, $d = -1/4$.

So using $\quad S_n = \frac{n}{2}[2a + (n - 1)d]$

we have $\quad S_{16} = \frac{16}{2}\left[2 \times 4\frac{3}{4} + 15(-1/4)\right]$

$$= 46$$

We could equally well have obtained S_{16} using the formula $S_n = \frac{n}{2}(a + l)$ where $a = 4\frac{3}{4}$, $l = 1$, and $n = 16$.

Example 3.5.4(5)

In an A.P. the fourth term is 13 and the seventh term is 22. Determine (a) the first term and the common difference (b) the value of n if the nth term is 100 (c) the value of m if the sum to m terms of the series is 175

Solution: (a) If the first term is a and the common difference is d,

$$u_4 = a + 3d = 13 \quad (1)$$

$$u_7 = a + 6d = 22 \quad (2)$$

from (2) – (1) $\quad 3d = 9$

i.e. $\quad d = 3$

(1) $\quad a + 3 \times 3 = 13$

i.e. $\quad a = 4$

so first term = 4 and common difference = 3

(b) If u_n is 100

then $\quad 4 + (n - 1)3 = 100$

i.e. $\quad 4 + 3n - 3 = 100$

i.e. $\quad 3n = 99$

i.e. $\quad n = 33$

(c) If the sum to m terms is 175 then $\frac{m}{2}[2 \times 4 + (m - 1)3] = 175$

i.e. $\quad m[8 + 3m - 3] = 350$

i.e. $\quad 8m + 3m^2 - 3m = 350$

i.e. $\quad 3m^2 + 5m - 350 = 0$

i.e. $\quad (3m + 35)(m - 10) = 0$

i.e. $\quad m = -35/3$ or $m = 10$

As the number of terms must be a positive integer then $m = 10$.

Example 3.5.4(6)

Show that the terms of $\sum_{r=1}^{n} \log 5^r$ are in A.P. Find the sum of the first twenty terms of the series correct to two decimal places and also the least value of n for which the sum to n terms exceeds 400.

Solution: If we put $r = 1, 2, 3, \ldots, n$ we have

$$\sum_{r=1}^{n} \log 5^r = \log 5 + \log 5^2 + \log 5^3 + \ldots + \log 5^n$$

$$= \log 5 + 2 \log 5 + 3 \log 5 + \ldots + n \log 5$$

From the last line it can be seen that there is a common difference of log 5 between successive terms, so the terms are in A.P.

Hence $$\sum_{r=1}^{20} \log 5^r = \log 5 + 2 \log 5 + 3 \log 5 + \ldots + 20 \log 5$$

$$= \frac{20}{2} (\log 5 + 20 \log 5)$$

i.e. the sum is $$\frac{20 \times 21 \log 5}{2}$$

$$\simeq 146.7837$$

$$\simeq 146.78 \text{ (to two decimal places)}$$

If the sum to n terms is made equal to 400 we have

$$\frac{n}{2} [\log 5 + n \log 5] = 400$$

i.e. $$n \log 5 + n^2 \log 5 = 800$$

i.e. $$n^2 \log 5 + n \log 5 - 800 = 0$$

$$\text{so } n = \frac{-\log 5 \pm \sqrt{[(\log 5)^2 - 4(\log 5)(-800)]}}{2 \log 5}$$

$$n = 33.33 \text{ or } -34.33$$

As $n > 0$ and n is a positive integer we take $n = 34$ for the sum to exceed 400 for the first time.

So the least value of n for which the sum exceeds 400 is 34.

Example 3.5.4(7)

The sum of the first n terms of a series S_n is given by $S_n = 2n(n + 6)$. Find the fifth term of the series and show that the terms are in A.P.

Solution: If the terms of the series are $u_1, u_2, u_3, \ldots, u_n$.

Then $$S_n = u_1 + u_2 + u_3 + \ldots + u_n$$
$$= 2n\,(n + 6)$$

So $$S_5 = u_1 + u_2 + u_3 + u_4 + u_5 = 2 \times 5\,(5 + 6) = 110$$

and $$S_4 = u_1 + u_2 + u_3 + u_4 = 2 \times 4\,(4 + 6) = 80$$

By subtraction $$u_5 = S_5 - S_4$$

i.e. $$u_5 = 110 - 80 = 30$$

So the fifth term is 30.

In the same way the nth term of the series u_n can be found from

$$S_n = u_1 + u_2 + u_3 + \ldots + u_{n-1} + u_n$$

$$S_{n-1} = u_1 + u_2 + u_3 + \ldots + u_{n-1}$$

$$= 2(n-1)(n-1+6)$$

(by replacing n by $n - 1$ in the formula for S_n)

Hence $\quad u_n = S_n - S_{n-1}$

i.e. $\quad u_n = 2n(n+6) - 2(n-1)(n+5)$

$$= 2n^2 + 12n - 2(n^2 + 4n - 5)$$

$$= 2n^2 + 12n - 2n^2 - 8n + 10$$

so $\quad u_n = 4n + 10$

By replacing n by $n - 1$ we have

$$u_{n-1} = 4(n-1) + 10$$

$$= 4n - 4 + 10$$

$$= 4n + 6$$

So $\quad u_n - u_{n-1} = 4n + 10 - (4n + 6)$

$$= 4n + 10 - 4n - 6$$

$$= 4$$

Since the difference between u_n and u_{n-1} is always 4 whatever the value of n, this is a common difference of 4 between successive terms, showing that the series has its terms in A.P.

3.5.5 Geometric Progressions

If we consider the sequence 36, 12, 4, 4/3, 4/9, 4/27, ... each term of the sequence is one third of the previous term, so we may write the sequence as

$$36,\ 36(1/3),\ 36(1/3)^2,\ 36(1/3)^3,\ 36(1/3)^4,\ 36(1/3)^5,\ \ldots\ldots$$

Such a sequence is known as a geometric progression (G. P.). A G. P. is a sequence in which each term is a constant times the previous term. This constant multiplying factor is known as the common ratio, and it may take any real value. If a G.P. has a first term 5 and a common ratio -4 the first four terms are:

$$5,\ 5(-4),\ 5(-4)^2,\ 5(-4)^3, \text{ or } 5,\ -20,\ 80,\ -320.$$

In general if a G.P. has a first term a and a common ratio r the first four terms will be a, ar, ar^2, ar^3, and the nth term u_n will be ar^{n-1}. So a G.P. having n terms can be written as a, ar, ar^2, ar^3, ... ar^{n-1}.

In the same way as we approached the problem of finding a formula for the sum of an A.P. we first explore the sum of a G.P. using first principles in order to see how the method of first principles will lead to the establishment of a formula.

Example 3.5.5(1)

Find using first principles the sum of the first 8 terms, S_8 of the G.P. with first term 3 and common ratio 1/2.

Solution:

Writing $$S_8 = 3 + 3\left(\frac{1}{2}\right) + 3\left(\frac{1}{2}\right)^2 + 3\left(\frac{1}{2}\right)^3 + \ldots\ldots + 3\left(\frac{1}{2}\right)^7$$

You will notice that if we multiply each of the terms of S_8 by $\frac{1}{2}$ we have

$$\frac{1}{2}S_8 = 3\left(\frac{1}{2}\right) + 3\left(\frac{1}{2}\right)^2 + 3\left(\frac{1}{2}\right)^3 + \ldots\ldots + 3\left(\frac{1}{2}\right)^7 + 3\left(\frac{1}{2}\right)^8$$

Setting like powers of $\left(\frac{1}{2}\right)$ below each other in the expressions for S_8, $\frac{1}{2}S_8$,

$$S_8 = 3 + 3\left(\frac{1}{2}\right) + 3\left(\frac{1}{2}\right)^2 + 3\left(\frac{1}{2}\right)^3 + \ldots + 3\left(\frac{1}{2}\right)^7$$

$$\frac{1}{2}S_8 = \quad 3\left(\frac{1}{2}\right) + 3\left(\frac{1}{2}\right)^2 + 3\left(\frac{1}{2}\right)^3 + \ldots + 3\left(\frac{1}{2}\right)^7 + 3\left(\frac{1}{2}\right)^8$$

then subtracting $$S_8 - \frac{1}{2}S_8 = 3 - 3\left(\frac{1}{2}\right)^8$$

i.e. $$\left(1 - \frac{1}{2}\right)S_8 = 3\left[1 - \left(\frac{1}{2}\right)^8\right]$$

i.e. $$S_8 = \frac{3\left[1 - \left(\frac{1}{2}\right)^8\right]}{1 - \frac{1}{2}} = \frac{3\left[1 - \left(\frac{1}{2}\right)^8\right]}{\frac{1}{2}}$$

$$= 6\left[1 - \left(\frac{1}{2}\right)^8\right] = 6 \times \frac{255}{256}$$

$$= \frac{765}{128}.$$

We can now use the same method to establish a formula for the sum to n terms of the G.P.

$$a + ar + ar^2 + ar^3 + \ldots\ldots + ar^{n-1}.$$

Proof

Let $$S_n = a + ar + ar^2 + ar^3 + \ldots + ar^{n-1}$$

and $$rS_n = \quad ar + ar^2 + ar^3 + \ldots + ar^{n-1} + ar^n$$

Subtracting $$S_n - rS_n = a - ar^n$$

i.e. $$(1 - r)S_n = a(1 - r^n)$$

i.e. $$S_n = \frac{a(1 - r^n)}{1 - r}$$

If $r > 1$ it is usually easier to use the equivalent formula

$$S_n = \frac{a(r^n - 1)}{r - 1}$$

This formula may be applied to find the sum to n terms of any G.P. whose first term and common ratio are known, provided the common ratio is not equal to 1.

Example 3.5.5(2)

The third term of a G.P. is $1\frac{1}{4}$ and the sixth term is $\frac{5}{32}$. Determine the first term, the common ratio and the sum of the first six terms of the series.

Solution:

If the series is $a + ar + ar^2 + ar^3 + \ldots$

then $$u_3 = ar^2 = \frac{5}{4} \quad (1)$$

$$u_6 = ar^5 = \frac{5}{32} \quad (2)$$

Divide (2) by (1) so that

$$\frac{ar^5}{ar^2} = \frac{\frac{5}{32}}{\frac{5}{4}}$$

i.e. $$r^3 = \frac{5}{32} \times \frac{4}{5}$$

i.e. $$r^3 = \frac{1}{8}$$

i.e. $$r = \frac{1}{2}$$

In (1) $$a\left(\frac{1}{2}\right)^2 = \frac{5}{4}$$

so $$a = 5$$

so the first term is 5, the common ratio is $\frac{1}{2}$ and the sum of the first six terms is

$$S_6 = \frac{5\left[1 - \left(\frac{1}{2}\right)^6\right]}{1 - \frac{1}{2}} = 10 \times \frac{63}{64}$$
$$= 315/32$$

Example 3.5.5(3)

What is the smallest number of terms of the G.P. 8 + 24 + 72 + ... which will give a total greater than one million?

Solution: The sum to n terms of the G.P. $8 + 24 + 72 + ... + 8(3)^{n-1}$ is given by

$$S_n = \frac{8(3^n - 1)}{3 - 1}$$

i.e. $$4(3^n - 1)$$

To find the required number of terms we equate S_n to one million

i.e. Solve $$4(3^n - 1) = 10^6$$

i.e. $$3^n - 1 = 250\ 000$$

i.e. $$3^n = 250\ 001$$

Taking logarithms, to base 10 of each side

$$n \log 3 = \log 250\ 001$$

i.e. $$n = \frac{\log 250\ 001}{\log 3}$$

$$= \frac{5.39794}{0.47712}$$

$$= 11.31$$

So the number of terms required to make a sum first exceed one million is therefore 12.

Example 3.5.5(4)

The sum of the first n terms of a series is $5^n - 1$. Show that the terms of the series are in G.P. and find the first term, the common ratio, and the ratio of the sum of the first $2n$ terms of the series to the sum of the first n terms of the series.

Solution:

If $$S_n = 5^n - 1$$

then $$S_{n-1} = 5^{n-1} - 1$$

and so
$$\begin{aligned} u_n &= S_n - S_{n-1} \\ &= 5^n - 1 - (5^{n-1} - 1) \\ &= 5^n - 1 - 5^{n-1} + 1 \\ &= 5^n - 5^{n-1} \\ &= 5^{n-1}(5 - 1) \\ &= 4(5^{n-1}) \end{aligned}$$

Similarly $$u_{n-1} = 4(5^{n-2})$$

and so
$$u_n \div u_{n-1} = \frac{4(5^{n-1})}{4(5^{n+2})}$$

$$= 5$$

So successive terms of the series have a constant ratio 5, irrespective of the value of n. Hence as $u_n = 4(5^{n-1})$ the series is a G.P. with first term 4 and common ratio 5.

So $$S_{2n} \div S_n = \frac{5^{2n}-1}{5^n-1} = \frac{(5^n-1)(5^n+1)}{5^n-1} = 5^n + 1.$$

Example 3.5.5(5)

A customer makes a single deposit of Sh 16 000 in an account which pays compound interest at a rate of 6% per annum.

(a) How much is the investment worth after twelve years?

(b) After how many years will the investment be worth three times its initial value?

Solution:

(a) If we apply a step by step solution initially, after one year the investment will be worth the initial amount plus the interest on the initial amount

i.e. $$\text{Sh } 16\,000 + \frac{6}{100} \times \text{Sh } 16\,000$$
$$= \text{Sh } 16\,000\ (1 + 6/100)$$

after two years it will be worth the amount after one year plus the interest on that amount

i.e. $$\text{Sh } 16\,000(1 + 6/100) + \text{Sh } 16\,000(1 + 6/100) \times 6/100$$
$$= \text{Sh } 16\,000(1 + 6/100)(1 + 6/100)$$
$$= \text{Sh } 16\,000(1 + 6/100)^2$$

after three years it will be worth the amount after two years plus the interest on that amount

i.e. $$\text{Sh } 16\,000(1 + 6/100)^2 + \text{Sh } 16\,000(1 + 6/100)^2 \times 6/100$$
$$= \text{Sh } 16\,000(1 + 6/100)^2\ (1 + 6/100)$$
$$= \text{Sh } 16\,000(1 + 6/100)^3$$

Looking at the pattern it can be seen that the investment after n years will be

$$\text{Sh } 16\,000(1 + 6/100)^n$$

and in particular after 12 years it will be

$$\text{Sh } 16\,000(1 + 6/100)^{12}$$
$$= \text{Sh } 16\,000(1.06)^{12}$$
$$= \text{Sh } 32\,195 \text{ (to the nearest shilling)}$$

(b) Using the formula we deduced for the value of the investment after n years it means that we have to determine n such that

$$\text{Sh } 16\,000(1 + 6/100)^n$$

or $$\text{Sh } 16\,000(1.06)^n = 3 \times \text{Sh } 16\,000$$

i.e. $$(1.06)^n = 3$$

i.e. $$n \log 1.06 = \log 3$$

i.e. $$n = \log 3/\log 1.06$$

$$\simeq 18.85$$

Since in most cases the interest is added after every complete year then the period required would have to be 19 years.

Example 3.5.5(6)

If a single deposit of Sh 150 000 is invested for four years at compound interest, at what rate will the investment be worth Sh 182 326?

Solution:

If the rate is R% per annum then after four years the investment will be worth

$$\text{Sh } 150\,000\,(1 + R/100)^4$$

and so $$\text{Sh } 150\,000(1 + R/100)^4 = \text{Sh } 182\,326$$

i.e. $$\log 150\,000 + 4 \log (1 + R/100)$$

$$= \log 182\,326$$

leading to $$\log (1 + R/100)$$

$$= \frac{\log 182\,326 - \log 150\,000}{4}$$

i.e. $$\log (1 + R/100) = 0.02119$$

or $$1 + R/100 = 1.05$$

hence $$R/100 = 0.05$$

or the rate is 5%.

In each of these examples we have applied the formula for the amount (or value) A of a single investment P at a rate of R% per annum for n years, which is

$$A = P(1 + R/100)^n$$

This same formula may be used for the value of say, a piece of land, which appreciates at a given rate R% per annum, and even for depreciation of the value of an item such as a car, where again the given rate per annum remains the same from year to year. However the value of R will be negative.

Example 3.5.5(7)

A car has an initial value of Sh 3 million when a Professor buys it. A student plans to buy the car from the Professor when the car is worth Sh 1 million.

If the rate of depreciation is 14% per annum, find after how many complete years the Professor will sell the car to the student.

Solution:

Since the value of the car is depreciating it will be worth only 86% of its value (100% –14%) at the end of each year when compared with its value at the start of the year.

If its initial value is Sh 3 million then after one year its value will be Sh 3 million (0.86). After two years its value will be Sh 3 million $(0.86)^2$. After n years its value will be Sh 3 million $(0.86)^n$. If the student wishes to buy when the car is worth Sh 1 million, then Sh 1 million $(0.86)^n$ = Sh 1 million

i.e. $$3(0.86)^n = 1$$

i.e. $$\log 3 + n \log 0.86 = 0$$

leading to
$$n = \frac{-\log 3}{\log 0.86}$$
$$= 7.28$$

If the Professor is willing to sell only after a number of complete years then the student will have to wait 8 years.

The same principles apply in accumulated investments, in which a constant amount is invested at the start of each year, as in the following two examples.

Example 3.5.5 (8)

A customer deposits Sh 10000 on 1 January every year for four years. How much is the total investment worth at the end of four years?

Solution: It is easy to see how the investment stands if we look at the first two years.

		Sh
Year 1	1 Jan	10 000
	31 Dec	$10\,000 + 10\,000 \times 12/100$
		$= 10\,000\ (1.12)$
Year 2	1 Jan	$10\,000\ (1.12) + 10\,000$
	31 Dec	$10\,000\ (1.12) + 10\,000$
		$+\ 10\,000 \times 1.12 \times \frac{12}{100} + 10\,000 \times \frac{12}{100}$
i.e. Year 2	31 Dec	$10\,000\ (1.12)\ (1 + 12/100) + 10\,000(1 + 12/100)$
		$= 10\,000\ (1.12)^2 + 10\,000\ (1.12)$

Continuing the pattern we see that at the end of year 4 the investment is

$$\text{Sh } \{10\,000(1.12)^4 + 10\,000(1.12)^3 + 10\,000(1.12)^2 + 10\,000(1.12)\}$$

If this is written the other way

i.e. $$\text{Sh } \{10\,000(1.12) + 10\,000(1.12)^2 + 10\,000\ (1.12)^3 + 10\,000(1.12)^4\}$$

it can be seen that this is the sum of a G.P. with first term 10 000(1.12) and common ratio 1.12. The total investment can then be worked out from the formula

$$\frac{10000\ (1.12)\left\{(1.12)^4 - 1\right\}}{(1.12 - 1)}$$

giving Sh 53 528 correct to the nearest shilling.

Example 3.5.5(9)

Sh 100 000 was invested on 1 January 1990. An additional Sh 6 000 was added to the investment at the beginning of each subsequent year. The investment returned compound interest at a rate of 8% per annum. Find the value of the investment on 31 December 2 000.

Solution:

Note that if the investment is held till 31 December 2 000 this is 31 December at the end of year 11.

		Sh
Year 1	1 Jan	100 000
	31 Dec	$100\ 000 + 100\ 000 \times 8/100$
		$= 100\ 00\ (1 + 8/100)$
		$= 100\ 000\ (1.08)$
Year 2	1 Jan	$100\ 000\ (1.08) + 6\ 000$
	31 Dec	$100\ 000\ (1.08) + 6\ 000 + 100\ 000(1.08) \times 8/100 + 6000 \times 8/100$
		$= 100\ 000\ (1.08)\ (1 + 8/100) + 6000\ (1 + 8/100)$
		$= 100\ 000\ (1.08)^2 + 6\ 000\ (1.08)$
Year 3	1 Jan	$100\ 000\ (1.08)^2 + 6\ 000\ (1.08) + 6\ 000$
	31 Dec	$100\ 000\ (1.08)^2 + 6\ 000\ (1.08) + 6\ 000 + 100\ 000(1.08)^2 \times 8/100 + 6\ 000(1.08) \times 8/100 + 6000 \times 8/100$
		$= 100\ 000\ (1.08)^2\ (1 + 8/100) + 6\ 000\ (1.08)\ (1 + 8/100) + 6\ 000(1 + 8/100)$
		$= 100\ 000\ (1.08)^3 + 6000(1.08)^2 + 6000(1.08)$

From the pattern we can see that at the end of year 11 the investment is worth

$$100\ 000(1.08)^{11} + 6\ 000(1.08)^{10} + 6\ 000(1.08)^9 + ... + 6\ 000(1.08)$$

$$100\ 000(1.08)^{11} + 6\ 000(1.08) + ... + 600(1.08)^9 + 6\ 000(1.08)^{10}$$

$= 100\ 000(1.08)^{11}$ + the sum of a G.P. with first term 6 000(1.08) and common ratio 1.08.

$$= 100\ 000(1.08)^{11} + \frac{6\ 000\ (1.08)\ [1.08^{10} - 1]}{(1.08 - 1)}$$

i.e. 233 163.9 + 93 872.9

giving Sh 327 037 as the value of the investment at the end of year 11.

3.5.6 Convergence of a Geometric Series

Suppose a frog has to move a distance along a straight line. In his first jump he covers exactly half the distance, in his second jumps he covers half the distance of his first jump, and so on, i.e. the distance for each new jump is exactly half the distance of the previous jump. Can we determine the total distance S?

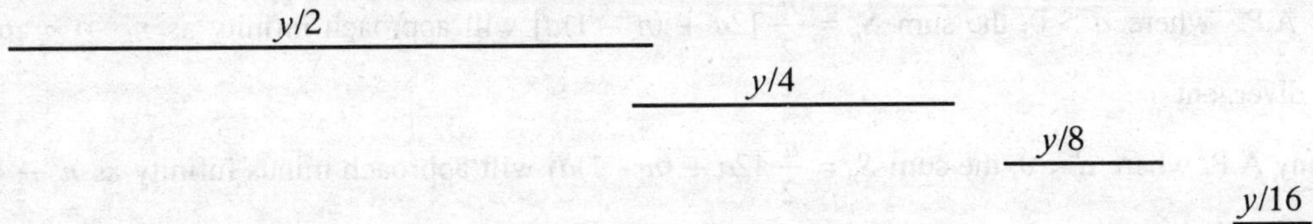

$$S = y/2 + y/4 + y/8 + y/16 + ...$$

In theory this process can go on indefinitely and the series formed is infinite.

After several jumps it is clear that the remaining distance to be covered will be very small, so that the total distance jumped will be very close to the total distance to be covered. The more jumps that are made the closer to y the sum becomes.

After n jumps the distance will be $y/2 + y/4 + y/8 + ... + y/2^n$

$$= \frac{y/2(1-1/2^n)}{(1-1/2)} = y(1-1/2^n)$$

As $n \to \infty$ (i.e. as n approaches infinity or n becomes very large) we can see that

$$y/2 + y/4 + y/8 + ... + y/2^n \to y$$

i.e. as $n \to \infty$

$$\lim_{n\to\infty} (y/2 + y/4 + y/8 + ... + y/2^n) = y$$

y is called the sum to infinity of the series.

In general if S_n is the sum of the first n terms of any series and if $\lim_{n\to\infty} (S_n)$ exists and is finite the series (S_n) is said to be *Convergent*.

In this case we can define the sum to infinity (often written as S or S_∞) = $\lim_{n\to\infty} S_n$.

The series we have been examining

$$y/2 + y/4 + y/8 + ... \text{ is convergent.}$$

$$S_n = y\left(1 - \frac{1}{2^n}\right)$$

and the sum to infinity is

$$\lim_{n\to\infty} y\left(1 - \frac{1}{2^n}\right) = y$$

N.B. as n gets large $\frac{1}{2^n}$ will become very small.

However the series $3 + 6 + 9 + ... + 3n$ has a sum to n terms given by

$$S_n = \frac{n}{2}(3 + 3n) = \frac{3n}{2}(1 + n).$$

As $n \to \infty$, $S_n \to \infty$. In this case the series does not converge, and we say that it is *Divergent.*

For any A.P. where $d > 0$, the sum $S_n = \frac{n}{2}[2a + (n-1)d]$ will approach infinity as $n \to \infty$ and will be divergent.

For any A.P. where $d < 0$, the sum $S_n = \frac{n}{2}[2a + (n-1)d]$ will approach minus infinity as $n \to \infty$ and will be divergent.

The series discussed earlier

$y/2 + y/4 + y/8 + \ldots$ is a G.P. with first term $y/2$ and common ratio $1/2$.

We have already seen that the sum to n terms

$$S_n = \frac{y/2\left[1-\frac{1}{2^n}\right]}{\left[1-\frac{1}{2}\right]}$$

$$= y\left[1-\frac{1}{2^n}\right]$$

We saw that as n increases

$$\frac{1}{2^n} = \left(\frac{1}{2}\right)^n \text{ decreases}$$

i.e. as $n \to \infty, \left(\frac{1}{2}\right)^n \to 0$

$$\lim_{n\to\infty} y\left[1-\left(\frac{1}{2}\right)^n\right] = y$$

So $S_n \to y$ as $n \to \infty$

i.e. S or $S_\infty = y$

We may apply the same idea for the sum to infinity of the general G.P.

$$a + ar + ar^2 + \ldots$$

The sum to n terms S_n

$$= \frac{a\,(1-r^n)}{1-r}$$

Provided $-1 < r < 1$ (i.e. $|r| < 1$) then the $\lim_{n\to\infty} r^n = 0$

and $$\lim_{n\to\infty} S_n = \lim_{n\to\infty} \frac{a(1-r^n)}{1-r}$$

$$= \frac{a}{1-r}$$

If $|r| > 1$ $\lim_{n\to\infty} |r|^n = \infty$ and the series *does not converge*.

So provided $-1 < r < 1$ (or $|r| < 1$) a G.P. converges to the sum

$$\frac{a}{1-r}$$

Example 3.5.6(1)

Determine whether the series below converge, and if so give the sum to infinity

(a) $$15 + 18 + 21 + ...$$

(b) $$1 - 1/6 + 1/36 - 1/216 + ...$$

(c) $$8 + 12 + 18 + 27 + ...$$

Solution:

(a) 15 + 18 + 21 + ... is an A.P. with common difference $d = 3$, and so it does not converge.

(b) 1 – 1/16 + 1/36 – 1/216 + ... is a G.P. with first term 1 and common ratio – 1/6. Since – 1 < 1/6 < 1, the series converges to a sum

$$\frac{1}{1-\left(-\frac{1}{6}\right)} = 6/7$$

(c) 8 + 12 + 18 + 27 + ... is a G.P. with first term 8, common ratio 3/2. As the common ratio > 1 the series does not converge.

Example 3.5.6(2)

Find the values of x for which the series $\sum_{r=0}^{\infty} \frac{x^{r+1}}{(1+x)^r}$ converges and find the sum of the series when it converges.

Solution:

The series $$= x + \frac{x^2}{1+x} + \frac{x^3}{(1+x)^2} + ...+$$

This is a G.P. with common ratio

$$\frac{x}{1+x}$$

and it will be convergent if

$$-1 < \frac{x}{1+x} < 1$$

for the case where $1 + x > 0$

i.e. $$x > -1$$

we multiply the inequality by $1 + x$ to arrive at $-1 - x < x < 1 + x$.
Solving separately: the left-hand inequality $-1 - x < x$

i.e. $$-1 < 2x$$

i.e. $$-1/2 < x$$

the right-hand inequality

$$x < 1 + x$$

gives $0 < 1$ which is true whatever the value of x.

So for the case where $x > -1$ we have shown (by the left-hand inequality) that $x > -1/2$. For the case where $1 + x < 0$

i.e.

we mult... of the inequality signs i.e. – 1

66 *Electromagnetic Field and Waves*

(*b*) a charge = 3 μC at $P_2(-1, 2, 3)$

(*c*) both the charges Q_1 and Q_2.

Solution: (*a*) The vector from poin...

g

w

$$x + \frac{x^2}{1+x} + \frac{x^3}{(1+x)^2} + \ldots\ldots$$

$$= \frac{x}{1 - \dfrac{x}{1+x}} = \frac{x}{\dfrac{1+x-x}{1+x}}$$

$$= \frac{x(1+x)}{1} = x\ (1 + x).$$

Example 3.5.6(3)

Express the recurring decimal $0.2\dot{4}5\dot{6}$ as a fraction in its lowest terms.

Solution:

$$0.2\dot{4}5\dot{6} = 0.2456\ 456\ 456\ \ldots.$$

$$= \frac{2}{10} + \frac{456}{10^4} + \frac{456}{10^7} + \frac{456}{10^{10}} + \ldots.$$

$$= \frac{2}{10} + \frac{456/10^4}{1 - 1/10^3} = \frac{2}{10} + \frac{456}{10^4} \times \frac{10^3}{999}$$

$$= \frac{2}{10} + \frac{456}{9990} = \frac{1998 + 456}{9990}$$

$$= \frac{2454}{9990} = \frac{409}{1665}.$$

Example 3.5.6(4)

The fifth term of a convergent G.P. is the arithmetic mean of the preceding two terms.

Find the common ratio, given that the common ratio and the first term are non-zero, and, if the first term is 12, find the sum to infinity.

Solution: If the series is $a + ar + ar^2 + ...$, the fifth term is ar^4 and the preceding two terms are ar^3, ar^2.

So
$$ar^4 = \frac{1}{2}(ar^3 + ar^2)$$
or
$$2ar^4 = ar^3 + ar^2$$

Since a and r are both non-zero we can cancel out by a factor ar^2 to arrive at

$$2r^2 = r + 1$$
i.e.
$$2r^2 - r - 1 = 0$$
i.e.
$$(2r + 1)(r - 1) = 0$$
i.e.
$$r = -1/2 \text{ or } 1$$

But the series is convergent so the only acceptable value of r is $r = -1/2$.

When $a = 12$, $r = -1/2$, the sum to infinity is $\dfrac{12}{1 - \frac{-1}{2}}$

$$= \frac{12}{3/2}$$
$$= 8.$$

3.5.7 Binomial Theorem

In Chapter 2 of the book (Sections 2.5.1, 2.5.2) we examined how to find the expansion of $(a + b)^n$ for a positive integer n, using Pascal's triangle, or by using a table of binomial coefficients. However if we wanted to expand $(1 + x)^{50}$ or $(2 - x)^{100}$ it would be likely that no table of coefficients would be available for such high powers and an extension of Pascal's triangle would be tedious.
We may now look at a different procedure:

e.g. $$(a + b)^5 = (a + b)\,(a + b)\,(a + b)\,(a + b)\,(a + b)$$

When five brackets are expanded each term is obtained by multiplying together an 'a' or a 'b' from each bracket.

Taking 'a' from each bracket we get a^5.

Taking 'a' from 4 brackets and 'b' from 1 bracket means that 'b' can be selected in ${}_5C_1$ ways.

We can choose the 'b' from each of the 5 brackets in turn. So the term in $a^4 b$ is $5a^4 b$.
Taking 'a' from 3 brackets and 'b' from the 2 remaining brackets we can obtain the term in $a^3 b^2$. But this can be done when we select 'b' from any 2 brackets out of 5 i.e. in $5^C 2$ ways.

Similarly it is possible to show that the coefficients of $a^2 b^3$, ab^4 and b^5 are ${}_5C_3$, ${}_5C_4$ and ${}_5C_5 = 1$, respectively.

Arranging in ascending powers of b we get

$$(a + b)^5 = a^5 + {}_5C_1\, a^4 b + {}_5C_2\, a^3 b^2 + {}_5C_3\, a^2 b^3 + {}_5C_4\, a\, b^4 + b^5$$

The right hand side of this identity is the binomial series expansion of $(a + b)^5$. Note that the expansion has 6 terms.

The argument can be generalised in the proof which follows.

Proof of Binomial Theorem: Expansion of $(a + b)^n$ where n is a positive integer.

$(a + b)^n$ can be expanded to give a series of terms in ascending powers of b where the term in $a^{n-r}\, b^r$ is obtained by multiplying together 'b' terms from r brackets and the remaining 'a' terms from $n - r$ brackets.

We may choose r brackets from n brackets in ${}_nC_r$ ways. Hence the term containing $a^{n-r}\, b^r$ is ${}_nC_r\, a^{n-r}\, b^r$

Hence $(a + b)^n = a^n + {}_nC_1\, a^{n-1}\, b^1 + {}_nC_2\, a^{n-2}\, b^2 + \ldots + {}_nC_r\, a^{n-r}\, b^r + \ldots + {}_nC_n\, b^n$

It may also be written in another form if we note that

$$ {}_nC_1 = \frac{n!}{1!\,(n-1)!} = \frac{n}{1} = n$$

$$ {}_nC_2 = \frac{n!}{2!\,(n-2)!} = \frac{n\,(n-1)}{2!}$$

$$ {}_nC_3 = \frac{n!}{3!\,(n-3)!} = \frac{n\,(n-1)\,(n-2)}{3!}$$

and

$$ {}_nC_r = \frac{n!}{(n-r)!\,r!}$$

$$= \frac{n\,(n-1)\,(n-2)\,\ldots.\,(n-r+1)\,(n-r)\,(n-r-1)\,\ldots.\,2\times 1}{(n-r)!\,r!}$$

$$= \frac{n\,(n-1)\,(n-2)\,\ldots.\,(n-r+1)}{r!}$$

and so $(a + b)^n = a^n + na^{n-1}\, b + \dfrac{n\,(n-1)}{2!}\, a^{n-2}\, b^2 + \dfrac{n\,(n-1)\,(n-2)}{3!}\, a^{n-3}\, b^3 + \ldots +$

$$\frac{n\,(n-1)\,(n-2)\,\ldots.\,(n-r+1)}{r!}\, a^{n-r}\, b^r + \ldots + b^n$$

or as

$$\sum_{r=0}^{n} {}_nC_r\, a^{n-r}\, b^r$$

Note also:

(i) this expansion is a *finite* series with $(n + 1)$ terms;

(ii) the coefficient of $a^{n-r}\, b^r$ is

$$ {}_nC_r = \frac{n\,(n-1)\,(n-2)\,\ldots.\,(n-r+1)}{r!}$$

which has r factors in both numerator and denominator;

(iii) the term containing b^1 is the *Second* term, the term containing b^2 is the *third* term, the term containing b^3 is the fourth term, and the term containing b^r is the $(r + 1)$th term, so the rth term is ${}_nC_r - 1\ a^{n-r+1}\ b^{r-1}$;

(iv) the 'combination' form of the expansion can be used when we require to find the coefficient of a large power of b or to find the general term, but it is more useful to keep the factor form of the expansion when we only require to find the first few terms of any expansion.

Example 3.5.7(1)

Find the term containing a^3 in the expansion of $(2a - b)^{50}$ and simplify your answer.

Solution:

In the expansion of $(2a - b)^{50}$

the term containing a^3 will be the term containing $(2a)^3\ (-b)^{47}$ and this is ${}_{50}C_{47}\ (2a)^3\ (-b)^{47}$

$$= \frac{50 \times 49 \times 48}{6}\ 8a^3\ (-1)^{47}\ b^{47}$$

$$= -156\ 800\ a^3\ b^{47}$$

Example 3.5.7(2)

Find in factor form the coefficient of x in the expansion of $(2x^2 - 1/x)^{50}$.

Solution:

The general term is ${}_{50}C_r\ (2\,x^2)^{50-r} \left(\dfrac{-1}{x}\right)^r$

i.e.
$$ {}_{50}C_r\ 2^{50-r}\ x^{100-2r}\ \frac{(-1)^r}{x^r}$$

The power of x is $x^{100-2r-r}$

$$= x^{100-3r}$$

and this should be x^1 for the term we require

hence
$$100 - 3r = 1$$

i.e.
$$r = 33$$

The required term is then

$$ {}_{50}C_{33}\ 2^{50-33}\ x^1\ (-1)^{33}$$

and the coefficient of x will be

$$\frac{-50!\,2^{17}}{33!\,17!}.$$

Example 3.5.7(3)

If x is so small that terms in x^2 and higher powers of x may be neglected show that

$$(1 - x)^6\ (2 + x/2)^8 \simeq 2^8\ (1 - 4x)$$

and using the approximation estimate the value of $0.999^6\ (2.0005)^8$ correct to two decimal places.

Solution:

$$(1 - x)^6\ (2 + x/2)^8$$

$$= (1 - 6x + ...)(2^8 + 8 \times 2^7 (x/2) + ...)$$

$$\simeq 2^8 - 6x\,(2^8) + 2^{10}\,(x/2) + ...$$

$$\simeq 2^8\,(1 - 6x + 2x)$$

or

$$2^8\,(1 - 4x)$$

From the required expression $0.999^6(2.0005)^8$ then we check the value of x which makes $1 - x = 0.999$ and also makes $2 + x/2 = 2.0005$ and the value of x, which makes *both* equations true, is 0.001. Using the approximation that

$$(1 - x)^6\,(2 + x/2)^8 \simeq 2^8(1 - 4x)$$

we have $0.999^6(2.0005)^8$

$$\simeq 2^8(1 - 4 \times 0.001)$$

or

$$256 \times 0.996 = 254.976$$

or

$$254.98 \text{ (two decimal places)}$$

We now wish to extend the binomial theorem to cover the case where the index n in the expansion $(a + b)^n$ could be a negative integer or a real number which is not an integer.
When n is a *positive* integer we already know that

$$(1 + x)^n = 1 + nx + \frac{n\,(n-1)}{2!}\,x^2 + ... + x^n$$

Consider the infinite series

$$1 - x + x^2 - x^3 + ...$$

which is a G.P. with first term 1 and common ratio $- x$. Provided $|\,x\,| < 1$ we know that the sum to infinity of this series is $\dfrac{1}{1-(-x)} = \dfrac{1}{1+x}$ or $(1 + x)^{-1}$

So

$$(1 + x)^{-1} = 1 - x + x^2 - x^3 + ...$$

If we go back to the binomial expansion for $(1 + x)^n$ and set $n = -1$ on the left hand side, and also set $n = -1$ on the first few terms of the right hand side, the left hand side becomes $(1 + x)^{-1}$ and the right hand side becomes

$$1 + (-1)\,x + \frac{(-1)(-2)}{2}\,x^2 + \frac{(-1)(-2)(-3)}{3!}\,x^3 + ...$$

$$= 1 - x + x^2 - x^3 + ...$$

However we already know that provided $|\,x\,| < 1$ then

$$(1 + x)^{-1} = 1 - x + x^2 - x^3 + ...$$

as an infinite series.

So the form of the binomial expansion quoted earlier appears to hold for a value of n which is not positive integer (in this case we took $n = -1$) but we must note the following points:

(i) the series expansion does not terminate but is an infinite series,
(ii) the values of x must lie in the range $|\,x\,| < 1$ or $-1 < x < 1$,
(iii) the expansion no longer provides an identity since the sum of the infinite series (in this case a G.P.) will never be exactly $(1 + x)^{-1}$.

It is true that for any real value of n, although we are unable to *prove* it at this stage, the general series

$$1 + nx + \frac{n(n-1)}{2}x^2 + \frac{n(n-1)(n-2)}{3!}x^3 + ...$$

has a sum to infinity of $(1 + x)^n$ where n is not a positive integer but subject to the condition that $-1 < x < 1$.

So for all values of x for which $-1 < x < 1$ we are able to expand $(1 + x)^n$, for any value of n, as a series of ascending powers of x in the form

$$(1 + x)^n = 1 + nx + \frac{n(n-1)}{2}x^2 + \frac{n(n-1)(n-2)}{3!}x^3 + ...$$

When using the general expression it should be noted that

(i) the expansion is valid for $(1 + x)^n$ but not for $(a + x)^n$, so to expand $(a + x)^n$ we should first express

$$(a + x)^n = (a(1 + x/a))^n$$
$$= a^n \ (1 + x/a)^n,$$

and then expand the expression in terms of powers of x/a,

(ii) the expansion is valid for a restricted set of values of x, and this set of values must always be stated.

Example 3.5.7(4)

Expand each of the following in ascending powers of x up to and including the term in x^3, stating the range of values of x for which each of the expansions is valid.

(a) $(1 + x)^{1/3}$ (b) $(1 - 4x)^{-3}$ (c) $(3 - x)^{-2}$

Solution:

(a)
$$(1 + x)^{1/3} = 1 + (1/3)x + \frac{1/3(-2/3)}{2}x^2 + \frac{\frac{1}{3}\left(\frac{-2}{3}\right)\left(\frac{-5}{3}\right)}{3!}x^3 + ...$$
$$= 1 + (1/3)x - (1/9)x^2 + (5/81)x^3 + ...$$

provided $| x | < 1$

(b)
$$(1 - 4x)^{-3}$$
$$= 1 + (-3)(-4x) + \frac{(-3)(-4)}{2}(-4x)^2 + \frac{(-3)(-4)(-5)}{3!}(-4x)^3 + ...$$
$$= 1 + 12x + 96x^2 + 640x^3 + ...$$

provided $-1 < -4x < 1$

i.e. $\frac{-1}{-4} > x > \frac{1}{-4}$

or $1/4 > x > -1/4$

(c) $(3 - x)^{-2}$ must first be expressed as $[3(1 - x/3)]^{-2}$

$$= 3^{-2}(1 - x/3)^{-2}$$

$$= 3^{-2}\left[1 + (-2)(-x/3) + \frac{(-2)(-3)}{2!}(-x/3)^2 + \frac{(-2)(-3)(-4)}{3!}(-x/3)^3 + ...\right]$$

$$= 1/9\,[1 + 2x/3 + x^2/3 + 4x^3/27]$$

$$= 1/9 + 2x/27 + x^2/27 + 4x^3/243 + ...$$

provided $\quad -1 < -x/3 < 1$

i.e. $\quad 1 > x/3 > -1$

i.e. $\quad 3 > x > -3$

Example 3.5.7(5)

Expand $\dfrac{4}{(1+4x)(1-2x)}$

in ascending powers of x as far as the term in x^3 and find the range of values of x for which the expansion is valid.

Solution: There is another method of starting this problem for those readers who are familiar with the method of resolving the given expression into partial fractions. However for the present we shall write the expansion as that for

$$\frac{4}{(1+4x)(1-2x)} = \frac{4}{1+2x-8x^2}$$

$$= 4(1 + 2x - 8x^2)^{-1}$$

If the reader wishes we can replace $2x - 8x^2$, by say y, as we are now going to expand the expression in powers of $(2x - 8x^2)$

Write $\quad 4(1 + y)^{-1}$

$$= 4[1 - y + y^2 - y^3 + ...]$$

$$= 4[1 - (2x - 8x^2) + (2x - 8x^2)^2 - (2x - 8x^2)^3 + ...]$$

$$= 4[1 - 2x + 8x^2 + 4x^2 - 32x^3 + 64\,x^4 - 8x^3 + ...]$$

since all other terms will involve powers of x^4 or higher powers.

i.e. $\quad 4[1 - 2x + 12x^2 - 40x^3] + ...$

$$= 4 - 8x + 48x^2 - 160x^3$$

To find the range of values for which the expansion is valid we need to look at the validity for expanding in powers of x from the point of view of both factors of the denominator i.e. $(1 + 4x)(1 - 2x)$ as we have effectively expanded $4(1 + 4x)^{-1}(1 - 2x)^{-1}$.

The first factor in the denominator results in the need to expand $(1 + 4x)^{-1}$ for which the range is

$$-1 < 4x < 1$$

or $\quad -1/4 < x < 1/4$

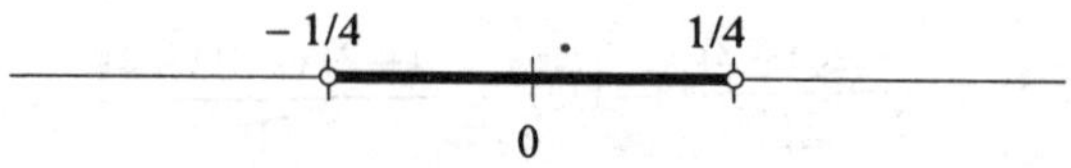

Fig. 3.5.7(a)

Considering the second factor in a similar manner we find the required range is

$$-1/2 < x < 1/2$$

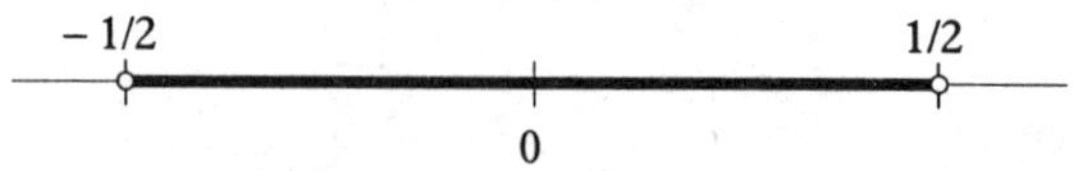

Fig. 3.5.7(b)

Now the expansion will be valid only when *both* the conditions $-1/4 < x < 1/4$ and $-1/2 < x < 1/2$ are satisfied. From the diagram below it is clear that x must satisfy

$$-1/4 < x < 1/4 \text{ [see Fig. 3.5.7(c)]}$$

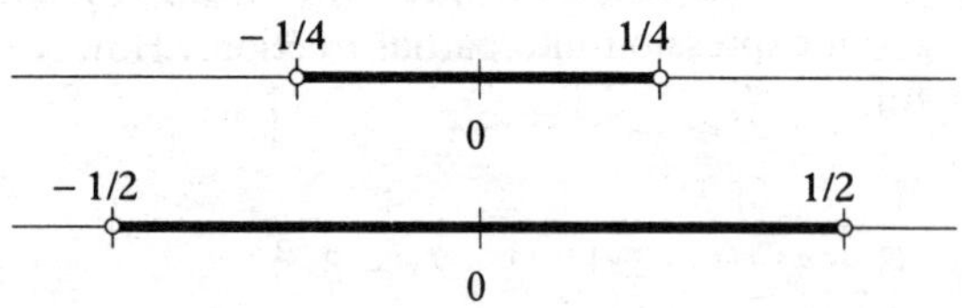

Fig. 3.5.7(c)

Example 3.5.7(6)

Expand
$$\sqrt{\left(\frac{2+x}{2-x}\right)}$$
in ascending powers of x up to and including the term in x^3. State the values of x for which the expansion is valid.

By putting $x = 0.04$ in your expansion determine the value of $\sqrt{51}$ correct to six decimal places.

Solution:

Write the expression

$$\sqrt{\left(\frac{2+x}{2-x}\right)} \text{ as } (2+x)^{1/2}\,(2-x)^{-1/2}$$
$$= [2(1+x/2)]^{1/2}\,[2(1-x/2)]^{-1/2}$$
$$= [2^{1/2}\,(1+x/2)^{1/2}\,2^{-1/2}\,(1-x/2)^{-1/2}]$$
$$= 2^{1/2}\,2^{-1/2}\,(1+x/2)^{1/2}\,(1-x/2)^{-1/2}$$

$$= 2^\circ \left[1 + 1/2\ (x/2) + \frac{1/2\,(-1/2)}{2}\,(x/2)^2 + \frac{1/2\,(-1/2)\,(-3/2)}{3!}\,(x/2)^3 + ... \right] \times$$

$$\left[1 + (-\ 1/2)(-\ x/2) + \frac{(-1/2)(-3/2)}{2}\,(-\ x/2)^2 + \frac{(-1/2)\,(-3/2)(-5/2)}{3!}\,(-\ x/2)^3 + ...\right]$$

$$= [1 + x/4 - x^2/32 + x^3/128 + ...]\ [1 + x/4 + 3x^2/32 + 5x^3/128 + ...]$$

Expanding the brackets term by term we get

$$1 + x/4 + x/4 - x^2/32 + x^2/16 + 3x^2/32 + x^3/128 - x^3/128 + 3x^3/128 + 5x^3/128 + ...$$

since all other terms will be in x^4 and higher powers of x.
and this will simplify to

$$1 + x/2 + x^2/8 + x^3/16$$

The reader should be able to check that the expansions for $(1 + x/2)^{1/2}$ and $(1 - x/2)^{-1/2}$ will be valid in the range $-\ 2 < x < 2$
In the final part if we write $x = 0.04$ the given expression is

$$\sqrt{\frac{(2+0.04)}{(2-0.04)}} = \sqrt{\left(\frac{2.04}{1.96}\right)}$$

$$= \sqrt{\frac{204}{196}} = \sqrt{\frac{51}{49}} = \frac{\sqrt{51}}{7}$$

and the expansion is

$$1 + 1/2 \times 0.04 + (0.04)^2/8 + (0.04)^3/16$$

which simplifies to 1.020204

hence $\qquad \dfrac{\sqrt{51}}{7} \simeq 1.020204$

i.e. $\qquad \sqrt{51} \simeq 7.141428$ to six decimal places.

3.6 FACTORISABLE POLYNOMIALS

Any expression of the form

$$a_n\, x^n + a_{n-1}\, x^{n-1} + ... + a_0$$

where $a_n, a_{n-1}, ...\ a_0$ are real numbers and n is a positive integer is called a *Polynomial* of degree n. For example the expression $7\ x^6 - 4\ x^5 + 3\ x^2 - 9$ is a polynomial of degree 6.

3.6.1 Long Division of a Polynomial Example 3.6.1(1)

Determine the quotient and the remainder when the polynomial $x^3 - 5\ x^2 + 7x - 1$ is divided by $x - 2$

Solution: We shall set out the working as for a long division, but we shall look at a method of division by inspection later.

$$\begin{array}{r|l} & x^2 - 3x + 1 \\ \hline x-2 & x^3 - 5x^2 + 7x - 1 \\ & -(x^3 - 2x^2) \\ \hline & -3x^2 + 7x \\ & -(-3x^2 + 6x) \\ \hline & x - 1 \\ & -(x-2) \\ \hline & 1 \end{array}$$

Here the quotient is $x^2 - 3x + 1$ and the remainder is 1.

We can set out the answer by writing

$$x^3 - 5x^2 + 7x - 1 = (x - 2)(x^2 - 3x + 1) + 1.$$

3.6.2 Remainder Theorem

The final form of the answer we obtained

$$x^3 - 5x^2 + 7x - 1 \equiv (x - 2)(x^2 - 3x + 1) + 1$$

is an identity which is true for all values of x.

In particular when we substitute the value $x = 2$

then $\quad \text{LHS} = 2^3 - 5 \times 2^2 + 7 \times 2 - 1$

and $\quad \text{RHS} = 1$

It would appear that if we had only required to find the remainder (and not the quotient) then the remainder could have been obtained by replacing x by 2 in the expression $x^3 - 5x^2 + 7x - 1$. This can be stated in general in the form shown below, which is the Remainder Theorem.

Before stating the Remainder Theorem it may be useful to study what is known as 'function notation', which will be referred to later in Chapter 6.

The phrase 'the value of the expression when $x = 2$' is too cumbersome, and an abbreviation can be used. When we draw graphs some expression in terms of x (function of x) is called y, and so we can now have some expression in x called $f(x)$ (or function of x). This does not mean f multiplied by x, but simply a function of x, e.g. when $f(x) = x^3 - 5x^2 + 7x - 1$

then $\quad f(2) = 2^3 - 5 \times 2^2 + 7 \times 2 - 1 = 1$

and $\quad f(-1) = (-1)^3 - 5(-1)^2 + 7 \times (-1) - 1 = -14.$

Statement and proof of the Remainder Theorem

The Remainder Theorem states: If a polynomial $f(x)$ is divided by $x - a$ until the remainder does not contain any power of x, then the remainder is $f(a)$.

Proof

If $Q(x)$ is the quotient and R the remainder when $f(x)$ is divided by $x - a$ until the remainder does not contain any powers of x (i.e. R is a constant) then we may write

$$f(x) = (x - a)\, Q(x) + R$$

and substituting $x = a$

then $$f(a) = R$$

hence the remainder R is $f(a)$.

Corollary 1 If a polynomial $f(x)$ is divided by $px + q$ until the remainder does not contain any powers of x then the remainder is $f(-q/p)$. This can be established in a similar manner to the result for the Remainder Theorem and is left as an exercise to the reader.

Corollary 2 If, for any polynomial $f(x)$, $f(a) = 0$ then $x - a$ is a factor of $f(x)$. This follows directly from the Remainder Theorem, since if $f(a) = 0$ then $f(x)$ is exactly divisible by $x - a$. In such case $x = a$ is said to be a root (or solution) of the equation $f(x) = 0$.

Example 3.6.2(1)

Find the remainder when

$$f(x) \equiv 4x^5 - x^3 + x^2 - 3$$

is divided by (a) $x - 1$ (b) $x + 2$ (c) $2x - 1$

Solution:

(a) The remainder when $f(x)$ is divided by $x - 1$

is $$f(1) = 4 \times 1^5 - 1^3 + 1^2 - 3$$
$$= 1$$

(b) The remainder when $f(x)$ is divided by $x + 2$ is

$$f(-2) = 4 \times (-2)^5 - (-2)^3 + (-2)^2 - 3$$
$$= 4 \times -32 - (-8) + 4 - 3$$
$$= -119$$

(c) The remainder when $f(x)$ is divided by $2x - 1$ is

$$f\left(\frac{1}{2}\right) = 4 \times (1/2)^5 - (1/2)^3 + 1/4 - 3$$
$$= -2\frac{3}{4}.$$

Example 3.6.2(2)

Factorise the expression

$$6x^3 - 17x^2 - 4x + 3$$

and hence solve the cubic equation

$$6x^3 - 17x^2 - 4x + 3 = 0.$$

Solution:

Let $$f(x) \equiv 6x^3 - 17x^2 - 4x + 3$$

We require to find a factor of the form $x - a$, but $x - a$ will be a factor of $f(x)$ only when there will be no remainder on division, i.e. $f(a) = 0$. It is necessary to try to find a factor by trial and error.

$$f(1) = 6 - 17 - 4 + 3 \neq 0 \text{ so } x - 1 \text{ is not a factor.}$$

$$f(-1) = 6(-1)^3 - 17(-1)^2 - 4(-1) + 3 \neq 0 \text{ so } x + 1 \text{ is not a factor.}$$

$x - 2$, $x + 2$ cannot be factors, since 2 is not a factor of constant term 3.

$$\begin{aligned} f(3) & \\ &= 6 \times 3^3 - 17 \times 3^2 - 4 \times 3 + 3 \\ &= 162 - 153 - 12 + 3 \\ &= 0 \end{aligned}$$

so $\qquad x - 3$ is a factor.

On division by inspection we know that when $6x^3 - 17x^2 - 4x + 3$ is divided by $x - 3$ the first term is $6x^2$ (when $6x^3$ is divided by x).

Write
$$\begin{aligned} & 6x^3 - 17x^2 - 4x + 3 \\ &= (x-3)(6x^2 + ?) \end{aligned}$$

On looking at the term in x^2 formed by the product of -3 and $6x^2$ the term is $-18x^2$, but in the final product obtained when the factors are multiplied the term in x^2 should be $-17x^2$.

So the terms in the question mark should include a term in x, which when multiplied by x, will be added to $-18x^2$ to give $-17x^2$. This term is x.

Thus
$$\begin{aligned} & 6x^3 - 17x^2 - 4x + 3 \\ &= (x-3)(6x^2 + x + ?) \end{aligned}$$

On looking at the term in x formed by the product of -3 and x the term is $-3x$, but in the final product obtained when the factors are multiplied the term in x should be $-4x$. So the term in the question mark should be a constant, which when multiplied by x, will be added to $-3x$ to give $-4x$. Clearly the constant is -1.

Hence
$$\begin{aligned} f(x) &\equiv 6x^3 - 17x^2 - 4x + 3 \\ &\equiv (x-3)(6x^2 + x - 1) \end{aligned}$$

A final check shows that when we multiply the constant terms in each of the factors we obtain 3. The second factor, a quadratic expression in x, may often be reduced to two linear factors in x, after we have carried out the first division of a cubic expression. In this case $6x^2 + x - 1$ can be written as

$$(3x - 1)(2x + 1)$$

so $\qquad f(x) \equiv (x - 3)(3x - 1)(2x + 1)$ in its factorised form.

In solving the cubic equation

$$6x^3 - 17x^2 - 4x + 3 = 0$$

this is equivalent to

$$(x - 3)(3x - 1)(2x + 1) = 0$$

i.e. $\qquad x = 3$ or $1/3$ or $-1/2$.

Example 3.6.2(3)

Solve by factors the cubic equation

$$18x^3 - 9x^2 - 5x + 2 = 0$$

Solution: We approach this example like the previous example by looking for a factor of the form $x - a$ for which $f(a) = 0$

if $$f(x) \equiv 18x^3 - 9x^2 - 5x + 2.$$

The factors of the constant term 2 are ± 1, ± 2, and the reader should check that in each case $f(1)$, $f(-1)$, $f(2)$, $f(-2)$ are all different from zero.

We now look for a factor of the form $px + q$, where p should be a factor of 18 and q should be a factor of 2, since the first term in $f(x)$ is $18x^3$ and the constant (last) term is 2. There is a possibility that $2x - 1$ could be a factor so we evaluate $f\left(\frac{1}{2}\right)$.

$$\begin{aligned} f\left(\frac{1}{2}\right) &= 18\left(\frac{1}{2}\right)^3 - 9\left(\frac{1}{2}\right)^2 - 5\left(\frac{1}{2}\right) + 2 \\ &= \frac{18}{8} - \frac{9}{4} - \frac{5}{2} + 2 \\ &= -\frac{1}{2} \\ &\neq 0 \end{aligned}$$

So $2x - 1$ is *not* a factor.

There is a possibility that $2x + 1$ could be a factor so we evaluate $f\left(-\frac{1}{2}\right)$.

$$\begin{aligned} f\left(-\frac{1}{2}\right) &= 18\left(-\frac{1}{2}\right)^3 - 9\left(-\frac{1}{2}\right)^2 - 5\left(-\frac{1}{2}\right) + 2 \\ &= \frac{-18}{8} - \frac{9}{4} + \frac{5}{2} + 2 \\ &= \frac{-9}{4} - \frac{9}{4} + \frac{10}{4} + \frac{8}{4} \\ &= 0 \end{aligned}$$

so $2x + 1$ is a factor

Dividing out by $2x + 1$

$$\begin{aligned} f(x) &\equiv 18x^3 - 9x^2 - 5x + 2 \\ &= (2x + 1)(9x^2 + ?) \end{aligned}$$

Following the arrows which give the term $-9x^2$ in the product, $9x^2 + 2x(?) = -9x^2$

$$2x\,(?) = -18x^2$$

$$2x(-9x) = -18x^2$$

so now the ? in the second bracket can be replaced by $-9x$

$$18x^3 - 9x^2 - 5x + 2$$

By a similar procedure, following the arrows which give the term $-5x$ in the product,

$$-9x + 2x\,(?) = -5x$$

$$2x\,(?) = 4x$$

so now the ? in the second bracket can be replaced by 2

So $$18x^2 - 9x^2 - 5x + 2 \equiv (2x + 1)\,(9x^2 - 9x + 2)$$

$$\equiv (2x + 1)(3x - 1)(3x - 2) \quad \text{on further factorisation.}$$

Solving $18x^3 - 9x^2 - 5x + 2 = 0$ is equivalent to solving in its factorised form

$$(2x + 1)(3x - 1)(3x - 2) = 0$$

i.e. $$x = -1/2 \text{ or } x = 1/3 \text{ or } x = 2/3.$$

Example 3.6.2(4)

A polynomial $F(x)$ has remainder 9 when divided by $x - 3$ and remainder -5 when divided by $2x + 1$. Find the remainder when $F(x)$ is divided by $(x - 3)(2x + 1)$.

Solution: When we divide by $(x - 3)(2x + 1)$ which is of degree 2 in x, then the remainder will be of degree 1, i.e. of the form $ax + b$.

Imagine that the division has been formed, i.e.

$$F(x) \equiv (x - 3)(2x + 1)\;Q(x) + ax + b,$$

where $Q(x)$ is the quotient.

Substituting $$x = 3$$

$$F(3) = 3a + b$$

But $F(3)$ is the remainder when $F(x)$ is divided by $x - 3$, so $F(3)$ is also equal to 9

hence $$3a + b = 9 \qquad (1)$$

Substituting $x = -1/2$ we obtain

$$F(-1/2) = -1/2a + b$$

hence $$-1/2a + b = -5 \qquad (2)$$

$$(1) - (2) \text{ gives } 3\frac{1}{2}a = 14$$

i.e. $$a = 4$$

Substituting in (1) $$3 \times 4 + b = 9$$

i.e. $$b = -3$$

so the remainder when $F(x)$ is divided by $(x - 3)(2x + 1)$ will be $4x - 3$.

EXERCISE 3

1. Express $\frac{3}{5}\sqrt{\left(\frac{6}{5}\right)}$ in the form $\sqrt{\left(\frac{p}{q}\right)}$ where p and q are integers.

2. Simplify $\dfrac{\sqrt{180}-\sqrt{80}}{\sqrt{125}}$.

3. Simplify:

(a) $\dfrac{\sqrt{35}-\sqrt{30}}{\sqrt{35}+\sqrt{30}}+\dfrac{\sqrt{35}+\sqrt{30}}{\sqrt{35}-\sqrt{30}}$;

(b) $\left(\dfrac{6\sqrt{5}-4\sqrt{3}}{2\sqrt{5}-\sqrt{3}}\right)\left(\dfrac{5\sqrt{5}+3\sqrt{3}}{3\sqrt{5}-2\sqrt{3}}\right)$.

4. Given $\sqrt{37} = 6.0827625$ and $\sqrt{35} = 5.9160798$, each correct to seven decimal places, find *without* tables or calculator the value of $\dfrac{1}{\sqrt{37}-\sqrt{35}}$ correct to six decimal places.

5. Given $\sqrt{2} = 1.4142136$ correct to seven decimal places, find *without* tables or calculator the value of $(\sqrt{3}+\sqrt{2})^3-(\sqrt{3}-\sqrt{2})^3-\sqrt{8}$ correct to six decimal places.

6. Simplify *without* tables or calculator

(a) $\tan^2 60° \sin 55° - (1+\tan^2 45°)\cos 35° - \tan 45° \cos 35°$;

(b) $\dfrac{\sin 315° (\cos^2 210° + 1)}{\cos 225° (\sin^2 150° + 1)}$.

7. (a) Simplify *without* use of tables or calculator and express your answer with a positive index

(i) $4^{-2} \times 16^2 \times 64^{-1} \times 8^{-3}$;

(ii) $6^{2/3} \times 2^{-1} \times 3^{-2} \times 2^{1/3} \div 3^{2/3}$.

(b) Evaluate $\log_5 12$ correct to two decimal places.

(c) (i) Solve without use of tables or calculator the equation $2^x = 32$;

(ii) Solve with use of tables or calculator the equation $2^x = 10$, and give your answer correct to two decimal places.

(iii) Hence find the possible values of x if $2^{2x} - 42(2^x) + 320 = 0$.

8. Solve the equation

$(2^x)(3^{x+1}) = 10$ giving the answer correct to three decimal places.

9. Simplify in each case *without* tables or calculator

(a) $\log 5 + \log 16 + \log 60 - \log 12 - 2 \log 2$,

(b) $\log 15 + \log 25 + 4 \log 3 + 3 \log 2 - \log 243$,

(c) $\log 243/\log 27$,

(d) $\log 3125/\log 125$.

10. (a) Solve *without* tables or calculator the equation $2^x = \dfrac{1}{32}$;

(b) Solve using tables or calculator the equation $2^x = 3$;

(c) Hence solve the equation

$$32(4^x) - 97(2^x) + 3 = 0.$$

11. (a) Determine $\log_7 2$ using tables or calculator, giving your answer correct to two decimal places.
(b) Solve for x *without* tables or calculator

$$\log_2 (10x + 2) - \log_2 (x + 1) = 3.$$

12. Solve the equation

$$2\log_5 x + 2 \log_x 5 = 5.$$

13. Solve the equation

$$2 \log_5 x + \log_x 5 = 3.$$

14. Solve for x where x is a real number

$$4 + 7 \ln e^{x^2} = 2e^{4 \ln x}$$

15. Do not use tables or calculator in this question.
(a) Given that $\log_{10} 3.142 = 0.4972$ write down the logarithms, to base 10 of
(i) 314200.
(ii) $(0.003142)^{3}$.
(b) Find the value of x which satisfies the equation

$$\left(\frac{4}{5}\right)^{5x-4} = \left(\frac{5}{4}\right)^{4x-5}$$

*16. (a) Show that $\log_{3^n} x = \frac{1}{n} \log_3 x$

(b) Hence solve the equation

$$\log_{81} x + \log_3 x + \log_{\sqrt{3}} x = 13.$$

*17. Solve for x *without* using tables or calculator

$$\frac{\log_{81} x}{\log_9 3} = \frac{\log_6 \frac{1}{216}}{\log_6 36}.$$

*18. (a) Show that $\log_8 x = \frac{1}{3} \log_2 x$

(b) Hence solve the equation, for $x > 0$,

$$\log_2 (3x + 1) + \log_8 (x - 1)^3 = 6$$

without using tables or calculator.

19. Solve each of the following quadratic equations, giving the roots if necessary to two decimal places:
(a) $x^2 + 15x + 54 = 0$;
(b) $2x^2 - 5x - 6 = 0$.

20. (a) On graph paper draw the graph of $y = x^2$ for the domain $-3 \leq x \leq 3$.
(b) By drawing a suitable straight line on the same diagram solve graphically the equation $4x^2 = 3 - 6x$.

21. Solve each of the following inequalities
(a) $x(x - 3) < 10$
(b) $16 x^2 - 8x - 15 > 0$
(c) $12 + x - x^2 \geq 0$
(d) $5x^2 - x - 6 \geq 0$

22. Determine the minimum value of the function $x^2 - 12x + 9$.

23. Determine the maximum value of the function $15 + 6x - 3x^2$.

24. Determine the values of k for which the equation $4x^2 - 2kx + (2k - 3) = 0$ has
   (a) equal roots
   (b) real distinct roots.

25. (a) Find the total number of arrangements if the letters of the word *Success* are arranged in a row.
   (b) Find the total number of selections of five letters from the letters of the word *Success*.

26. In how many ways can the coach select a team of 4 players for tennis if two of the players, A and B, play well together and should either both play or both be excluded, and two others X and Y have recently quarreled and should not both be included? Eight players are available.

27. How many different arrangements of three balls in a row can be made from one red ball, two green balls and three blue balls?

28. In how many ways can a committee of 7 be selected from 10 men and 7 women if there must be a majority of women serving?

29. In a hotel the menu for lunch is always a choice of three types of meat dish, or irio, or rice. A man arranges to have lunch at the hotel for three consecutive days. In how many ways can he arrange his lunch over the three days
   (a) if he makes no restriction,
   (b) if he does not take a meat dish two days running nor repeat any dish?

30. (a) Show that ${}_nC_r + {}_nC_{r+1} = {}_{n+1}C_{r+1}$
   (b) Hence obtain an expression for ${}_2C_0 \times {}_nC_7 + {}_2C_1 \times {}_nC_8 + {}_2C_2 \times {}_nC_9$ in the form ${}_aC_b$.

31. A box contains 5 red, 2 white and 3 blue balls.
   How many arrangements are possible if
   (a) all ten balls are arranged in a row?
   (b) any four are taken and arranged in a row?

32. There are 8 persons, including a married couple Mr and Mrs Sharma, from whom a committee of 4 has to be chosen. In how many ways can the committee be chosen.
   (a) if both Mr and Mrs Sharma are excluded,
   (b) if Mrs Sharma is included but Mr Sharma is excluded,
   (c) if both Mr and Mrs Sharma are included?

33. At a school open day there are 12 special people for whom the cateress has prepared supper. Unfortunately she can only fit 9 people into the dining room, and of these the Headmaster, the Board Chairman and the Guest of Honour must be included. In how many ways can the six remaining guests be chosen, if Mr and Mrs A must both be included or both excluded, and if Miss X and Miss Y have recently quarreled and must not both be invited?

*34. Five digit numbers greater than 10000 are to be formed using the digits 0, 1, 2, 3, 4 and 5.
   (a) How many such numbers can be formed if no repetition of digits is allowed?
   (b) How many of these numbers (without repetition) are even?
   (c) Show that if repetitions are allowed then the number of even numbers which can be formed is less than 3240.

*35. Five married couples are sitting on a bench. In how many ways can the ten people be arranged if
   (a) each man sits next to his wife.
   (b) men may sit next to another man's wife but no two men may sit next to each other.

*36. Twenty-two people are called to form two teams of eleven for a hockey match. In how many ways can the two teams be selected if there are five married couples and twelve single people, if the married couples are selected so that the husband and wife have to play together in the same team?

37. An arithmetic progression has fourth term 22 and eighth term 46.
    (a) Find the first term and the common difference.
    (b) Find the sum of the first twelve terms.
    (c) Find the number n such that the sum of the first n terms is 310.
38. In an arithmetic progression the first term is 6 and the common difference is 4. Find the tenth term and the sum of the first ten terms.
39. In a geometric progression the third term is 81 and the sixth term is 3. Determine the first term, the common ratio, and the sum to infinity of the series.
40. Show that the sum to n terms of the series

$$\log a + \log 2a + \log 4a + \log 8a + ... \text{ is } \log [a^n\, 2^{n(n-1)/2}].$$

*41. S_n denotes the sum of the first n terms of a geometric progression whose first term is a and whose common ratio is r.

Show that
$$\frac{S_{3n} - S_{2n}}{S_n} = r^{2n}.$$

*42. If
$$S = 1 + 2 + 2^2 + 2^3 + ... + 2^{n-1}$$
$$T = 1 + 3 + 3^2 + 3^3 + ... + 3^{n-1},$$
$$U = 1 + 6 + 6^2 + 6^3 + ... + 6^{n-1}$$

Show that
$$2ST + S + 2T = 5U.$$

43. Determine the sum of all integers between 1 and 1000 inclusive which are multiples of 5 or 8 or both.
44. Determine the sum of all natural numbers between 1 and 200 inclusive which are not divisible by 9 or 12.
45. For what values of x is there a sum to infinity of the series

$$x + \frac{x}{1+x} + \frac{x}{(1+x)^2} + ...$$

and find in simplified form the sum when it exists.
46. Ali has a starting salary of 5000 Buranian dollars with annual increments of 350 Buranian dollars. After how many years will his total earnings be 265750 Buranian dollars?
47. Determine the number of complete years for which a single investment should be held at 7% compound interest per annum in order to realise double its original value.
48. Mrs Prudence deposits 2000 Buranian dollars in a savings account on 1 January each year for 20 years. If the account gives 3% compound interest per annum, what will be the total value of her savings at the end of 20 years? (Answer correct to the nearest Buranian dollar).
49. The king of Burania has a beautiful daughter aged 10 years. A prospective bridegroom agrees to deposit 1000 Buranian dollars in the king's account at the beginning of each year for each of 10 years. The account gains 6% compound interest per annum.
    (a) Determine the total value of the account after one year.
    (b) Determine the total value of the account after two years.
    (c) Show that the total value of the account after n years is $\frac{106\,000}{6}\{1.06^n - 1\}$ Buranian dollars
    (d) Determine the total value of the account at the end of ten years. (answer to the nearest Buranian dollar)

*50. Show that the term independent of x in the expansion $\left(3x - \dfrac{2}{x}\right)^{50}$ is $\dfrac{-50!}{25!\,25!}\,(6^{25})$.

51. Expand $(1 - x)^{1/3}$ up to and including the term in x^3 in ascending powers of x and simplify your answer. Use your expansion with $x = 0.005$ to evaluate $995^{1/3}$ correct to four decimal places.

52. Expand $(1 - x^2)^{1/4}$ in ascending powers of x up to and including the term in x^6 and simplify your answer. Use the first three terms of your expansion to show that

$$\left(\frac{3}{4}\right)^{1/4} \simeq \frac{477}{512}$$

53. Expand $(1 + x)^{1/2}$ in ascending powers up to and including the term in x^3, and simplify your answer. By putting $x = 0.08$ in your expansion evaluate $\sqrt{27}$ correct to four decimal places.

54. Show that if x^4 and higher powers of x are so small that they may be neglected then

$$(1 + 2x + x^2)^{10} \simeq 1 + 20x + 190x^2 + 1140x^3$$

By putting $x = 0.01$ in your expansion estimate 1.0201^{10} correct to three decimal places.

55. Show that if x^3 and higher powers of x may be neglected then $(1 + x)^{1/2}\,(1 - x)^{-1/2} \simeq 1 + x + \dfrac{1}{2}x^2$. By putting $x = 0.1$ in your expansion obtain an estimate of $\sqrt{11}$.

56. (a) Use the binomial theorem to show that if x^4 and higher powers are neglected then

$$(1 + 2x - x^2)^{1/2} \simeq 1 + x - x^2 + x^3$$

for small x.

(b) By putting $x = 0.1$ in your expansion estimate $1.19^{1/2}$ correct to three decimal places.

57. Given that $x + 1$ is a factor of $f(x) = px^3 + qx^2 + px + q$ show that $p = q$. Deduce that if $p = q = 1$ then the equation $f(x) = 0$ has only one real root.

58. The function $f(x) = ax^3 + bx^2 + 7x + 6$ has remainder 6 if divided by $x - 1$, and is exactly divisible by $x - 2$. Find the values of a and b and hence factorise $f(x)$ completely.

59. When the polynomial $f(x)$ is divided by $x - 2$ the remainder is 7 and when $f(x)$ is divided by $x + 1$ the remainder is -5. When $f(x)$ is divided by $(x - 2)(x + 1)$ the remainder is $ax + b$. Determine the values of a and b.

60. A polynomial $f(x)$ when divided by $x + 2$ leaves remainder -13 and when divided by $2x - 1$ leaves remainder $5\dfrac{1}{2}$. Find the remainder when $f(x)$ is divided by $(x + 2)(2x - 1)$.

61. If $3x - 1$ is a factor of $6x^3 + ax^2 + x - 2$, find the value of a and hence solve the equation

$$6x^3 + ax^2 + x - 2 = 0.$$

62. If $x - 2$ is a factor of the polynomial $f(x) \equiv ax^3 - 9x^2 + 12x - a^2$, find the possible values of a and the other factors of $f(x)$ when a has the smaller value.

*63. When a polynomial $f(x)$ is divided by $x - 1$ the remainder is 3, when $f(x)$ is divided by $x + 1$ the remainder is 5, and when $f(x)$ is divided by $x - 2$ the remainder is 20. When $f(x)$ is divided by $(x^2 - 1)(x - 2)$ the remainder is $ax^2 + bx + c$. Determine the values of a, b, c.

*64. The polynomial $Ax^3 + Bx^2 + Cx + D$ leaves a remainder $x^2 - 3x + 2$ when divided by $(x + 2)^3$ and a remainder $ax + b$ when divided by $(x + 2)^2$. Determine the values of a and b.

4

Trigonometry

INTRODUCTION

Trigonometry is the science of measuring (or determining) the sides and angles of a triangle by means of information given about some of the sides and angles.

An angle can be defined as the amount of rotation between one staright line and another. We shall have first to consider how this rotation can be measured.

4.1 MEASUREMENT OF ROTATION

When a line OP is pivoted at O and rotates from an initial position OA to a new position OB then the angle AOB is a measure of the rotation of the line OP.
See Fig. 4.1.1(a).

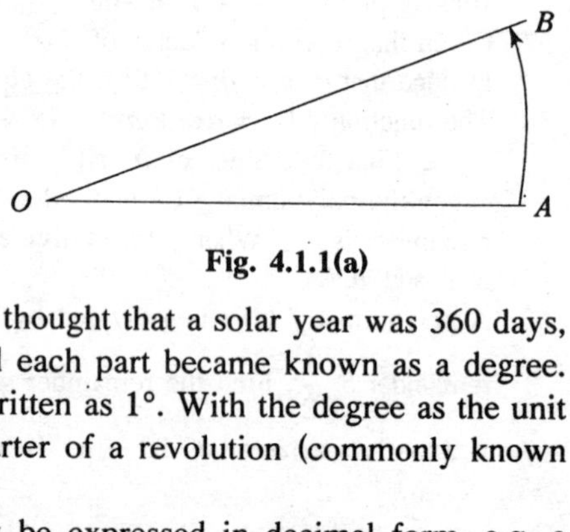

Fig. 4.1.1(a)

4.2 DEGREES AND RADIANS

4.2.1 The Degree

The unit of measurement most commonly used for an angle is the degree. The ancient Babylonian Mathematicians thought that a solar year was 360 days, so they divided one complete revolution into 360 parts, and each part became known as a degree. The symbol for degree measure is °, thus 1 degree can be written as 1°. With the degree as the unit of rotation half a revolution corresponds to 180° and a quarter of a revolution (commonly known as a right angle) corresponds to 90°. See Fig. 4.2.1(a).

Any angle which is smaller than a degree can normally be expressed in decimal form, e.g. a quarter of a degree is 0.25° but in some fields of study such as navigation a degree may be divided into 60 minutes (abbreviated ′) and each minute into 60 seconds (abbreviated ″),

So
$$1^\circ = 60'$$
$$1' = 60''$$

4.2.2 The Radian

If the arc $P_0 P_1$ of a circle is drawn in such a way that it is the same length as the radius of the circle the angle $P_0 OP_1$ is called one *Radian* (abbreviated to 1^c) See Fig. 4.2.2(a).

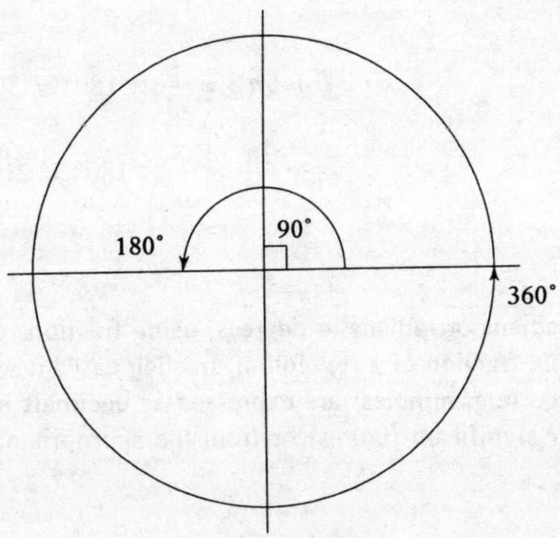

Fig. 4.2.1(a)

The number of radians in a complete revolution can then be given by the ratio

$$\frac{\text{Circumference of Circle}}{\text{Radius of Circle}}$$

However we know that the circumference of a circle of radius r is $2\pi r$ and so there will be

$$\frac{2\pi r}{r} = 2\pi \text{ radians in one revolution.}$$

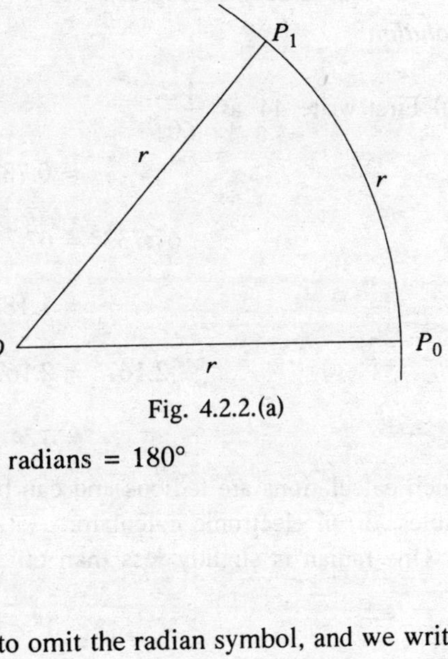

Fig. 4.2.2.(a)

i.e. 2π radians = 360° or (often used as the conversion factor from degrees to radians or radians to degrees) half a revolution = π radians = 180°

$$\text{One right angle} = \frac{\pi}{2} \text{ radians}$$

$$= 90°$$

Normally when an angle is quoted in terms of π it is normal to omit the radian symbol, and we write 180° = π rather than π^c.

Any other angle which is a simple fraction of π is normally expressed in radians in terms of π, using the relationship,

$$180° = \pi.$$

Example 4.2.2(1)

(a) Express 18° and 320° in radians.

(b) Express $5\pi/3$ and $7\pi/6$ in degrees.

Solution:

(a) $180° = \pi$

so $18° = \dfrac{\pi}{10}$

$320° = \dfrac{320}{180}\pi = \dfrac{16\pi}{9}$

(b) $5\pi/3 = \dfrac{5}{3} \times 180° = 300°$

$\dfrac{7\pi}{6} = \dfrac{7}{6} \times 180° = 210°$

Conversion from degrees to radians or radians to degrees, using fractions of π, may not be easy for any angle which is not a simple fraction of a revolution. In such cases it will be necessary to ensure that smaller units than a degree (e.g. minutes) are expressed as decimals of a degree, and the value of π be read as 3.142 (to four significant figures) or from the appropriate function on an electronic calculator.

Example 4.2.2(2)

Express (a) 67° 44′ in radians
(b) 2.167^c in degrees.

Solution:

(a) First write 44′ as $\dfrac{44°}{60}$

$$= 0.733°$$

so

$$67.733° = 67.733 \times \frac{\pi}{180}$$

$$= 1.182^c$$

(b)

$$2.167^c = 2.167 \times \frac{180}{\pi}$$

$$= 124.16°$$

Such calculations are tedious and can be handled by use of logarithm tables, or suitable conversion tables, or an electronic calculator.

One radian is slightly less than 60°, as can be seen by converting

$$1^c = \frac{180°}{\pi} \simeq 57.296° \quad \text{(to five significant figures)}$$

4.3 MENSURATION OF A CIRCLE

4.3.1 Arc Length

We are already familiar with the formula for measuring the circumference C of a circle of radius r,

$$C = 2\pi r.$$

See Fig. 4.3.1(a).

If an arc subtends an angle θ (measured in *radians*) at the centre of a circle, then from the definition of a radian we know that the length of an arc which subtends 1 radian at the centre is of length r.

So an arc which subtends an angle θ radians at the centre has length = $r\theta$.

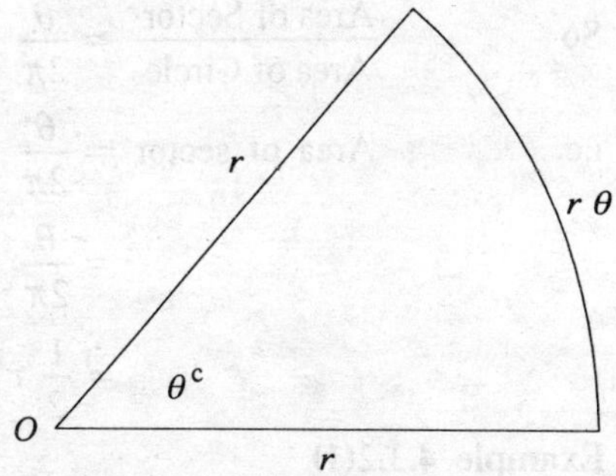

Fig. 4.3.1(a)

Example 4.3.1(1)

Express in terms of π the arc length AB where angle AOB = 300° and the radius of the circle, centre O *is* 5 cm.

Solution:

See Fig. 4.3.1(b)

$$300° = \frac{300\pi}{180} = \frac{5\pi}{3}$$

and the arc length of AB is

$$5 \times \frac{5\pi}{3} \text{ cm}$$

or

$$\frac{25\pi}{3} \text{ cm.}$$

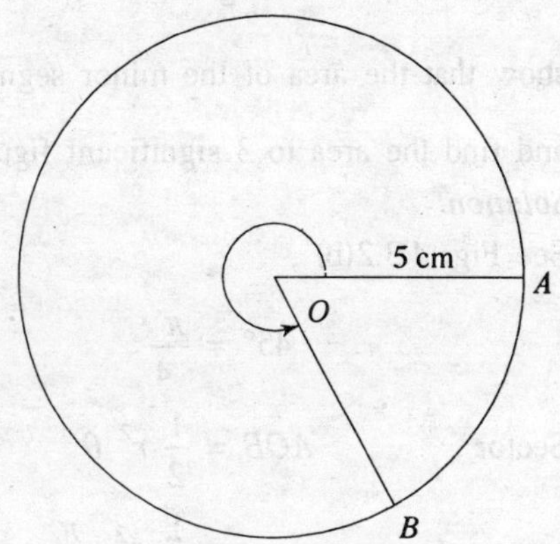

Fig. 4.3.1(b)

Example 4.3.1(2)

A track AB in the shape of an arc of a circle, centre O, is 400 metres long. If the angle AOB is 45°, calculate the radius of the arc correct to the nearest metre.

Solution:

See Fig. 4.3.1(c)

Using length of arc = $r\theta$ (θ in radians)

$$r\frac{\pi}{4} = 400$$

i.e.

$$r = \frac{4 \times 400}{\pi}$$

$$= 509 \text{ metres (correct to the nearest metre)}$$

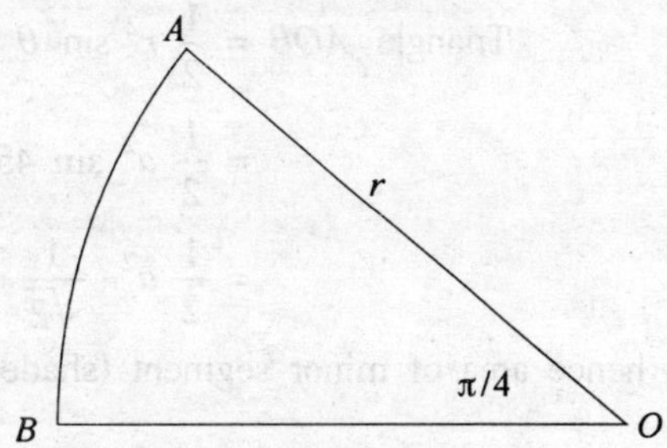

Fig. 4.3.1(c)

The reader may try working out this problem using the formula $\frac{45°}{360°} \times 2\pi r$ for the length of the arc, which should give the same answer.

4.3.2 Area of a Sector and of a Segment

The area of a sector of a circle containing an angle θ radians at the centre can be determined by considering the sector as a fraction of a circle. See Fig. 4.3.2(a).

The ratio of the area of the sector to the area of the whole circle is the same ratio as that of the angle θ contained in the sector AOB to the angle 2π contained in the whole circle.

So $\qquad \dfrac{\text{Area of Sector}}{\text{Area of Circle}} = \dfrac{\theta}{2\pi}$

i.e. $\qquad \text{Area of sector} = \dfrac{\theta}{2\pi} \times \text{Area of circle}$

$$= \frac{\theta}{2\pi}\pi r^2$$

$$= \frac{1}{2} r^2\theta$$

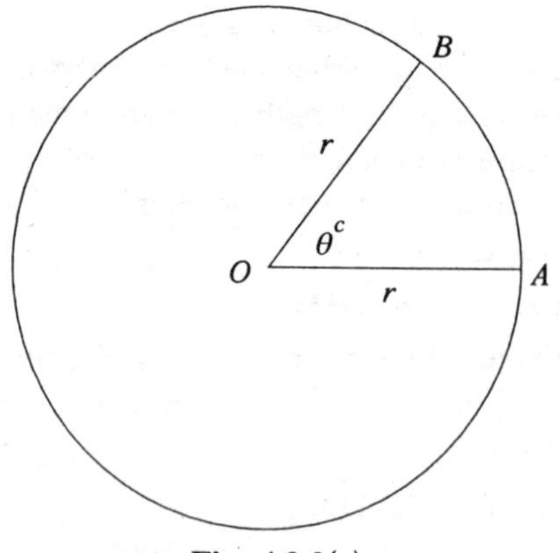

Fig. 4.3.2(a)

Example 4.3.2(1)

A chord AB divides a circle of radius a into two segments. If AB subtends an angle of 45° at the centre of the circle show that the area of the minor segment in $\dfrac{a^2}{8}(\pi - 2\sqrt{2})$ and find the area to 3 significant figures when $a = 4$ cm.

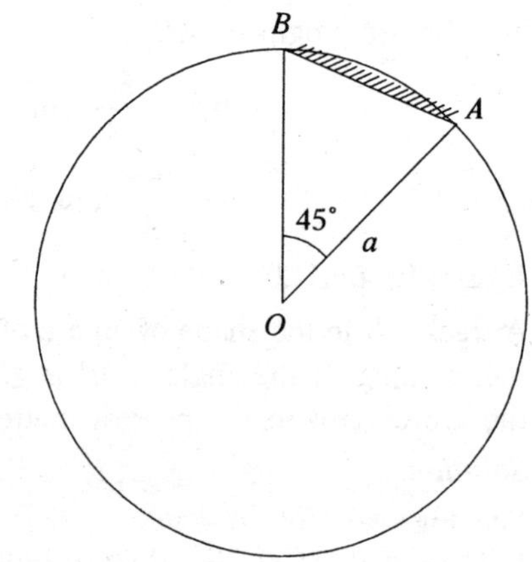

Fig. 4.3.2(b)

Solution:

See Fig. 4.3.2(b)

$$45° = \frac{\pi^c}{4}$$

Sector $\qquad AOB = \dfrac{1}{2} r^2 \theta$

$$= \frac{1}{2} a^2 \cdot \frac{\pi}{4} = \frac{\pi}{8} a^2$$

Triangle $AOB = \dfrac{1}{2} r^2 \sin \theta$

$$= \frac{1}{2} a^2 \sin 45°$$

$$= \frac{1}{2} a^2 \cdot \frac{1}{\sqrt{2}}$$

hence area of minor segment (shaded)

$$= \frac{\pi}{8} a^2 - \frac{a^2}{2\sqrt{2}}$$

$$= \frac{a^2}{8} (\pi - 2\sqrt{2})$$

The reader who wishes to find out the derivation of the formula for the area of triangle AOB is referred to 4.6.3.

If $a = 4$ cm

the area of the minor segment $\qquad = \dfrac{16}{8}(\pi - 2\sqrt{2})\ \text{cm}^2$

$$= 0.626\ \text{cm}^2 \text{ (3 significant figures)}$$

Example 4.3.2(2)

The two circles shown in Fig. 4.3.2(c) have centres A and B and radii 24 cm and 7 cm respectively. Given that AB = 25 cm determine the area of the shaded region and the perimeter $CEDFC$.

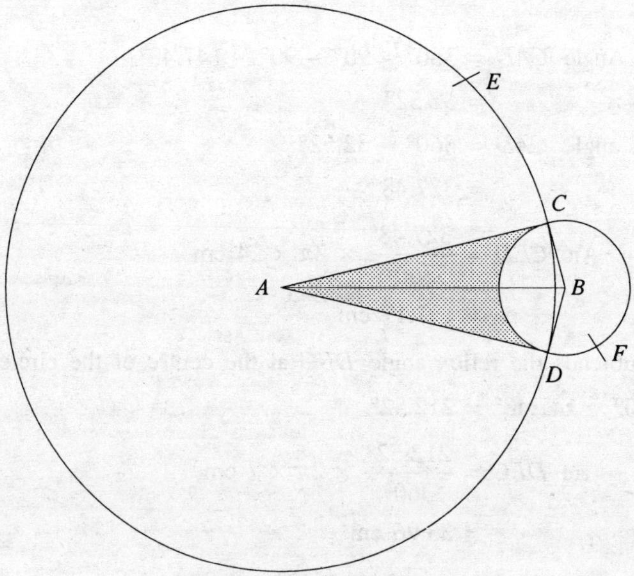

Fig. 4.3.2(c)

Solution:

In triangle ABC

$$AB^2 = 25^2 = 625 \text{ cm}^2$$

and $$AC^2 + CB^2 = (24^2 + 7^2)$$

$$= (576 + 49) \text{ cm}^2$$

$$= 625 \text{ cm}^2$$

As $$AB^2 = AC^2 + CB^2$$

$$\text{angle } ACB = 90°$$

$$\text{and similarly angle } ADB = 90°$$

$$\text{Area of Kite } ACBD = 2 \times \frac{1}{2} \times 24 \times 7 \text{ cm}^2$$

$$= 168 \text{ cm}^2$$

In triangle ACB

$$\text{sine } \angle AB\dot{C} = 24/25 = 0.96$$

$$\angle ABC = 73.74°$$

$$\text{so area of sector } BCD = \frac{2 \times 73.74°}{360°} \times \pi \times 7^2$$

$$= 63.06 \text{ cm}^2$$

$$\text{So shaded area} = \text{area of kite } ACBD - \text{area of sector } BCD$$
$$= 104.9 \text{ cm}^2 \text{ (to four significant figures)}$$

The major arc CED subtends the reflex angle CAD at the centre of the circle.
In kite $ACBD$

$$\text{Angle } CAD = 360° - 90° - 90° - 147.48°$$
$$= 32.52°$$
$$\text{So reflex angle } CAD = 360° - 32.52°$$
$$= 327.48°$$
$$\text{Arc } CED = \frac{327.48°}{360°} \times 2\pi \times 24 \text{ cm}$$
$$= 137.17 \text{ cm}$$

The major arc DFC subtends the reflex angle DBC at the centre of the circle, and this angle is

$$360° - 147.48° = 212.52°$$
$$\text{arc } DFC = \frac{212.52°}{360°} \times 2\pi \times 7 \text{ cm}$$
$$= 25.96 \text{ cm}$$

So the perimeter of the figure is

$$(137.17 + 25.96) \text{ cm}$$
$$= 163.1 \text{ cm (to four significant figures)}$$

4.4 CIRCULAR FUNCTIONS

4.4.1 Definitions of Circular Functions in a Right Angled Triangle

For a right angled triangle we can define *six* trigonometric ratios using an angle θ in the triangle [See Fig. 4.4.1(a)]

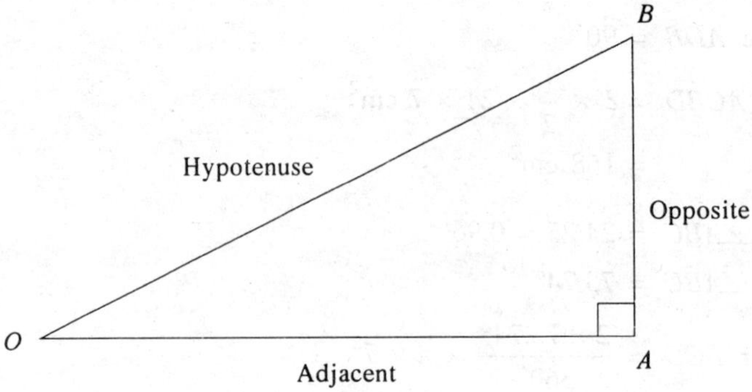

Fig. 4.4.1(a)

$$\text{sine } \theta \text{ (or sin } \theta \text{ in abbreviated form)} = \frac{AB}{OB} = \frac{\text{opposite}}{\text{hypotenuse}}$$

$$\text{cosine } \theta \text{ (or cos } \theta \text{ in abbreviated form)} = \frac{OA}{OB} = \frac{\text{adjacent}}{\text{hypotenuse}}$$

$$\text{tangent } \theta \text{ (or tan } \theta \text{ in abbreviated form)} = \frac{AB}{OA} = \frac{\text{opposite}}{\text{adjacent}}$$

Two helpful ways of remembering these trigonometric ratios are:

(i) SOH; CAH; TOA

(i.e. Sine equals Opposite over Hypotenuse, Cosine equals Adjacent over Hypotenuse, Tangent equals Opposite over Adjacent)

(ii) Starting with the tangent ratio

The Old Ant Sat On Her Coat And Hat.

(i.e. Tangent equals Opposite over Adjacent, Sine equals Opposite over Hypotenuse, Cosine equals Adjacent over Hypotenuse).

There are also the ratios

$$\text{Cosecant } \theta \text{ (or cosec } \theta \text{ in abbreviated form)} = \frac{1}{\sin\theta} = \frac{OB}{AB} = \frac{\text{hypotenuse}}{\text{opposite}}$$

$$\text{Secant } \theta \text{ (or sec } \theta \text{ in abbreviated form)} = \frac{1}{\cos\theta} = \frac{OB}{OA} = \frac{\text{hypotenuse}}{\text{adjacent}}$$

$$\text{Cotangent } \theta \text{ (or cot } \theta \text{ in abbreviated form)} = \frac{1}{\tan\theta} = \frac{OA}{AB} = \frac{\text{adjacent}}{\text{opposite}}$$

In any problem involving a right angled triangle where the lengths of two of the sides have been given we are able to use one of the six trigonometric ratios to calculate the size of one of the unknown angles, and where the length of one side and the size of one of the acute angles have been given we are able to use one of the six trigonometric ratios to calculate the length of a second side.

4.4.2 Special Angles

A lot of theoretical work in trigonometry, geometry, calculus and physics, makes use of the exact values of the trigonometric ratios of the angles 30°, 45°, 60°.

These can be easily determined using Figs. 4.4.2(a) and (b) on any occasion they may be required.

Using an equilateral triangle of side 2 units with one altitude drawn it is easily shown, using the symmetry of the triangle and applying Pythagoras, that the length of the altitude is $\sqrt{3}$ units.

So $\sin 30°$ or $\sin \frac{\pi}{6} = \frac{1}{2}$

$\cos 30°$ or $\cos \frac{\pi}{6} = \frac{\sqrt{3}}{2}$

$\tan 30°$ or $\tan \frac{\pi}{6} = \frac{1}{\sqrt{3}}$

$\sin 60°$ or $\sin \frac{\pi}{3} = \frac{\sqrt{3}}{2}$

$\cos 60°$ or $\cos \frac{\pi}{3} = \frac{1}{2}$

$\tan 60°$ or $\tan \frac{\pi}{3} = \sqrt{3}$

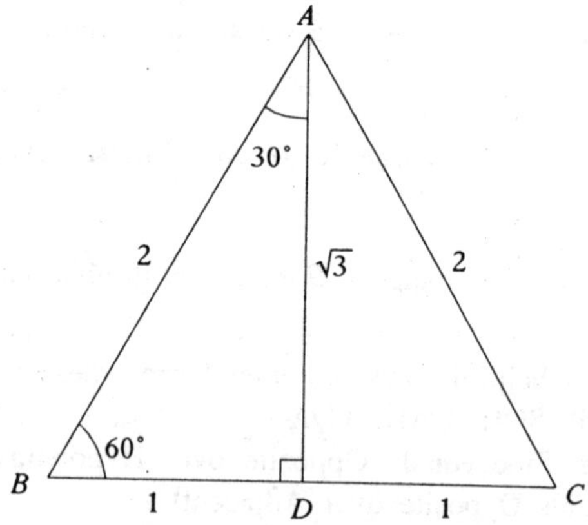

Fig. 4.4.2(a)

Using the isosceles right angled triangle in Fig. 4.4.2(b), where the two sides containing the right angle are chosen to be each 1 unit, we can show that

$\sin 45°$ or $\sin \frac{\pi}{4} = \frac{1}{\sqrt{2}}$

$\cos 45°$ or $\cos \frac{\pi}{4} = \frac{1}{\sqrt{2}}$

$\tan 45°$ or $\tan \frac{\pi}{4} = 1.$

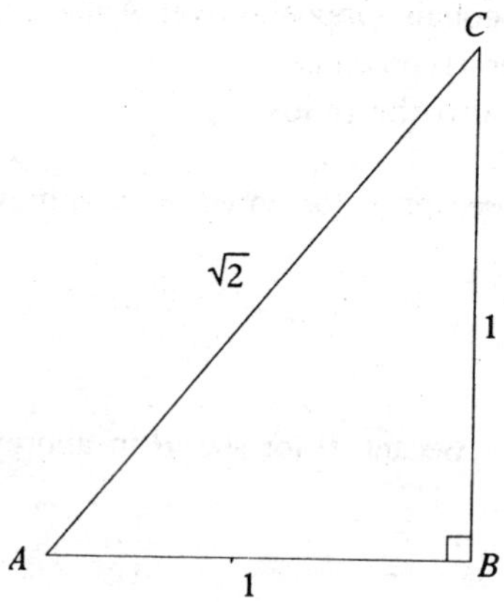

Fig. 4.4.2(b)

4.4.3 Complementary Angles

If the sum of two acute angles [such as α, β in Fig. 4.4.3(a)] is $90° \left(\text{or } \frac{\pi}{2}\right)$ then the angles are said to be *Complementary*.

In Fig. 4.4.3(a) we can say that α is the complement of β, and β is the complement of α.

Using the right angled triangle *OAB* shown in Fig. 4.4.3(a).

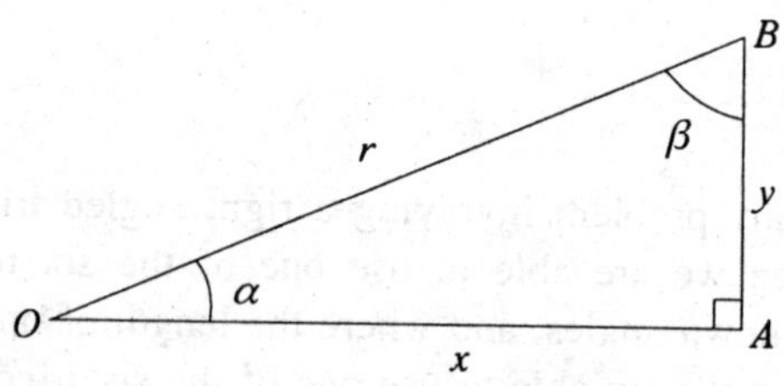

Fig. 4.4.3(a)

$$\sin \alpha = \frac{y}{r} = \cos \beta$$

$$\cos \alpha = \frac{x}{r} = \sin \beta$$

$$\tan \alpha = \frac{y}{x} = \cot \beta$$

$$\cot \alpha = \frac{x}{y} = \tan \beta$$

Because of this property the sine and cosine ratios of the same angle are called complementary ratios. In the same way the tangent and cotangent ratios are said to be complementary.

4.5 CIRCULAR FUNCTIONS FOR A GENERAL ANGLE

So far we have considered the circular functions as defined in a right angled triangle. These definitions only cover acute angles. We now extend these definitions so that we may be able to find any trigonometric ratio (or trigonometric function) for an angle of any magnitude.

4.5.1 The Four Quadrants

Consider a line OP which rotates about the origin O from the initial direction of the positive x-axis [labelled OX in Figs. 4.5.1(a) (b) and (c)]. Any angle which is formed by rotating the line *Anticlockwise* is taken to be *Positive*, and any angle which is formed by rotating the line *Clockwise* is taken to be *Negative*.

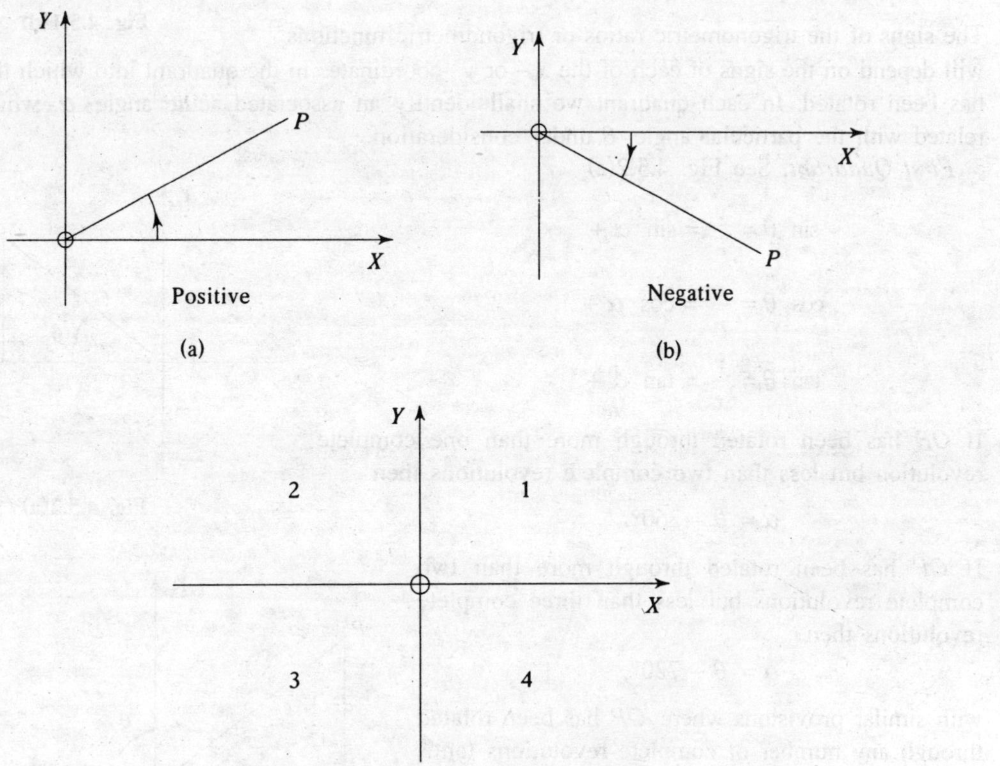

Fig. 4.5.1 (a, b and c)

The pair of Cartesian axes will divide this plane of rotation into four quadrants numbered 1, 2, 3, 4 as shown in Fig. 4.5.1(c).

As the line OP is rotated, P moves round the first quadrant for which both coordinates are positive. When OP moves into the second quadrant its x-coordinate becomes negative. When OP

moves into the third quadrant both coordinates become negative. When OP moves into the fourt (or last) quadrant its x-coordinate is positive but its y-coordinate is negative.

Denoting the distance OP as $r > 0$, and denoting the angle, which is formed by rotating from the positive direction of the x-axis (OX) to the direction of OP as θ, we *define* the trigonometric ratios of θ as

$$\sin\ \theta = \frac{y}{r}$$

$$\cos\ \theta = \frac{x}{r}$$

$$\tan\ \theta = \frac{y}{x}, \text{ as in Fig. 4.5.1(d)}$$

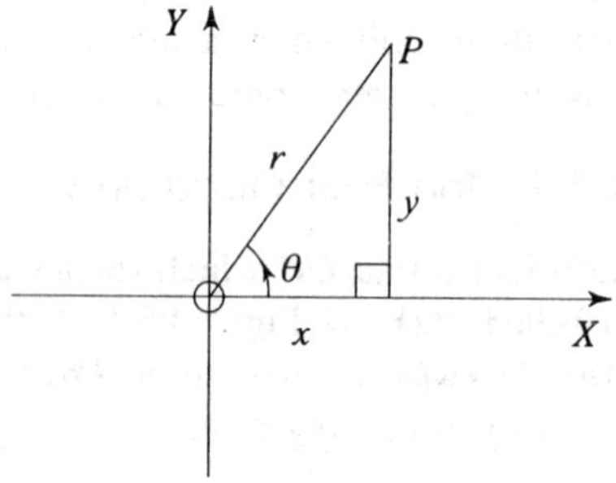

Fig. 4.5.1(d)

4.5.2 The Signs of the Trigonometric Ratios in Each of the Four Quadrants

The signs of the trigonometric ratios or trigonometric functions will depend on the signs of each of the x – or y –coordinates in the quadrant into which the line OP has been rotated. In each quadrant we shall identify an associated acute angle α, which will be related with the particular angle θ under consideration.

First Quadrant, See Fig. 4.5.2(a)

$$\sin\ \theta = \frac{y}{r} = \sin\ \alpha +$$

$$\cos\ \theta = \frac{x}{r} = \cos\ \alpha +$$

$$\tan\ \theta = \frac{y}{x} = \tan\ \alpha +$$

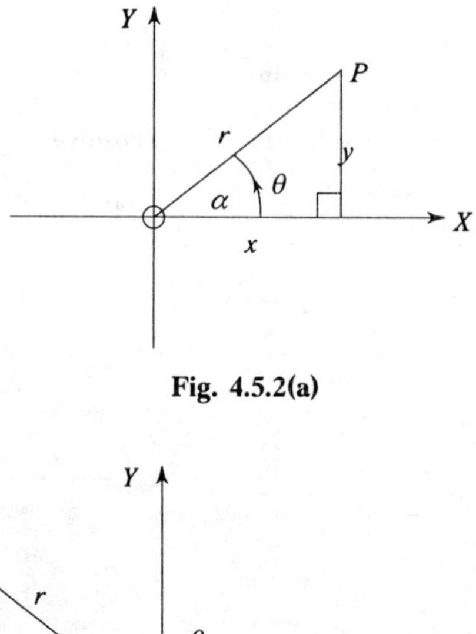

Fig. 4.5.2(a)

If OP has been rotated through more than one complete revolution but less than two complete revolutions then

$$\alpha = \theta - 360°.$$

If OP has been rotated through more than two complete revolutions but less than three complete revolutions then

$$\alpha = \theta - 720°,$$

with similar provisions where OP has been rotated through any number of complete revolutions (anti-clockwise or clockwise).

Second Quadrant, See Fig. 4.5.2(b)

The associated acute angle α is $180° - \theta$

or $$\theta = 180° - \alpha$$

$$\sin\ \theta = \frac{y}{r} = \sin\ \alpha$$

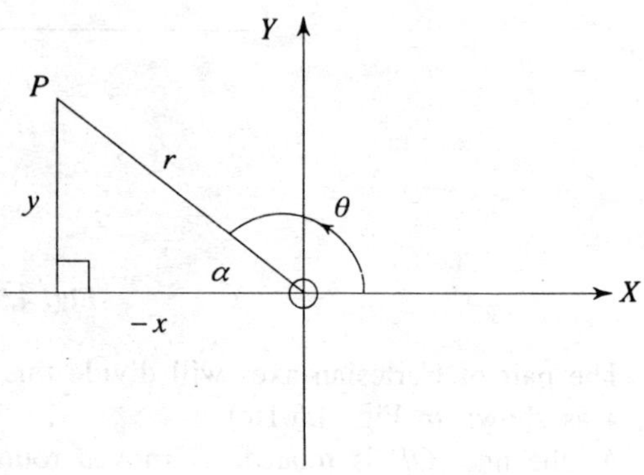

Fig. 4.5.2(b)

or $\sin(180° - \alpha) = +\dfrac{y}{r} = \sin\alpha$ $\quad +$

$$\cos\theta = \frac{-x}{r} = -\cos\alpha$$

or $\cos(180° - \alpha) = -\cos\alpha$ $\quad -$

$$\tan\theta = \frac{y}{-x} = -\tan\alpha$$

or $\tan(180° - \alpha) = -\tan\alpha$ $\quad -$

Third Quadrant. See Fig. 4.5.2(c). The associated angle α is $\theta - 180°$.

or $\theta = 180° + \alpha$

$$\sin\theta = \frac{-y}{r} = -\sin\alpha$$

or $\sin(180° + \alpha) = -\sin\alpha$ $\quad -$

$$\cos\theta = \frac{-x}{r} = -\cos\alpha$$

or $\cos(180° + \alpha) = -\cos\alpha$ $\quad -$

$$\tan\theta = \frac{-y}{-x} = \frac{y}{x} = +\tan\alpha$$

or $\tan(180° + \alpha) = \tan\alpha$ $\quad +$

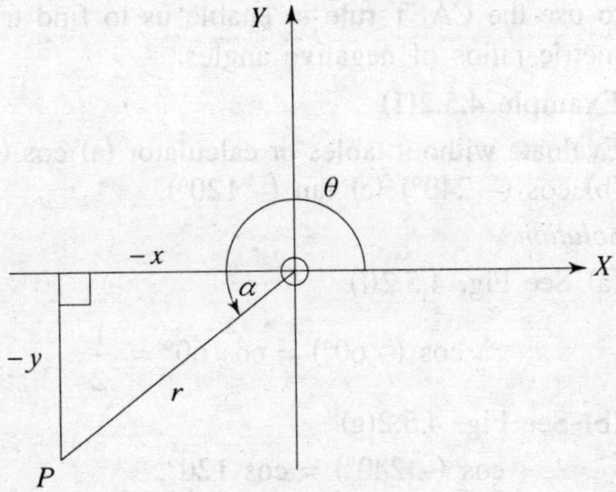

Fig. 4.5.2(c)

Fourth Quadrant. See Fig. 4.5.2 (d). Here the final direction OP may be reached by an anticlockwise rotation of θ from OX or by a clockwise rotation (of $-\alpha$) from OX.

However $\alpha = 360° - \theta$

or $\theta = 360° - \alpha$

$$\sin\theta = \frac{-y}{r} = -\sin\alpha$$

or $\sin(360° - \alpha) = \sin(-\alpha)$
(anticlockwise) (clockwise)

$= -\sin\alpha$ $\quad -$

$$\cos\theta = \frac{x}{r} = \cos\alpha$$

or $\cos(360° - \alpha) = \cos(-\alpha)$

$= \cos\alpha$ $\quad +$

$$\tan\theta = \frac{-y}{x} = -\tan\alpha$$

or $\tan(360° - \alpha) = \tan(-\alpha)$

$= -\tan\alpha$ $\quad -$

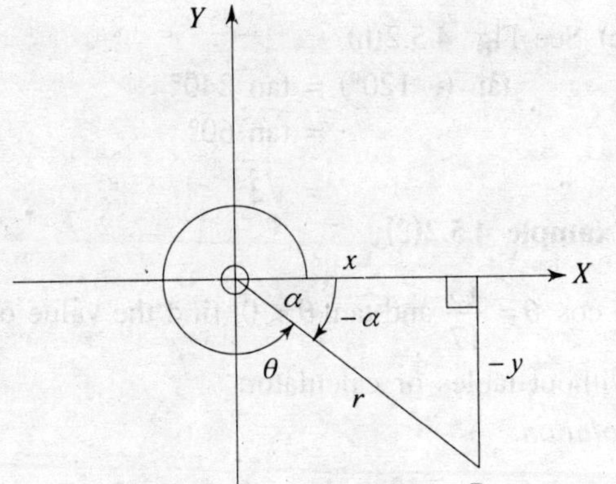

Fig. 4.5.2(d)

Studying the summary in Fig. 4.5.2(e) we can see that in each of the four quadrants we may use this quadrant rule, together with the associated acute angle α (which should always be measured from the directions of 0°, 180° or 360°) to give the trigonometric ratios (trigonometric functions) of *Any Angle*.

There are several useful ways in which we can remember the quadrant rule, such as cast into the Last,

or all students take care,

or all silly tom cats,

or acts Clockwise.

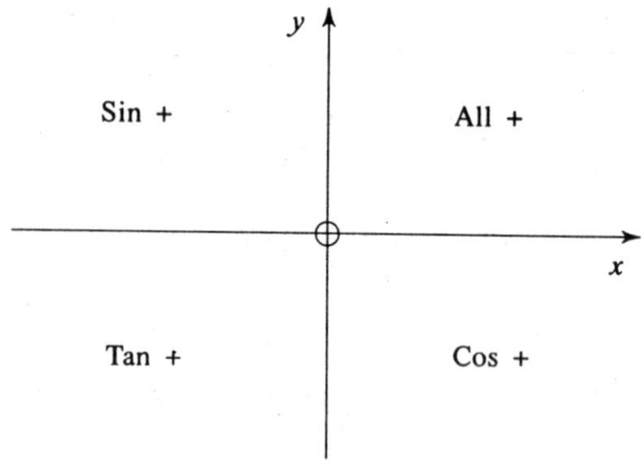

Fig. 4.5.2.(e)

As has been seen from Fig. 4.5.2(d) it is possible to use the CAST rule to enable us to find trigonometric ratios of negative angles.

Example 4.5.2(1)

Evaluate without tables or calculator (a) cos (– 60°) (b) cos (– 240°) (c) tan (– 120°).

Solution:

(a) See Fig. 4.5.2(f)

$$\cos(-60°) = \cos 60° = \frac{1}{2}.$$

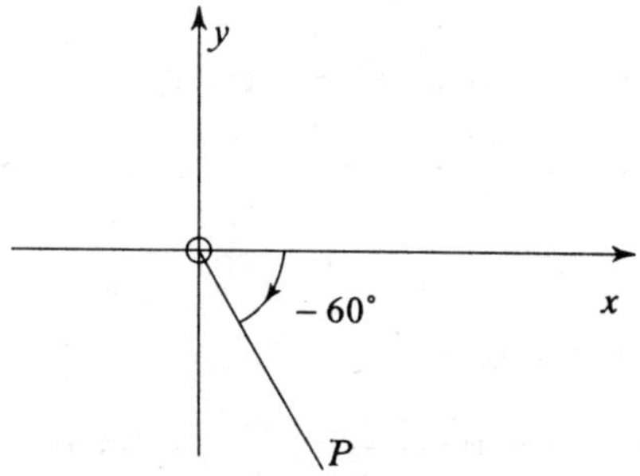

Fig. 4.5.2(f)

(b) See Fig. 4.5.2(g)

$$\begin{aligned}\cos(-240°) &= \cos 120°\\ &= -\cos 60°\\ &= -\frac{1}{2}.\end{aligned}$$

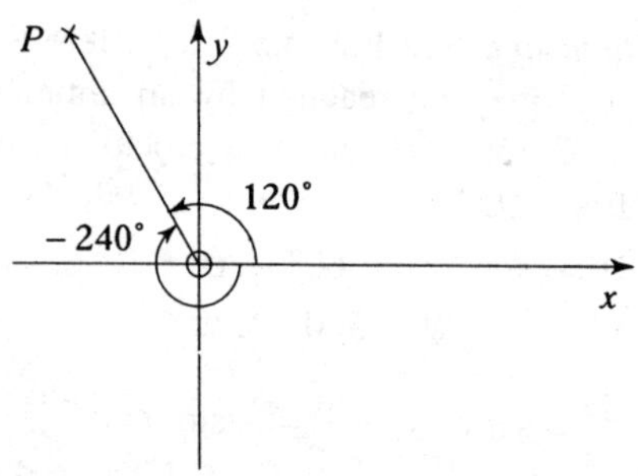

Fig. 4.5.2(g)

(c) See Fig. 4.5.2(h)

$$\begin{aligned}\tan(-120°) &= \tan 240°\\ &= \tan 60°\\ &= \sqrt{3}.\end{aligned}$$

Example 4.5.2(2)

If $\cos\theta = \dfrac{15}{17}$ and $\tan\theta < 0$, find the value of $\sin\theta$ without tables or calculator.

Solution:

Since $\cos\theta = \dfrac{15}{17} > 0$ and $\tan\theta < 0$ then θ lies in the fourth quadrant. Construct a triangle OAB in the fourth quadrant taking $x = 15$ and $r = 17$ as in Fig. 4.5.2(i).

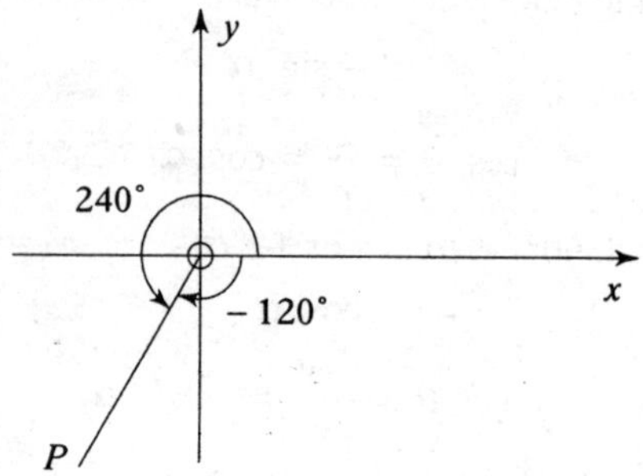

Fig. 4.5.2(h)

Then $y = -\sqrt{(17^2 - 15^2)} = -8$ (by Pythagoras)

From Fig. 4.5.2(i) it should be clear that

$$\sin \theta = -\frac{-8}{17}.$$

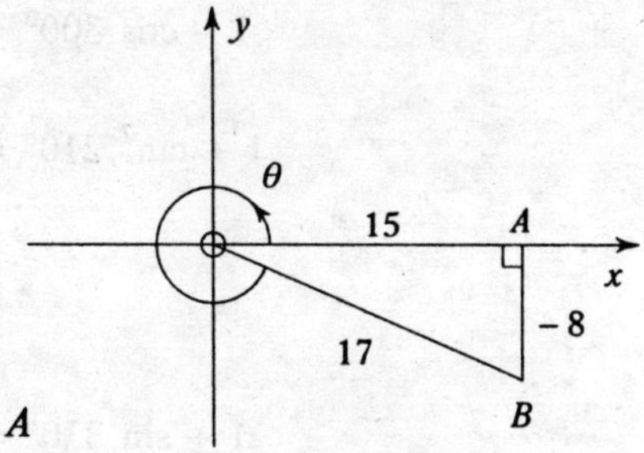

Fig. 4.5.2(i)

Example 4.5.2(3)

If sin A is $\frac{3}{4}$ and A is obtuse find the values of sec A and cot A without tables or calculator.

Solution:

Construct a triangle as in Fig. 4.5.2(j), in the second quadrant, taking $y = 3$, $r = 4$.

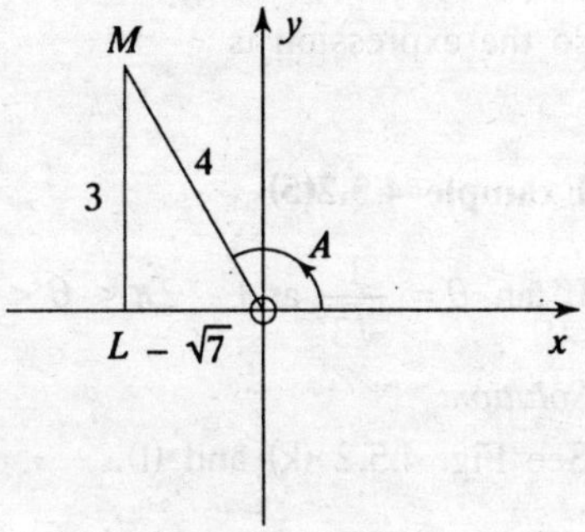

Fig. 4.5.2(j)

Then $$x = -\sqrt{(4^2 - 3^2)}$$

$$= -\sqrt{7} \text{ by Pythagoras}$$

From Fig. 4.5.2(j), we can read $\sec A = \dfrac{1}{\cos A}$

$$-\frac{1}{\sqrt{7}/4} = \frac{-4}{\sqrt{7}}$$

and $$\cot A = \frac{1}{\tan A}$$

$$= \frac{1}{3/-\sqrt{7}}$$

$$= \frac{-\sqrt{7}}{3}.$$

Example 4.5.2(4)

Evaluate $\dfrac{\sec^2 150° (1 - \cos 300°)}{(1 + \tan^2 210°)(1 + \sin 330°)}$ without tables or calculator.

Solution:

Taking each factor in turn

$$\sec^2 150° = \frac{1}{\cos^2 150°} = \frac{1}{(-\cos 30°)^2}$$

$$= \frac{1}{\left(-\frac{\sqrt{3}}{2}\right)^2} = \frac{1}{\frac{3}{4}} = \frac{4}{3}$$

$$1 - \cos 300° = 1 - \cos 60° = 1 - \frac{1}{2} = \frac{1}{2}$$

$$1 + \tan^2 210° = 1 + \tan^2 30°$$

$$= 1 + \left(\frac{1}{\sqrt{3}}\right)^2 = 1 + \frac{1}{3} = \frac{4}{3}$$

$$1 + \sin 330° = 1 - \sin 30° = 1 - \frac{1}{2} = \frac{1}{2}$$

so the expression is $$= \frac{\frac{4}{3} \times \frac{1}{2}}{\frac{4}{3} \times \frac{1}{2}} = 1.$$

Example 4.5.2(5)

If $\tan \theta = \frac{1}{\sqrt{3}}$ and $-2\pi < \theta < 2\pi$ give four possible values of θ.

Solution:

See Fig. 4.5.2 (k) and (l).

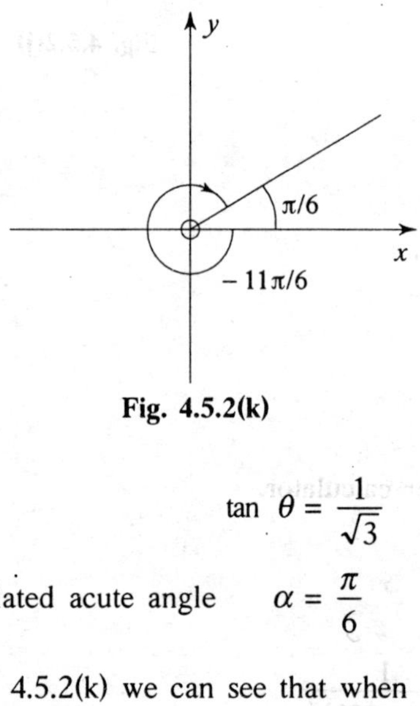

Fig. 4.5.2(k)

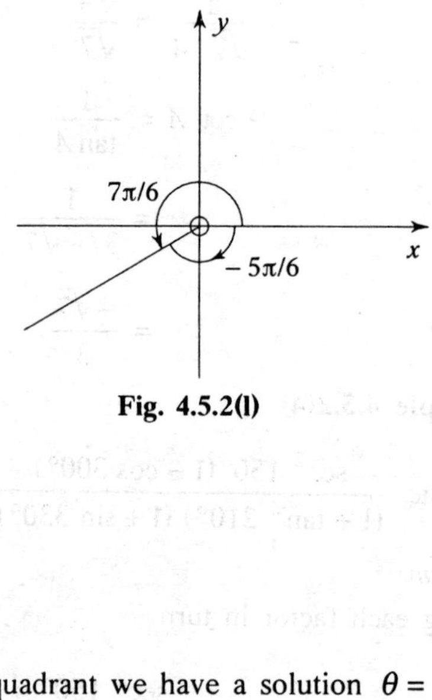

Fig. 4.5.2(l)

When $$\tan \theta = \frac{1}{\sqrt{3}}$$

the associated acute angle $$\alpha = \frac{\pi}{6}$$

In Fig. 4.5.2(k) we can see that when θ is in the first quadrant we have a solution $\theta = \frac{\pi}{6}$ for $0 < \theta < 2\pi$ and also a solution $\theta = \frac{-11\pi}{6}$ (rotating clockwise) for $-2\pi < \theta < 0$

However $\tan \theta = \frac{1}{\sqrt{3}}$ has a solution in the third quadrant i.e. $$\pi + \frac{\pi}{6} = \frac{7\pi}{6} \text{ for } 0 < \theta < 2\pi$$

and also $\theta = -5\frac{\pi}{6}$ (rotating clockwise) for $-2\pi < \theta < 0$.

so the four possible values for θ in $-2\pi < \theta < 2\pi$ are

$$\theta = \frac{\pi}{6} \text{ or } \theta = \frac{-11\pi}{6} \text{ or } \theta = \frac{7\pi}{6} \text{ or } \theta = \frac{-5\pi}{6}.$$

4.6 NUMERICAL TRIGONOMETRY

In many exercises, such as surveying or map reading, practical measurements will normally include the measurements of sides and angles of a triangle.

A triangle has 3 sides and 3 angles and can be specified provided we are given the values of any three members from three sides and two angles (remember that the sum of the angles of a triangle is always 180°).

Given any set of data which will be sufficient to define a triangle it is possible to calculate the remaining sides and angles. This exercise is known as *solving* the triangle and may require the use of one or both of the *Sine Rule* and *Cosine Rule*.

4.6.1 The Sine Rule

Example 4.6.1(1)

Prove the formula for the sine rule

i.e. $$\frac{a}{\sin A} = \frac{b}{\sin B} = \frac{c}{\sin C} = 2R,$$

where R is the radius of circumcircle of triangle ABC, considering the cases where C is (i) acute (ii) obtuse

Proof:

Let the sides opposite the angles A, B, C be denoted a, b, c respectively (this is a normal convention in solving triangles)

In Fig. 4.6.1(a) (i) C is acute and in Fig. 4.6.1(a) (ii) C is obtuse.

In both figures join diameter AD.

In Figs. 4.6.1(a)(i) and (ii) angle $ADB = 90°$ (angle in a semi-circle)

and $$c = 2R \sin \angle ADB$$

In Fig. 4.6.1 (a) (i)

$$\angle ACB = \angle ADB \text{ (angles in the same segment)}$$

so $$c = 2R \sin C$$

In Fig. 4.6.1(a)(ii)

$$\angle ACB + \angle ADB = 180° \text{ (opposite angles of a cyclic quadrilateral)}$$

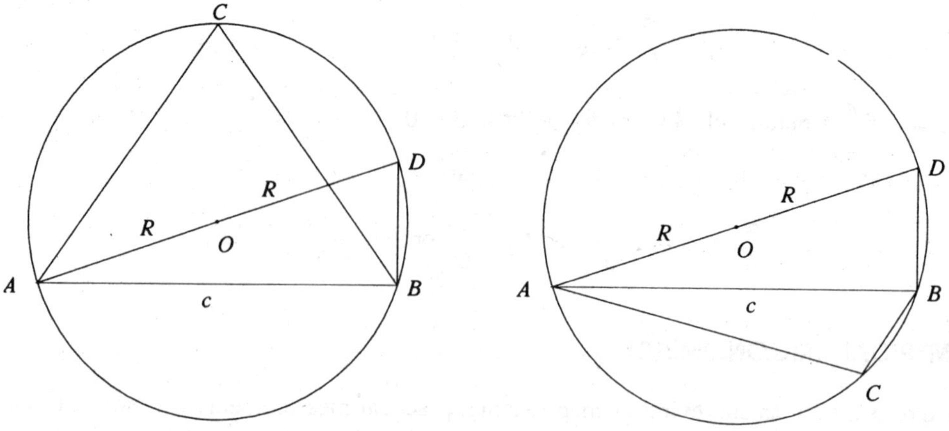

(i) C acute (ii) C obtuse

Fig. 4.6.1(a)

so

$$c = 2R \sin (180° - C)$$

hence in both cases

$$c = 2R \sin C$$

Similarly it may be shown that

$$\frac{a}{\sin A} = 2R, \; \frac{b}{\sin B} = 2R$$

Hence

$$\frac{a}{\sin A} = \frac{b}{\sin B} = \frac{c}{\sin C} = 2R$$

The sine rule can be used immediately to solve a triangle where we are given either (i) two angles and one side, or (ii) two sides and an angle opposite one of these sides.

Example 4.6.1(2)

Calculate the remaining angle and sides of triangle ABC in which $A = 73°$, $B = 49°$ and $a = 6.1$ cm.

Solution:

From the information shown in Fig. 4.6.1(b)

$$C = 180° - 73° - 49° = 58°$$

Now calculate b using

$$\frac{b}{\sin 49°} = \frac{6.1}{\sin 73°}$$

i.e.

$$b = \frac{6.1 \sin 49°}{\sin 73°} = 4.81 \text{ cm (3sf)}$$

also

$$\frac{c}{\sin 58°} = \frac{6.1}{\sin 73°}$$

i.e. $c = \dfrac{6.1 \sin 58^\circ}{\sin 73^\circ} = 5.41 \text{ cm (3sf)}$

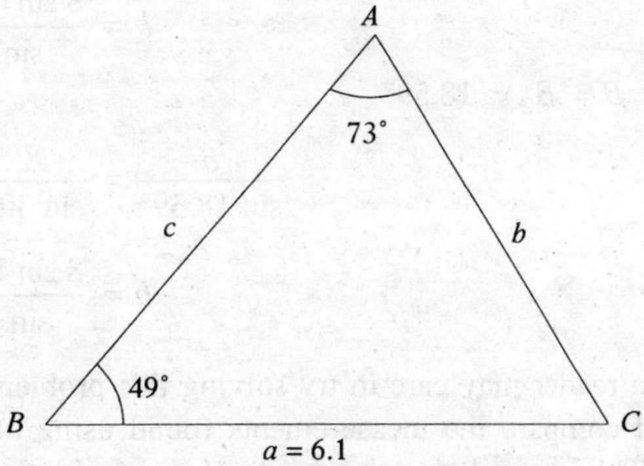

Fig. 4.6.1 (b)

Example 4.6.1(3)

Calculate the remaining side and angles of triangle *ABC* in which $c = 12$ cm, $a = 8$ cm and angle $A = 30°$.

Solution: See Fig. 4.6.1(c).

We can determine C using

$$\frac{\sin C}{12} = \frac{\sin 30^\circ}{8}$$

i.e. $\sin C = \dfrac{12 \sin 30^\circ}{8} = 0.75$

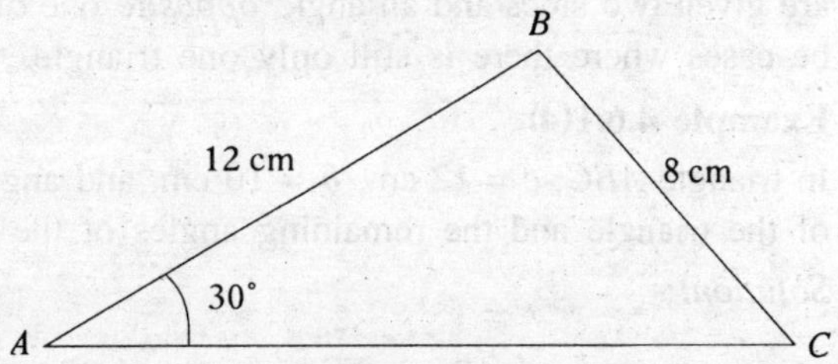

Fig. 4.6.1(c)

Now $C = 48.59°$

or $180° - 48.59°$ (two possible angles)

i.e. $C = 48.59°$ or $131.41°$

If $C = 48.59°$

$$B = 180^\circ - 30^\circ - 48.59^\circ = 101.41^\circ$$

If $C = 131.41°$

$$B = 180^\circ - 30^\circ - 131.41^\circ = 18.59^\circ$$

In this case there are two triangles as shown in Fig. 4.6.1(d).

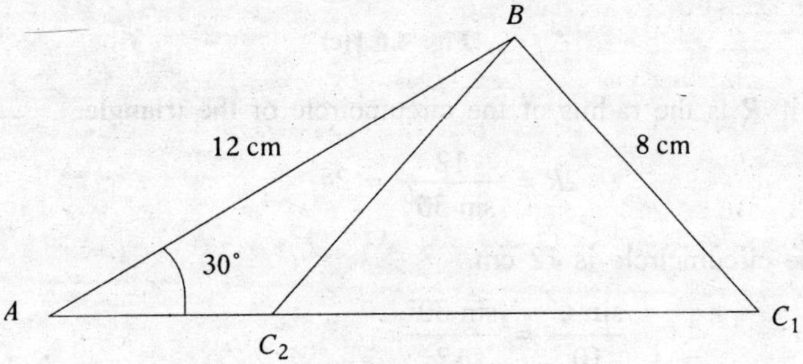

Fig. 4.6.1(d)

In Fig. 4.6.1(d) $C_1 = 48.59°$ and B_1 (i.e. $\angle ABC_1$) is $101.41°$

$C_2 = \angle AC_2B = 131.41°$ and B_2 (i.e. $\angle ABC_2$) is $18.59°$.

This in turn means that there will be two possible lengths [AC_1 and AC_2 in Fig. 4.6.1(d)] for b.
For $B = B_1 = 101.41°$

$$\frac{b}{\sin 101.41^\circ} = \frac{8}{\sin 30^\circ}$$

i.e. $$b = \frac{8 \sin 101.41°}{\sin 30°} = 15.68 \text{ cm}$$

For $B = B_2 = 18.59°$

$$\frac{b}{\sin 18.59°} = \frac{8}{\sin 30°}$$

i.e. $$b = \frac{8 \sin 18.59°}{\sin 30°} = 5.10 \text{ cm}$$

The reader may care to try solving this problem by drawing, using a ruler and a pair of compasses, and compare the measurements found using both methods.

This example is an example of the *ambiguous case* where there are *two* possible triangles which can fit the specification. It is not always true that there will be two possible triangles whenever we are given two sides and an angle *opposite* one of the sides. The next example shows that there may be cases where there is still only one triangle.

Example 4.6.1(4)

In triangle ABC, $c = 12$ cm, $b = 10$ cm, and angle $A = 30°$. Calculate the radius of the circumcircle of the triangle and the remaining angles of the triangle.

Solution:

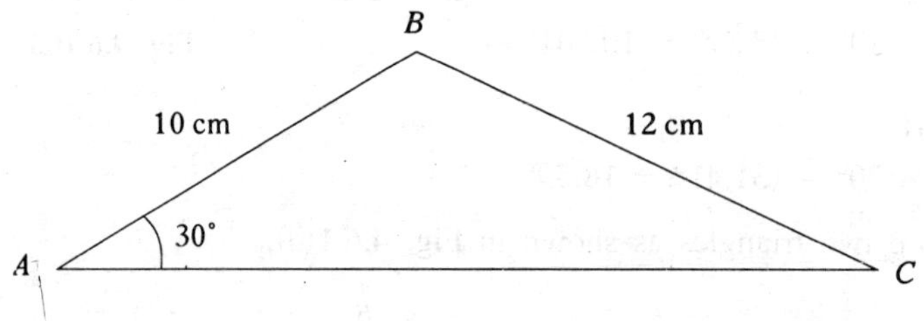

Fig. 4.6.1(e)

Using Fig. 4.6.1(e) if R is the radius of the circumcircle of the triangle,

$$2R = \frac{12}{\sin 30°} = 24$$

i.e. the radius of the circumcircle is 12 cm.

Also $$\frac{\sin C}{10} = \frac{\sin 30°}{12}$$

i.e. $$\sin C = \frac{10 \sin 30°}{12} = 0.4167$$

$$C = 24.62° \text{ or } 180° - 24.62°$$

i.e. $$C = 24.62 \text{ or } 155.38°$$

Clearly one possibility is

$$A = 30°,\ C = 24.62°$$

then $$B = 180° - 30° - 24.62° = 125.38°.$$

There is no other possibility because

(i) for $C = 155.38°$ we see that $A + C = 185.38° > 180°$ (no triangle is possible)

(ii) Angle C is opposite side $c = 10$ cm and $c < a = 12$ cm, so the size of angle C *must* be less than the size of angle A.

So in this case there is only *one* possible triangle with $A = 30°$, $B = 125.38°$, $C = 24.62°$.

We now examine how many triangles can be found, based on the way the lengths c, a and $c \sin A$ are related to each other in order of magnitude.

This is shown in the following example.

Example 4.6.1(5)

Given a triangle is which the lengths a, c, and the angle A are specified, show using suitable diagrams how there may be no triangle, one triangle or two triangles possible, based on the relative magnitudes of the lengths c, a and $c \sin A$.

Solution:

Case (i) $a < c \sin A$ no triangle can be drawn see Fig. 4.6.1(f), hence no possible triangle.

Case (ii) $a = c \sin A$ the side BC will touch the line AC at C and one triangle may be drawn. See Fig. 4.6.1(g), hence one possible triangle.

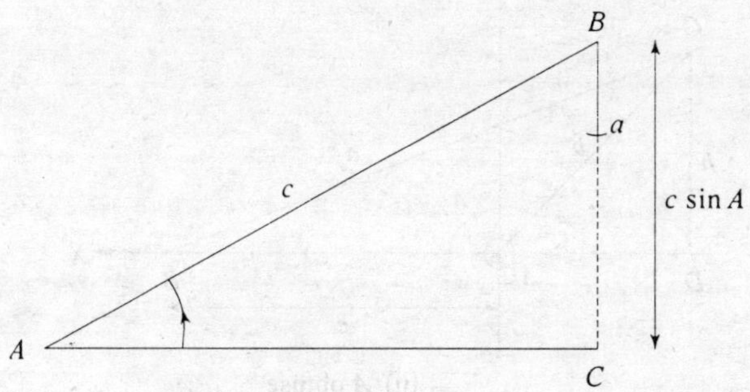

Fig. 4.6.1 (f)

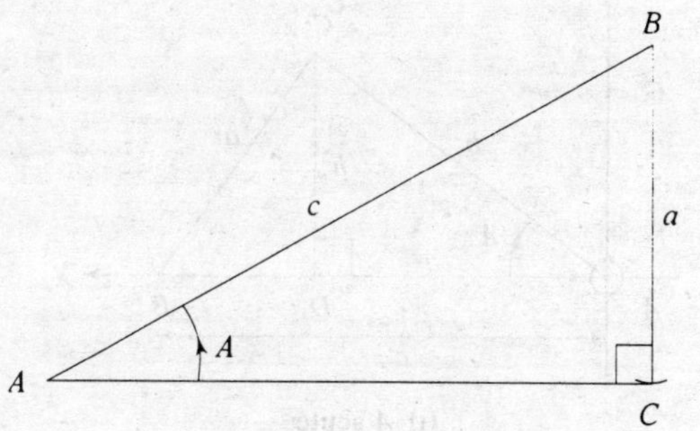

Fig. 4.6.1 (g)

Case (iii) $c > a > c \sin A$ the side BC can take up either of positions BC_1 or BC_2 so two triangles may be drawn, see Fig. 4.6.1(h), hence two possible triangles.

Case (iv) a > c, only one triangle is possible (in the 'second' triangle the sides c and a (indicated by a broken line) will be the specifications but the angle at vertex A will not be the given angle A but 180°—the given angle A) [see Fig. 4.6.1(i)]

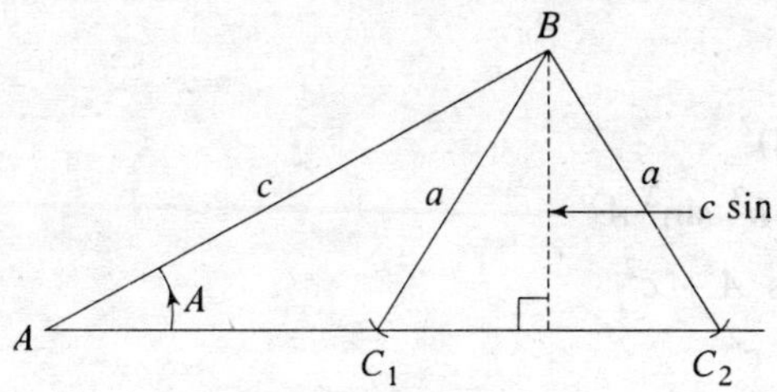

Fig. 4.6.1 (h)

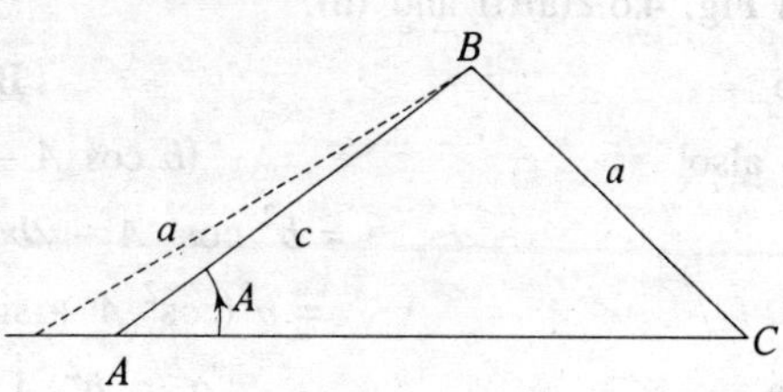

Fig. 4.6.1 (i)

In other problems we may be given either the lengths of all three sides of a triangle or the lengths of two sides and the included angle of a triangle. In both cases it is *not* possible with only the given information to determine a known value of any of the ratios, $\dfrac{a}{\sin A}$, $\dfrac{b}{\sin B}$, or $\dfrac{c}{\sin C}$.

In both cases it is necessary to use another rule to determine the first unknown element (side or angle), and then the sine rule can be used to calculate an angle which is *known* to be acute (i.e. an angle which is not the angle opposite the longest side).

4.6.2 The Cosine Rule

Example 4.6.2(1)

Prove that in any triangle ABC $a^2 = b^2 + c^2 - 2\,bc \cos A$ (the cosine rule) considering the cases where A is (i) acute (ii) obtuse.

Proof:

Set triangle ABC with A at the origin and B on the x-axis. See Fig. 4.6.2(a).

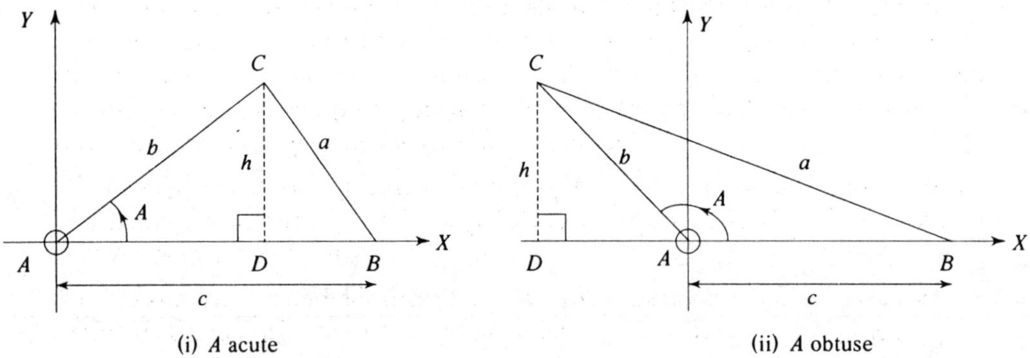

Fig. 4.6.2(a)

In Fig. 4.6.2(a) A is (0, 0), B is (c, 0) and C (using triangle ADC) is ($b \cos A$, $b \sin A$).

Remember that, in Fig. 4.6.2(a)(ii), $\cos A$ is negative.

hence
$$\mathbf{BC} = \begin{pmatrix} b\cos A - c \\ b \sin A \end{pmatrix}$$

in Fig. 4.6.2(a)(i) and (ii).

so
$$|\mathbf{BC}|^2 = a^2$$

is also
$$(b\cos A - c)^2 + (b \sin A)^2$$
$$= b^2\cos^2 A - 2bc\cos A + c^2 + b^2\sin^2 A$$
$$= b^2(\cos^2 A + \sin^2 A) - 2bc\cos A + c^2$$

i.e.
$$a^2 = b^2 \cdot 1 - 2bc\cos A + c^2$$

i.e.
$$a^2 = b^2 + c^2 - 2bc\cos A$$

In a similar manner it may be shown that

$$b^2 = a^2 + c^2 - 2ac \cos B$$

$$c^2 = a^2 + b^2 - 2ab \cos C$$

Example 4.6.2(2)

Calculate the size of the largest angle in triangle *ABC* where a = 35.0 cm, b = 16.8 cm, c = 23.8 cm.

Solution:

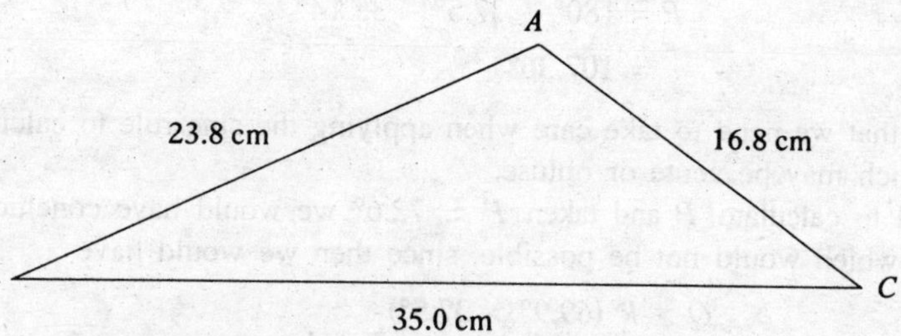

Fig. 4.6.2(b)

The largest angle will be opposite the longest side and so we have to calculate the size of angle *A*. Using the alternative form of the cosine rule,

$$\cos A = \frac{b^2 + c^2 - a^2}{2bc} \quad \text{(make cos A the subject of the formula)}$$

i.e.

$$\cos A = \frac{16.8^2 + 23.8^2 - 35.0^2}{2 \times 16.8 \times 23.8}$$

$$= -0.4706$$

$$A = 180° - 61.9° = 118.1°$$

A is obtuse since $\cos A < 0$.

Whenever a problem requiring that the triangle be solved has to be started by one application of the cosine rule any subsequent sides or angles can then be determined using the sine rule, as the next example shows.

Example 4.6.2(3)

Solve the triangle *PQR*, given p = 6.05 cm, q = 3.65 cm, and R = 37.5°.

Using cosine rule to calculate r we have

$$r^2 = 6.05^2 + 3.65^2 - 2 \times 6.05 \times 3.65 \cos 37.5°$$

$$= 14.89$$

i.e. r = 3.86 cm (3 sf)

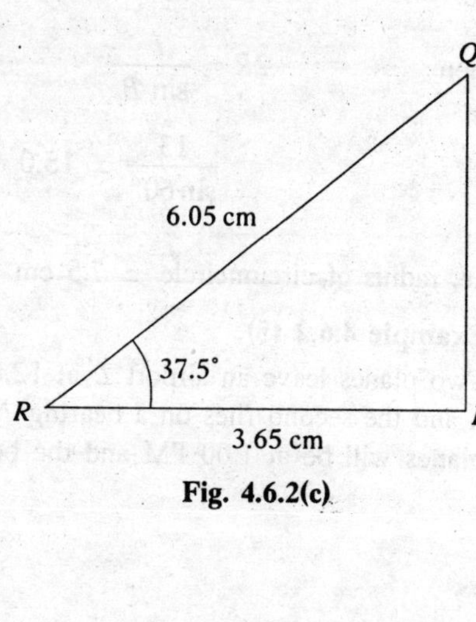

Fig. 4.6.2(c)

Now use the sine rule to calculate Q as we know r, R, p and q.

$$\frac{\sin Q}{3.65} = \frac{\sin 37.5°}{3.86}$$

i.e. $$\sin Q = \frac{3.65 \sin 37.5°}{3.86} = 0.57\ 56$$

i.e. $$Q = 35.14°$$

NB Q *cannot* be obtuse since it is *not* opposite the longest side.

then $$P = 180° - 37.5° - 35.14°$$

$$= 107.36°$$

This example shows that we need to take care when applying the sine rule to calculate the size of the largest angle, which may be acute or obtuse.

Had we proceeded to calculate P and taken $P \simeq 72.6°$ we would have concluded that Q was approximately 69.9° which would not be possible, since then we would have

$$Q > R\ (69.9° > 37.5°)$$

but $$q < r\ (3.65 < 3.86)$$

and we would not have the smallest angle opposite the shortest side.

Example 4.6.2(4)

Calculate the size of the angle B and the radius of the circumcircle of triangle ABC in which a = 8 cm, b = 13 cm, c = 15 cm.

Solution: Using the cosine rule

$$\cos B = \frac{a^2 + c^2 - b^2}{2ac}$$

$$= \frac{8^2 + 15^2 - 13^2}{2 \times 8 \times 15} = 0.5$$

i.e. $$B = 60°$$

then $$2R = \frac{b}{\sin B}$$

$$\frac{13}{\sin 60°} \simeq 15.0 \text{ cm}$$

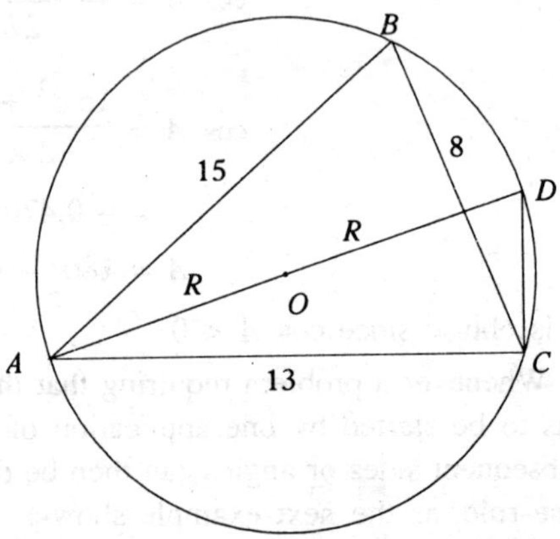

Fig. 4.6.2(d)

i.e. radius of circumcircle $\simeq$ 7.5 cm (exact value $\frac{13\sqrt{3}}{3}$ cm)

Example 4.6.2 (5)

Two planes leave an airport L at 12.0 noon. The first plane flies due West at a speed of 600 km/h and the second flies on a bearing $N\ 30°\ E$ at a speed of 1000 km/h. Calculate how far apart the planes will be at 1.00 PM and the bearing of the second from the first at that time.

Solution:

See Fig. 4.6.2(e). After one hour the first plane will be at M, 600 km due West of L, and the second will be at P, 1000 km on a bearing of $N\ 30°\ E$ from L. Using triangle LMP we have

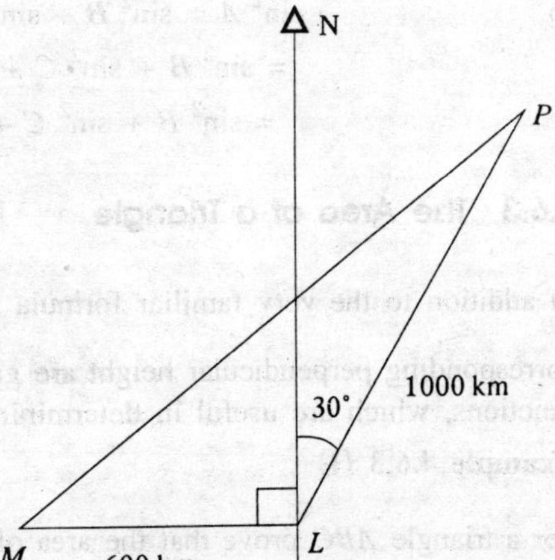

Fig. 4.6.2(e)

$$p = 600,\ m = 1000 \text{ and } \angle MLP = 120°$$

and so $l^2 = 600^2 + 1000^2 - 2 \times 600 \times 1000 \cos 120°$

$$= 360000 + 1000000 + 2 \times 600 \times 1000 \cos 60°$$

$$= 1960000$$

i.e. $\quad l = 1400$

so the planes are 1400 km apart at 1.00 PM.

Also $\quad \dfrac{\sin \angle PML}{1000} = \dfrac{\sin 120°}{1400}$

i.e. $\quad \sin \angle PML = \dfrac{1000 \sin 120°}{1400} = 0.6186$

i.e. $\quad \angle PML = 38.21°$

See Fig. 4.6.2(f). Since angle $PML = 38.21°$ made with the East direction (ML), then the bearing of P from M will be $N\ (90° - 38.21°)\ E$

i.e. $\quad N\ 51.79°\ E$

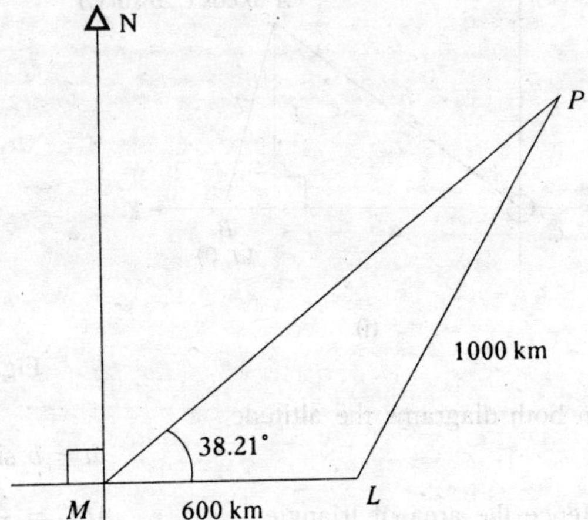

Fig. 4.6.2(f)

Example 4.6.2(6)

Using the sine and cosine rules show that in any triangle ABC,

$$\sin^2 A = \sin^2 B + \sin^2 C + 2 \sin B \sin C \cos (B + C)$$

Solution:

Using $a^2 = b^2 + c^2 - 2bc \cos A$ and re-writing

$$a = 2R \sin A,\ b = 2R \sin B,\ c = 2R \sin C,$$

then

$$4R^2 \sin^2 A = 4R^2 \sin^2 B + 4R^2 \sin^2 C - 2 \cdot 2R \sin B \cdot 2R \sin C \cos A$$

Dividing throughout by $4R^2$ then

$$\sin^2 A = \sin^2 B + \sin^2 C - 2 \sin B \sin C \cos A$$

but $\quad \cos (180° - A) = \cos (B + C)$ in triangle ABC

and so $\quad -\cos A = \cos (B + C)$

so $$\sin^2 A = \sin^2 B + \sin^2 C + 2 \sin B \sin C\ (-\cos A)$$
$$= \sin^2 B + \sin^2 C + 2 \sin B \sin C\ (\cos (B + C))$$
$$= \sin^2 B + \sin^2 C + 2 \sin B \sin C \cos (B + C)$$

4.6.3 The Area of a Triangle

In addition to the very familiar formula $\frac{1}{2}$ base × height for the area of triangle where base and corresponding perpendicular height are given there are two other formulae, involving trigonometric functions, which are useful in determining the area of a triangle.

Example 4.6.3 (1)

For a triangle ABC prove that the area of the triangle is $\frac{1}{2}\, a\, b \sin C$, considering the cases where C is (i) acute (ii) obtuse. Hence deduce that in any triangle ABC

$$\frac{a}{\sin A} = \frac{b}{\sin B} = \frac{c}{\sin C}.$$

Proof:
Set triangle ABC with C at the origin and side CB along the x-axis as in Figs 4.6.3(a)(i) and (ii) Then B will have coordinate (a, o) and A will have coordinates $(b \cos C, b \sin C)$

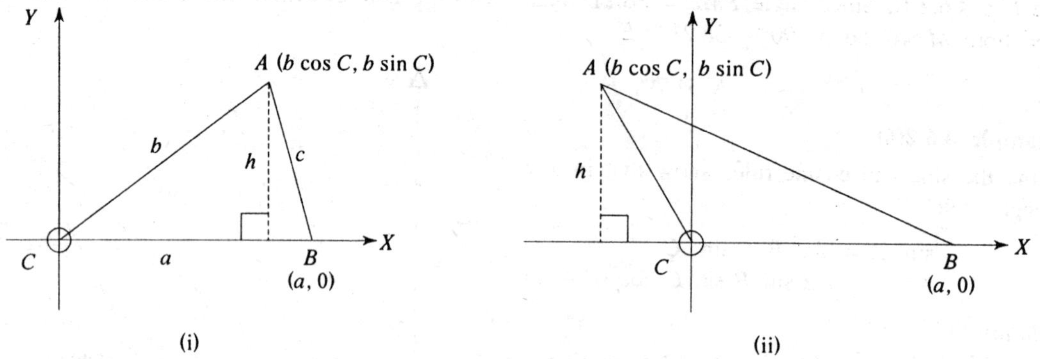

Fig. 4.6.2(a)

In both diagrams the altitude

$$h = b \sin C$$

Hence the area of triangle $$ABC = \frac{1}{2}\, a\, h$$
$$= \frac{1}{2}\, a\, (b \sin C)$$
$$= \frac{1}{2}\, a\, b \sin C$$

In a similar manner it may be proved that the area of triangle ABC is $\frac{1}{2}\ bc\ \sin A = \frac{1}{2}\ ca\ \sin B$

Hence we may write the area of the triangle ABC in three different ways using the formulae just established

i.e. $$\frac{1}{2}\ a\ b\ \sin C = \frac{1}{2}\ b\ c\ \sin A = \frac{1}{2}\ c\ a\ \sin B$$

or $$a\ b\ \sin C = bc\ \sin A = ca\ \sin B$$

Dividing each expression by abc

Gives $$\frac{\sin C}{c} = \frac{\sin A}{a} = \frac{\sin B}{b}$$

or $$\frac{a}{\sin A} = \frac{b}{\sin B} = \frac{c}{\sin C}.$$

Example 4.6.3(2)

A prism is on a triangular base ABC where $AB = 5$ cm, $BC = 10$ cm, and angle $B = 120°$.

If the perpendicular height of the prism is $4\ \sqrt{3}$ cm calculate the volume of the prism without tables or calculator.

Solution: See Fig. 4.6.3(b)

Triangle ABC has area $\frac{1}{2}\ ca\ \sin B$

$$= \frac{1}{2} \times 5 \times 10 \sin 120°$$

$$= \frac{1}{2} \times 5 \times 10 \times \frac{\sqrt{3}}{2}\ \text{cm}^2$$

So the volume of the prism is

$$\frac{1}{2} \times 5 \times 10 \times \frac{\sqrt{3}}{2} \times 4\ \sqrt{3}\ \text{cm}^3 = 150\ \text{cm}^3.$$

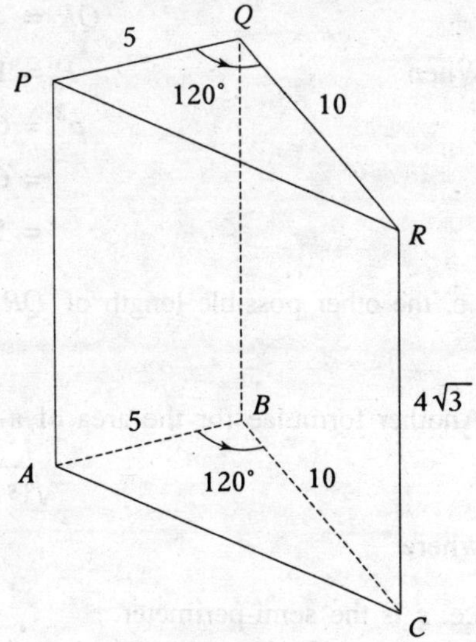

Fig. 4.6.3(b)

Example 4.6.3 (3)

In triangle PQR the side PQ is 6 cm and the side PR is 8 cm. If the area of the triangle is 9 cm^2 calculate the possible sizes of angle P and the possible lengths of side QR.

Solution:

Using area of triangle PQR

$$= \frac{1}{2}\ qr\ \sin P$$

then $$\frac{1}{2} \times 8 \times 6 \sin P = 9$$

i.e. $$\sin P = \frac{9}{24} = 0.375$$

and $$P = 22.02° \text{ or } 157.98°$$

i.e. there are two possible angles P [see Figs. 4.6.3(c)]

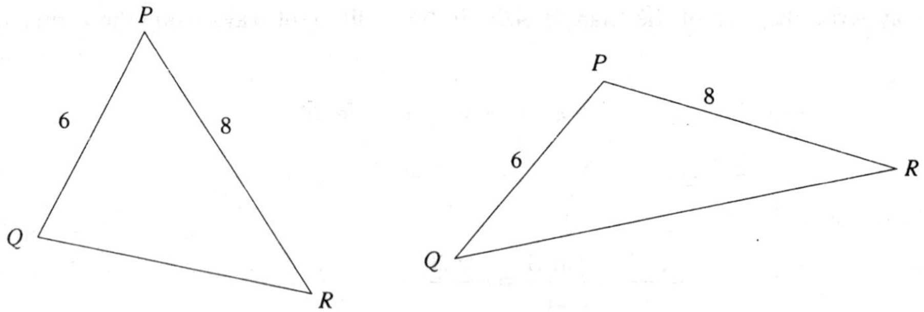

Fig. 4.6.3(c)

When $$P = 22.02°$$

$$p^2 = 6^2 + 8^2 - 2 \times 6 \times 8 \cos 22.02° = 11.00$$

i.e. one possible length of

$$QR = \sqrt{11.00} = 3.32 \text{ cm (3 sf)}$$

When $$P = 157.98°$$

$$p^2 = 6^2 + 8^2 - 2 \times 6 \times 8 \cos 157.98°$$
$$= 6^2 + 8^2 + 2 \times 6 \times 8 \cos 22.02°$$
$$= 189.00$$

i.e. the other possible length of $QR = \sqrt{189.00}$ cm.

$$= 13.7 \text{ cm (3sf)}$$

Another formulae for the area of a triangle, whose three sides are known, is given by

$$\sqrt{\{s(s-a)(s-b)(s-c)\}}$$

where $$2s = a + b + c$$

i.e. s is the semi-perimeter.

The proof of the formula is one of the problems in exercise IV. However it will be possible meanwhile for the reader to make applications of the formula when appropriate.

Example 4.6.3(4)

A surveyor measures the three sides of a triangular field XYZ as XY = 65.5 metres, YZ = 56.7 metres and ZX = 48.9 metres. Calculate the area of the field correct to the nearest square metre.

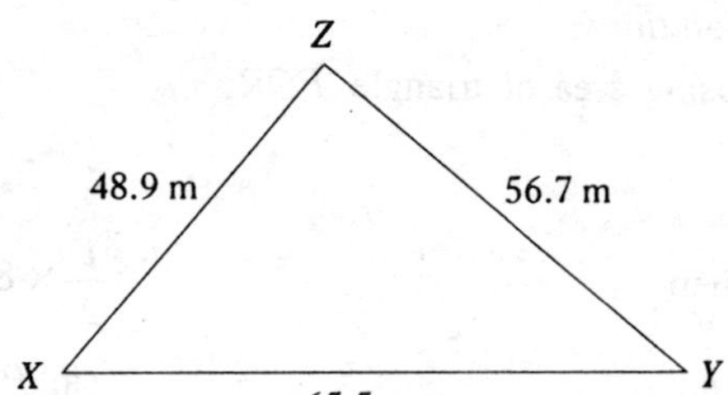

Fig. 4.6.3(d)

Solution:

From $z = 65.5$

$x = 56.7$

$y = 48.9$

$2s = x + y + z = 171.1$

so $s = 85.55$

required area

$$= \sqrt{\{85.55\ (85.55 - 65.5)\ (85.55 - 56.7)\ (85.55 - 48.9)\}}$$

$$= \sqrt{\{85.55 \times 20.05 \times 28.85 \times 36.65\}}$$

$$= \sqrt{1813\ 653}$$

$$= 1346.7\ \text{m}^2 \simeq 1347\ \text{metre}^2\ (\text{to nearest m}^2).$$

Example 4.6.3(5)

A triangle has one side of length 5 cm, perimeter 18 cm and area $6\sqrt{6}$ cm^2. Determine the lengths of the other two sides.

Solution:

If side $a = 5$ cm and side $b = x$ cm,

then as $a + x + c = 18.$

$5 + x + c = 18$

i.e. $c = (13 - x)$ cm

[see Fig. 4.6.3(e)]

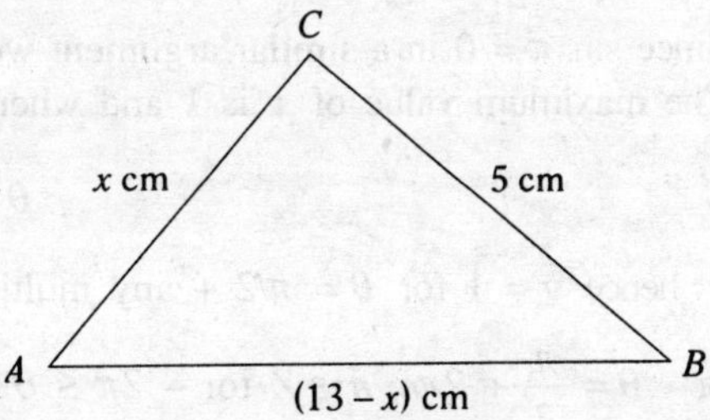

Fig. 4.6.3(e)

The area is given as $6\sqrt{6}$ cm^2.

so $6\sqrt{6} = \sqrt{\{9\ (9 - 5)\ (9 - x)\ (9 - (13 - x))\}}$

i.e. $6\sqrt{6} = \sqrt{\{9 \times 4\ (9 - x)\ (x - 4)\}}$

i.e. $216 = 36\ (9 - x)\ (x - 4)$

i.e. $6 = 9x - x^2 - 36 + 4x$

i.e. $x^2 - 13x + 42 = 0$

i.e. $(x - 6)\ (x - 7) = 0$

i.e. $x = 6$ or 7

If $b = 6$ cm, $c = 13 - 6 = 7$ cm

If $b = 7$ cm, $c = 13 - 7 = 6$ cm

So the other two sides are 6 cm, 7 cm respectively.

4.7 GRAPHS OF THE FUNCTIONS SIN θ, COS θ, TAN θ, AND OF THEIR INVERSE FUNCTIONS

Using the CAST diagram (as in 4.5.2) we know that there is a unique value for each trigonometric ratio of any angle. Of particular importance are the graphs of the three circular functions $y = \sin\theta$, $y = \cos\theta$ and $y = \tan\theta$. These will now be examined, mainly with θ measured in radians, but in later problems the degree measure will be used.

4.7.1 The Graph of the Function Sin θ

Example 4.7.1(1)

Sketch the graph of $y = \sin\theta$ for $-2\pi \le \theta \le 3\pi$.

Solution:

Since a sketch is required it will not be necessary to plot many points on graph paper and then join them. We shall concentrate on the points where the graph meets the axes, the points where the graph has a maximum or a minimum, and also where appropriate, any asymptotes (this will be explained when we look at example 4.7.3 (1)

When $y = 0$ then $\sin\theta = 0$

i.e. $\theta = \ldots, -2\pi, -\pi, 0\ \pi, 2\pi, 3\pi, \ldots$

This is based on the knowledge that $\sin\theta = 0$ and so $\theta = 0 +$ any multiple of 2π

e.g.

$$\sin 2\pi = 0$$

$$\sin 4\pi = 0 \qquad \text{(see Fig. 4.7.1(a))}$$

Since $\sin\pi = 0$ in a similar argument we may say that $\sin\theta = 0$ when $\theta = \pi +$ any multiple of 2π.

The maximum value of y is 1 and when $\sin\theta = 1$,

$$\theta = \frac{\pi}{2} \text{ for } 0 \le \theta \le 2\pi$$

hence $y = 1$ for $\theta = \pi/2 +$ any multiple of 2π

or $\theta = \dfrac{\pi}{2} + 2\pi n,\ n \in Z$ for $-2\pi \le \theta \le 3\pi$

$y = 1$ for $\theta = \dfrac{-3\pi}{2}, \dfrac{\pi}{2}, \dfrac{5\pi}{2}$ (See Fig. 4.7.1(b))

The minimum value of y is -1 and when

$$\sin\theta = -1$$

$$\theta = \frac{3\pi}{2} \text{ for } 0 \le \theta \le 2\pi$$

hence $y = -1$ for $\theta = \dfrac{3\pi}{2} +$ any multiple of 2π

or $\dfrac{3\pi}{2} + 2\pi n,\ n \in Z$

for $-2\pi \le \theta \le 3\pi$ $y = -1$ for $\theta = -\dfrac{\pi}{2}, \dfrac{3\pi}{2}$

(See Fig. 4.7.1(c))

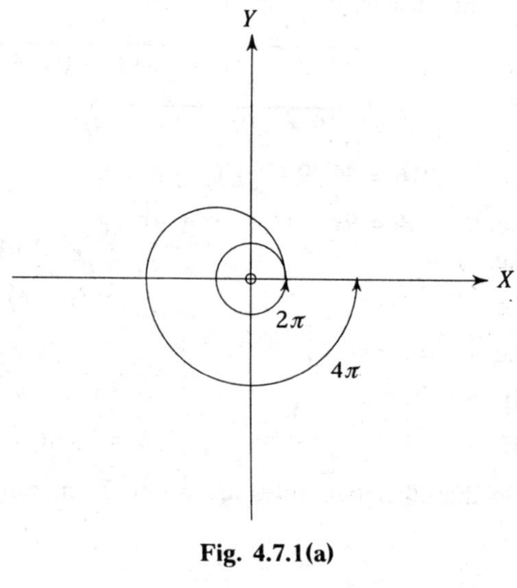

Fig. 4.7.1(a)

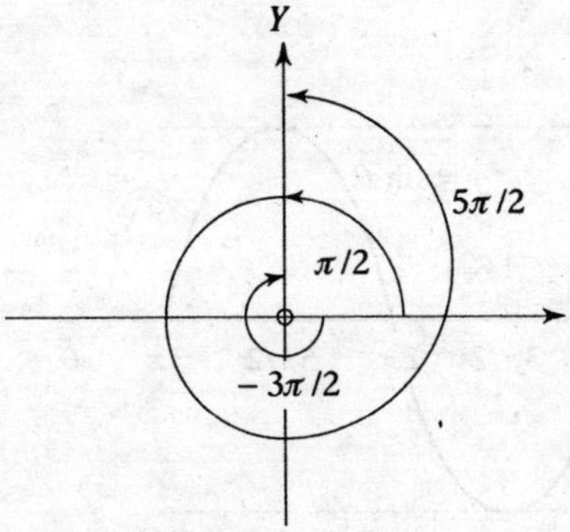

Fig. 4.7.1(b)

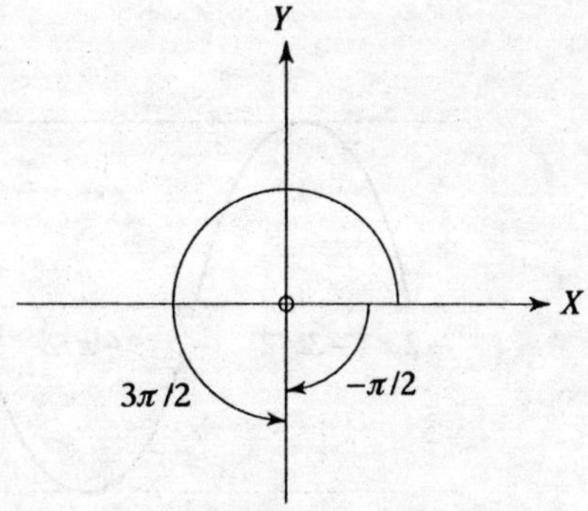

Fig. 4.7.1(c)

For the graph of $y = \sin\theta$, θ will be on the horizontal axis (x-axis) and we know that there are certain points through which the graph will pass. These are the points of the θ-axis

i.e. $(-2\pi, 0), (-\pi, 0), (0, 0), (\pi, 0), (2\pi, 0), (3\pi, 0)$

and the points where y has a maximum

i.e. $$\left(-\frac{3\pi}{2}, 1\right), \left(\frac{\pi}{2}, 1\right), \left(\frac{5\pi}{2}, 1\right)$$

and the points where y has a minimum

i.e. $$\left(-\frac{\pi}{2}, -1\right), \left(\frac{3\pi}{2}, -1\right)$$

Mark all the points where the graph crosses the axes with a cross and the maximum and minimum points with a short horizontal line segment (as the gradient of the graph at these points will be zero). See Fig. 4.7.1(d).

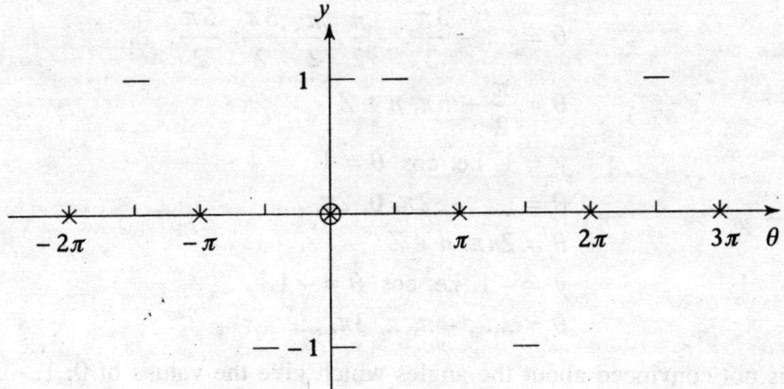

Fig. 4.7.1(d)

Now follow with a pencil or a pen the shape of the curve and the sketch will appear as in Fig. 4.7.1(e).

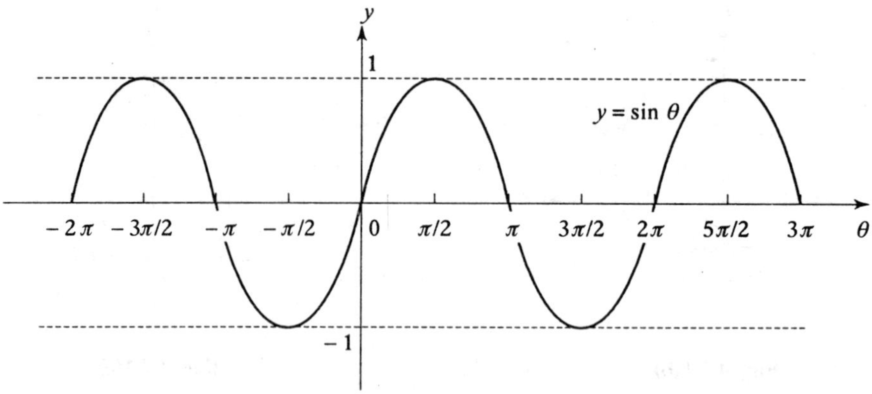

Fig. 4.7.1(e)

Note (i) the graph is *continuous* i.e. there are *no* breaks in the graph,

(ii) the range of values of y is $-1 \leq y \leq 1$ (or $-1 \leq \sin\theta \leq 1$)

(iii) the slope repeats itself after every 2π radians along the graph and this repeating width (in the case of the graph of $y = \sin\theta$ the width $= 2\pi$) is called the *Period* of the graph.

4.7.2 The Graph of the Function cos θ

Example 4.7.2(1)

Sketch the graph of $y = \cos\theta$ for $-2\pi \leq \theta \leq 3\pi$

Solution:

As for the graph of $y = \sin\theta$ we concentrate on the points where the graph meets the axes, and also the points where the graph has maximum or a minimum.

When $y = 0$, i.e. $\cos\theta = 0$

$$\theta = \ldots\ldots -\frac{3\pi}{2}, -\frac{\pi}{2}, \frac{\pi}{2}, \frac{3\pi}{2}, \frac{5\pi}{2}, \ldots\ldots$$

i.e. $\theta = \frac{\pi}{2} + n\pi,\ n \in Z$

When $y = 1$, i.e. $\cos\theta = 1$

$$\theta = \ldots\ldots, -2\pi, 0, 2\pi, \ldots\ldots$$

or $\theta = 2n\pi,\ n \in Z$

When $y = -1$, i.e. $\cos\theta = -1$,

$$\theta = \ldots\ldots, -\pi, \pi, 3\pi, \ldots\ldots$$

The reader who is not convinced about the angles which give the values of 0, 1, − 1 respectively to $\cos\theta$ should sketch axes as in Fig. 4.7.1(a), (b) and (c).

Marking the points where the graph crosses the horizontal axis and also the points which give maximum and minimum values of $\cos\theta$ (1 and − 1 respectively) and then joining the points in a curve will produce a sketch graph as in Fig. 4.7.2(a).

NB as for the graph of $y = \sin\theta$, it should be observed that the graph is *continuous*, the range of values *is* $-1 \leq y \leq 1$ (or $-1 \leq \cos\theta \leq 1$), and it is periodic with period 2π. Note also that the graph

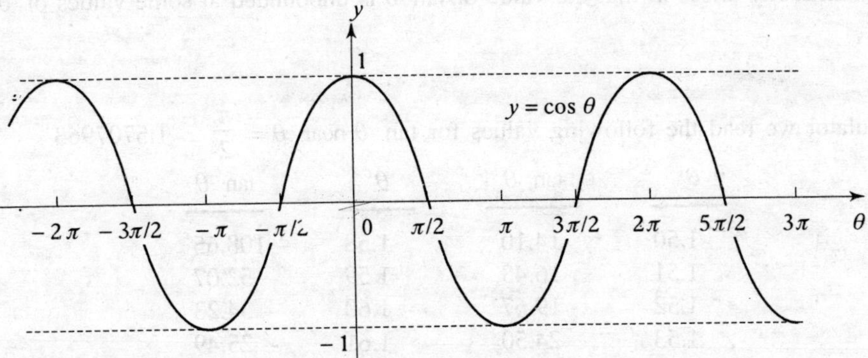

Fig. 4.7.2(a)

of $y = \cos\theta$ has the same slope as the graph of $y = \sin\theta$ [shown in Fig. 4.7.1(e)] but that the graph of $y = \sin\theta$ has been translated a distance of $\frac{\pi}{2}$ to the left, or given the translation $\begin{pmatrix} -\pi/2 \\ 0 \end{pmatrix}$ to give the graph of $y = \cos\theta$.

Such a displacement is known as a phase angle or phase difference or phase shift.

Some readers might have observed that other translations, such as $\begin{pmatrix} 3\pi/2 \\ 0 \end{pmatrix}$ or $\begin{pmatrix} -5\pi/2 \\ 0 \end{pmatrix}$ would also take the graph of $y = \sin\theta$ onto the graph of $y = \cos\theta$ and in general a translation $\begin{pmatrix} -\pi/2 + 2n\pi \\ 0 \end{pmatrix}$ where $n \in Z$.

In fact reflections in lines such as $x = \frac{\pi}{4}$, $x = \frac{5\pi}{4}$, or in general $x = \frac{\pi}{4} + n\pi$, would also take the graph of $y = \sin\theta$ onto the graph of $y = \cos\theta$.

4.7.3 The Graph of the Function Tan θ

Example 4.7.3(1)

Sketch the graph of $y = \tan\theta$ for $-7\pi/4 \le \theta \le 3\pi$

Solution:

We may note a few points which lie on the curve

θ	$-\pi$	$-3\pi/4$	$-\pi/4$	0
$\tan\theta$	0	1	–1	0

θ	$\pi/4$	$3\pi/4$	π	$5\pi/4$
$\tan\theta$	1	–1	0	1

However a difficulty arises in that the value of tan θ is unbounded at some values of θ, such as $\theta = \frac{\pi}{2}$.

Using a calculator we read the following values for tan θ near $\theta = \frac{\pi}{2} \simeq 1.5707963$

θ^c	tan θ	θ^c	tan θ
1.50	14.10	1.58	– 108.65
1.51	16.43	1.59	– 52.07
1.52	19.67	1.60	– 34.23
1.53	24.50	1.61	– 25.49
1.54	32.46	1.62	– 20.31
1.55	48.08	1.63	– 16.87
1.56	92.62	1.64	– 14.43
1.57	1255.77	1.65	– 12.60
1.571	– 4909.83		
1.575	– 237.89		

As θ approaches $\frac{\pi}{2}$ on the left i.e. $\theta \to \frac{\pi}{2}-$ we can make θ as close to $\frac{\pi}{2}$ as we like but a little less than $\frac{\pi}{2}$, and we observe that tan θ becomes larger and larger on the positive side, i.e. we can make tan θ as large as we like by taking θ as close to $\frac{\pi}{2}$ (on the left) as necessary,

i.e. as $\theta \to \frac{\pi}{2}-$ we say that tan $\theta \to \infty$.

However when $\pi/2 < \theta < \pi$, θ is in the second quadrant where tan $\theta < 0$.

If we take values of θ just greater than $\frac{\pi}{2}$ we observe that the values of tan θ are very large on the negative side. We say that as $\theta \to \frac{\pi}{2}+$ (i.e. θ approaches $\frac{\pi}{2}$ on the right) then tan θ becomes larger and larger on the negative side, and we say that : as $\theta \to \frac{\pi}{2}+$, tan $\theta \to -\infty$.

It will be helpful to complete a table of values for points which lie on the curve and also to note what is happening at points such as $\theta = \ldots, \frac{-3\pi}{2}, -\frac{\pi}{2}, \frac{\pi}{2}, \frac{3\pi}{2}, \frac{5\pi}{2}, \ldots$ at which tan θ is unbounded.

θ	$-7\pi/4$	$-3\pi/2$	$-3\pi/2+$	$-5\pi/4$	$-\pi$
tan θ	1	$+\infty$	$-\infty$	-1	0

θ	$-3\pi/4$	$-\pi/2-$	$-\pi/2+$	$-\pi/4$	0
tan θ	1	$+\infty$	$-\infty$	-1	0

θ	$\pi/4$	$\pi/2-$	$\pi/2+$	$3\pi/4$	π
$\tan\ \theta$	1	$+\infty$	$-\infty$	-1	0

θ	$5\pi/4$	$3\pi/2-$	$3\pi/2+$	$7\pi/4$	2π
$\tan\ \theta$	1	$+\infty$	$-\infty$	-1	0

θ	$9\pi/4$	$5\pi/2-$	$5\pi/2+$	$11\pi/4$	3π
$\tan\ \theta$	1	$+\infty$	$-\infty$	-1	0

The graph is now shown in Fig. 4.7.3(a).

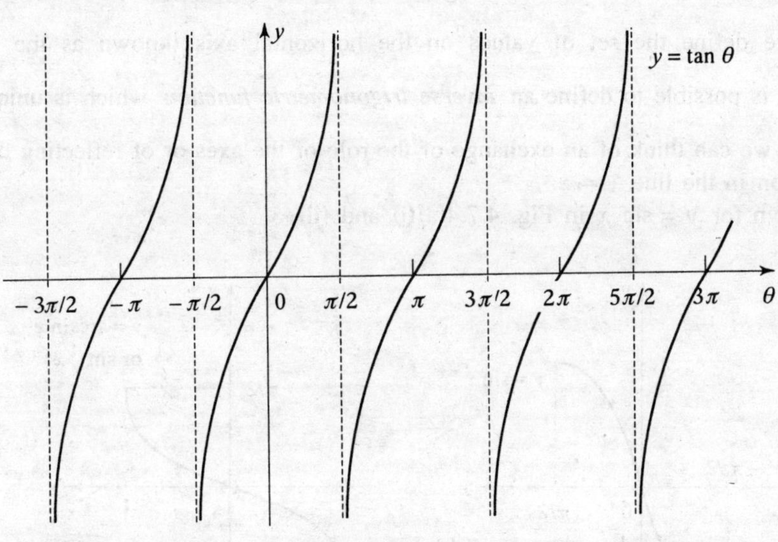

Fig. 4.7.3(a)

N.B. for the curve $y = \tan\ \theta$

(i) the curve is *not* continuous being *undefined* when $\theta = ..., -\dfrac{\pi}{2}, \dfrac{\pi}{2}, \dfrac{3\pi}{2}, ...,$ and the vertical lines $.... x = -\dfrac{\pi}{2}, x = \dfrac{\pi}{2}, x = \dfrac{3\pi}{2}$, are known as *asymptotes* in the sense that the branches of the curves get closer to the lines but never actually touch the lines (if you like these lines could be tangent lines to the branches at $y = -\infty$ or $y = +\infty$),

(ii) when sketching the curve no attempt should be made to join adjacent branches, which should all be kept separate,

(iii) the range of possible values for $\tan\ \theta$ is unlimited, i.e. $-\infty < \tan\ \theta < \infty$,

(iv) the function is periodic but in this case the period is π (*Not* 2π as in the case of the graphs of $y = \sin\ \theta$ and $y = \cos\ \theta$).

The reader will have observed that tan $\theta = 0$ at $\theta = -3\pi, -2\pi, -\pi, 0, \pi, 2\pi, 3\pi$, or at $\theta = n\pi$, $n \varepsilon Z$ and that tan $\theta \to +\infty$

$$\text{as } \theta \to (2n + 1)\ \pi/2\ -,\ n\ \varepsilon\ Z$$

$$\tan\ \theta \to -\infty \text{ as } \theta \to (2n + 1)\pi/2\ +,\ n\ \varepsilon\ Z.$$

The reader who is still uncertain about the behaviour of tan θ near $\theta = \pi/2$ may prefer to tabulate values for tan 89.95°, tan 89.96°, tan 89.97°, tan 89.98°, tan 89.99°, tan 89.999°, tan 89.9999°, and corresponding readings on the other side of 90° if further investigation is desired.

4.7.4 Graphs of Inverse Trigonometric Functions

As the reader will have observed from the graph of $y = \sin\ \theta$ in part 4.7.1 there are many angles for which the value of sin θ is the same e.g. for sin $\theta = -1$ we saw that $\theta = -\dfrac{\pi}{2} + 2n\pi,\ n\ \varepsilon\ Z$. However if we define the set of values on the horizontal axis (known as the domain) to be $-\dfrac{\pi}{2} \le \theta \le \dfrac{\pi}{2}$ it is possible to define an *inverse trigonometric function* which is unique.

Graphically we can think of an exchange of the role of the axes or of reflecting the graph of the original function in the line $y = x$.

This is shown for $y = \sin\ x$ in Fig. 4.7.4(a)(i) and (ii).

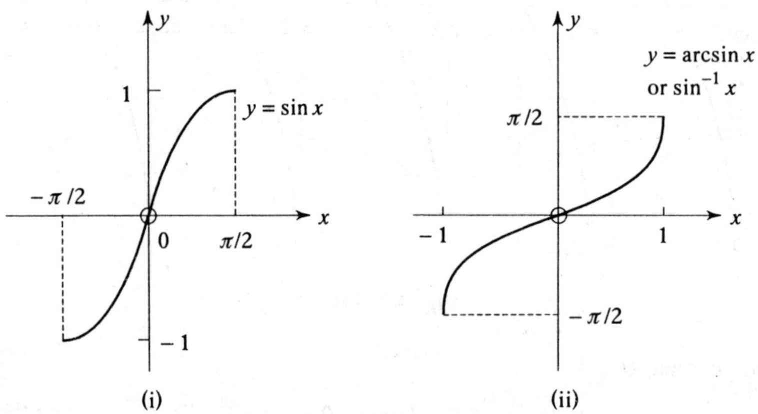

Fig. 4.7.4(a)

For $y = \sin x$ we can think of the input as an angle (here it is measured in radians) and the output is a number (the value of the sine ratio for that angle).

If we define $y = f(x) = \sin\ x$ for domain $-\dfrac{\pi}{2} \le x \le \dfrac{\pi}{2}$ which has a range (on the readings on the vertical axis) $-1 \le y \le 1$ then the inverse function denoted by $f^{-1}(x)$ or arc sin x or $\sin^{-1} x$ (the angle whose sine is x) has a domain for x of $-1 \le x \le 1$ and a range for y of $-\dfrac{\pi}{2} \le y \le \dfrac{\pi}{2}$ (the reader should check Fig. 4.7.4(a)(i) and (ii) again for the positions of the axes).

N.B: The use of the '– 1' is a special notation for an inverse function in the present context, where arc sin x or $\sin^{-1} x$ denotes the angle whose sine is x, and should *not* confused with cosec $x = \dfrac{1}{\sin x}$ which must be written as $(\sin\ x)^{-1}$ if the index notation is used.

Example 4.7.4(1)

Determine a suitable domain for the function $f(x) = \cos\ x$ so as to determine $f^{-1}(x)$. Sketch the graphs of the function and its inverse, showing clearly the domain and range in each case.

Solution:

Take $y = \cos\ x$ restricted to domain $0 \le x \le \pi$ and for which the range is $-1 \le y \le 1$ but note that $\cos 0 = 1$ and $\cos\ \pi = -1$.

As will be seen in Fig. 4.7.4(b)(i) given any value of $y = \cos\ x$, say $\cos\ x = -\dfrac{1}{2}$, there will be one and only value of x, i.e. $x = \dfrac{2\pi}{3}$, for which $y = -1/2$, on the domain which has been restricted to $0 \le x \le \pi$.

For $f(x) = \cos\ x$ on domain $0 \le x \le \pi$.

We can then define $f^{-1}(x)$, denoted by arc cos x or $\cos^{-1} x$ (i.e. the angle whose cosine is x).

If $f(x) = \cos\ x$ for domain $0 \le x \le \pi$ and range $-1 \le y \le 1$

then $f^{-1}(x) = \cos^{-1} x$ for domain $-1 \le x \le 1$ and range $0 \le y \le \pi$

As before it can be thought of as an exchange of the role of the axes, or of reflecting the graph of the original function in the line $y = x$, to obtain the graph of $f^{-1}(x)$.

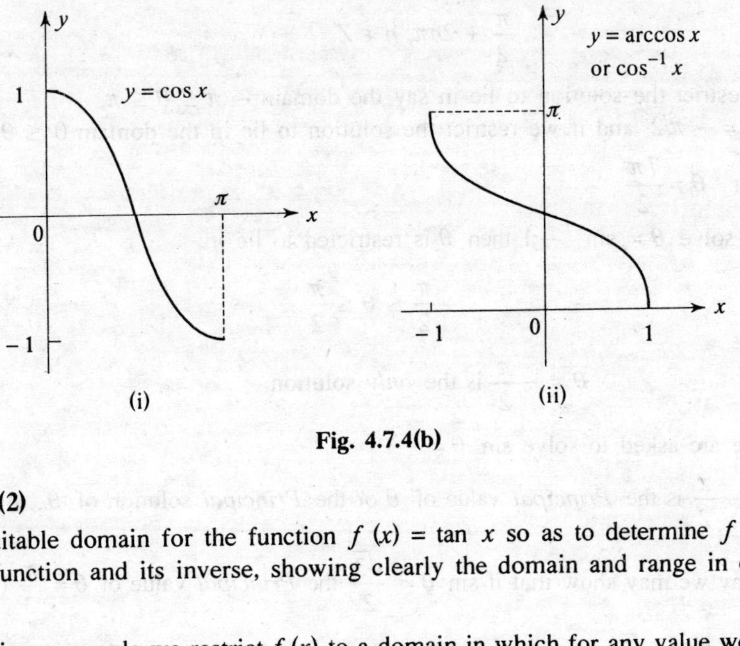

Fig. 4.7.4(b)

Example 4.7.4(2)

Determine a suitable domain for the function $f(x) = \tan\ x$ so as to determine $f^{-1}(x)$. Sketch the graphs of the function and its inverse, showing clearly the domain and range in each case.

Solution:

As for the previous example we restrict $f(x)$ to a domain in which for any value we choose for $f(x)$ there will be one and only one value of x. If we restrict $f(x)$ to $-\pi/2 < x < \pi/2$ there will be a *unique* inverse $f^{-1}(x)$ = arc tan x or $\tan^{-1} x$ (the angle whose tangent is x). The graphs of $y = \tan x$ and of $y = \tan^{-1} x$ are shown in Fig. 4.7.4(c).

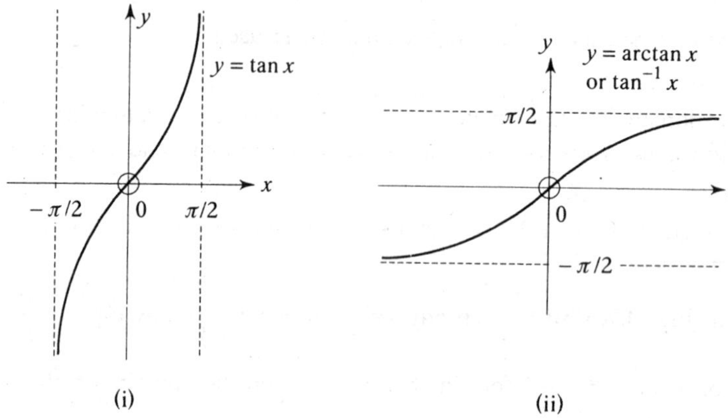

Fig. 4.7.4(c)

4.8 INTRODUCTION TO SOLUTION OF TRIGONOMETRIC EQUATIONS

We have already observed from the graph of $y = \sin\theta$ that the equation $\sin\theta = 1$ has infinitely many solutions ..., $\frac{-5\pi}{2}, \frac{-3\pi}{2}, \frac{\pi}{2}, \frac{3\pi}{2}, \frac{7\pi}{2}$

i.e. $$\frac{\pi}{2} + 2n\pi,\ n \in Z.$$

However if we restrict the solution to lie in say the domain $-\pi \le \theta \le \pi$.
We have only $\theta = -\pi/2$, and if we restrict the solution to lie in the domain $0 \le \theta \le 4\pi$, we have only $\theta = \frac{3\pi}{2}$ or $\theta = \frac{7\pi}{2}$.

If we wish to solve $\theta = \sin^{-1} -1$ then θ is restricted to lie in

$$-\frac{\pi}{2} \le \theta \le \frac{\pi}{2}$$

and $$\theta = -\frac{\pi}{2} \text{ is the } \textit{only} \text{ solution.}$$

In this case if we are asked to solve $\sin\theta = -1$

we say that $\theta = -\frac{\pi}{2}$ is the *Principal* value of θ or the *Principal* solution of θ.

In a similar way we may show that if $\sin\theta = \frac{\sqrt{3}}{2}$ the *Principal* value of $\theta = \frac{\pi}{3}$ (this is the only value for which $\sin\theta = \frac{\sqrt{3}}{2}$ and $-\frac{\pi}{2} \le \theta \le \frac{\pi}{2}$) and that of $\cos\theta = -\frac{1}{\sqrt{2}}$ the *Principal* value of

$\theta = \dfrac{3\pi}{4}$ (as it must lie in the domain $0 \le \theta \le \pi$) and that if $\tan\theta = -1$ the *Principal* value of θ $= -\dfrac{\pi}{4}$ (as it must lie in the domain $-\dfrac{\pi}{2} \le \theta \le \dfrac{\pi}{2}$).

4.8.1 Secondary Values in Solution of Trigonometric Equations

Now we shall solve the same three equations in the domain $-\pi \le \theta \le \pi$

(i) $\sin\theta = \dfrac{\sqrt{3}}{2}$.

See Fig. 4.8.1(a). There are two possible angles $\theta = \dfrac{\pi}{3}$, $\theta = \dfrac{2\pi}{3}$ using the *Cast* diagram.

The *Principal* value is $\theta = \dfrac{\pi}{3}$ and the other value (here $\theta = \dfrac{2\pi}{3}$) is known as the *Secondary* value.

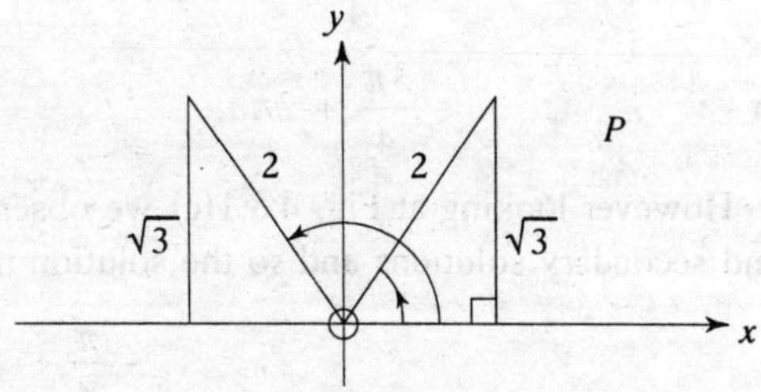

Fig. 4.8.1(a)

Clearly either of these angles plus any multiple of 2π will give the same value $\dfrac{\sqrt{3}}{2}$ to $\sin\theta$, so the general solution of $\sin\theta = \dfrac{\sqrt{3}}{2}$ is

$$\frac{\pi}{3} + 2n\pi$$

$$\text{or} \quad \frac{2\pi}{3} + 2n\pi, \qquad n \in Z.$$

(ii) $\cos\theta = -\dfrac{1}{\sqrt{2}}$

See Fig. 4.8.1(b).

There is a secondary value in the third quadrant using the *Cast* diagram. So there are *two* possible values in the domain $-\pi \le \theta \le \pi$

i.e. $\dfrac{3\pi}{4}$ or $\dfrac{-3\pi}{4}$

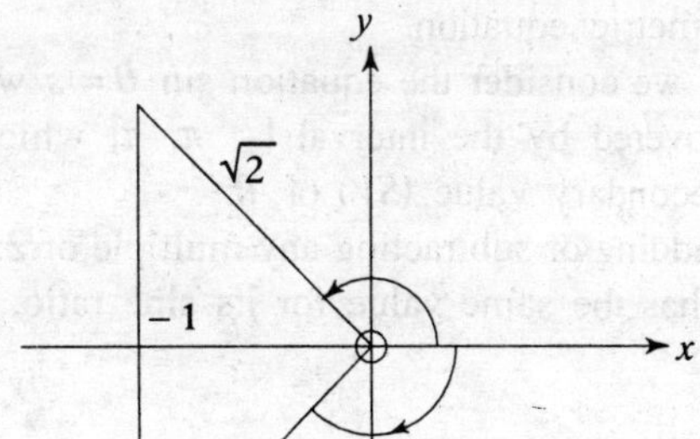

Fig. 4.8.1(b)

This can be written as $\pm\dfrac{3\pi}{4}$. In the same way as in (i) we may add multiples of 2π to either angle and find the cosine function will still take the value $-\dfrac{1}{\sqrt{2}}$, hence we may give a general solution in the form

$$\pm\frac{3\pi}{4} + 2\pi n, \; n \in Z.$$

(iii) $\tan\ \theta = -1$. See Fig. 4.8.1(c).

There are *two* possible values in the domain $-\pi \le \theta \le \pi$,

of which $\dfrac{3\pi}{4}$ is the secondary value.

These two values are $-\dfrac{\pi}{4}$ and $\dfrac{3\pi}{4}$. Adding suitable multiples of 2π the general solution can be given as

$$-\frac{\pi}{4} + 2\pi n$$

or $$\frac{3\pi}{4} + 2\pi n, \qquad n \in Z$$

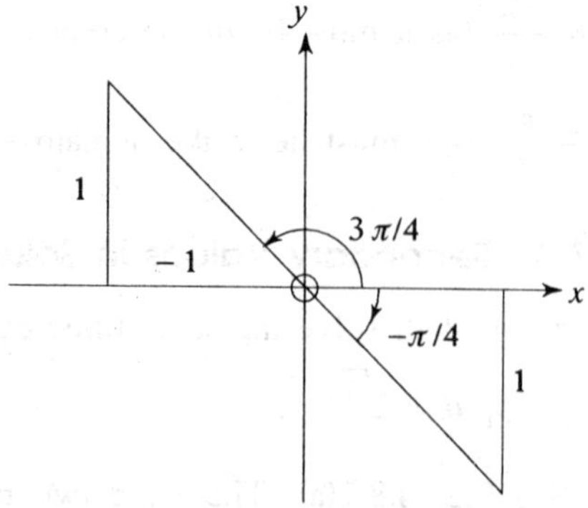

Fig. 4.8.1(c)

However looking at Fig. 4.8.1(c) we observe that there is a difference of π between the principal and secondary solutions and so the solution may be combined to give a general solution in the form

$$-\frac{\pi}{4} + \pi k,\ k \in Z.$$

4.8.2 Graphical Representation of the Principal and Secondary Solutions of Trigonometric Equations

We may use the sketch graphs of $y = \sin\ \theta$, $y = \cos\ \theta$, $y = \tan\ \theta$, to show how to use the principal value, and if necessary the secondary value, to enable us to find the general solution of a trigonometric equation.

(i) If we consider the equation $\sin\ \theta = s$ where $|s| \le 1$ the period 2π of the sine function will be covered by the interval $[-\pi, \pi]$ which will include both the principal value (*PV*) and the secondary value (*SV*) of θ.

By adding or subtracting any multiple of 2π to either the *PV* or the *SV* we can obtain another angle which has the same value for its sine ratio. See Fig. 4.8.2(a).

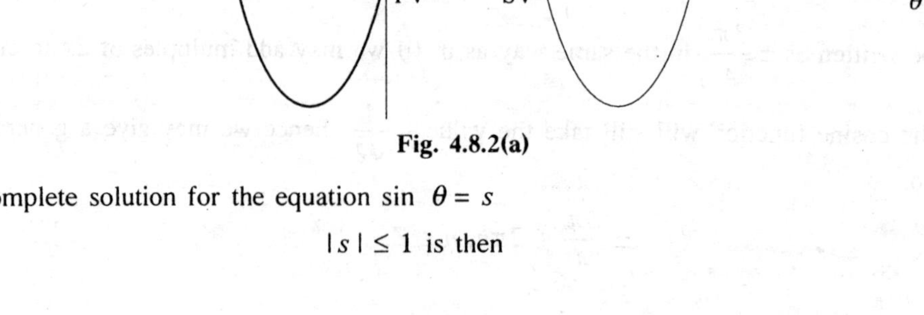

Fig. 4.8.2(a)

The complete solution for the equation $\sin\ \theta = s$

where $\qquad |s| \le 1$ is then

$$\theta = PV + 2n\pi$$

or
$$\theta = SV + 2n\pi, \quad n \in Z$$

or in degrees

$$\theta = PV + 360n°$$

or
$$\theta = SV + 360n°, \quad n \in Z$$

(ii) The same situation arises in considering the equation cos $\theta = c$ where $|c| \le 1$, since both the PV and the SV of the cosine function will occur within one period, which is 2π. See Fig. 4.8.2(b).

The complete solution of the equation cos $\theta = c$ where $|c| \le 1$ is also given by adding or subtracting multiples of 2π to either the PV or the SV. But for cosine the PV and the SV are numerically equal but of opposite sign (i.e. $SV = -PV$).

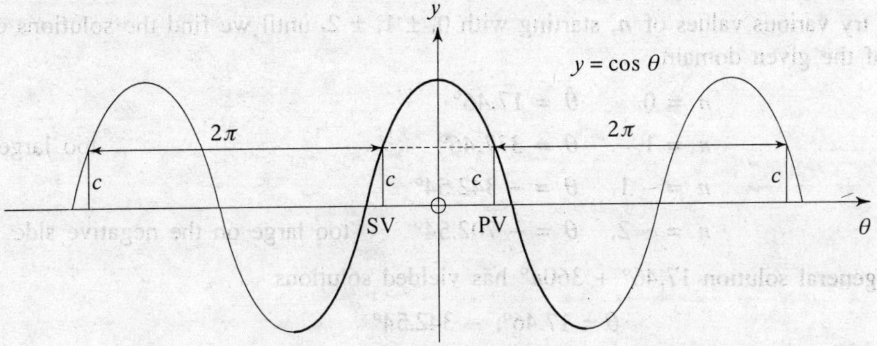

Fig. 4.8.2(b)

The general solution of the equation cos $\theta = c$ can be given in the form

$$\theta = \pm PV + 2n\pi \text{ or } \pm PV + 360n°, \qquad n \in Z.$$

(iii) For the equation tan $\theta = t$, we already know that only the PV will be included in the domain $\left(-\dfrac{\pi}{2}, \dfrac{\pi}{2}\right)$, which covers a complete period.

But all further angles θ for which the tangent value is t can be found by combining multiples of the period π, with the PV. See Fig. 4.8.2(c).

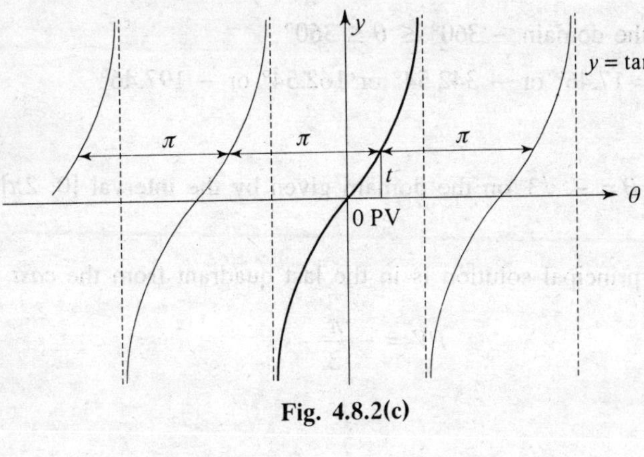

Fig. 4.8.2(c)

The general solution of the equation tan $\theta = t$ is

$$\theta = PV + n\pi \text{ or } PV + 180n°, \qquad n \varepsilon Z.$$

We shall now apply the above principles to particular trigonometric equations.

Example 4.8.2(1)

Find all angles θ in the domain $-360° \le \theta \le 360°$ for which sin $\theta = 0.3$.

Solution:

The Principal solution is 17.46°. By *Cast* diagram the secondary solution is in the second quadrant

i.e. $$180° - 17.46° = 162.54°.$$

One general solution is $17.46° + 360n°$

We now try various values of n, starting with 0, ± 1, ± 2, until we find the solutions exceeding the limits of the given domain.

$n = 0, \quad \theta = 17.46°$

$n = 1, \quad \theta = 377.46°$ too large

$n = -1, \quad \theta = -342.54°$

$n = -2, \quad \theta = -702.54°$ too large on the negative side

So far the general solution $17.46° + 360n°$ has yielded solutions

$$\theta = 17.46°, \ -342.54°$$

The other general solution is $162.54° + 360n°$

$n = 0, \quad \theta = 162.54°$

$n = 1, \quad \theta = 522.54°$ too large

$n = -1, \quad \theta = -197.46°$

$n = -2, \quad 1 = -557.46°$ too large on the negative side

So the other general solution $162.54° + 360n°$ has yielded solutions

$$\theta = 162.54°, \ -197.46°.$$

Hence the solution on the domain $-360° \le \theta \le 360°$

is $$\theta = 17.46° \text{ or } -342.54° \text{ or } 162.54° \text{ or } -197.46°$$

Example 4.8.2(2)

Solve the equation tan $\theta = -\sqrt{3}$ on the domain given by the interval $[0, 2\pi]$.

Solution:

For tan $\theta = -\sqrt{3}$ the principal solution is in the last quadrant from the *cast* diagram.

i.e. $$PV = -\frac{\pi}{3}$$

and the general solution will be

$$-\frac{\pi}{3}+n\pi, \qquad n\ \varepsilon\ Z$$

for $n=0,\ \theta=-\dfrac{\pi}{3}<0$ not in domain

$$n=1,\ \theta=-\frac{\pi}{3}+\pi=\frac{2\pi}{3}$$

$$n=2,\ \theta=-\frac{\pi}{3}+2\pi=\frac{5\pi}{3}$$

$$n=3,\ \theta=-\frac{\pi}{3}+3\pi=\frac{8\pi}{3}>2\pi,\ \text{not in domain}$$

hence

$$\theta=\frac{2\pi}{3}\ \text{or}\ \frac{5\pi}{3}.$$

Example 4.8.2(3)

Find the angles in the domain given by the interval [$-360°, 0°$] which satisfy the equation $\cos\theta = 0.6$.

Solution:

Using the *Cast* diagram the Principal Solution will be in the first quadrant and the secondary solution will be in the last quadrant.

The *PV* is 53.13°

and the *SV* is $-53.13°$

with a general solution

$$\pm 53.13+360n°$$

$n=0,\ \theta=\pm 53.13°$ but $53.13°>0°$ not in domain

$n=-1,\ \theta=-306.87°$ or $-413.13°$ but $-413.13°<-360°$ not in domain.

$n=-2,\ \theta=-666.87°$ or $-773.13°$

both of these angles are less than $-360°$, so the required angles are

$$\theta=-53.13° \text{ or } \theta=-306.87°.$$

Example 4.8.2(4)

Solve for $0° \le \theta \le 360°$ the equation

$$\sin\theta+4\sin\theta\cos\theta=0$$

Solution:

Factorise the equation as

$$\sin\theta\,(1+4\cos\theta)=0$$

which means $\sin\theta=0$ (do *not* omit this possibility otherwise some solutions will be missing) or $$\cos\theta = -\frac{1}{4}$$

$\sin\theta = 0$ has a general solution $180n°$ which gives 0°, 180°, 360° in the given domain.

$$\cos\theta = -\frac{1}{4}$$

has principal solution 104.48° in the second quadrant and secondary solution – 104.48° in the third quadrant so the general solution is

$$\pm 104.48° + 360n°$$

$$n = 0,\ \theta = 104.48° \text{ or } -104.48° < 0° \quad \text{not in domain}$$

$$n = 1,\ \theta = 464.48° > 360° \quad \text{not in domain}$$

or $$\theta = 255.52°$$

$$n = 2,\ \theta = 824.48° > 360°$$

or $$\theta = 615.52° > 360° \quad \text{both of these are not in the domain}$$

hence the required solutions are $\theta = 0°$ or 180° or 360° or 104.48° or 255.52°.
N.B: If we had cancelled out the factor $\sin\theta$ without considering the possibility of $\sin\theta = 0$ we would have missed the first three solutions in our answer.

Example 4.8.2(5)
Find the general solution of

(a) $$\tan\theta = \frac{1}{\sqrt{3}}$$

(b) $$\tan\theta = -0.5$$

Solution:
(a) The *PV* from

$$\tan\theta = \frac{1}{\sqrt{3}} \text{ is } \theta = \frac{\pi}{6}$$

so the general solution is

$$\theta = \frac{\pi}{6} + \pi n, \quad n \in Z$$

(b) Using degree measure the *PV* from $\tan\theta = -0.5$ is $-26.57°$
so the general solution is

$$\theta = -26.57° + 180n°. \quad n \in Z$$

Example 4.8.2(6)
Find the general solution of

(a) $$\cos\theta = \frac{\sqrt{3}}{2}$$

(b) $$\cos\theta = -\frac{1}{2}$$

Solution:

(a) The *PV* from the equation $\cos\,\theta = \dfrac{\sqrt{3}}{2}$ is $\theta = \dfrac{\pi}{6}$

and hence the general solution is

$$\pm\, PV + 2\pi n$$

i.e.

$$\theta = \pm\,\frac{\pi}{6} + 2\pi n, \qquad n \,\varepsilon\, Z$$

(b) The *PV* from the equation $\cos\,\theta = -\dfrac{1}{2}$ is $\theta = \dfrac{2\pi}{3}$ and hence the general solution is $\pm\, PV + 2\pi n$

i.e.

$$\theta = \pm\,\frac{2\pi}{3} + 2\pi n, \qquad n \,\varepsilon\, Z.$$

Example 4.8.2(7)

Find the general solution of the equation $\sin\,\theta = -\dfrac{1}{2}$

Solution:

The *PV* from the equation $\sin\,\theta = -\dfrac{1}{2}$ is $-\dfrac{\pi}{6}$

And for $-\pi \le \theta \le \pi$ the *SV* is $\dfrac{-5\pi}{6}$

hence the general solution will be

$$\theta = -\frac{\pi}{6} + 2\pi n$$

or

$$\theta = \frac{-5\pi}{6} + 2\pi n, \qquad n \,\varepsilon\, Z$$

N.B: In this case if the reader chose to take angles in $[0, 2\pi]$ as the starting angles the solutions would have read

$$\theta = \frac{7\pi}{6} + 2\pi n \text{ or } \theta = \frac{11\pi}{6} + 2\pi n,\; n \,\varepsilon\, Z.$$

and these solutions would have given the same set of angles.

Example 4.8.2(8)

Find the general solution of the equation

$$8\sin\,\theta\,\tan\,\theta + 4\tan\,\theta + 6\sin\,\theta + 3 = 0.$$

Solution:

Using factors by grouping

$$8\sin\,\theta\,\tan\,\theta + 4\tan\,\theta + 6\sin\,\theta + 3 = 0$$

gives

$$4\tan\,\theta\,(2\sin\,\theta + 1) + 3(2\sin\,\theta + 1) = 0$$

i.e.

$$(2\sin\,\theta + 1)(4\tan\,\theta + 3) = 0$$

i.e.

$$\sin\,\theta = -\frac{1}{2} \text{ or } \tan\,\theta = -\frac{3}{4}$$

For

$$\sin\,\theta = -\frac{1}{2}$$

PV is – 30° (last quadrant) and *SV* is – 150° (third quadrand) so the general solution is

$$\theta = -30° + 360n° \quad \text{or} \quad \theta = -150° + 360°\,n, \qquad n \in Z$$

For $\tan\theta = -\frac{3}{4}$ *PV* is – 37.87°, so the general solution is

$$\theta = -36.87° + 180n°, \qquad n \in Z$$

hence the complete general solution is

$$\theta = -30° + 360n°$$

or
$$\theta = -150° + 360n°$$

or
$$\theta = -36.87° + 180n°, \quad n \in Z$$

as an alternative, the solution

$$\theta = 210° + 360n°$$

or
$$\theta = 330° + 360n°$$

or
$$\theta = 143.13° + 180n°$$

would give the same set of angles.

4.8.3 Trigonometric Equations Involving Multiple Angles

Often we meet equations of the form $\cos 3\theta = \frac{\sqrt{3}}{2}$ or $\tan 4\theta = -3$, and the methods used in subsection 4.8.2 will now be applied first to the multiple angle, from which it will be possible to determine all the values of the angle required.

Example 4.8.3(1)

Find the general solution of the equation $\cos 3\theta = \frac{\sqrt{3}}{2}$.

Solution:

A common mistake is to think that $\cos 3\theta$ is the same as $3\cos\theta$ and so the equation can be written as $\cos\theta = \frac{\sqrt{3}}{6}$. *This is wrong,*

$\cos 3\theta$ is not the same as $3\cos\theta$ e.g. if $\theta = 30°$

$$\cos 3\theta = \cos 90° = 0$$

and
$$3\cos\theta = 3\cos 30° = 3 \times \frac{\sqrt{3}}{2} \neq 0$$

For $\cos 3\theta = \frac{\sqrt{3}}{2}$ the *PV* for $\phi = 3\theta$ is $\frac{\pi}{6}$, so the general solution for ϕ is given by

$$\phi = \pm\frac{\pi}{6} + 2\pi n, \qquad n \in Z$$

so
$$3\theta = \pm\frac{\pi}{6} + 2\pi n$$

i.e.
$$\theta = \pm\frac{\pi}{18} + \frac{2}{3}\pi n, \qquad n \in Z$$

Another common mistake is to ignore the general solution at the time when we identify the multiple angle ϕ and simply take ϕ as $\pm \dfrac{\pi}{6}$ leading to $\theta = \pm \dfrac{\pi}{18}$ then hopefully adding $2\pi n$ to the final solution to give

$$\theta = \pm \frac{\pi}{18} + 2\pi n$$

The latter solution *does not* give all the angles.

Example 4.8.3(2)

Find all the angles in the domain $-180° \leq \theta \leq 180°$ for which $\tan 4\theta = -3$.

Solution:

Again the reader is warned that $\tan 4\theta$ is *not* the same as $4 \tan \theta$ and so writing $\tan \theta = -\dfrac{3}{4}$ is *wrong*.

We proceed by saying that the *PV* for $\phi = 4\theta$ is $-71.57°$ and so the general solution for $\phi = 4\theta = -71.57 + 180n°$

so $$\theta = -17.89° + 45n°, \quad n \in Z$$

$n = 0$, $\theta = -17.89°$
$n = 1$, $\theta = 27.11°$
$n = -1$, $\theta = -62.89°$
$n = 2$, $\theta = 72.11°$
$n = -2$, $\theta = -107.89°$
$n = 3$, $\theta = 117.11°$
$n = -3$, $\theta = -152.89°$
$n = 4$, $\theta = 162.11°$
$n = -4$, $\theta = -197.89° < -180°$ outside domain
$n = 5$, $\theta = 197.11° > 180°$ outside domain

It should be clear that values of n which are greater than 5 or which are less than -4 will all give angles which are outside the given domain.

So for the domain $-180° \leq \theta \leq 180°$, the solutions are

$$\theta = -17.89° \text{ or } \theta = 27.11°$$

or $$\theta = -62.89° \text{ or } \theta = 72.11°$$

or $$\theta = -107.89° \text{ or } \theta = 117.11°$$

or $$\theta = -152.89° \text{ or } \theta = 162.11°$$

Example 4.8.3(3)

Find all solutions to the equation $\sin \dfrac{\theta}{2} = 0.8$ where $0° \leq \theta \leq 1080°$.

Solution:

Set $$\frac{\theta}{2} = \phi$$

i.e. $$\sin \phi = 0.8$$

so PV for $\phi = \dfrac{\theta}{2}$ is 53.13°

SV for $\phi = \dfrac{\theta}{2}$ is 126.87°

so the general solution is

$$\frac{\theta}{2} = 53.13° + 360n°$$

or
$$\frac{\theta}{2} = 126.87° + 360n°$$

i.e.
$$\theta = 106.26° + 720n°$$

or
$$\theta = 253.74° + 720n°$$

For
$$0° \le \theta \le 1080°$$

it is clear that the solutions are

$$\theta = 106.26°$$

or
$$\theta = 826.26°$$

or
$$\theta = 253.74°$$

or
$$\theta = 973.74°.$$

4.8.4 Trigonometric Equations of Form cos A = cos B

The type of equation where we have cos A = cos B where A, B are simple functions of θ, or equivalent forms using one of the other trigonometric functions, can be solved later using a factor formula. However at present we shall make use of the format of the general solution as applied in the previous two parts.

Example 4.8.4(1)

Solve the equation

$$\cos 5\theta = \cos 2\theta,$$

giving the general solution.

Solution:

From
$$\cos 5\theta = \cos 2\theta$$

we can write
$$5\theta = \pm\, 2\theta + 2\pi n, \quad n \in Z$$ (if we treat this initially as an equation in 5θ)

if
$$5\theta = 2\theta + 2\pi n$$

then
$$3\theta = 2\pi n$$

i.e.
$$\theta = \frac{2\pi n}{3}, \qquad n \in Z$$

if
$$5\theta = -\,2\theta + 2\pi n$$

then
$$7\theta = 2\pi n$$

i.e.
$$\theta = \frac{2\pi n}{7}, \qquad n \in Z$$

hence the general solution is

$$\theta = \frac{2\pi n}{3}$$

or

$$\theta = \frac{2\pi n}{7}, \quad n \in Z$$

Example 4.8.4(2)

Find the values of θ in the domain [0°, 360°] for which tan $(4\theta - 40°] = \tan 2\theta$.

Solution:

The general solution is

$$4\theta - 40° = 2\theta + 180n°, \qquad n \in Z$$

i.e.

$$2\theta = 40° + 180n°, \qquad n \in Z$$

i.e.

$$2\theta = 40° + 180n°$$

i.e.

$$\theta = 20° + 90n°$$

$$n = 0, \quad \theta = 20°$$

$$n = 1, \quad \theta = 110$$

$$n = -1, \ \theta = -70° < 0° \text{ outside domain}$$

$$n = 2, \quad \theta = 200°$$

$$n = 3, \quad \theta = 290°$$

$$n = 4, \quad \theta = 380° \quad > 360° \text{ outside domain}$$

so solutions are

$$\theta = 20°$$

or

$$\theta = 110°$$

or

$$\theta = 200°$$

or

$$\theta = 290°.$$

Example 4.8.4(3)

Find the general solution of the equation cos $3\theta = \sin 2\theta$.

Solution: This is not of the required form

$$\cos A = \cos B$$

However we know that

$$\sin 2\theta = \cos\left(\frac{\pi}{2} - 2\theta\right)$$

using the property of complementary angles in part 4.4.3.

So $\cos 3\theta = \cos\left(\frac{\pi}{2} - 2\theta\right)$ where now both sides have been written in the terms of cosine.

i.e. $$3\theta = \pm\left(\frac{\pi}{2} - 2\theta\right) + 2\pi n, n \varepsilon Z$$

If $$3\theta = \frac{\pi}{2} - 2\theta + 2\pi n$$

then $$5\theta = \frac{\pi}{2} + 2\pi n$$

i.e. $$\theta = \frac{\pi}{10} + \frac{2\pi n}{5}$$

If $$3\theta = -\frac{\pi}{2} + 2\theta + 2\pi n$$

then $$\theta = -\frac{\pi}{2} + 2\pi n$$

so the general solution is

$$\theta = \frac{\pi}{10} + \frac{2\pi n}{5} \quad \text{or} \quad \theta = -\frac{\pi}{2} + 2\pi n, \qquad n \varepsilon Z$$

4.9 GRAPHS OF CIRCULAR FUNCTIONS INVOLVING MULTIPLE ANGLES AND PHASE ANGLES

We shall be finding out how to sketch graphs of circular functions of multiple angles, and later some graphs where phase angles may be involved.

As in Section 4.7, we shall concentrate on finding out the points where the graph meets the horizontal axis, the points at which the graph has its maximum and minimum values, and if appropriate the symptotes. The knowledge of how to obtain the general solution of a trigonometric equation involving multiple angles, as was used in sub-sec. 4.8.3, will be used extensively.

4.9.1 Graphs of Circular Functions of Multiple Angles

Example 4.9.1(1)

Sketch the graph of $y = \sin 3\theta$ on the domain $0 \leq \theta \leq 2\pi$.

Solution:

$$\sin 3\theta = 0 \text{ when } 3\theta = n\pi$$

i.e. $$\theta = \frac{n\pi}{3}$$

i.e. $$\theta = 0, \frac{\pi}{3}, \frac{2\pi}{3}, \pi, \frac{4\pi}{3}, \frac{5\pi}{3}, 2\pi \text{ in } [0, 2\pi]$$

$$\sin 3\theta = 1 \text{ when } 3\theta = \frac{\pi}{2} + 2\pi n$$

i.e. $$\theta = \frac{\pi}{6} + \frac{2n\pi}{3}$$

i.e. $$\theta = \frac{\pi}{6}, \frac{5\pi}{6}, \frac{3\pi}{2} \text{ in } [0, 2\pi]$$

$$\sin 3\theta = -1 \text{ when } 3\theta = -\frac{\pi}{2} + 2n\pi$$

i.e. $$\theta = -\frac{\pi}{6} + \frac{2n\pi}{3}$$

i.e. $$\theta = \frac{\pi}{2}, \frac{7\pi}{6}, \frac{11\pi}{6} \text{ in } [0, 2\pi]$$

The graph is shown in Fig. 4.9.1(a).

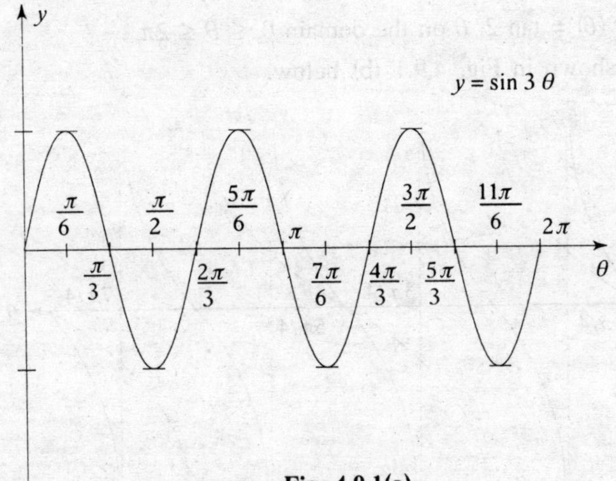

Fig. 4.9.1(a)

N.B: (i) the graph is *continuous*
(ii) the function is *cyclic* or *periodic* but with period of $2\pi/3$,
(iii) the range is $-1 \leq y \leq 1$,
(iv) the shape is that of a sine wave,
(v) within the given domain $0 \leq \theta \leq 2\pi$ there are *three* complete curve patterns compared with only one for the basic function $y = \sin\theta$ sketched in example 4.7.1, i.e. the complete cycle occurs with three times the frequency.

As an exercise the reader is encouraged to sketch the graphs of the equations $y = \sin 2\theta$ and $y = \sin 4\theta$, each of the graphs having the domain $0 \leq \theta \leq 2\pi$.

The reader should observe that with the graph of $y = \sin 2\theta$ the function is cyclic with period π, and with the graph of $y = \sin 4\theta$ the function is cyclic with period $\dfrac{\pi}{2}$.

It seems likely that the graph of $y = \sin k\theta$ will be a sine wave of period $2\pi/k$ and frequency k times that of $y = \sin\theta$.

In fact this is true and can be demonstrated by considering the value of $y = \sin k\theta$ at values $\theta = \theta_0$, $\theta = \theta_0 + 2\pi/k$

At $\theta = \theta_0, y = \sin k\theta_0$

At $\theta = \theta_0 + 2\pi/k, y = \sin (k(\theta_0 + 2\pi/k))$

$$= \sin (k\theta_0 + 2\pi)$$

$$= \sin k\theta_0$$

i.e. for any θ_0, these two values are equal, hence the graph of $y = \sin k\theta$ will be repeated after a period $2\pi/k$.

These properties are valid for all values of k, and similar properties are true for $y = \cos k\theta$, $y = \tan k\theta$. In the case of $y = \tan k\theta$ the period will be π/k, *not* $2\pi/k$.

As examples of other graphs of multiple angles three further examples are shown below.

Example 4.9.1(2)

Sketch the graph of $y = f(\theta) = \tan 2\ \theta$ on the domain $0 \le \theta \le 2\pi$.

Solution: The sketch is shown in Fig. 4.9.1 (b) below.

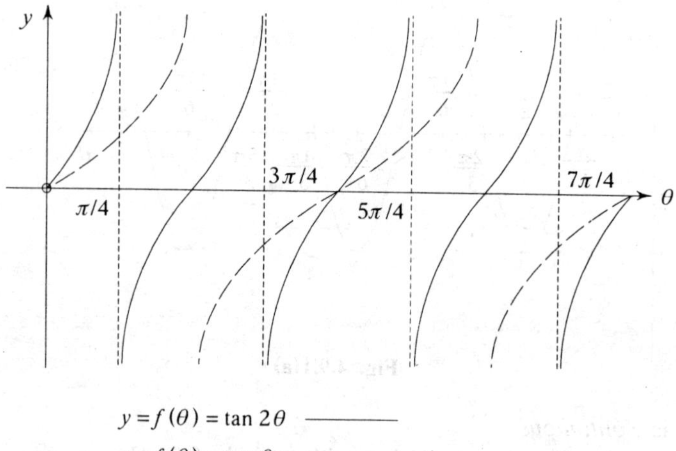

Fig. 4.9.1 (b)

Note that the period of the graph is $\dfrac{\pi}{2}$.

Example 4.9.1(3)

Sketch the graph of $y = \cos 5\theta$ on the domain $0 \le \theta \le 2\pi$.

Solution:

The sketch is shown in Fig. 4.9.1 (c) below.

The reader who has doubts about the points where the graph crosses the horizontal axis and has its maximum and minimum points should solve the equations.

$$\cos 5\theta = 0, \cos 5\theta = 1, \cos 5\theta = -1, \quad \text{on the given domain*}$$

The period of the graph is $2\pi/5$.

* as a check, see next page

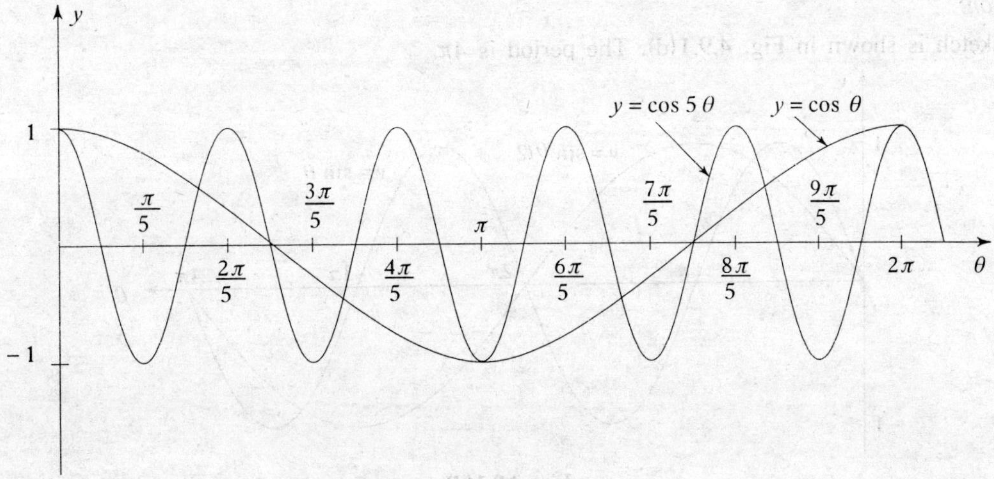

Fig. 4.9.1(c)

Check

$\cos 5\theta = 0$

means $5\theta = \pm\dfrac{\pi}{2} + 2\pi n,\ n \ \varepsilon\ Z$

i.e. $\theta = \pm\dfrac{\pi}{10} + \dfrac{2\pi n}{5}$

giving $\theta = \pm\dfrac{\pi}{10}, \dfrac{3\pi}{10}, \dfrac{5\pi}{10}$, or $\dfrac{\pi}{2}, \dfrac{7\pi}{10}, \dfrac{9\pi}{10}, \dfrac{11\pi}{10}, \dfrac{13\pi}{10}, \dfrac{15\pi}{10}$, or $\dfrac{3\pi}{2}, \dfrac{17\pi}{10}, \dfrac{19\pi}{10}$

$\cos 5\theta = 1$

means $5\theta = 2\pi n,\ n \ \varepsilon\ Z$

i.e. $\theta = \dfrac{2\pi n}{5}$

giving $\theta = \dfrac{2\pi}{5}, \dfrac{4\pi}{5}, \dfrac{6\pi}{5}, \dfrac{8\pi}{5}, 2\pi$

$\cos 5\theta = -1$

means $5\theta = \pi + 2\pi n,\ n \ \varepsilon\ Z$

i.e. $\theta = \dfrac{\pi}{5} + \dfrac{2\pi n}{5}$

giving $\theta = \dfrac{\pi}{5}, \dfrac{3\pi}{5}, \pi, \dfrac{7\pi}{5}, \dfrac{9\pi}{5}.$

Example 4.9.1(4)

Sketch the graph of $y = \sin\dfrac{\theta}{2}$ on domain $0 \le \theta \le 4\pi$.

Solution:

The sketch is shown in Fig. 4.9.1(d). The period is 4π.

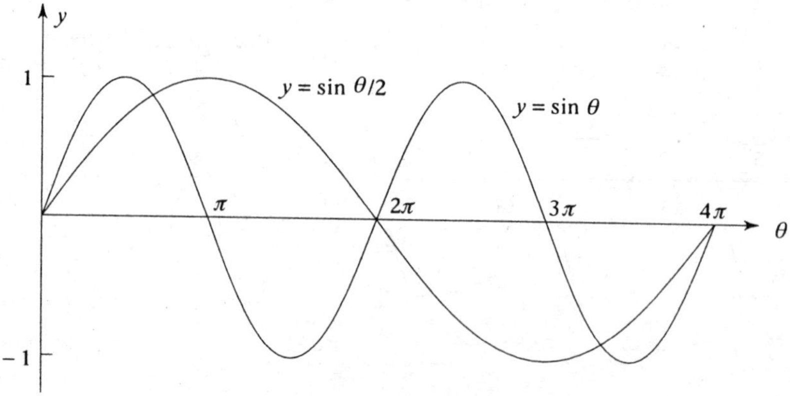

Fig. 4.9.1(d)

4.9.2 Graphs of Circular Functions Involving Phase Angles

Examples 4.9.2(1)

Sketch the graph of $y = \cos\left(\theta - \frac{\pi}{4}\right)$

on the domain $-2\pi \le \theta \le 2\pi$.

Solution:

As in previous examples we shall determine the points at which the graph meets the horizontal axis and also the points at which the maximum and minimum values occur.

$$y = 0 \text{ when } \cos\left(\theta - \frac{\pi}{4}\right) = 0$$

i.e. $$\theta - \frac{\pi}{4} = \pm\frac{\pi}{2} + 2\pi n,\ n \,\varepsilon\, Z$$

i.e. $$\theta = \frac{3\pi}{4} + 2\pi n$$

or $$\frac{-\pi}{4} + 2\pi n$$

i.e. $$\theta = \frac{-5\pi}{4}, \frac{-\pi}{4}, \frac{3\pi}{4} \text{ or } \frac{7\pi}{4}$$

$$y = 1 \text{ when } \cos\left(\theta - \frac{\pi}{4}\right) = 1$$

i.e. $$\theta - \frac{\pi}{4} = 2\pi n, \quad n \,\varepsilon\, Z$$

i.e. $\theta = \frac{\pi}{4} + 2\pi n$

i.e. $\theta = \frac{-7\pi}{4}$ or $\frac{\pi}{4}$

$$y = -1 \text{ when } \cos\left(\theta - \frac{\pi}{4}\right) = -1$$

i.e. $\theta - \frac{\pi}{4} = \pi + 2\pi n$

i.e. $\theta = \frac{5\pi}{4} + 2\pi n$

i.e. $\theta = -\frac{3\pi}{4}$ or $\frac{5\pi}{4}$

The sketch is shown in Fig. 4.9.2(a).

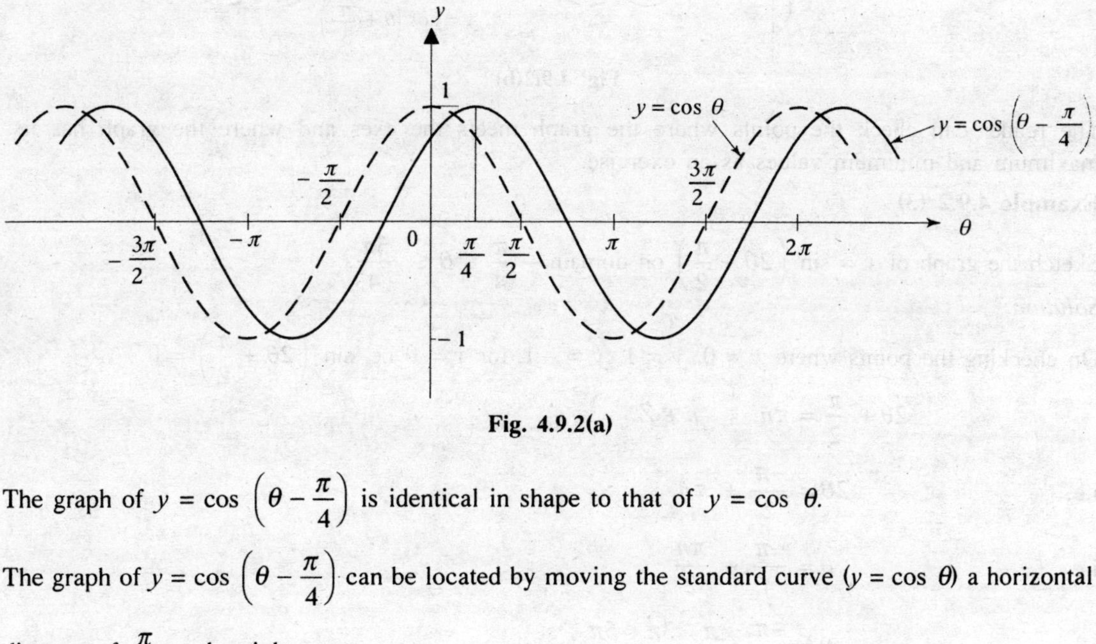

Fig. 4.9.2(a)

The graph of $y = \cos\left(\theta - \frac{\pi}{4}\right)$ is identical in shape to that of $y = \cos\ \theta$.

The graph of $y = \cos\left(\theta - \frac{\pi}{4}\right)$ can be located by moving the standard curve ($y = \cos\ \theta$) a horizontal distance of $\frac{\pi}{4}$ to the right.

The angle $\frac{\pi}{4}$ is known as the phase angle.

Example 4.9.2(2)

Sketch the graph of $y = \cos\left(\theta + \frac{\pi}{3}\right)$ on domain $-\frac{\pi}{3} \leq \theta \leq \frac{8\pi}{3}$

Solution:

The graph of $y = \cos\left(\theta + \frac{\pi}{3}\right)$ can be located by moving the standard curve ($y = \cos\,\theta$) a horizontal distance of $\frac{\pi}{3}$ to the left as in Fig. 4.9.2(b).

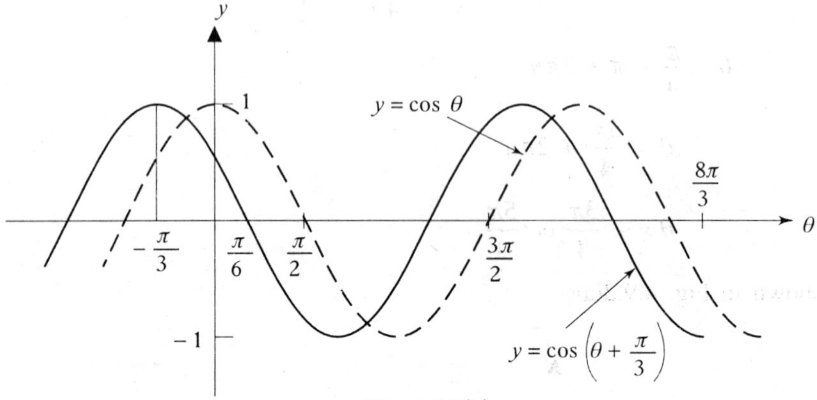

Fig. 4.9.2(b)

The reader can check the points where the graph meets the axes and where the graph has its maximum and minimum values as an exercise.

Example 4.9.2 (3)

Sketch the graph of $y = \sin\left(2\theta + \frac{\pi}{2}\right)$ on domain $-\frac{\pi}{4} \leq \theta \leq \frac{5\pi}{4}$.

Solution:

On checking the points where $y = 0$, $y = 1$, $y = -1$, for $y = 0$ i.e. $\sin\left(2\theta + \frac{\pi}{2}\right) = 0$

$$2\theta + \frac{\pi}{2} = \pi n, \qquad n \in Z$$

i.e. $$2\theta = \frac{-\pi}{2} + \pi n$$

i.e. $$\theta = \frac{-\pi}{4} + \frac{\pi n}{2}$$

i.e. $$\theta = \frac{-\pi}{4}, \frac{\pi}{4}, \frac{3\pi}{4}, \frac{5\pi}{4}$$

$$y = 1 \text{ i.e. } \sin\left(2\theta + \frac{\pi}{2}\right) = 1$$

$$2\theta + \frac{\pi}{2} = \frac{\pi}{2} + 2\pi n, \qquad n \in Z$$

i.e. $$\theta = \pi n$$

i.e. $$\theta = 0, \pi, 2\pi$$

$$y = -1 \text{ i.e. } \sin\left(2\theta + \frac{\pi}{2}\right) = -1$$

$$2\theta + \frac{\pi}{2} = -\frac{\pi}{2} + 2\pi n,\ n \varepsilon Z$$

i.e.
$$2\theta = -\pi + 2\pi n$$

$$\theta = \frac{-\pi}{2} + \pi n$$

i.e.
$$\theta = \frac{\pi}{2}, \frac{3\pi}{2}$$

The sketch is shown in Fig. 4.9.2 (c)

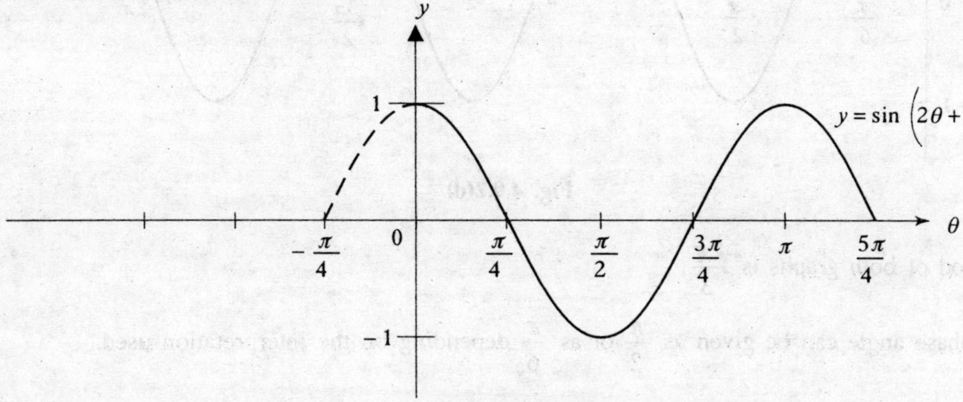

Fig. 4.9.2(c)

In comparison with the graph of $y = \sin 2\theta$ the graph of $y = \sin\left(2\theta + \frac{\pi}{2}\right)$ has been located by moving the graph of $y = \sin 2\theta$, a horizontal distance of $\frac{\pi}{4}$ *to the left.*

In graphs involving a phase angle the literature is divided on the definition of the phase angle. Some texts define the phase angle to be $\frac{\pi}{2}$ as given in the original equation $y = \sin\left(2\theta + \frac{\pi}{2}\right)$, whereas others would define the phase angle to be $\frac{\pi}{4}$, the actual horizontanl distance moved (in this case to the left) to locate the given curve from the standard curve.

Example 4.9.2 (4)

Sketch the grapah of $y = \cos\left(3\theta - \frac{\pi}{2}\right)$ on the domain $0 \le \theta \le 2\pi$.

Solution:

The graph can be compared with the graph of $y = \cos 3\theta$. The reader can check the points where the graph meets the horizontal axis and where the graph has its maximum and minimum values and should then recognise that the graph of $y = \cos\left(3\theta - \dfrac{\pi}{2}\right)$ can be located by moving the graph of $y = \cos 3\theta$.

a horizontal distance of $\dfrac{\pi}{6}$ to the right. [See Fig. 4.9.2 (d)]

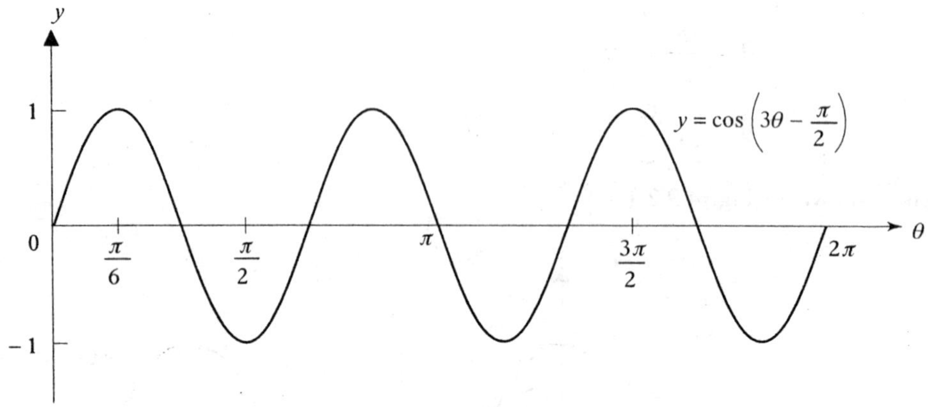

Fig. 4.9.2(d)

The period of both graphs is $\dfrac{2\pi}{3}$.

The phase angle can be given as $\dfrac{\pi}{2}$ or as $\dfrac{\pi}{6}$ depending on the interpretation used.

4.10 TRIGONOMETRIC IDENTITIES

In sections 4.4 and 4.5 we defined the six trigonometric ratios (or functions) sin θ, cos θ, tan θ, sec θ, cosec θ, cot θ, first of all from an acute angle θ in a right angled triangle, and later for a general angle. Apart from the reciprocal relations.

$$\sec\theta = 1/\cos\theta,\ \text{cosec}\ \theta = 1/\sin\theta,\ \cot\theta = 1/\tan\theta,$$

there are certain other relationships between two or more members of this set of six trigonometric ratios, and these will be derived in this section.

4.10.1 Relationships between Trigonometric Ratios

In this part we shall prove four important identities which can help in solving problems involving trigonometric functions.

In Fig. 4.10. 1(a) we have that, for any angle θ,

$$\sin\theta = y/r,\ \cos\theta = x/r$$

$$\tan\theta = y/x$$

But $$\frac{\sin\theta}{\cos\theta} = \frac{y/r}{x/r} = \frac{y}{r} \times \frac{r}{x} = \frac{y}{x} = \tan\theta$$

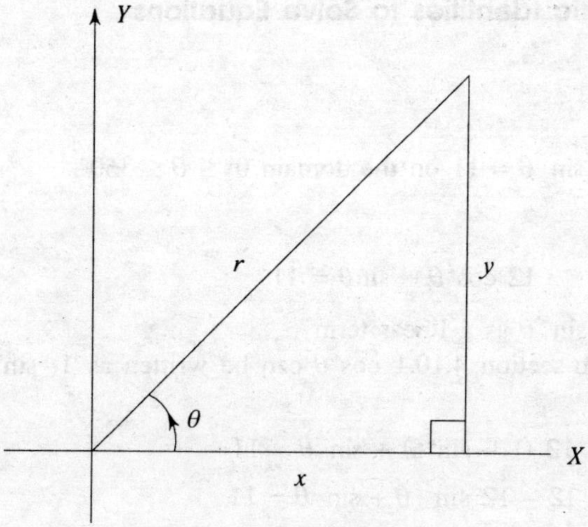

Fig. 4.10.1(a)

i.e. $$\frac{\sin\theta}{\cos\theta} = \tan\theta \quad \text{(i)}$$

and in a similar manner we may also show that

$$\frac{\cos\theta}{\sin\theta} = \cot\theta \quad (2)$$

Also the following identities may be proved using Pythagoras

$$(\cos\theta)^2 + (\sin\theta)^2$$

which may commonly be written as $\cos^2\theta + \sin^2\theta$

is $$\frac{x^2}{r^2} + \frac{y^2}{r^2}$$

$$= \frac{x^2 + y^2}{r^2} = \frac{r^2}{r^2} = 1$$

so $$\cos^2\theta + \sin^2\theta = 1 \quad (3)$$

Dividing (3) by $\cos^2\theta$

we have $$1 + \frac{\sin^2\theta}{\cos^2\theta} = \frac{1}{\cos^2\theta}$$

i.e. $$1 + \tan^2\theta = \sec^2\theta \quad (4)$$

Dividing (3) by $\sin^2\theta$

$$\frac{\cos^2\theta}{\sin^2\theta} + 1 = \frac{1}{\sin^2\theta}$$

i.e. $$\cot^2\theta + 1 = \text{cosec}^2\theta \quad (5)$$

4.10.2 Use of Trigonometric Identities to Solve Equations

Examples 4.10.2(1)

Solve the equation

$$12 \cos^2 \theta + \sin \theta = 11 \text{ on the domain } 0° \le \theta \le 360°.$$

Solution:

$$12 \cos^2\theta + \sin\theta = 11$$

has $\cos^2\theta$ as a quadratic term $\sin \theta$ as a linear term

By using identity (3) in sub-section 4.10.1 $\cos^2\theta$ can be written as $1-\sin^2 \theta$ so as to give an equation in $\sin \theta$.

i.e. $$12 (1 - \sin^2\theta) + \sin \theta = 11$$

i.e. $$12 - 12 \sin^2 \theta + \sin \theta = 11$$

i.e. $$0 = 12 \sin^2\theta - \sin \theta - 1$$

i.e. $$0 = (3 \sin \theta - 1)(4 \sin \theta + 1)$$

i.e. $$\sin \theta = \frac{1}{3} \text{ or } \sin \theta = -\frac{1}{4}$$

for $$\sin \theta = \frac{1}{3}$$

$$PV = 19.47°$$

$$SV = 160.53°$$

These are the only values in domain $0° \le \theta \le 360°$

for $$\sin\theta = -\frac{1}{4}$$

$$PV = -14.48°$$

$$SV = -165.52°$$

and general solution is

$$\theta = -14.48° + 360n°$$

or $$-165.52 + 360n°$$

on domain $0° \le \theta \le 360°$

$$\theta = 345.52° \quad \text{or} \quad 194.48°$$

so the complete solution on domain $0° \le \theta \le 360°$ is $\theta = 19.47°$ or $\theta = 160.53°$

or $$\theta = 194.48° \quad \text{or} \quad \theta = 345.52°.$$

Example 4.10.2(2)

Find the general solution of the equation

$$12 \sec^2\theta - 13 \tan\theta - 9 = 0$$

Solution:

$12\sec^2\theta - 13\tan\theta - 9 = 0$ has $\sec^2\theta$ as a quadratic term and $\tan\theta$ as a linear term

By using identity (4) in sub-section 4.10.1 it is easy to replace $\sec^2\theta$ by $1 + \tan^2\theta$, giving an equation in $\tan\theta$

i.e. $$12(1 + \tan^2\theta) - 13\tan\theta - 9 = 0$$

i.e. $$12 + 12\tan^2\theta - 13\tan\theta - 9 = 0$$

i.e. $$12\tan^2\theta - 13\tan\theta - 3 = 0$$

i.e. $$(4\tan\theta - 3)(3\tan\theta - 1) = 0$$

i.e. $$\tan\theta = \frac{3}{4} \text{ or } \tan\theta = \frac{1}{3}$$

giving a general solution

$$\theta = 36.87° + 180\, n° \text{ or } \theta = 18.43° + 180n°,\ n \in Z$$

Example 4.10.2(3)

Determine in radians the general solution of the equation $\sqrt{3}\sec\theta = \operatorname{cosec}\theta$

Solution:

Writing $$\sec\theta = 1/\cos\theta$$

and $$\operatorname{cosec}\theta = \frac{1}{\sin\theta}$$

$$\sqrt{3} \times \frac{1}{\cos\theta} = \frac{1}{\sin\theta}$$

and cross–multiplying

$$\sqrt{3}\sin\theta = \cos\theta$$

which still has two different trigonometric ratios in the equation, however if we use the fact that

$$\frac{\sin\theta}{\cos\theta} = \tan\theta$$

we can write the equation as

$$\sqrt{3}\,\frac{\sin\theta}{\cos\theta} = 1$$

i.e. $$\sqrt{3}\tan\theta = 1$$

i.e. $$\tan\theta = \frac{1}{\sqrt{3}}$$

and the general solution is

$$\theta = \frac{\pi}{6} + \pi n, \qquad n \in Z$$

4.10.3 Proving Trigonometric Identities

It is possible to use the identities which were established in 4.10.1 to prove identities involving quite complicated expressions involving trigonometric functions. Unfortunately there is no set routine which can be followed when proving identities, but two helpful suggestions would be

(i) aim to simplify the more complicated side first,

(ii) aim to write the expressions on the side which you are simplifying in terms of the trigonometric functions given on the other side.

Example 4.10.3 (1)

$$\text{Prove that } \frac{\tan\theta - \cot\theta}{\tan\theta + \cot\theta} = (\sin\theta - \cos\theta)\,(\sin\theta + \cos\theta).$$

Solution:

The left hand side looks more complicated so we shall aim to simplify it, putting the expressions in terms of $\sin\theta$ or $\cos\theta$

i.e.

$$\text{LHS} = \frac{\tan\theta - \cot\theta}{\tan\theta + \cot\theta}$$

$$= \frac{\dfrac{\sin\theta}{\cos\theta} - \dfrac{\cos\theta}{\sin\theta}}{\dfrac{\sin\theta}{\cos\theta} + \dfrac{\cos\theta}{\sin\theta}} = \frac{\dfrac{\sin^2\theta - \cos^2\theta}{\cos\theta\sin\theta}}{\dfrac{\sin^2\theta + \cos^2\theta}{\cos\theta\sin\theta}}$$

$$\text{LHS} = \frac{(\sin^2\theta - \cos^2\theta)}{\cos\theta\sin\theta} \times \frac{\cos\theta\sin\theta}{(\sin^2\theta + \cos^2\theta)}$$

$$= \sin^2\theta - \cos^2\theta$$ (cancelling out the factors $\cos\theta$ and $\sin\theta$ and using the fact that $\sin^2\theta + \cos^2\theta = 1$)

$$\text{LHS} = \sin^2\theta - \cos^2\theta$$
$$= (\sin\theta - \cos\theta)\,(\sin\theta + \cos\theta) = \text{RHS}.$$

Example 4.10.3(2)

Prove that

$$\frac{1}{(\operatorname{cosec}\theta - \sin\theta)(\sec\theta - \cos\theta)} = \tan\theta + \cot\theta.$$

Solution:

Writing all the expressions in the left hand side in terms of $\sin\theta$ and $\cos\theta$

$$\text{LHS} = \frac{1}{\left(\dfrac{1}{\sin\theta} - \sin\theta\right)\left(\dfrac{1}{\cos\theta} - \cos\theta\right)}$$

$$= \frac{1}{\dfrac{(1-\sin^2\theta)}{\sin\theta}\,\dfrac{(1-\cos^2\theta)}{\cos\theta}}$$

$$= \frac{\sin\theta \qquad \cos\theta}{(1-\sin^2\theta)\,(1-\cos^2\theta)}$$

$$= \frac{\sin\theta}{\cos^2\theta}\,\frac{\cos\theta}{\sin^2\theta} = \frac{1}{\sin\theta\ \cos\theta}$$

Now writing the expressions on the right hand side in terms of $\sin\theta$ and $\cos\theta$

$$\text{RHS} = \tan\theta + \cot\theta$$

$$= \frac{\sin\theta}{\cos\theta} + \frac{\cos\theta}{\sin\theta}$$

$$= \frac{\sin^2\theta + \cos^2\theta}{\cos\theta\sin\theta}$$

$$\text{RHS} = \frac{1}{\cos\theta\sin\theta} = \text{LHS.}$$

N.B: There is a difference between an identity and an equation.

An identity will be true for all values of the angle θ on a given domain (in example 4.10.3(1) it would be true for all possible values of θ but in example 4.10.3(2) we would have to exclude all multiples of 90° for which neither the left hand side nor the right hand side could be defined).

An equation will only be true for some values of θ on a given domain.

In an identity it is wrong to approach the problem as if we have to solve an equation. It is essential to prove an identity by simplifying each side separately.

The key word which will distinguish an identity from an equation is that for an identity the instruction will normally include 'show that' or 'prove' whereas for an equation the instruction will normally include 'solve'.

4.11 COMPOUND ANGLE IDENTITIES

A very important set of trigonometric identities, often known as compound angle identities or addition formulae, will now be proved, as these identities are very important both in solving certain types of equation involving trigonomatric functions, and also in many applications in calculus.

4.11.1 Establishment of Formulae for sin (A ± B), cos (A ± B), tan (A ± B)

First we shall consider the matrix, $M(\theta)$, which represents a rotation of angle θ about the origin O. See Fig. 4.11.1 (a)

Clearly under this rotation

$$I\,(1,0) \longrightarrow I_1\,(\cos\theta, \sin\theta)$$

and

$$J\,(0,1) \longrightarrow J_1\,(-\sin\theta, \cos\theta)$$

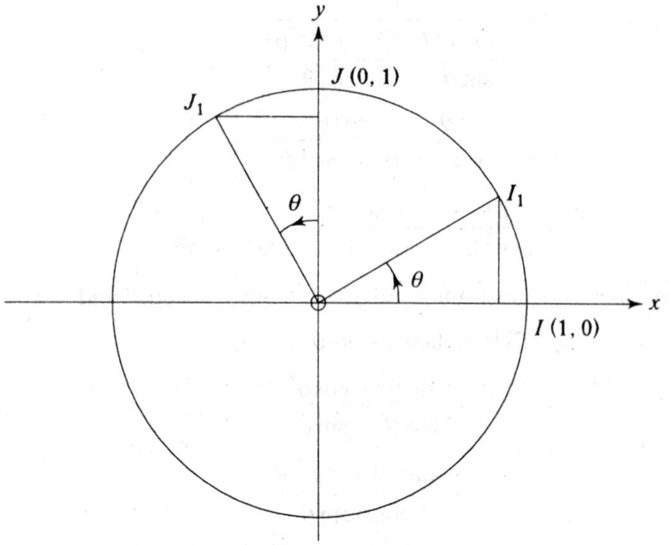

Fig. 4.11.1(a)

since the image points I_1, J_1 of I, J respectively must lie on the unit circle, being also a distance of 1 unit from the origin.

If $$M(\theta) = \begin{pmatrix} a & b \\ c & d \end{pmatrix}$$

then $$\begin{pmatrix} a & b \\ c & d \end{pmatrix} \overset{\quad I \quad J}{\begin{pmatrix} 1 & 0 \\ 0 & 1 \end{pmatrix}} = \overset{\quad I_1 \quad\quad J_1}{\begin{pmatrix} \cos\theta & -\sin\theta \\ \sin\theta & \cos\theta \end{pmatrix}}$$

but on the left hand side the right hand matrix is the identity matrix for multiplication

hence $$M(\theta) = \begin{pmatrix} a & b \\ c & d \end{pmatrix} = \begin{pmatrix} \cos\theta & -\sin\theta \\ \sin\theta & \cos\theta \end{pmatrix}$$

In the same way the matrix $M(\phi)$ which represents a rotation of angle ϕ about O will be given by

$$M(\phi) = \begin{pmatrix} \cos\phi & -\sin\phi \\ \sin\phi & \cos\phi \end{pmatrix}$$

Now a rotation of ϕ about O followed by a rotation of θ about O is equivalent to a roatioin of $\theta + \phi$ about O and this may be represented by the matrix

$$M(\theta + \phi) = \begin{pmatrix} \cos(\theta + \phi) & -\sin(\theta + \phi) \\ \sin(\theta + \phi) & \cos(\theta + \phi) \end{pmatrix}$$

However the rotation of ϕ about O, followed by a rotation of θ about O, may also be represented by the matrix product.

$$M(\theta)\,M(\phi) = \begin{pmatrix} \cos\theta & -\sin\theta \\ \sin\theta & \cos\theta \end{pmatrix}\begin{pmatrix} \cos\phi & -\sin\phi \\ \sin\phi & \cos\phi \end{pmatrix}$$

$$= \begin{pmatrix} \cos\theta\cos\phi - \sin\theta\sin\phi & -\cos\theta\sin\phi - \sin\theta\cos\phi \\ \sin\theta\cos\phi + \cos\theta\sin\phi & -\sin\theta\sin\phi + \cos\theta\cos\phi \end{pmatrix}$$

The matrix $M(\theta + \phi)$ and the product matrix $M(\theta)\,M(\phi)$ must be equal, so on equating the entries in the first row and the first column

$$\cos(\theta + \phi) = \cos\theta\cos\phi - \sin\theta\sin\phi$$

and in the second row and first column

$$\sin(\theta + \phi) = \sin\theta\cos\phi + \cos\theta\sin\phi$$

In the same way, for angles A, B we may write

$$\cos(A + B) = \cos A\cos B - \sin A\sin B \qquad (1)$$

$$\sin(A + B) = \sin A\cos B + \cos A\sin B \qquad (2)$$

in (1) write $-B$ for B to give

$$\begin{aligned}\cos(A - B) &= \cos A\cos(-B) - \sin A\sin(-B)\\ &= \cos A\cos B - \sin A(-\sin B)\\ &= \cos A\cos B + \sin A\sin B\end{aligned}$$

i.e. $$\cos(A - B) = \cos A\cos B + \sin A\sin B \qquad (3)$$

in (2) write $-B$ for B to give

$$\begin{aligned}\sin(A - B) &= \sin A\cos(-B) + \cos A\sin(-B)\\ &= \sin A\cos B + \cos A(-\sin B)\\ &= \sin A\cos B - \cos A\sin B\end{aligned}$$

i.e. $$\sin(A - B) = \sin A\cos B - \cos A\sin B \qquad (4)$$

Grouping the two cosine results (1) and (3) together

$$\left.\begin{aligned}\cos(A + B) &= \cos A\cos B - \sin A\sin B\\ \cos(A - B) &= \cos A\cos B + \sin A\sin B\end{aligned}\right\}$$ like ratios together, unlike connecting signs

Grouping the two sine results (2) and (4) together

$$\left.\begin{aligned}\sin(A + B) &= \sin A\cos B + \cos A\sin B\\ \sin(A - B) &= \sin A\cos B - \cos A\sin B\end{aligned}\right\}$$ unlike ratios together, like connecting signs.

NB: $\sin(A + B) \neq \sin A + \sin B$

e.g. $$\sin(60° + 30°) \neq \sin 60° + \sin 30°$$

as $$\sin(60° + 30°) = \sin 90° = 1$$

but $$\sin 60° + \sin 30° = \frac{\sqrt{3}}{2} + \frac{1}{2} = \frac{\sqrt{3}+1}{2}$$

However we may evaluate sin (60° + 30°) using the formula for sin $(A + B)$ where $A = 60°$, $B = 30°$ to give sin (60° + 30°)

$$= \sin 60° \cos 30° + \cos 60° \sin 30°$$
$$= \frac{\sqrt{3}}{2} . \frac{\sqrt{3}}{2} + \frac{1}{2} . \frac{1}{2}$$
$$= \frac{3}{4} + \frac{1}{4}$$
$$= 1$$

using the formulae (1) and (2) it is now possible to find an expansion for tan $(A + B)$ and this can be done using

$$\tan (A + B) = \frac{\sin(A + B)}{\cos (A + B)}$$

$$\frac{\sin(A + B)}{\cos (A + B)} = \frac{\sin A \cos B + \cos A \sin B}{\cos A \cos B - \sin A \sin B}$$

$$= \frac{\dfrac{\sin A \cos B}{\cos A \cos B} + \dfrac{\cos A \sin B}{\cos A \cos B}}{\dfrac{\cos A \cos B}{\cos A \cos B} - \dfrac{\sin A \sin B}{\cos A \cos B}}$$

(on dividing each term in the expressions by cos A cos B)

$$= \frac{\dfrac{\sin A \cancel{\cos B}^{1}}{\cos A \cancel{\cos B}_{1}} + \dfrac{\cancel{\cos A}^{1} \sin B}{{}_{1}\cancel{\cos A} \cos B}}{\dfrac{{}^{1}\cancel{\cos A}\ \cancel{\cos B}^{1}}{\cancel{\cos A}_{1}\ \cancel{\cos B}_{1}} - \dfrac{\sin A \sin B}{\cos A \cos B}}$$

$$= \frac{\tan A + \tan B}{1 - \tan A \tan B}$$

i.e. tan $(A + B)$

$$= \frac{\tan A + \tan B}{1 - \tan A \tan B} \quad (5)$$

If we replace B by $-B$ in (5) this gives tan $(A - B)$

$$= \frac{\tan A + \tan (-B)}{1 - \tan A \tan (-B)}$$

$$= \frac{\tan A - \tan B}{1 - \tan A\,(-\tan B)}$$

$$= \frac{\tan A - \tan B}{1 + \tan A \tan B)} \qquad (6)$$

These results (1) to (6) may now be used together with results for special angles to determine the exact values of trigonometeic ratios of other angles, as can be shown in example 4.11.1 (1).

Example 4.11.1 (1)

Without using tables or calculator evaluate (a) sin 105° (b) cos 255° (c) tan (–75°).

Solution:

(a) $\sin 105° = \sin (60° + 45°)$

$$= \sin 60° \cos 45° + \cos 60° \sin 45°$$

$$= \frac{\sqrt{3}}{2}.\frac{1}{\sqrt{2}} + \frac{1}{2}.\frac{1}{\sqrt{2}} = \frac{\sqrt{3}+1}{2\sqrt{2}}$$

$$= \frac{\sqrt{2}\left(\sqrt{3}+1\right)}{2\sqrt{2}.\sqrt{2}}$$

$$= \frac{\sqrt{2}\left(\sqrt{3}+1\right)}{4} \text{ or } \frac{\sqrt{6}+\sqrt{2}}{4}$$

(b) $\cos 255°$

$$= \cos (225° + 30°)$$

$$= \cos 225° \cos 30° - \sin 225° \sin 30°$$

$$= -\cos 45° \cos 30° - (-\sin 45°) \sin 30°$$

$$= -\frac{1}{\sqrt{2}}.\frac{\sqrt{3}}{2} - \left(\frac{-1}{\sqrt{2}}\right).\frac{1}{2} = \frac{-\sqrt{3}+1}{2\sqrt{2}}$$

$$= \frac{\sqrt{2}\,(-\sqrt{3}+1)}{2\sqrt{2}.\sqrt{2}}$$

$$= \frac{\sqrt{2}\,(-\sqrt{3}+1)}{4} \text{ or } \frac{-\sqrt{6}+\sqrt{2}}{4}$$

(c) $\tan (-75°)$

$$= -\tan 75°$$

$$= -\tan (45° + 30°)$$

$$= -\left(\frac{\tan 45° + \tan 30°}{1 - \tan 45° \tan 30°}\right)$$

$$= -\left(\frac{1+\frac{1}{\sqrt{3}}}{1-1.\frac{1}{\sqrt{3}}}\right)$$

tan (– 75°)

$$= -\left(\frac{\frac{\sqrt{3}+1}{\sqrt{3}}}{\frac{\sqrt{3}-1}{\sqrt{3}}}\right)$$

$$= -\frac{(\sqrt{3}+1)}{\sqrt{3}} \times \frac{\sqrt{3}}{(\sqrt{3}-1)}$$

$$= -\frac{(\sqrt{3}+1)(\sqrt{3}+1)}{(\sqrt{3}-1)(\sqrt{3}+1)}$$

$$= -\frac{(3+2\sqrt{3}+1)}{(3-1)}$$

$$= -\frac{(4+2\sqrt{3})}{2} = -\frac{2(2+\sqrt{3})}{2}$$

$$= -(2+\sqrt{3}) \text{ or } -2-\sqrt{3}$$

N.B: In each of parts (a), (b), (c) we could have used alternative compound angles e.g. in (a) 105° = 135° – 30°, in (b) 255° = 210° + 45°, in (c) – 75° = 60° – 135°.

The reader could try as an exercise each of these parts using different compound angles.

The results (1) to (6) can also be used to determine compound angle formulae without the use of tables or calculator for angles other than the special angles, as is illustrated in the following example.

Example 4.11.1 (2)

A is obtuse and $\sin A = \frac{4}{5}$, B is reflex and $\sin B = \frac{-5}{13}$ and $\cos B$ is negative.

Without using tables or calculator evaluate (a) $\cos (A + B)$ (b) $\tan (A - B)$.

Solution:

In order to use the relevant formulae we have to determine $\cos A$, $\tan A$, $\cos B$, $\tan B$, and these can be found by using Pythagoras in *Two* different right angled trianges as in Fig. 4.11.1 (b) and (c).

(a) Reading the relevant values of the trigonometric ratios from Fig. 4.11.1 (b) and (c)

$$\cos (A + B) = \cos A \cos B - \sin A \sin B$$

$$= \frac{-3}{5} \times \frac{-12}{13} - \frac{4}{5} \times \frac{-5}{13}$$

$$= \frac{36+20}{65} = \frac{56}{65}$$

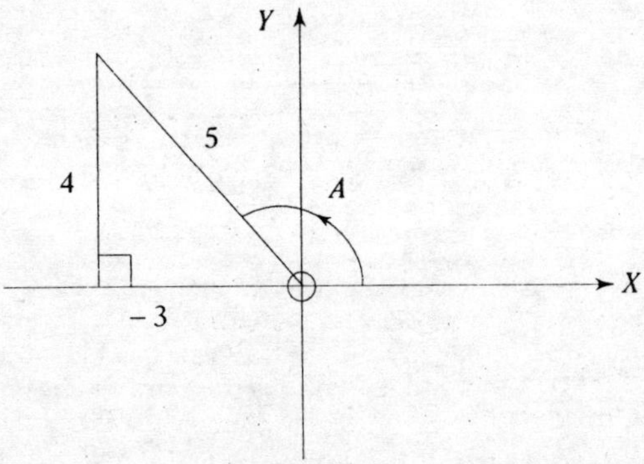

Fig. 4.11.1(b)

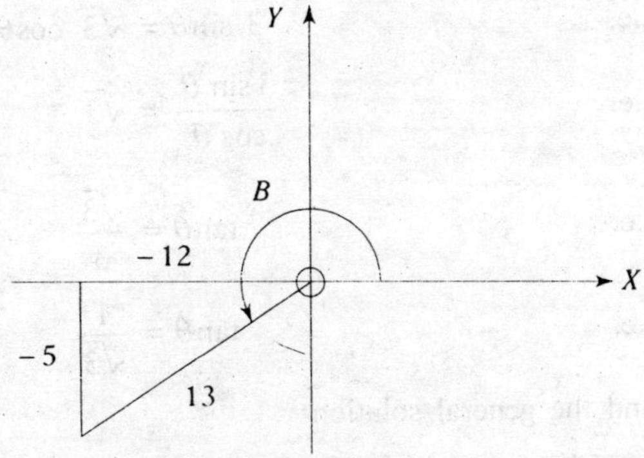

Fig. 4.11.1(c)

(b) In a similar manner

$$\tan(A - B) = \frac{\tan A - \tan B}{1 + \tan A \tan B}$$

$$= \frac{\frac{-4}{3} - \frac{-5}{-12}}{1 + \left(\frac{-4}{3}\right)\left(\frac{-5}{-12}\right)}$$

$$= \frac{\frac{-4}{3} - \frac{5}{12}}{1 - \frac{4}{3}.\frac{5}{12}} = \frac{\frac{-48-15}{36}}{\frac{36-20}{36}}$$

$$\tan(A - B) = \frac{-63}{36} \times \frac{36}{16} = \frac{-63}{16}$$

Example 4.11.1 (3)

Solve the equation

$$2 \sin\theta = \cos(30° - \theta)$$

giving the general solution.

Solution:

Note that this is *not* an identity to be proved, it is an equation to be solved, and therefore only a limited number of values of θ will satisfy the equation.

$$2 \sin \theta = \cos(30° - \theta)$$

i.e. $$2 \sin\theta = \cos 30° \cos\theta + \sin 30° \sin \theta$$

i.e. $$2 \sin\theta = \frac{\sqrt{3}}{2}\cos \theta + \frac{1}{2} \sin \theta.$$

i.e. $$4 \sin\theta = \sqrt{3} \cos\theta + \sin\theta$$

i.e. $$3 \sin\theta = \sqrt{3} \cos\theta$$

i.e. $$\frac{3 \sin \theta}{\cos \theta} = \sqrt{3}$$

i.e. $$\tan\theta = \frac{\sqrt{3}}{3}$$

i.e. $$\tan\theta = \frac{1}{\sqrt{3}}.$$

and the general solution

$$\theta = \frac{\pi}{6} + \pi n, \qquad n \in Z$$

Example 4.11.1 (4)

If $0 < x < \pi$ and $\tan(x - A) = 3$ where $\tan A = 2$ show that $x = \dfrac{3\pi}{4}$ without using tables or calculator.

Solution:

Since it is given that $\tan(x - A) = 3$

then $$\frac{\tan x - \tan A}{1 + \tan x \tan A} = 3$$

i.e. $$\frac{\tan x - 2}{1 + 2 \tan x} = 3$$

i.e. $$\tan x - 2 = 3 + 6 \tan x$$

i.e. $$-5 = 5 \tan x$$

i.e. $$-1 = \tan x$$

but since $$0 < x < \pi$$

then $$x = \frac{3\pi}{4}$$

4.11.2 Multiple Angle Formulae

The compound angle formulae, which were established and applied in part 4.11.1, are true for any two angles A and B, and in particular they are true for two equal angles, i.e. when $B = A$ Replacing B by A in the compound angle formulae then $\sin 2A = \sin(A + A)$

$$= \sin A \cos A + \cos A \sin A$$

$$= 2 \sin A \cos A$$

i.e. $$\sin 2A = 2 \sin A \cos A \qquad (1)$$

also $$\cos 2A = \cos(A + A)$$

$$= \cos A \cos A - \sin A \sin A$$

i.e. $$\cos 2A = \cos^2 A - \sin^2 A \qquad (2)$$

The formula in (2) can easily be rewritten so as to give cos 2 A in terms of only one of cos A or sin A

From $\cos 2A = \cos^2 A - \sin^2 A$

$$= \cos^2 A - (1 - \cos^2 A)$$

$$= \cos^2 A - 1 + \cos^2 A$$

$$= 2\cos^2 A - 1$$

i.e. $$\cos 2A = 2\cos^2 A - 1 \qquad (3)$$

also from $\cos 2A = \cos^2 A - \sin^2 A$

$$= 1 - \sin^2 A - \sin^2 A$$

$$= 1 - 2\sin^2 A$$

i.e. $$\cos 2A = 1 - 2\sin^2 A \qquad (4)$$

The expressions (3) and (4) may be rewritten to give

$$2\cos^2 A = 1 + \cos 2A \qquad (5)$$

and $$2\sin^2 A = 1 - \cos 2A \qquad (6)$$

Results (5) and (6) are applied in calculus, especially in the integration of even powers of sines and cosines, and examples are given in Chapter 6.

Also $\tan 2A = \tan (A + A)$

$$= \frac{\tan A + \tan A}{1 - \tan A \tan A}$$

$$= \frac{2\tan A}{1 - \tan^2 A}$$

i.e. $$\tan 2A = \frac{2\tan A}{1 - \tan^2 A} \qquad (7)$$

Examples 4.11.2 (1)

Find the general solution of the equation $\cos 2x - 3\cos x + 2 = 0$.

Solution:

The presence of a term in cos x suggests that it may be possible to rearrange the equation as a quadratic in cos x by writing $\cos 2x = 2\cos^2 x - 1$

so $$\cos 2x - 3\cos x + 2 = 0$$

becomes $$2\cos^2 x - 1 - 3\cos x + 2 = 0$$

i.e. $$2\cos^2 x - 3\cos x + 1 = 0$$

i.e. $$(2\cos x - 1)(\cos x - 1) = 0$$

i.e. $$\cos x = \frac{1}{2} \text{ or } \cos x = 1$$

From $$\cos x = \frac{1}{2}$$

$$x = \pm \frac{\pi}{3} + 2\pi n, \qquad n \ \varepsilon \ Z$$

From $$\cos x = 1$$

$$x = 2\pi n, \qquad n \ \varepsilon \ Z$$

so the complete general solution is

$$x = \pm \frac{\pi}{3} + 2\pi n$$

or $$x = 2\pi n, \qquad n \ \varepsilon \ Z$$

Example 4.11.2(2)

If $\tan \theta = \frac{4}{3}$ and θ is acute find without tables or calculator the values of $\tan 2\theta$, $\tan 4\theta$, and $\tan \frac{\theta}{2}$.

Solution:

In each case we use result (7) of this sub-section i.e.

$$\tan 2A = \frac{2 \tan A}{1 - \tan^2 A}$$

and replace A in turn by θ, then by 2θ, then by $\frac{1}{2}\theta$

If $A = \theta$

then $$\tan 2\theta = \frac{2 \tan \theta}{1 - \tan^2 \theta} = \frac{2 \times 4/3}{1 - (4/3)^2}$$

$$= \frac{\frac{8}{3}}{1 - \frac{16}{9}} = \frac{\frac{8}{3}}{\frac{-7}{9}}$$

$$= \frac{8}{3} \times \frac{9}{-7} = \frac{-24}{7}$$

If $A = 2\theta$

then $$\tan 4\theta = \frac{2 \tan 2\theta}{1 - \tan^2 2\theta}$$

$$= \frac{2 \times \left(\frac{-24}{7}\right)}{1 - \left(-\frac{24}{7}\right)^2} = \frac{-\frac{48}{7}}{1 - \frac{576}{49}}$$

$$\tan 4\theta = \frac{\frac{-48}{7}}{\frac{-527}{49}} = \frac{-48}{7} \times \frac{49}{-527} = \frac{336}{527}$$

If $\quad A = \theta/2$

then $2A = \theta$

and write the known value of $\tan\theta$ $\left(\text{here } \frac{4}{3}\right)$ in the formula

$$\tan\theta = \frac{2\tan\frac{\theta}{2}}{1-\tan^2\frac{\theta}{2}}$$

i.e. $\quad \frac{4}{3} = \frac{2t}{1-t^2} \text{ if } t = \tan\frac{\theta}{2}$

i.e. $\quad 4 - 4t^2 = 6t$

i.e. $\quad 0 = 4t^2 + 6t - 4$

i.e. $\quad 0 = 2t^2 + 3t - 2$

i.e. $\quad 0 = (2t-1)(t+2)$

i.e. $\quad t = \frac{1}{2} \text{ or } -2$

but as θ is acute then $\frac{\theta}{2}$ is also acute

so $\quad \tan\frac{\theta}{2} > 0$

i.e. $\quad \tan\frac{\theta}{2} = \frac{1}{2}.$

Some problems using the compound and double angle formulae may appear in terms of inverse trigonometric functions. In such cases this should be re-set in the direct form, bearing in mind the quadrants in which the angles may be.

Example 4.11.2 (3)

Show that $\tan^{-1} 2 + 2\tan^{-1} 3 = \pi + \cot^{-1} 2$

Solution:

Let $\quad \tan^{-1} 2 = A$

So $\quad \tan A = 2$

Let $\quad \tan^{-1} 3 = B$

so $\qquad \tan B = 3$

Let $\qquad \cot^{-1} 2 = C$

so $\qquad \cot C = 2$

The left hand side can now be expressed as $A + 2B$ and we can use

$$\tan(A + 2B) = \frac{\tan A + \tan 2B}{1 - \tan A \tan 2B}$$

but
$$\tan 2B = \frac{2\tan B}{1 - \tan^2 B}$$

$$= \frac{2 \times 3}{1 - 3^2} = \frac{-3}{4}$$

Since $\qquad \tan B = 3 > 1$

then $\qquad \frac{\pi}{4} < B < \pi/2$

and so $\frac{\pi}{2} < 2B < \pi$ i.e. $2B$ is obtuse hence $\tan 2B < 0$

$$\tan(A + 2B) = \frac{2 + \left(\frac{-3}{4}\right)}{1 - 2\left(-\frac{3}{4}\right)} = \frac{\frac{5}{4}}{\frac{10}{4}} = \frac{1}{2}$$

Now it should be clear that we have added an acute angle A to an obtuse angle $2B$ and the value of $\tan(A + 2B) = \frac{1}{2} > 0$ hence $A + 2B$ must be a reflex angle in the third quadrant.

Now $\quad \cot C = 2$

so $\quad \tan C = \frac{1}{2}$

therefore, $A + 2B = \pi + C$ or $\tan^{-1} 2 + 2\tan^{-1} 3$

$$= \pi + \cot^{-1} 2$$

Examples 4.11.2(4)

Solve the equation

$$\sin^{-1} x + \sin^{-1} 3x = \frac{\pi}{2}.$$

Solution:

Let $A = \sin^{-1} x$ so $\sin A = x$

Let $B = \sin^{-1} 3x$ so $\sin B = 3x$

and we know that $A + B = \frac{\pi}{2}$.

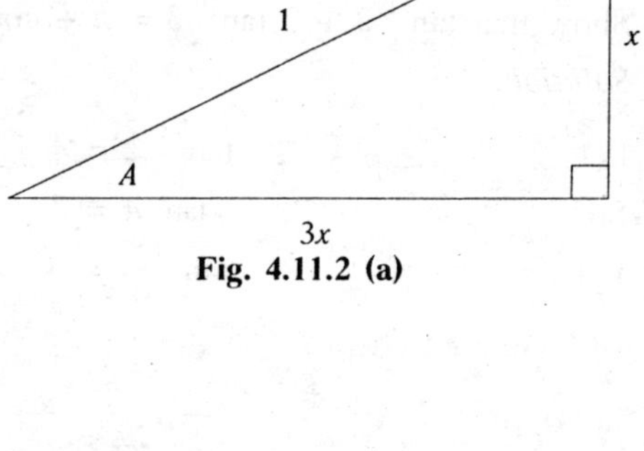

Fig. 4.11.2 (a)

See Fig. 4.11.2(a)

We can set for A an opposite side x and a hypotenuse 1, and also for B an opposite side $3x$ and a hypotenuse 1,

since $\quad A + B = \dfrac{\pi}{2}$

so $\quad x^2 + 9x^2 = 1$

i.e. $\quad 10x^2 = 1$

i.e. $\quad x^2 = \dfrac{1}{10}$

i.e. $\quad x = \pm \dfrac{1}{\sqrt{10}}$

but $\sin^{-1}x > 0$ if $\sin^{-1}x + \sin^{-1}3x = \dfrac{\pi}{2}$

so $\quad x = \dfrac{1}{\sqrt{10}}.$

Examples 4.11.2(5)

Prove that $\cos 3A = 4\cos^3 A - 3\cos A$

Solution:

Expand $\cos(3A)$ as $\cos(2A + A)$

$$= \cos 2A \cos A - \sin 2A \sin A$$
$$= (2\cos^2 A - 1)\cos A - 2\sin A \cos A \sin A$$
$$= 2\cos^3 \text{A} - \cos A - 2\sin^2 A \cos A$$
$$= 2\cos^3 A - \cos A - 2(1 - \cos^2 A)\cos A$$
$$= 2\cos^3 A - \cos A - 2\cos A + 2\cos^3 A$$
$$= 4\cos^3 A - 3\cos A$$

A similar expansion can be used to show that $\sin 3A = 3\sin A - 4\sin^3 A$ and this is left as an exercise to the reader.

4.11.3 Half angle '*t*' Formulae

As we saw in example 4.11.2(2) using the formula

$$\tan 2A = \frac{2\tan A}{1 - \tan^2 A}$$

and writing $\quad 2A = \theta$ and so $A = \dfrac{\theta}{2}$

we have $$\tan\,\theta = \frac{2\tan\dfrac{\theta}{2}}{1-\tan^2\dfrac{\theta}{2}} \quad \text{or} \quad \frac{2t}{1-t^2} \qquad \text{where } t = \tan\frac{\theta}{2}$$

hence $$\tan\,\theta = \frac{2t}{1-t^2} \tag{1}$$

It is possible to find formulae for $\sin\,\theta$ and $\cos\,\theta$ in terms of t.

Set up a right angle triangle where $y = 2t$ and $x = 1 - t^2$ as shown in Fig. 4.11.3(a)

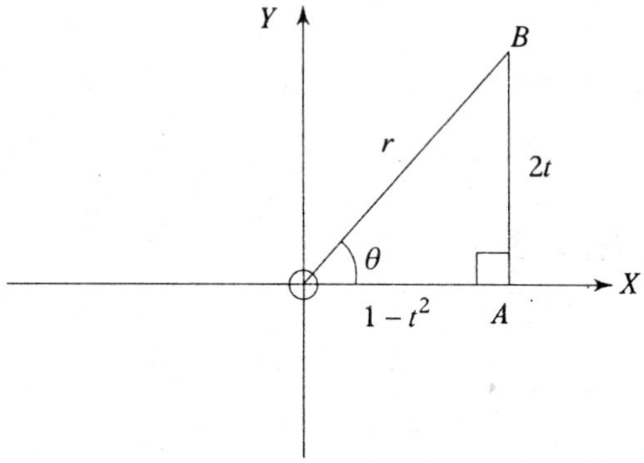

Fig. 4.11.3(a)

From Fig. 4.11.3(a) it can be seen that

$$\begin{aligned} r^2 &= (1 - t^2)^2 + (2t)^2 \\ &= 1 - 2t^2 + t^4 + 4t^2 \\ &= 1 - 2t^2 + t^4 = (1 + t^2)^2 \end{aligned}$$

so $$r = 1 + t^2 \text{ as } r > 0$$

hence we may write

$$\sin\,\theta = \frac{2t}{1+t^2}$$

$$\cos\,\theta = \frac{1-t^2}{1+t^2}$$

These formulae are important in calculus, especially in the integration of some trigonometric functions.

Example 4.11.3(1)

If $\cos\,\theta = \dfrac{5}{13}$ and θ is acute determine the value of $\tan\,\dfrac{\theta}{2}$.

Solution:

If $\tan\,\dfrac{\theta}{2} = t$

then $$\frac{5}{13} = \frac{1-t^2}{1+t^2}$$

i.e. $$5 + 5t^2 = 13 - 13t^2$$

i.e. $$18t^2 = 8$$

i.e. $$9t^2 = 4$$

i.e. $$t^2 = \frac{4}{9}$$

so $$t = \pm\ \frac{2}{3}$$

but θ is acute

so $$\tan\ \frac{\theta}{2} = \frac{2}{3}.$$

Example 4.11.3(2)

Express $$\frac{2\cos\theta}{5\cos\theta + 3\sin\theta + 4}$$

in terms of $t = \tan\ \dfrac{\theta}{2}$.

Solution:

$$\frac{2\cos\theta}{5\cos\theta + 3\sin\theta + 4} = \frac{\dfrac{2\,(1-t^2)}{(1+t^2)}}{\dfrac{5\,(1-t^2)}{(1+t^2)} + \dfrac{3\times 2t}{(1+t^2)} + 4}$$

$$= \frac{\dfrac{2\,(1-t^2)}{(1+t^2)}}{\dfrac{5-5t^2+6t+4+4t^2}{(1+t^2)}}$$

$$= \frac{2\,(1-t^2)}{(1+t^2)} \times \frac{(1+t^2)}{9+6t-t^2}$$

$$= \frac{2\,(1-t^2)}{9+6t-t^2}.$$

Example 4.11.3(3)

Solve the equation $4\sin\ \theta + 3\cos\ \theta = 4$ for the domain $0° \le \theta \le 360°$.

It is not possible in this case to form a simple equation in tan θ, since in this case dividing both sides by cos θ would bring a term in sec θ on the right hand side.

However we can write the left hand side in terms of $t = \tan\frac{\theta}{2}$

i.e. $4\sin\theta + 3\cos\theta = 4$

becomes $\frac{4\times 2t}{1+t^2} + \frac{3(1-t^2)}{1+t^2} = 4$

i.e. $8t + 3 - 3t^2 = 4 + 4t^2$

i.e. $0 = 7t^2 - 8t + 1$

i.e. $0 = (7t - 1)(t - 1)$

i.e. $t = \frac{1}{7}$ or 1

so $\tan\frac{\theta}{2} = \frac{1}{7}$ or $\tan\frac{\theta}{2} = 1$

i.e. $\frac{\theta}{2} = 8.13° + 180n°$

or $\frac{\theta}{2} = 45° + 180n°\ n\ \varepsilon\ Z$

i.e. $\theta = 16.26° + 360n°$

or $\theta = 90° + 360\ n°$

on domain $0° \le \theta \le 360°$

$\theta = 16.26$ or $90°$

4.12 THE EXPRESSION *A* cos θ + *B* sin θ

4.12.1 The Graphs of $y = k\sin\theta$ and $y = k\cos\theta$

Fig. 4.12.1 (a) shows on the same axes the graphs of

$$y = \frac{1}{2}\sin\theta,\ y = \sin\theta,$$

$y = 2\sin\theta$, $y = 3\sin\theta$, for the domain $-36\ 0° \le \theta \le 360°$.

Clearly all the graphs have the same period, which is 360°. However, the maximum and minimum values of y are different for each graph.

Since $-1 \le \sin\theta \le 1$

then $-\frac{1}{2} \le \frac{1}{2}\sin\theta \le 1/2$

$-2 \le 2\sin\theta \le 2$

$-3 \le 3\sin\theta \le 3$

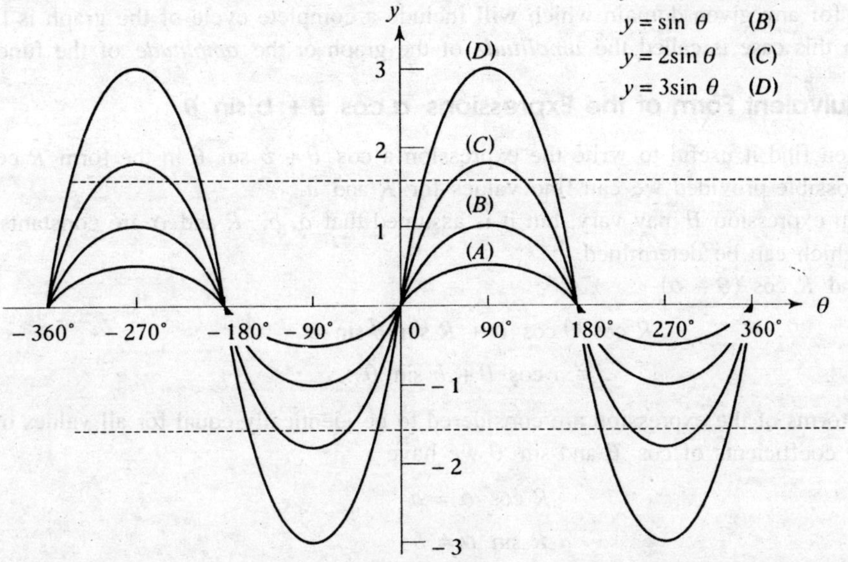

Fig. 4.12.1(a)

and so for the graph of $y = k \sin \theta$ the range of values for y will be $-k \le y \le k$ for $k > 0$.

Geometrically the graph of $y = k \sin \theta$ can be obtained from the graph of $y = \sin \theta$ by a stretch of scale factor k parallel to the y-axis with the x-axis invariant, since any point $(\theta, \sin \theta)$ on the graph of $y = \sin \theta$ is mapped to a point $(\theta, k \sin \theta)$ on the graph of $y = k \sin \theta$.

In a similar way the graph of $y = k \cos \theta$ can be obtained from the graph of $y = \cos \theta$ by a stretch of scale factor k parallel to the y-axis with the x-axix invariant, as shown in Fig. 4.12.1(b).

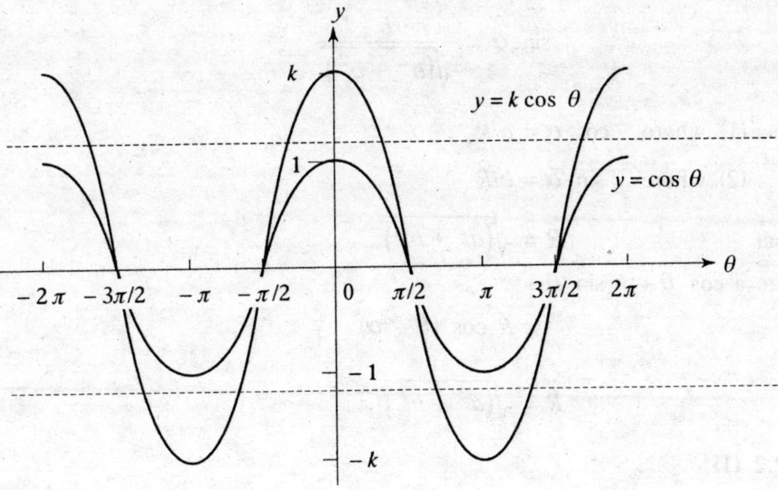

Fig. 4.12.1(b)

The maximum distance from the horizontal (or central) axis of the graphs of $y = k \sin \theta$ and of $y = k \cos \theta$ for any given domain which will include a complete cycle of the graph is $|k|$ and the value $|k|$ in this case is called the *amplitude* of the graph or the *amplitude* of the function.

4.12.2 Equivalent Form of the Expressions $a \cos \theta + b \sin \theta$

We may often find it useful to write the expression a $\cos \theta + b \sin \theta$ in the form $R \cos (\theta - \alpha)$.

This is possible provided we can find values for R and a.

In such an expression θ may vary, but it is assumed that a, b, R and α are constants which are known or which can be determined.

We expand $R \cos (\theta - \alpha)$

as $$R \cos \theta \cos \alpha + R \sin \theta \sin \alpha$$
$$\equiv a \cos \theta + b \sin \theta$$

i.e. the two forms of the expression are considered to be identically equal for all values of θ). If we compare the coefficients of $\cos \theta$ and $\sin \theta$ we have

$$R \cos \alpha = a \qquad (1)$$
$$R \sin \alpha = b \qquad (2)$$

if we divide (2) by (1)

we have $\tan \alpha = \dfrac{b}{a}$

using Fig. 4.12.2 (a)

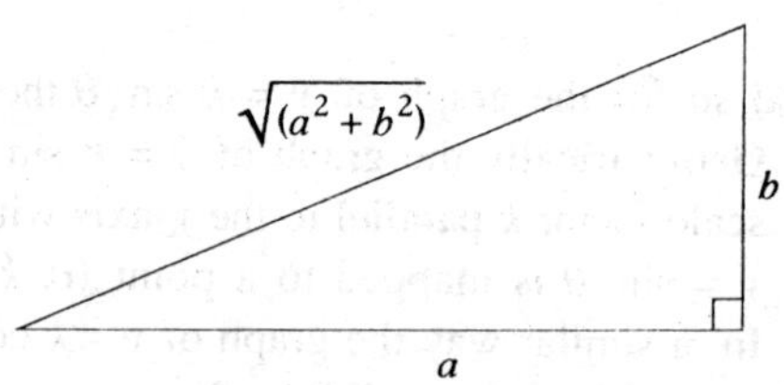

Fig. 4.12.2(a)

we recognise $\cos \alpha = \dfrac{a}{\sqrt{(a^2 + b^2)}}$ and

$$\sin \alpha = \frac{b}{\sqrt{(a^2 + b^2)}}$$

and using either (1) where $\cos \alpha = a/R$

or (2) where $\sin \alpha = b/R$

we observe that $R = \sqrt{(a^2 + b^2)}$

so we can write a $\cos \theta + b \sin \theta$

$$\equiv R \cos (\theta - \alpha)$$

where $R = \sqrt{(a^2 + b^2)}$.

Example 4.12.2 (1)

Express $8 \cos \theta + 15 \sin \theta$ in the form $R \cos (\theta - \alpha)$ giving the values of R and α.

Solution:

Let $R \cos(\theta - \alpha)$

$$\equiv R(\cos\theta\cos\alpha + \sin\theta\sin\alpha)$$

$$\equiv R\cos\theta\cos\alpha + R\sin\theta\sin\alpha$$

$$\equiv 8\cos\theta + 15\sin\theta$$

so $R\cos\alpha = 8$, $R\sin\alpha = 15$

and $\tan\alpha = \dfrac{15}{8}$

using Fig. 4.12.2 (b)

$$R = \sqrt{(8^2 + 15^2)} = 17$$

hence $R = 17$, $\alpha = \tan^{-1}\dfrac{15}{8} = 61.93°$

and so $8\cos\theta + 15\sin\theta$

$$\equiv 17\cos(\theta - 61.93°)$$

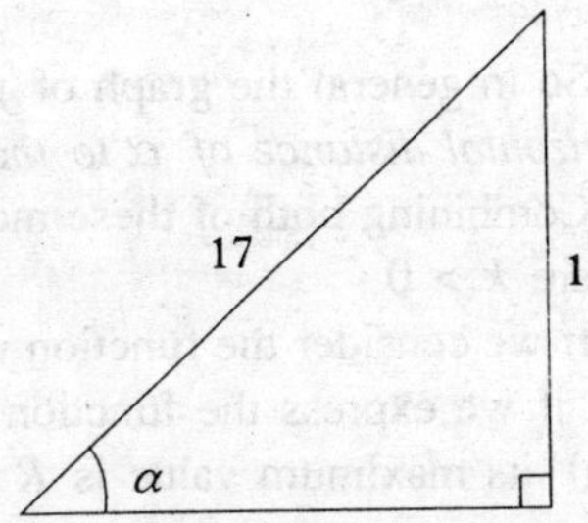

Fig. 4.12.2 (b)

In some problems it may be easier to use the expression of one of the other compound angle formulae, say expressing $a\cos\theta + b\sin\theta$ in the form $R\sin(\theta + \alpha)$ so that $R\sin(\theta + \alpha)$

$$\equiv R(\sin\theta\cos\alpha + \cos\theta\sin\alpha)$$

$$\equiv R\sin\theta\cos\alpha + R\cos\theta\sin\alpha$$

$$\equiv a\cos\theta + b\sin\theta$$

Comparing coefficients of $\sin\theta$ and of $\cos\theta$,

$$R\cos\alpha = b$$
$$R\sin\alpha = a$$

and this leads to $\tan\alpha = a/b$

$$R = \sqrt{(a^2 + b^2)}$$

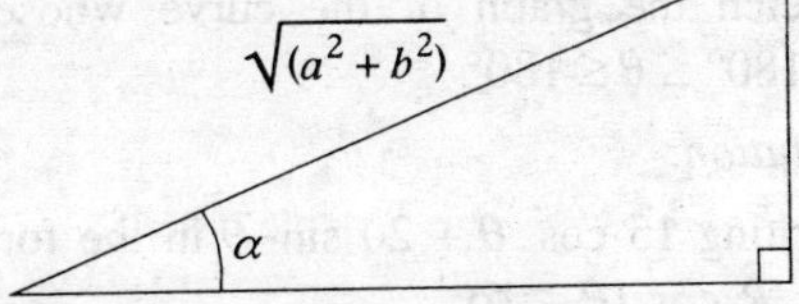

Fig. 4.12.2(c)

as in Fig. 4.12.2 (c)

N.B: α is not the same angle as the α which we found earlier when we used the form $R\cos(\theta - \alpha)$.

It is possible to use further variations of the form, such as $R\cos(\theta + \alpha)$, $R\sin(\theta - \alpha)$. In such cases it is better to work from a comparison of the coefficients of $\cos\theta$ and $\sin\theta$ as was done in example 4.12.1(1) rather than to quote R and α.

For example the expression $7\cos\theta + 24\sin\theta$ could be written in either of the forms $R\cos(\theta - \alpha)$ or $R\sin(\theta + \alpha)$.

The expression $3\cos\theta - 4\sin\theta$ could be written in either of the form $R\cos(\theta + \alpha)$ or $R\sin(\theta - \alpha)$.

The expression in the latter case would have angle α in the third quadrant as both $R\sin\alpha$ and $R\cos\alpha$ would be negative. Further investigation is left as an exercise to the reader.

4.12.3 The Graph of the Equation $y = a \cos \theta + b \sin \theta$

We saw in subsection 4.12.1 that the graph of $y = k \cos \theta$ where $k > 0$ is similar to a standard cosine curve but that the graph has a maximum value of k and a minimum value of $-k$, i.e. the curve will have amplitude k.

We saw earlier in example 4.9.2 (1) of sub-section 4.9.2 that the graph of $y = \cos\left(\theta - \frac{\pi}{4}\right)$ can be obtained from the graph of $y = \cos \theta$ by moving the graph of $y = \cos \theta$ a horizontal distance of $\frac{\pi}{4}$ to the right.

So in general the graph of $y = \cos(\theta - \alpha)$ can be obtained by moving the graph of $y = \cos \theta$ *a horizontal distance of* α *to the right.*

Combining both of these modifications it is then possible to sketch the graph of $y = k \cos(\theta - \alpha)$ where $k > 0$.

If we consider the function $y = a \cos \theta + b \sin \theta$ the graph for this equation is not easy to visualise but if we express the function $a \cos \theta + b \sin \theta$ in the form $R \cos(\theta - \alpha)$ then

(a) its maximum value is R when $\theta = \alpha + 2\pi n,\ n \varepsilon Z$

(or $\alpha + 360n°$ if α is in degrees)

(b) its minimum value is $-R$ when $\theta = \alpha + \pi + 2\pi n,\ n \varepsilon Z$

(or $\alpha + 180° + 360n°$ if α is in degrees)

(c) its position is at a distance α to the right of the standard curve.

Example 4.12.3 (1)

Sketch the graph of the curve whose equation is $y = 15 \cos \theta + 20 \sin \theta$ on the domain $-180° \le \theta \le 180°$

Solution:

Writing $15 \cos \theta + 20 \sin \theta$ in the form $R \cos(\theta - \alpha)$

i.e. $R \cos(\theta - \alpha)$

$$\equiv R(\cos \theta \cos \alpha + \sin \theta \sin \alpha)$$
$$\equiv R \cos \theta \cos \alpha + R \sin \theta \sin \alpha$$
$$\equiv 15 \cos \theta + 20 \sin \theta$$

then $\quad R \cos \alpha = 15$

$\quad R \sin \alpha = 20$

leading to $\quad \tan \alpha = \frac{20}{15} = \frac{4}{3}$

i.e. $\quad \alpha = 53.13°$

See Fig. 4.12.3 (a)

$$R = \sqrt{(15^2 + 20^2)} = 25$$

so the expression $15 \cos \theta + 20 \sin \theta$ can be written as $25 \cos(\theta - 53.13°)$ and the graph of the equation

$$y = 15 \cos \theta + 20 \sin \theta$$
$$= 25 \cos (\theta - 53.13°)$$

is a cosine curve with amplitude 25 and phase angle 53.13° to the right.

The sketch for the domain $-180° \le \theta \le 180°$ is shown on Fig. 4.12.3 (b)

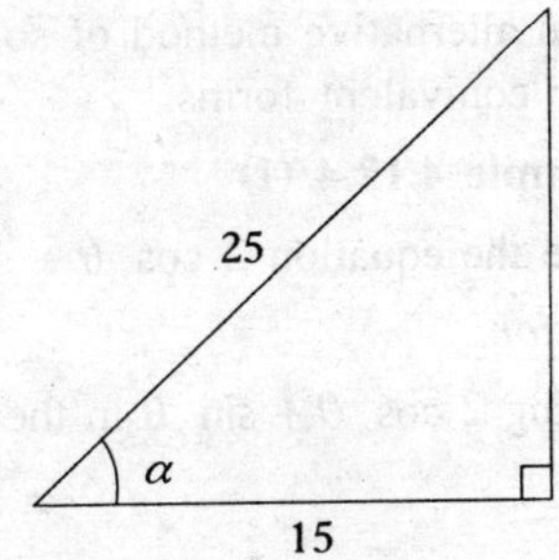

Fig. 4.12.3(a)

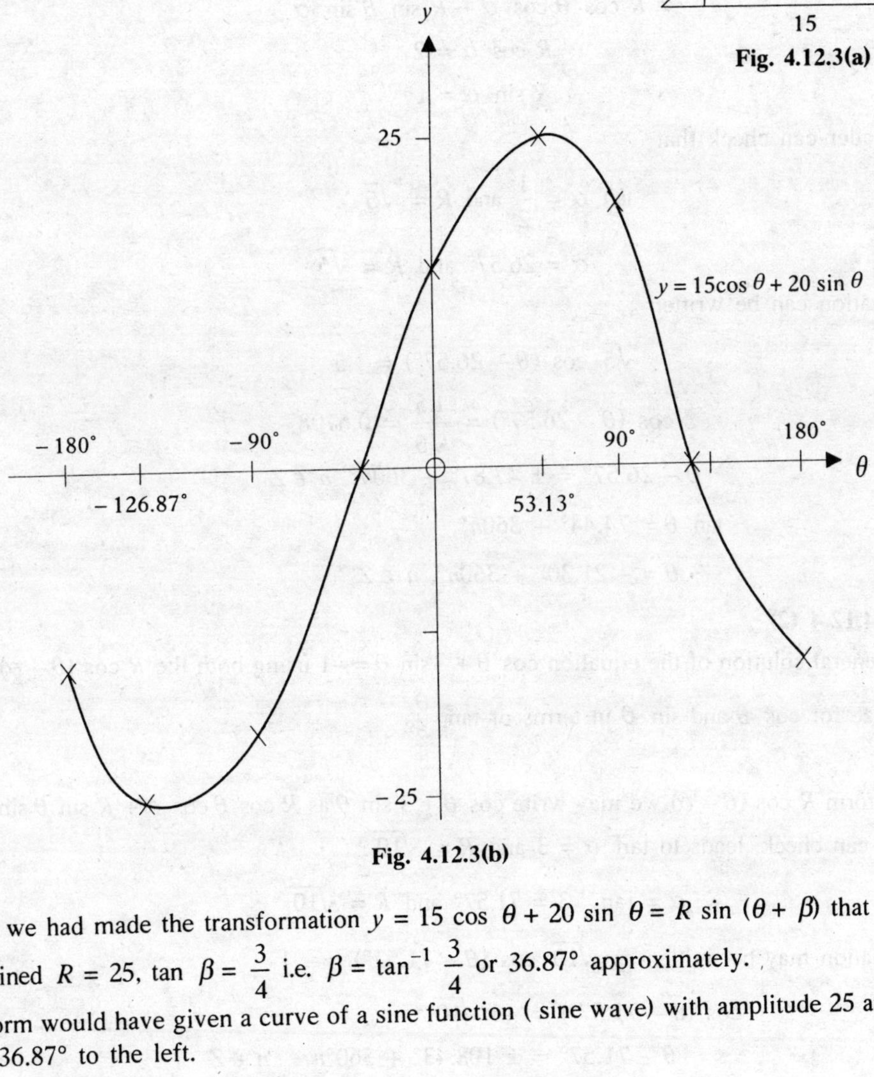

Fig. 4.12.3(b)

If instead we had made the transformation $y = 15 \cos \theta + 20 \sin \theta = R \sin (\theta + \beta)$ that we would have obtained $R = 25$, $\tan \beta = \frac{3}{4}$ i.e. $\beta = \tan^{-1} \frac{3}{4}$ or 36.87° approximately.

This form would have given a curve of a sine function (sine wave) with amplitude 25 and a phase angle of 36.87° to the left.

4.12.4 Solutions of the Equation $a \cos \theta + b \sin \theta = c$

We have already seen in example 4.11.3 (3) how such an equation may be solved using the formulae for $\cos \theta$ and $\sin \theta$ in terms of $t = \tan \frac{\theta}{2}$.

An alternative method of solution is now available usisng the form $R\cos(\theta-\alpha)$ or one of the other equivalent forms.

Example 4.12.4 (1)

Solve the equation $2\cos\theta+\sin\theta=1.5$, giving the general solution.

Solution:

Writing $2\cos\theta+\sin\theta$ in the form $R\cos(\theta-\alpha)$

$$\equiv R\cos\theta\cos\alpha+R\sin\theta\sin\alpha$$

and so
$$R\cos\alpha=2$$
$$R\sin\alpha=1$$

and the reader can check that

$$\tan\alpha=\frac{1}{2}\text{ and }R=\sqrt{5}$$

i.e.
$$\alpha=26.57°\text{ and }R=\sqrt{5}$$

so the equation can be written

as
$$\sqrt{5}\cos(\theta-26.57°)=1.5$$

i.e.
$$\cos(\theta-26.57°)=\frac{1.5}{\sqrt{5}}=0.6708$$

i.e.
$$\theta-26.57°=\pm 47.87°+360n°,\ n\ \varepsilon\ Z$$

leading
$$\tan\theta=74.44°+360n°,$$

or
$$\theta=-21.30°+360n°,\ n\ \varepsilon\ Z$$

Example 4.12.4 (2)

Find the general solution of the equation $\cos\theta+3\sin\theta=-1$ using both the $R\cos(\theta-\alpha)$ form and the formulae for $\cos\theta$ and $\sin\theta$ in terms of $\tan\frac{\theta}{2}$.

Solution:

Using the form $R\cos(\theta-\alpha)$ we may write $\cos\theta+3\sin\theta$ as $R\cos\theta\cos\alpha+R\sin\theta\sin\alpha$ which, the reader can check, leads to $\tan\alpha=3$ and $R=\sqrt{10}$.

i.e.
$$\alpha=\tan^{-1}3=71.57°\text{ and }R=\sqrt{10}$$

so the equation may be written as $\sqrt{10}\cos(\theta-71.57°)=-1$

i.e.
$$\cos(\theta-71.57°)=-1/\sqrt{10}$$

i.e.
$$\theta-71.57°=\pm 108.43°+360°n,\quad n\ \varepsilon\ Z$$

i.e.
$$\theta=-36.86°+360n°$$

or
$$\theta=180°+360n°,\quad n\ \varepsilon\ Z$$

and the equation becomes in $t = \tan\dfrac{\theta}{2}$

$$\frac{1-t^2}{1+t^2} + \frac{3\times 2t}{1+t^2} = -1$$

i.e. $$1 - t^2 + 6t = -1 - t^2 \;*$$

i.e. $$6t = -2$$

i.e. $$t = -1/3$$

so as $$\tan\frac{\theta}{2} = -1/3$$

$$\frac{\theta}{2} = -18.43° + 180n° \qquad n \in Z$$

i.e. $$\theta = -36.86° + 360n°, \qquad n \in Z$$

We appear to have lost one set of solutions, and this happened at the line marked *.

In some problems cancelling out a term of the form $-t^2$ on each side of an equation may not affect the solution, but here we know that there is a possibility that $t = \pm\,\infty$ could be a solution.

It can be checked rigorously even from the first line of the equation, since on the left hand side we can take the limit as $t^2 \to \infty$ (equivalent to $t \to \infty$ or $t \to -\infty$) on each expression,

i.e. $$\lim_{t^2\to\infty}\frac{1-t^2}{1+t^2} + \lim_{t^2\to\infty}\frac{6t}{1+t^2}$$

$$= \lim_{t^2\to\infty}\frac{1/t^2-1}{1/t^2+1} + \lim_{t^2\to\infty}\frac{6/t}{1/t^2+1}$$

$$= -1/1 + \frac{0}{0+1} = -1$$

and this is the right hand side.

The reader who finds this kind of argument difficult to understand may omit it for the time being, and proceed on with the rest of this section.

The solution $$t = \pm\,\infty$$

or $$\tan\frac{\theta}{2} = \pm\,\infty$$

gives $$\frac{\theta}{2} = 90° + 180n°$$

i.e. $$\theta = 180° + 360n°,$$

and so the complete general solution (as found by the earlier method)

is $$\theta = -36.86° + 360n°$$

or $$\theta = 180° + 360n°, \; n \in Z$$

Example 4.12.4 (3)

Express 7 sin θ + 24 cos θ in the form $R \sin(\theta + \alpha)$, giving the values of R and α, and show that if $f(\theta) = 5 + 7\sin\theta + 24\cos\theta$

then $$-20 \le f(\theta) \le 30$$

Solution:

Setting 7 sin θ + 24 cos θ in the form $R \sin(\theta + \alpha)$

$$\equiv R(\sin\theta\cos\alpha + \cos\theta\sin\alpha)$$

$$\equiv R\sin\theta\cos\alpha + R\cos\theta\sin\alpha$$

so $$R\cos\alpha = 7$$

$$R\sin\alpha = 24$$

See Fig. 4.12.4 (a)

$$\tan\alpha = \frac{24}{7} \text{ and } R = \sqrt{(24^2 + 7^2)} = 25$$

i.e. $$\alpha = \tan^{-1}\frac{24}{7} = 73.74° \text{ and } R = 25$$

so $$7\sin\theta + 24\cos\theta \equiv 25\sin(\theta + 73.74°)$$

and $$f(\theta) = 5 + 7\sin\theta + 24\cos\theta$$

$$= 5 + 25\sin(\theta + 73.74°)$$

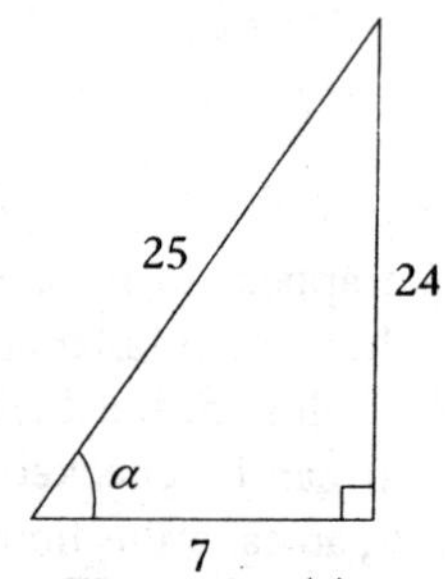

Fig. 4.12. 4(a)

We know that the maximum value of sin $(\theta + 73.74°)$ is 1, and to the maximum value of $f(\theta)$ is $5 + 25 \times 1 = 30$

and the minimum value of sin $(\theta + 73.74°)$ is -1,

and so the minimum value of

$$f(\theta) \text{ is } 5 + 25 \times -1 = -20$$

so $$-20 \le f(\theta) \le 30.$$

4.13 FACTOR FORMULAE

To factorise means expressing a sum or difference of terms as a product. The set of identities described as factor formulae (the last to be learnt in this section) will convert expressions of the form sin A + sin B into a product, so expressing the same in factor form.

4.13.1 From Products to Sums

First we shall use the original compound angle formulae

$$\sin A\cos B + \cos A\sin B = \sin(A + B)$$

$$\sin A\cos B - \cos A\sin B = \sin(A - B)$$

$$\text{Adding } 2 \sin A \cos B = \sin (A + B) + \sin (A - B) \quad (1)$$

$$\text{Subtracting } 2 \cos A \sin B = \sin (A + B) - \sin (A - B) \quad (2)$$

$$\text{similarly } \cos A \cos B - \sin A \sin B = \cos (A + B)$$

$$\cos A \cos B + \sin A \sin B = \cos (A - B)$$

$$\text{Adding } 2 \cos A \cos B = \cos (A + B) + \cos (A - B) \quad (3)$$

$$\text{Subtracting } -2 \sin A \sin B = \cos (A + B) - \cos (A - B) \quad (4)$$

Example 4.13.1 (1)

Show that $$2 \sin 75° \cos 45° = \frac{\sqrt{3}+1}{2}.$$

Solution:

This problem could be tackled by writing sin 75° as sin (45° + 30°) and then expanding using the formula sin $(A + B)$

However if we use (1) above with A = 75° cos 45°

we have 2 sin 75° cos 45°

$$= \sin (75° + 45°) + \sin (75° - 45°)$$

$$= \sin 120° + \sin 30°$$

$$= \frac{\sqrt{3}}{2} + \frac{1}{2} = \frac{\sqrt{3}+1}{2}.$$

Example 4.13.1 (2)

Given that $\cos 72° = \dfrac{\sqrt{5}+1}{4}$

show that

$$\sin 81° \sin 9° = \frac{\sqrt{5}-1}{8}.$$

Solution:

Using (4)

$$-2 \sin 81° \sin 9° = \cos (81° + 9°) - \cos (81° - 9°)$$

i.e. $$-2 \sin 81° \sin 9° = \cos 90° - \cos 72°$$

i.e. $$\sin 81° \sin 9° = -\frac{1}{2} \cos 90° - \frac{1}{2} (-\cos 72°)$$

$$= -\frac{1}{2} \times 0 - \frac{1}{2} \left(\frac{-(\sqrt{5}+1)}{4} \right)$$

i.e. $$\sin 81° \sin 9° = \frac{\sqrt{5}-1}{8}.$$

Example 4.13.1 (3)

Show that $\cos\alpha\cos(\alpha+\beta)\cos(\alpha+2\beta)$

$$= \cos(\alpha+\beta)[\cos(\alpha+\beta)-\sin\beta][\cos(\alpha+\beta)+\sin\beta].$$

Solution:

Since the factor $\cos(\alpha+\beta)$ appears on both sides of the identity we shall take the other factors of the left hand side and write their product as a sum of cosines using (3)

i.e.

$$\cos\alpha\cos(\alpha+2\beta)$$
$$= \cos(\alpha+2\beta)\cos\alpha$$
$$= \frac{1}{2}[\cos(2\alpha+2\beta)+\cos 2\beta]$$

so the left hand side can be written as

$$\frac{1}{2}\cos(\alpha+\beta)[\cos(2\alpha+2\beta)+\cos 2\beta]$$
$$= \frac{1}{2}\cos(\alpha+\beta)[2\cos^2(\alpha+\beta)-1+1-2\sin^2\beta]$$

using the appropriate double angle formulae
so the left hand side is

$$2\times\frac{1}{2}\cos(\alpha+\beta)[\cos^2(\alpha+\beta)-\sin^2\beta]$$
$$= \cos(\alpha+\beta)[\cos(\alpha+\beta)-\sin\beta][\cos(\alpha+\beta)+\sin\beta]$$
$$= \text{the right hand side.}$$

4.13.2 Sums to Products

Referring back to the four formulae (1) to (4) at the beginning of sub-section 4.13.1 we had shown that

$$2\sin A\cos B = \sin(A+B)+\sin(A-B) \quad (1)$$
$$2\cos A\sin B = \sin(A+B)-\sin(A-B) \quad (2)$$
$$2\cos A\cos B = \cos(A+B)+\cos(A-B) \quad (3)$$
$$-2\sin A\sin B = \cos(A+B)-\cos(A-B) \quad (4)$$

If we now write $A+B=P$

and $A-B=Q$

adding $2A=P+Q$

or $A=\frac{1}{2}(P+Q)$

and subtracting $2B=P-Q$

or $$B = \frac{1}{2}(P - Q)$$

So that (1) can now be written as

$$2 \sin \frac{1}{2}(P + Q) \cos \frac{1}{2}(P - Q) = \sin P + \sin Q$$

or taking the right hand side first,

$$\sin P + \sin Q = 2 \sin \frac{1}{2}(P + Q) \cos \frac{1}{2}(P - Q) \quad (5)$$

(2) can now be written as

$$\sin P - \sin Q = 2 \cos \frac{1}{2}(P + Q) \sin \frac{1}{2}(P - Q) \quad (6)$$

(3) can now be written as

$$\cos P + \cos Q = 2 \cos \frac{1}{2}(P + Q) \cos \frac{1}{2}(P - Q) \quad (7)$$

(4) can now be written as

$$\cos P - \cos Q = -2 \sin \frac{1}{2}(P + Q) \sin \frac{1}{2}(P - Q) \quad (8)$$

For example if we want to express $\sin 10\theta - \sin 6\theta$ as a product we use (6) and write it as

$$2 \cos \frac{10\theta + 6\theta}{2} \sin \frac{10\theta - 6\theta}{2}$$

$$= 2 \cos 8\theta \sin 2\theta.$$

We can often remember these eight formulae best in words as

(1) 2 sin cos = sin (sum) + sin (difference)

(2) 2 cos sin = sin (sum) – sin (difference)

(3) 2 cos cos = cos (sum) + cos (difference)

(4) –2 sin sin = cos (sum) – cos (difference)

(5) sin + sin = 2 sin (half sum) cos (half difference)

(6) sin – sin = 2 cos (half sum) sin (half difference)

(7) cos + cos = 2 cos (half sum) cos (half difference)

(8) cos – cos = – 2 sin (half sum) sin (half difference)

The reader may have difficulty with the formulae (4) and (8) because of the presence of the negative sign before the factor form.

However, if all the angles are acute it should be remembered that as an acute angle becomes larger its cosine becomes smaller and so in (8) if P and Q are both acute and $P > Q$ then $\cos P < \cos Q$ so that $\cos P - \cos Q$ will be negative.

Example 4.13.2 (1)

Solve using two methods the equation sin 6 x – sin 2 x = 0 giving the general solution.

Solution:

This equation can be solved by writing $\sin 6x = \sin 2x$

and hence $6x = 2x + 2\pi n$

or $6x = \pi - 2x + 2\pi n \quad n \in Z$

leading to $x = \dfrac{\pi n}{2}$

or $x = \dfrac{\pi}{8} + \dfrac{\pi n}{4} \quad n \in Z$

Alternatively using a difference of sines

we have $\sin 6x - \sin 2x = 0$

i.e. $2 \cos \dfrac{1}{2} (6x + 2x) \sin \dfrac{1}{2} (6x - 2x) = 0$

i.e. $2 \cos 4x \sin 2x = 0$

i.e. $\cos 4x = 0$ or $\sin 2x = 0$

From $\cos 4x = 0$

$$4x = \pm \frac{\pi}{2} + 2\pi n, \quad n \in Z$$

or $x = \pm \dfrac{\pi}{8} + \dfrac{\pi n}{2} \quad n \in Z$

From $\sin 2x = 0$

$$2x = \pi n, \quad n \in Z$$

or $x = \dfrac{\pi n}{2}, \quad n \in Z$

So the complete general solution is

$$x = \pm \frac{\pi}{8} + \frac{\pi n}{2}$$

or $x = \dfrac{\pi n}{2}, \quad n \in Z$

The reader may be curious to know whether the solutions we obtained using these two different methods would actually give us the same set of angles as formulae for the general solution appear to be different in each case. The explanation is quite involving, and the reader may omit it if it is difficult to understand, and proceed to study example 4.13.2 (2). The solution $x = \dfrac{n\pi}{2}, n \in Z$ has appeared in both sets but the form of the other solution appears as $x = \frac{\pi}{8} + \frac{\pi n}{4}$, $n \varepsilon Z$ using the first method, and we call this solution set A, whereas the form of the other solution appear as $x = \pm\frac{\pi}{8} + \frac{\pi n}{2}$, $n \varepsilon Z$, using the second method, and we call this solution set B.

It is required to show that the sets A and B are equal, i.e. that every angle in set A also belongs to set B and that every angle in set B also belongs to set A.

Taking the solution in set A where solutions are of the form $x = \frac{\pi}{8} + \frac{\pi n}{4}$, $n \varepsilon Z$ then we may take n as either an even integer, say $n = 2p$ where p is an integer or $n = 2q - 1$ where q is an integer. If n is an even integer, say $n = 2p$, for a typical angle in the set, we can write the solution as

$$\frac{\pi}{8} + \frac{\pi n}{4} = \frac{\pi}{8} + \frac{\pi(2p)}{4} = \frac{\pi}{8} + \frac{\pi p}{2}$$

and this angle belongs to set B.

If n is an odd integer, say $n = 2q - 1$ for a typical angle in the set, we can write the solution as

$$\begin{aligned}\frac{\pi}{8} + \frac{\pi n}{4} &= \frac{\pi}{8} + \frac{(2q-1)\pi}{4} \\ &= \frac{\pi}{8} + \frac{\pi(2q)}{4} - \frac{\pi.1}{4} \\ &= \frac{\pi}{8} + \frac{\pi(2q)}{4} - \frac{2\pi}{8} \\ &= -\frac{\pi}{8} + \frac{\pi q}{2}\end{aligned}$$

and again this angle belongs to set B.

Now consider all angles in set B which have the form $\pm\frac{\pi}{8} + \frac{\pi n}{2}$

If we take $n = k$, say, where $k \varepsilon Z$, two typical angles would be

$$\frac{\pi}{8} + \frac{\pi k}{2} \quad \text{and} \quad -\frac{\pi}{8} + \frac{\pi k}{2}$$

The first angle could be written as

$$\frac{\pi}{8} + \frac{\pi k}{2} = \frac{\pi}{8} + \frac{\pi(2k)}{4}$$

and since 2k is an integer.
and this angle is of the form

$$\frac{\pi}{8} + \frac{\pi n}{4}, \quad n \varepsilon Z$$

and so belongs to set A

The second angle would be written as

$$\frac{-\pi}{8}+\frac{\pi k}{2}$$

and rewritten as

$$\frac{\pi}{8}-\frac{\pi}{8}-\frac{\pi}{8}+\frac{\pi k}{2}$$

$$=\frac{\pi}{8}-\frac{2\pi}{8}+\frac{\pi(2k)}{4}$$

$$=\frac{\pi}{8}-\frac{\pi}{4}+\frac{\pi(2k)}{4}$$

$$=\frac{\pi}{8}+\frac{\pi(2k-1)}{4}$$

but $2k - 1$ is an integer so this angle is also of the form

$$\frac{\pi}{8}+\frac{\pi n}{4},\ n\ \varepsilon\ Z$$

and so belongs to set A.

We have now shown that any angle in set A also belongs to set B, and that any angle in set B also belongs to set A, and so the two solution sets A and B are equal.

Example 4.13.2 (2)

Show that

$$\cos\theta-\cos 5\theta+\cos 9\theta-\cos 13\theta = 4\sin 2\theta\sin 7\theta\cos 4\theta$$

Solution:

We shall try to express the left hand side in factor form using some of the results (5), (6),(7), (8) established at the beginning of 4.13.2.

$$\text{L.H.S.} = \cos\theta-\cos 5\theta+\cos 9\theta-\cos 13\theta$$

$$=-2\sin\frac{\theta+5\theta}{2}\sin\frac{\theta-5\theta}{2}-2\sin\frac{9\theta+13\theta}{2}\sin\frac{9\theta-13\theta}{2}$$

$$=-2\sin 3\theta\sin(-2\theta)-2\sin 11\theta\sin(-2\theta)$$

$$=-2\sin 3\theta(-\sin 2\theta)-2\sin 11\theta(-\sin 2\theta)$$

$$=2\sin 3\theta\sin 2\theta+2\sin 2\theta\sin 11\theta$$

$$=2\sin 2\theta(\sin 3\theta+\sin 11\theta)$$

$$\text{L.H.S.} =2\sin 2\theta\left(2\sin\frac{3\theta+11\theta}{2}\cos\frac{3\theta-11\theta}{2}\right)$$

$$=2\sin 2\theta(2\sin 7\theta\cos(-4\theta))$$

$$=2\sin 2\theta(2\sin 7\theta\cos 4\theta)$$

$$=4\sin 2\theta\sin 7\theta\cos 4\theta$$

$$=\text{R.H.S.}$$

The reader could try to solve this problem by starting with the right hand side and using results (1), (2), (3), (4) established in sub-section 4.13.1 to arrive at the left hand side. This is left as an exercise to the reader.

Some problems using the factor formulae involve identities in which the trigonometric ratios are those of the angles of a triangle. In such cases the best method of attack is to start by changing one of the angles, say C, to $180° - (A + B)$, simplifying either the left hand side or the right hand side in terms of A and B only, and then re-writing one or more trigonometric functions in terms of C at the end to arrive at the expression given on the other side. The method is illustrated in examples 4.13.2(3) and (4).

Example 4.13.2(3)

If A, B, C are angles of a triangle prove the identity.

$$\text{Cos } A + \cos B - \cos C = 4 \cos \frac{A}{2} \cos \frac{B}{2} \sin \frac{C}{2} - 1.$$

Solution:

$$\text{L.H.S.} = \cos A + \cos B - \cos C$$

$$= \cos A + \cos B - \cos (180° - (A + B))$$ (switching out of C and expressing all the functions in terms of A and B)

$$= \cos A + \cos B - -\cos (A + B)$$

$$\text{L.H.S.} = 2 \cos \frac{A+B}{2} \cos \frac{A-B}{2} + \cos (A + B)$$

$$= 2 \cos \frac{A+B}{2} \cos \frac{A-B}{2} + 2 \cos^2 \frac{A+B}{2} - 1$$ (using $A + B$ as 2θ in the double angle formula for $\cos 2\theta$)

$$= 2 \cos \frac{A+B}{2} \left(\cos \frac{A-B}{2} + \cos \frac{A+B}{2} \right) - 1$$

$$= 2 \cos \frac{A+B}{2} \left(2 \cos \frac{A}{2} \cos \frac{-B}{2} \right) - 1$$

$$= 2 \cos \frac{180° - C}{2} \left(2 \cos \frac{A}{2} \cos \frac{B}{2} \right) - 1$$ (switching back to have a function of angle C)

$$= 2 \cos \left(90° - \frac{C}{2} \right) \left(2 \cos \frac{A}{B} \cos \frac{B}{2} \right) - 1$$

$$= 4 \sin \frac{C}{2} \cos \frac{A}{2} \cos \frac{B}{2} - 1$$

$$= 4 \sin \frac{C}{2} \cos \frac{A}{2} \cos \frac{B}{2} - 1$$

$$= 4 \cos \frac{A}{2} \cos \frac{B}{2} \sin \frac{C}{2} - 1$$

$$= \text{R.H.S.}$$

Example 4.13.2 (4)

If A, B, C are angles of a triangle, prove the identity

$\sin 2A + \sin 2B - \sin 2C = 4 \cos A \cos B \sin C$

Solution:

L.H.S. $= \sin 2A + \sin 2B - \sin 2C$

$= 2 \sin (A + B) \cos (A - B) - \sin (360° - (2A + 2B))$ (switching out of C and using the fact that if $A + B + C = 180°$ then $2A + 2B + 2C = 360°$)

$= 2 \sin (A + B) \cos (A - B) - - \sin (2A + 2B)$

$= 2 \sin (A + B) \cos (A - B) + \sin (2A + 2B)$

$= 2 \sin (A + B) \cos (A - B) + 2 \sin (A + B) \cos (A + B)$

L.H.S. $= 2 \sin (A + B) [\cos (A - B) + \cos (A + B)]$

$= 2 \sin (A + B) [2 \cos A \cos (-B)]$

$= 2 \sin (180° - C) \cdot (2 \cos A \cos B)$ (switching one function back to be a function of C)

$= 2 \sin C (2 \cos A \cos B)$

$= 4 \cos A \cos B \sin C$

$=$ R.H.S.

EXERCISE 4

Problems on Chapter IV

1. Express in degrees and decimals of a degree

   (a) $28° 33'$ (b) $35° 37'$ (c) $56° 41' 32''$

2. Express in degrees, minutes and seconds

   (a) $63.36°$ (b) $75.4291°$

3. Express, as fractions of π, in radians

   (a) $135°$ (b) $67.5°$ (c) $110°$ (d) $320°$

   (e) $1020°$.

4. Express in degrees

   (a) $\frac{11\pi}{8}$ (b) $\frac{13\pi}{6}$ (c) 1.15^c (d) 0.2874^c

5. In Fig. 4(a) O is the centre of the circle, radius 12 cm, and angle $AOB = 36°$. Calculate

   (a) the length of arc APB,

   (b) area of sector OAPB,

   (c) the shaded area. (all answers to two decimal places)

6. In Fig 4(b) O is the centre of the circle, and angle $AOB = 10°$, arc $APB = 5\pi$ cm, calculate

   (a) the exact value of the radius of the circle,

   (b) the shaded area correct to two decimal places

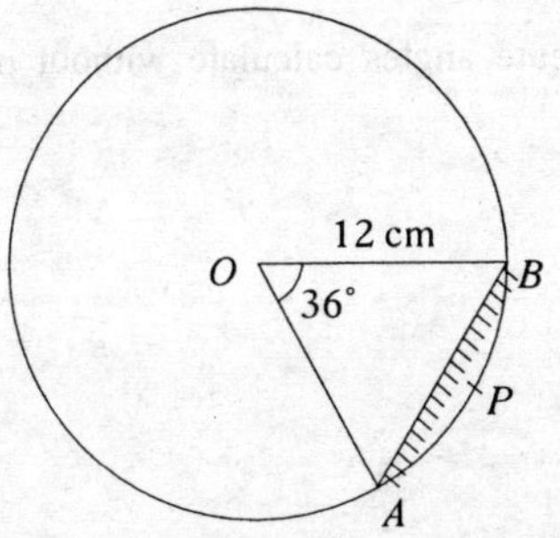

Fig. 4(a)

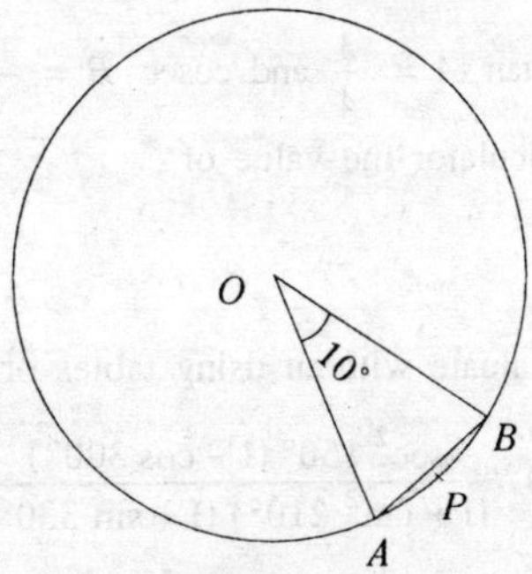

Fig. 4 (b)

7. In Fig. 4 (c) two equal circles of radius 10 cm intersect at S and T. Their centres are at P and Q, where P, S, Q, T be on a third circle with PQ as diameter. Show that $PSQT$ is a square and calculate the area of the shaded region.

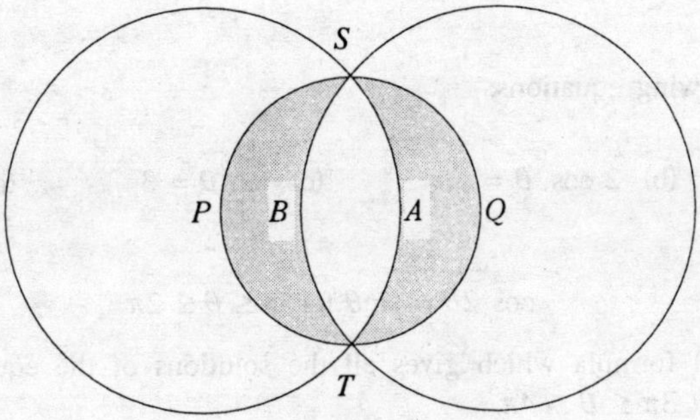

Fig. 4(c)

8. In Fig. 4 (d) the circle centre O has radius 8 cm and AB, CB are tangents to the circle at A and C respectively. Length AB is 15 cm. Calculate correct to three significant figures
   (a) the shaded area
   (b) the perimeter of the figure bounded by the lines AB, CB and the arc CPA.

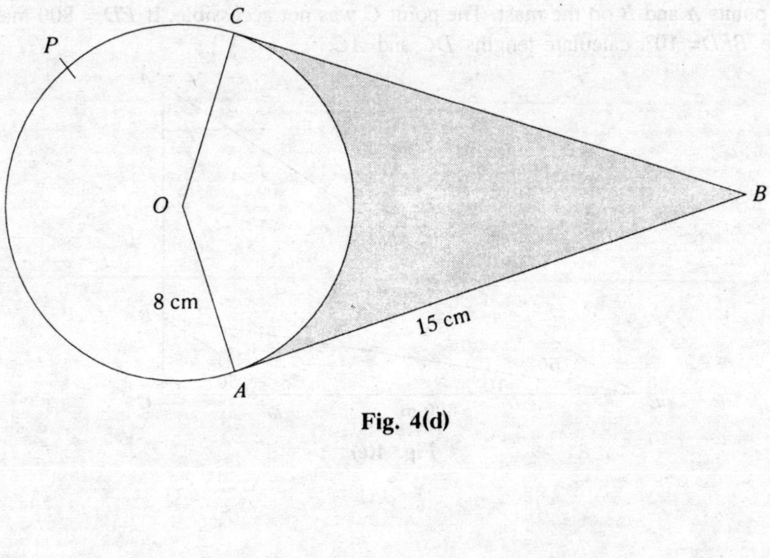

Fig. 4(d)

9. If $\tan A = \frac{3}{4}$ and $\operatorname{cosec} B = \frac{17}{8}$ where A and B are acute angles calculate without using tables or calculator the value of

$$\frac{3\sin A + 2\cos A}{\sec B}.$$

10. Evaluate without using tables or calculator

(a) $\dfrac{\sec^2 150° (1 - \cos 300°)}{(1 + \tan^2 210°)(1 + \sin 330°)}$

(b) $\tan^2 60° \cos 32° - \sec^2 300° \sin 122° + \tan 225° \sin 302°$

(c) $\sin \frac{\pi}{10} \cos \frac{3\pi}{10} \tan \frac{7\pi}{10} \operatorname{cosec} \frac{9\pi}{10} \operatorname{cosec} \frac{13\pi}{10}$

11. Solve for the domain

$O \leq \theta \leq 2\pi$

each of the following equations:

(a) $\sin \theta = \frac{\sqrt{3}}{2}$ (b) $2 \cos \theta = -1$ (c) $\tan^2\theta = 3$ (d) $\sec^2\theta = 2$

*12. Solve the equations

$$\cos 2\theta = \sin\theta \text{ for } 0 \leq \theta \leq 2\pi$$

Give *one* general formula which gives all the solutions of the equations. Find also a particular solution such that $3\pi \leq \theta \leq 4\pi$.

13. Calculate the remaining sides and angle of triangle ABC, and also the radius of the circumcircle, in which $a = 10$ cm, $A = 30°$, $B = 42°$, (lengths to be given correct to two decimal places)

14. Triangle ABC has side $a = 3$ cm, angle $C = 120°$, and area $\frac{15\sqrt{3}}{4}$ cm^2. Calculate the side b, and hence the remaining side and angles of the triangle.

15. In Fig. 4 (e) ABC is a flagpole and EDC is the ground level. To find AC observations were taken from D and E to points A and B on the mast. The point C was not accessible. If $ED = 800$ metres, angle $BDC = 50°$, angle $BED = 10°$, calculate lengths DC and AC.

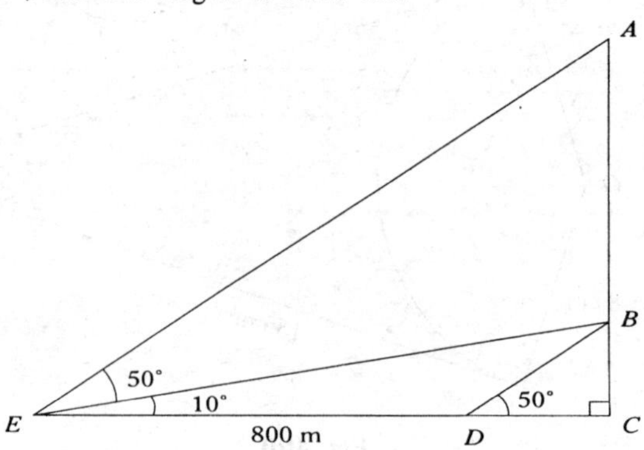

Fig. 4(e)

16. Calculate the angles of triangle ABC whose sides are $a = 14$ cm, $b = 10$ cm, $c = 6$ cm and also the area of the triangle.
17. In triangle XYZ, $XY = 14$ cm, $XZ = 10$ cm, and angle $Y = 30°$, calculate the possible values of the side YZ.
18. In triangle ABC, $c = 20$ cm angle $A = 30°$ and a is given as (i) 15 cm, (ii) 25 cm, (iii) 5 cm, (iv) 10 cm. State how many possible triangles theres are in each of cases (i) to (iv).
19. Calculate the radius of the circumcircle and the possible values of the remaining angles of triangle ABC where $a = 12$ cm, $c = 8$ cm, $C = 30°$.

*20. Use no tables or calculator in the question.
In triangle ABC, $a = 5k$, $b = 6k$, $c = 7k$.
(a) Determine the values of $\cos A$ and $\cos C$.
(b) Using the appropriate double angle formulae determine the exact values of $\cos \frac{A}{2}$ and $\sin \frac{A}{2}$ and hence show that $\sin\left(C + \frac{A}{2}\right) = \frac{13}{5\sqrt{7}}$
(c) The bisector of angle A meets BC at D. Use the Sine Rule to determine the length BD in terms of k.

*21. Use no tables, or calculator in this question.
(a) In triangle ABC $\tan A = \frac{1}{2}$, $\tan B = \frac{1}{3}$. Find the value of C.
(b) Deduce that if a, b, c are the lengths of the sides opposite angles A, B, C respectively, then $a\sqrt{5} = b\sqrt{10} = c\sqrt{2}$.
(c) If $a = 4$. Calculate the area of the triangle.

22. Determine the general solution of each of the following equations in x
(a) $\cos x = \frac{\sqrt{3}}{2}$
(b) $\cos 3x = \frac{\sqrt{3}}{2}$
(c) $\tan x = \sqrt{3}$
(d) $\tan 3x = \sqrt{3}$
(e) $\cos 2x = -0.5$
(f) $\cos 3x = \cos 2x$
(g) $\cos 4x = \sin 2x$

23. Solve each of the equations on the domain $0 \le x \le 360°$
(a) $6\cos^2 x + \cos x - 1 = 0$
(b) $6\sin^2 x - \sin x - 1 = 0$
(c) $\sec^2 x + 5(1 - \tan x) = 0$
(d) $7 + 5\cos x = 6\sin^2 x$

24. Solve each of the following equations for $0° \le \theta \le 360°$
(a) $\cos 2\theta - 3\cos\theta - 1 = 0$
(b) $4\cos 2\theta - 2\cos\theta + 1 = 0$
(c) $\cos 2\theta = \cos\theta + 2$
(d) $7\sec 2\theta + 3\,\text{cosec}\,\theta + 4 = 0$

25. Prove each of the following trigonometric identities:

(a) $\dfrac{\sec\theta}{\sec\theta - \cos\theta} = \text{cosec}^2\theta$ (b) $\dfrac{\cos(A+B)}{\cos A\ \cos B} = 1 - \tan A \tan B$

26. If A, B, C, D are acute angles in which $\tan A = \dfrac{1}{2}$, $\tan B = \dfrac{1}{3}$, $\tan C = \dfrac{1}{7}$, $\tan D = \dfrac{4}{3}$. Show without tables or calculator that $A + B + C = D$.

27. If $\sin A = \dfrac{3}{5}$ and $\cos B = \dfrac{12}{13}$ where A and B are both acute angles determine without tables or calcularor the value of each of $\sin(A - B)$ and $\cos(2A - B)$.

28. Find the general solutions of the equation

$$4\cos 3\theta + 3\cos\theta = 0$$

*29. Show that $3\tan^{-1} 2 - \pi = \tan^{-1}\dfrac{2}{11}$.

30. Two equal circles of radius r intersect and the common chord subtends an angle of 2θ radians at either centre. Find an expression for the area common to both circles.
If this area is equal to half the area of either circles show that θ is a root of the equation

$$\theta - \frac{\pi}{4} = \sin\theta\cos\theta.$$

*31. Given that $x = 18°$ is a root of the equation $\sin 3x = \cos 2x$ use the appropriate multiple angle formulae to show that $\sin 18° = \dfrac{\sqrt{5}-1}{4}$.

32. Prove that if $\theta \neq \dfrac{n\pi}{2}$ where $n \varepsilon Z$, then $\dfrac{\sin 3\theta}{\sin\theta} - \dfrac{\cos 3\theta}{\cos\theta} = 2.$

*33. Simplify without tables or calculator

$$\frac{\sin 18°\ \cos 12° + \cos 162°\ \cos 102°}{\sin 22°\ \cos 8° + \cos 158°\ \cos 98°}.$$

34. Find one single formula for the general solution of the equation $5\cos 2\theta + 8\sin^4\theta - 3 = 0$ and a particular solution for which $\dfrac{7\pi}{2} < \theta < 4\pi$.

*35. Given that $\tan\ \alpha = \dfrac{r\sin\theta}{1 - r\cos\theta}$.

$\sin\ \theta \neq 0$

and that $\alpha = \pi - 2\ \theta$, prove that $r = 2\sin\dfrac{1}{2}\alpha$.

*36. (a) Given that $\sin^{-1}x = A$ and $\cos^{-1}x = B$ where A, B are acute angles show that $\sin^{-1}x + \cos^{-1}x = \dfrac{\pi}{2}$

(b) Hence solve the equation $\cos^{-1}x - \sin^{-1}x = \dfrac{\pi}{3}$.

37. (a) Express $4\sin x + 3\cos x$ in the form $R\sin(x + \alpha)$, stating clearly the values of R and α.
(b) Hence determine the value of x on the domain $0° \leq x \leq 360°$ for which $4\sin x + 3\cos x$ is a minimum.

(c) Solve the equation $4 \sin x + 3 \cos x = 3$ on the domain $0° \le x \le 360°$

38. Solve for $-\pi < x < 2\pi$ each of the equations.

(a) $\cos 2x - \sqrt{3} \cos x - 2 = 0$ (b) $\cos x - \sin x = \sqrt{2}$

39. Solve for $0° \le \theta \le 360°$ the equation $2 \cos \frac{1}{2}\theta + 4 \sin \frac{1}{2}\theta = \sqrt{5}$

40. Given that $7 \sin x - \alpha \cos x \equiv 25 \sin (x - b)$ where α is a positive integer and $1^C < b < 2^C$ determines the values of α and b, giving yoiur answer for b correct to three decimal places.

41. If $5 \cos x + K \sin x \equiv 13 \cos (x - \alpha)$ determine the exact value of K if $K > 0$ and also α correct to the nearest 0.01 radians if $0 < \alpha < \pi/2$. Hence solve the equation

$$5 \cos x + K \sin x = 5$$

for $0 \le x \le 2\pi$, using the values of K, as found earlier, giving your answer correct to two decimal places where appropriate.

42. Without using tables or calculator determine the value of $\tan \theta$ given that $\tan\left(\theta - \frac{3\pi}{4}\right) = -\frac{4}{3}$

43. Using the general expression $K \cos (3x + A)$ solve the equation $2 \sec 3x + 3 \tan 3x = 2$, on the domain $0° \le x \le 360°$.

44. The graph for the function $y = R \sin Kx$ is shown on Fig. 4 (f) for domain $0° \le x \le 240°$ where R and K are constants.

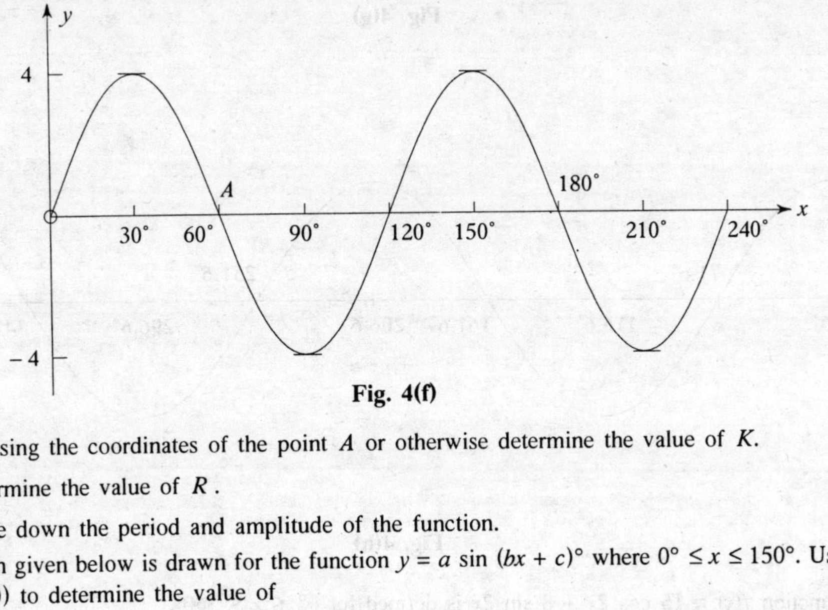

Fig. 4(f)

(a) By using the coordinates of the point A or otherwise determine the value of K.

(b) Determine the value of R.

(c) Write down the period and amplitude of the function.

45. The graph given below is drawn for the function $y = a \sin (bx + c)°$ where $0° \le x \le 150°$. Use the graph (Fig. 4(g)) to determine the value of

(i) a,

(ii) b, and

(iii) c such that $-180° < c < 150°$.

Find the three values of x in the range $0° \le x \le 180°$ when $y = 1$.

46. The graph shows the function $y = R \cos (2x - A)$ where R, A are constants. (See Fig. 4(h))

(a) By making suitable substitutions determine the values of R, A.

(b) State the period, amplitude and phase constant of the function.

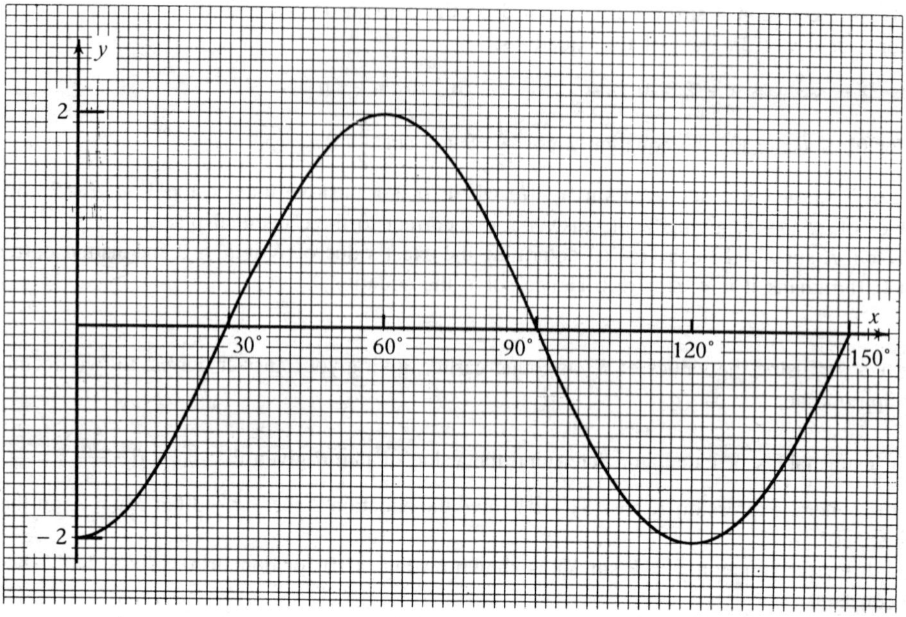

Fig. 4(g)

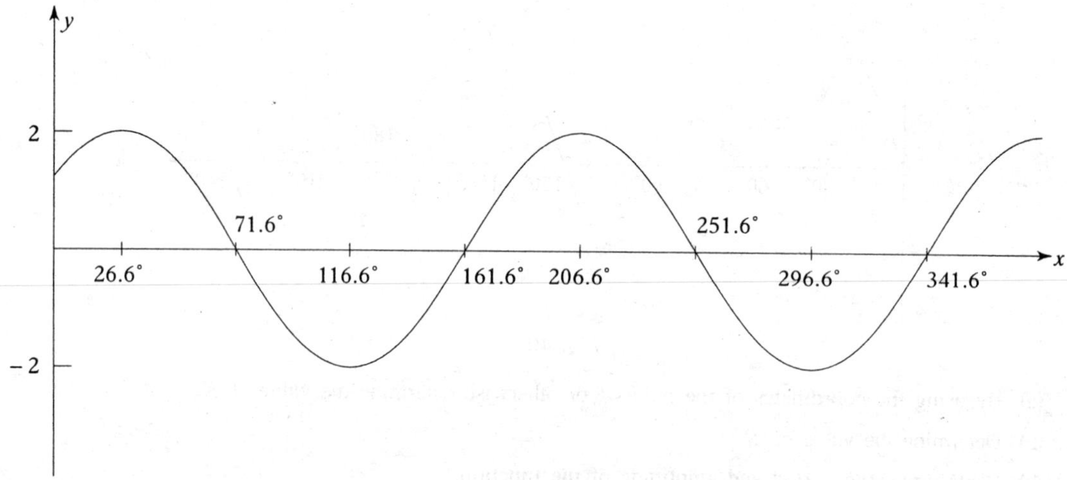

Fig. 4(h)

47. The function $f(x) \equiv 15 \cos 2x + 8 \sin 2x$ is defined for $0° \leq x \leq 360°$.

(a) Express $f(x)$ in the form $R \cos (2x - \alpha)$ stating clearly your values of R and α.

(b) Determine the maximum and minimum values of $f(x)$ and the values of x at which they occur.

(c) Deternine the values of x at which $f(x) = 0$.

(d) Sketch the graph of $f(x)$.

(e) State the period, amplitude and phase constant of $f(x)$.

48. (a) Express $f(x) = \frac{\sqrt{3}}{2} \sin 2x - \frac{1}{2} \cos 2x$ in the form $R \sin (2x - A)$, stating clearly your values of R and A.

(b) If in addition $g(x) = 2 \cos (3x + 60°)$ copy and complete the tables for the graphs of $f(x)$ and $g(x)$ for the domain $0° \leq x \leq 90°$ using two decimal places.

x	0°	10°	15°	30°	40°	45°	60°	70°	75°	90°
$2x$	0°		30°	60°	80°		120°	140°	150°	
$2x - A$	− 30°		0°	30°	50°		90°	110°	120°	
$f(x)$	−0.50		0.00	0.50	0.77		1.00	0.94	0.87	
$3x$	0°		45°	90°	120°		180°	210°	225°	
$3x+60$	60°		105°	150°	180°		240°	270°	285°	
$\cos(3x + 60)$	0.50		−0.26	− 0.87	− 1.00		− 0.50	0.00	0.26	
$g(x)$	1.00		− 0.52	− 1.74	− 2.00		− 1.00	0.00	0.52	

(c) Draw the graphs of $f(x)$ and $g(x)$ on the same axes for $0° \leq x \leq 90°$

(d) Solve graphically the equation

$$\frac{\sqrt{3}}{2} \sin 2x - \frac{1}{2} \cos 2x$$

$$= 2 \cos (3x + 60°) \text{ for } 0 \leq x \leq 90°$$

49. Solve the equation

$$\sin 4x - \sin 2x + 2 \cos 3x = 0$$

for

$$0° \leq x \leq 360°$$

50. (a) With the usual notation for a triangle ABC show that

$$\frac{a+b+c}{\sin A + \sin B + \sin C} = \frac{a}{\sin A}.$$

(b) Given that two angles of a triangle are 45° and 60° and the perimeter of the triangle is 100 cm. Calculate to one decimal place the length of the longest side of the triangle.

*51. (a) In any triangle ABC show that $\sin 2A + \sin 2B + \sin 2C = 4 \sin A \sin B \sin C$

(b) If R is the radius of the circumcircle of triangle ABC use part (a) to show that the area of triangle ABC is given by $2R^2 \sin A \sin B \sin C$.

*52. (a) In triangle ABC angle $A = 2\alpha$, angle $B = 2\beta$, angle $C = 2\gamma$.

(i) Prove that $\sin (\alpha + \beta) = \cos \gamma$

(ii) Using the Sine Rule prove that

$$\frac{a+b}{c} = \frac{\cos (\alpha - \beta)}{\sin \gamma}$$

(b) (i) Obtain a similar expression for $\frac{a-b}{c}$

(ii) Hence prove that

$$\frac{a^2 - b^2}{c^2} = \frac{\sin (A - B)}{\sin C}$$

*53. In triangle *ABC* the semi-perimeter, $s = \frac{1}{2}(a + b + c)$

(a) using the Cosine Rule show that

$$2\,bc\,(1 + \cos A) = 4s\,(s - a)$$

(b) using the Cosine Rule show that

$$2\,bc\,(1 - \cos A) = 4\,(s - b)\,(s - c)$$

(c) Deduce a formula for $4b^2 c^2 \sin^2 A$ in term of s, a, b, c, and hence show that the area of triangle *ABC* is given by

$$\sqrt{\{s\,(s - a)\,(s - b)\,(s - c)\}}$$

5

Coordinate Geometry

5.1 COORDINATES AND THE STRAIGHT LINE

5.1.1 Coordinates

In coordinate geometry we are looking for a way in which we can describe exactly and briefly the position of a point in a plane (a flat surface) in two dimensions, or the position of a point in space in three dimensions.

There was once a university student who felt that the safest place to keep his money was underneath a loose floorboard in his study cubicle. He knew how to locate the loose floorboard whenever he needed to take his money out, because he measured 90 centimetres along the wall by the hinge of the door when he came into the room, and from there he measured 150 centimetres away from the wall in a direction at right angles to the wall.

This illustrates one method we use to determine the position of a point on a plane. We can choose two straight lines which meet at right angles to determine our directions, and then start our measurements from O, the point of intersection. The point O is normally called the *origin*, and the two straight lines meeting at right angles at O are called the x-axis (drawn across the page), and the y-axis (drawn up the page). Units of distance can then be shown on each of the axes, and by convention these are positive to the right of O and negative to the left of O on the x-axis, and are positive above O and negative below O on the y-axis. The plane containing these axes is called the Cartesian plane, named in honour of René Descartes (1588-1648) who made a major contribution in laying the foundations of the subject which is now known as Coordinate Geometry.

Often the axes have to be drawn on a vertical plane such as a blackboard, and in such cases the x-axis is always drawn as a horizontal line and the y-axis as a vertical line. Hence the axes are often referred to as the horizontal and vertical axes respectively.

Consider the position of the point A in Fig. 5.1.1(a) below. To reach A from O we travel 5 units in the direction Ox and then 3 units in the direction Oy. The x-coordinate (known also as the abscissa) of A is 5, and the y-coordinate (known also as the ordinate) of A is 3. We can then say that the coordinates of A are (5, 3) or that A is the point (5, 3). By convention the x-coordinate is always given first, hence distinguishing A (5, 3) from B (3, 5). By use of the negative sign we distinguish between the points $A(5, 3)$ and $C(-5, -3)$.

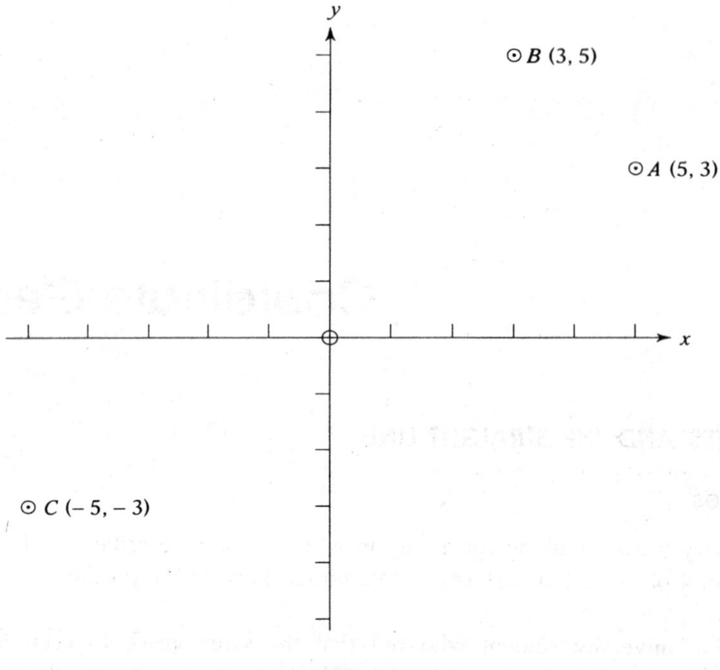

Fig. 5.1.1(a)

5.1.2 The Length of a Straight Line

Example 5.1.2(1)

Calculate the length of the straight line joining $A(-2, 4)$ and $B(3, -2)$.

Solution:

The diagram is shown in Fig. 5.1.2(a).

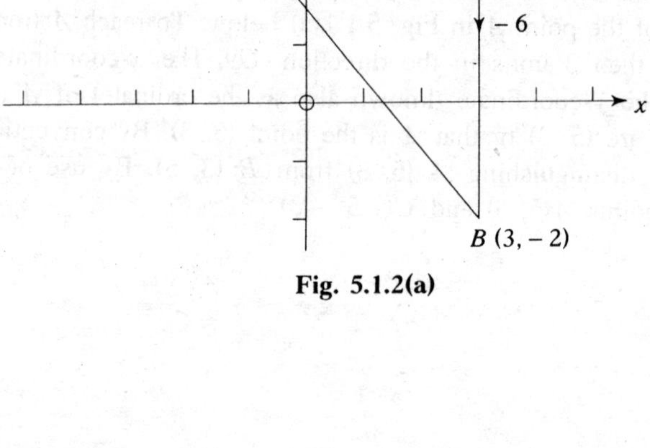

Fig. 5.1.2(a)

$$\text{Vector } \mathbf{AB} = \begin{pmatrix} 5 \\ -6 \end{pmatrix}$$

and $\qquad$ hence $|\mathbf{AB}| = \sqrt{(5^2 + (-6)^2)}$

$$= \sqrt{61}$$

or $\qquad$ 7.81 $\qquad$ (two decimal places)

This approach can also be used to find the distance formula for the length of a straight line joining two points $P_1(x_1, y_1)$ and $P_2(x_2, y_2)$.

The suffix notation (or subscript notation) is very useful when referring to points whose coordinates may not be known. Remember to write the suffix below the letter to avoid confusing between say x_3 and x^3. As can be seen below and in further results to be proved in this section the notation makes it easy to summarise results.

Proof of Distance Formula

Let $P_1(x_1, y_1)$ and $P_2(x_2, y_2)$ be two distinct points in the Cartesian plane, as in Fig. 5.1.2(b).

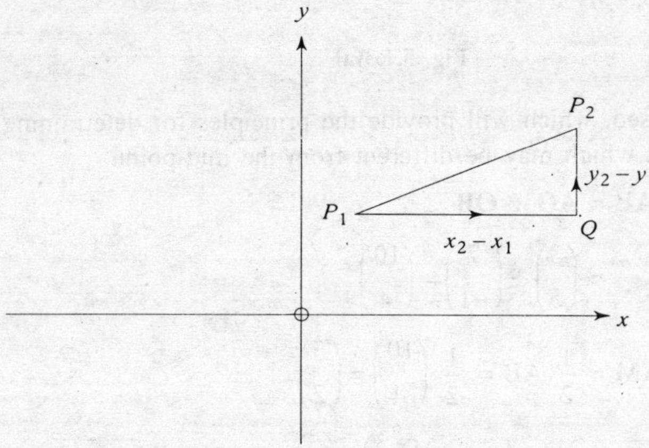

Fig. 5.1.2(b)

From the diagram it can be seen that the distance P_1Q is $x_2 - x_1$ and that the distance QP_2 is $y_2 - y_1$. Using Pythagoras theorem in triangle P_1QP_2 or using the vector

$$\mathbf{P_1 P_2} = \begin{pmatrix} x_2 - x_1 \\ y_2 - y_1 \end{pmatrix}$$

We find that $\qquad |\mathbf{P_1 P_2}| = \sqrt{\{(x_2 - x_1)^2 + (y_2 - y_1)^2\}}$.

5.1.3 The Coordinates of a Point Dividing a Line Segment in a given Ratio

Example 5.1.3(1)

Determine the coordinates of the mid-point M of the line segment joining A and B where A is $(-3, -5)$ and B is $(7, -1)$.

Solution:

See Fig. 5.1.3(a)

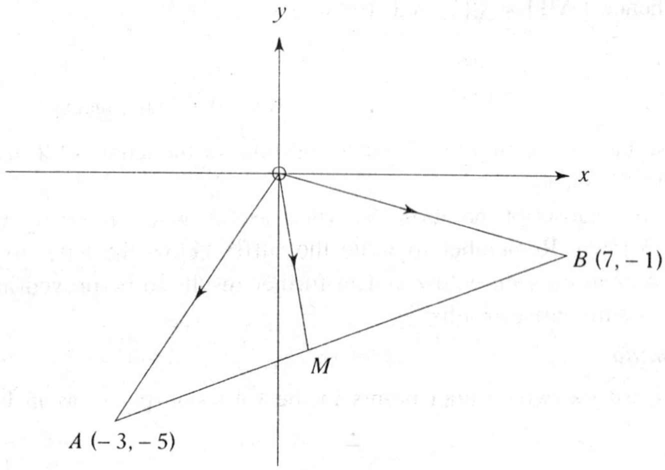

Fig. 5.1.3(a)

A vector approach will be used, which will provide the principles for determining the coordinates of a point on a line segment which may be different from the mid-point.

We can determine $\mathbf{AB} = \mathbf{AO} + \mathbf{OB}$

$$= \begin{pmatrix} 3 \\ 5 \end{pmatrix} + \begin{pmatrix} 7 \\ -1 \end{pmatrix} = \begin{pmatrix} 10 \\ 4 \end{pmatrix}$$

$$\mathbf{AM} = \frac{1}{2}\,\mathbf{AB} = \frac{1}{2}\begin{pmatrix} 10 \\ 4 \end{pmatrix} = \begin{pmatrix} 5 \\ 2 \end{pmatrix}$$

$$\mathbf{OM} = \mathbf{OA} + \mathbf{AM} = \begin{pmatrix} -3 \\ -5 \end{pmatrix} + \begin{pmatrix} 5 \\ 2 \end{pmatrix}$$

$$= \begin{pmatrix} 2 \\ -3 \end{pmatrix}$$

so M is the point (2, – 3).

An observant reader will have noticed that the coordinates of the mid-point in this case are given by averaging the x– and y–coordinates respectively, and this will be proved in the general case in the next example.

Example 5.1.3(2)

Determine the coordinates of the mid-point M of the line segment $P_1 P_2$ where P_1 is (x_1, y_1) and P_2 is (x_2, y_2).

Solution:

See Fig. 5.1.3(b)

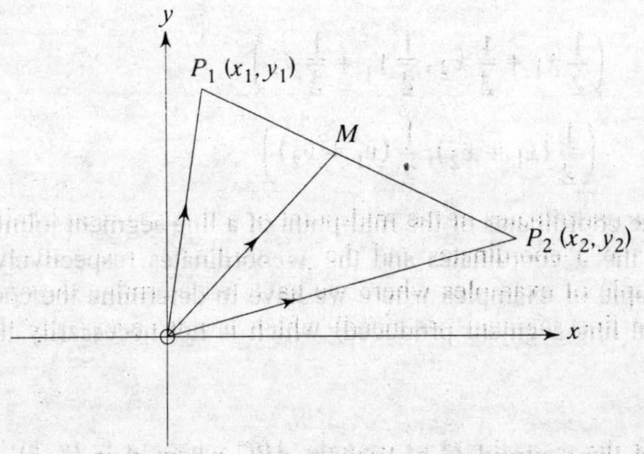

Fig. 5.1.3(b)

Following the approach of the previous example, we first determine the vector $\mathbf{P_1 P_2} = \mathbf{P_1}\, O + \mathbf{OP_2}$

$$= \begin{pmatrix} -x_1 \\ -y_1 \end{pmatrix} + \begin{pmatrix} x_2 \\ y_2 \end{pmatrix}$$

$$= \begin{pmatrix} -x_1 + x_2 \\ -y_1 + y_2 \end{pmatrix}$$

$$\mathbf{P_1\, M} = \frac{1}{2}\ \mathbf{P_1\, P_2}$$

$$= \frac{1}{2} \begin{pmatrix} -x_1 + x_2 \\ -y_1 + y_2 \end{pmatrix}$$

$$= \begin{pmatrix} -\frac{1}{2} x_1 + \frac{1}{2} x_2 \\ -\frac{1}{2} y_1 + \frac{1}{2} y_2 \end{pmatrix}$$

$$\mathbf{OM} = \mathbf{OP_1} + \mathbf{P_1\, M} = \begin{pmatrix} x_1 \\ y_1 \end{pmatrix} + \begin{pmatrix} -\frac{1}{2} x_1 + \frac{1}{2} x_2 \\ -\frac{1}{2} y_1 + \frac{1}{2} y_2 \end{pmatrix}$$

$$= \begin{pmatrix} x_1 - \frac{1}{2} x_1 + \frac{1}{2} x_2 \\ y_1 - \frac{1}{2} y_1 + \frac{1}{2} y_2 \end{pmatrix}$$

$$= \begin{pmatrix} \frac{1}{2} x_1 + \frac{1}{2} x_2 \\ \frac{1}{2} y_1 + \frac{1}{2} y_2 \end{pmatrix}$$

so M is the point

$$\left(\frac{1}{2}x_1 + \frac{1}{2}x_2, \frac{1}{2}y_1 + \frac{1}{2}y_2\right)$$

or

$$\left(\frac{1}{2}(x_1 + x_2), \frac{1}{2}(y_1 + y_2)\right)$$

It is true in general that the coordinates of the mid-point of a line segment joining two given points will be found by averaging the x-coordinates and the y-coordinates respectively.

We shall now study a couple of examples where we have to determine the coordinates of a point on a line segment (or on the line segment produced) which is not necessarily the mid-point of the line segment.

Example 5.1.3(3)

Determine the coordinates of the centroid G of triangle ABC where A is (3, 5), B is (– 2, – 7) and C is (14, – 19)

Solution:

See Fig. 5.1.3(c)

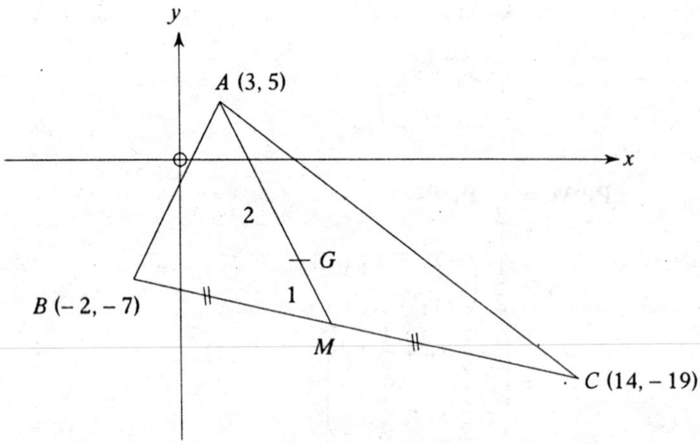

Fig. 5.1.3(c)

G is the point of intersection of the medians of the triangle and is a point of trisection of each median.

If M is the mid-point of BC then $AG : GM = 2 : 1$

Since M is the mid-point of the join of $B(-2, -7)$ and $C(14, -19)$ then M has coordinates

$$\left(\frac{1}{2}(-2 + 14), \frac{1}{2}(17 + 19)\right)$$

or

$$(6, -13)$$

$$\mathbf{AM} = \mathbf{AO} + \mathbf{OM}$$

$$= \begin{pmatrix} -3 \\ -5 \end{pmatrix} + \begin{pmatrix} 6 \\ -13 \end{pmatrix}$$

$$= \begin{pmatrix} 3 \\ -18 \end{pmatrix}$$

$$\mathbf{AG} = \frac{2}{3}\ \mathbf{AM}$$

$$= \frac{2}{3}\begin{pmatrix} 3 \\ -18 \end{pmatrix} = \begin{pmatrix} 2 \\ -12 \end{pmatrix}$$

$$\mathbf{OG} = \mathbf{OA} + \mathbf{AG}$$

$$= \begin{pmatrix} 3 \\ 5 \end{pmatrix} + \begin{pmatrix} 2 \\ -12 \end{pmatrix} = \begin{pmatrix} 5 \\ -7 \end{pmatrix}$$

so G is the point (5, – 7)

Example 5.1.3(4)

Determine the coordinates of the point R on AB which divides AB *externally* in the ratio 5 : 3 if A is (3, 1) and B is (7, 9).

Solution:

See Fig. 5.1.3(d)

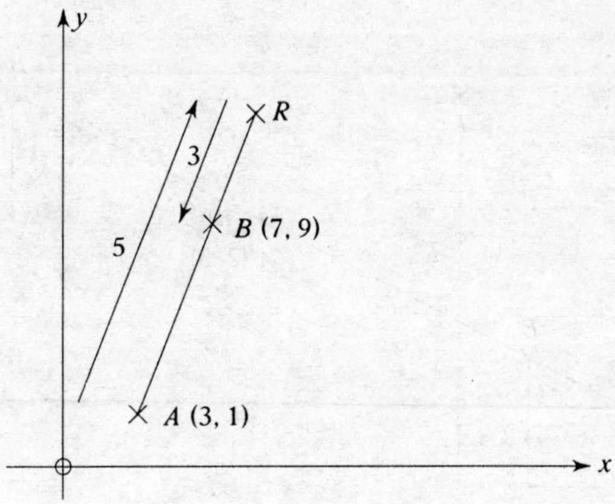

Fig. 5.1.3 (d)

As in the previous examples we determine a column vector for the join of the given line segment, in this case **AB**

$$\mathbf{AB} = \mathbf{AO} + \mathbf{OB}$$

$$= \begin{pmatrix} -3 \\ -1 \end{pmatrix} + \begin{pmatrix} 7 \\ 9 \end{pmatrix} = \begin{pmatrix} 4 \\ 8 \end{pmatrix}$$

Now AR is 5 parts and RB is 3 parts (back) [see Fig. 5.1.3(d)] and so AB must be 2 parts

Hence $$\mathbf{AR} = \frac{5}{2}\,\mathbf{AB}$$

$$= \frac{5}{2}\begin{pmatrix}4\\8\end{pmatrix} = \begin{pmatrix}10\\20\end{pmatrix}$$

$$\mathbf{OR} = \mathbf{OA} + \mathbf{AR} = \begin{pmatrix}3\\1\end{pmatrix} + \begin{pmatrix}10\\20\end{pmatrix} = \begin{pmatrix}13\\21\end{pmatrix}$$

so R is (13, 21)

5.1.4 The Gradient of a Straight Line

If we consider the straight line passing through $A(2, 3)$ and $B(7, 4)$ as in Fig. 5.1.4(a), and think of the x-axis as horizontal with the line passing through A and B as a pathway up a hill, then a person walking from A to B would move up a vertical distance CB at the same time as he has covered a horizontal distance AC. The gradient of the pathway would be CB/AC

$$= (4 - 3)/(7 - 2) = 1/5.$$

Instead of the points A and B, we could have taken any other two points on the line such as D and E, The gradient would have been expressed as $FE/DF = CB/AC$, since the two triangles ABC, DEF are similar. The gradient can also be calculated using the tangent of the angle formed by rotating from the direction of the positive x-axis to the direction of the line by an anticlockwise rotation.

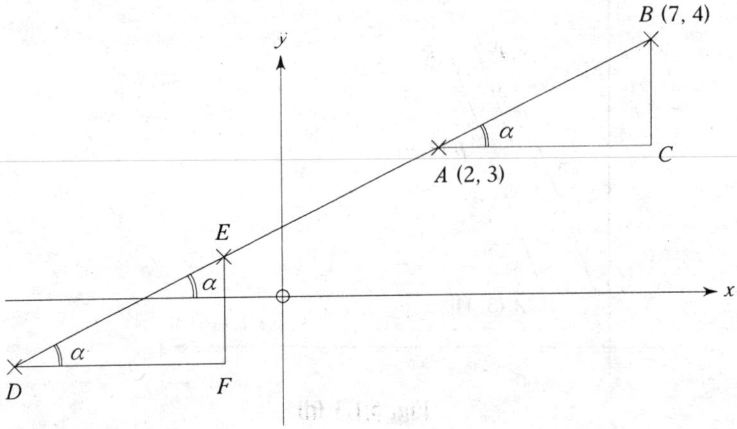

Fig. 5.1.4(a)

Definition The gradient of a straight line is $\dfrac{\text{the increase in } y}{\text{the increase in } x}$

in moving from one point on the line to another.

In the case of Fig. 5.1.4(a), when moving from A to B, both x and y increase by a positive amount so the gradient is positive.

Now consider the gradient of PQ in Fig. 5.1.4(b). When we move from P to Q the increase in x is +3 but the increase in y is – 4 (y is decreasing). Hence the gradient of PQ is $\frac{-4}{3}$.

In travelling along a line with x increasing when going uphill the gradient is positive, but when going downhill the gradient is negative.

It is not necessary to draw a figure, although a figure may help in working out the first few exercises.

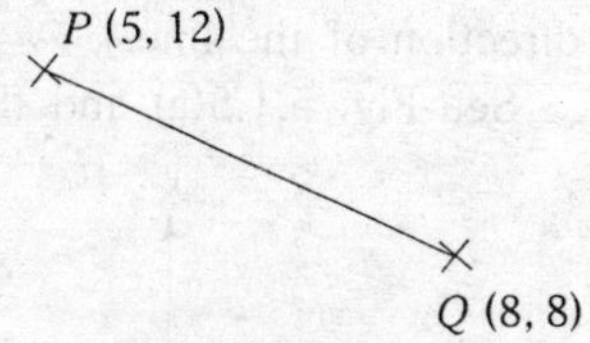

Fig. 5.1.4(b)

Example 5.1.4(1)

Determine the gradient of the line joining $R(4.5, 9.5)$ and $S(5.5, – 2.5)$.

Solution:

See Fig. 5.1.4(c)

The gradient of RS (or grad RS for short) can be found using either

$$\frac{-2.5 - 9.5}{5.5 - 4.5} = -12$$

or

$$\frac{9.5 - -2.5}{4.5 - 5.5} = \frac{12}{-1} = -12$$

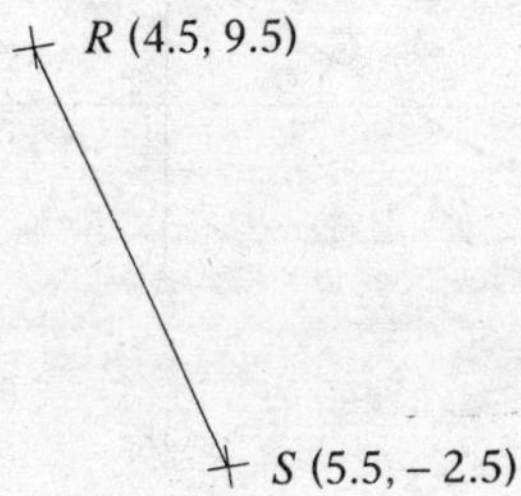

Fig. 5.1.4(c)

N.B: If you select the coordinates of R minus the coordinates of S then this must be written in *both* numerator and denominator.

Writing $\frac{9.5 - -2.5}{5.5 - 4.5}$

(where on the numerator you are writing the y-coordinates at R minus the y-coordinate at S but on the denominator you are writing the x-coordinate at S minus the x-coordinate at R) is wrong.

Example 5.1.4(2)

Determine if possible the gradient of the line joining $A(3, 1)$ and $B(3, 6)$.

Solution:

We could use either $\frac{6-1}{3-3} = 5/0$

or $\frac{1-6}{3-3} = -5/0$

What meaning can we give to two different expressions like 5/0 and – 5/0 which appear to be giving two different gradients? In algebra there are always difficulties when we try to divide by zero, and normally division by zero is not allowed in Mathematics. We therefore conclude that expressions such as 5/0 and – 5/0 do not exist.

Fig. 5.1.4(d) shows the graph of the line AB, which is a vertical line (i.e. parallel to the y-axis).

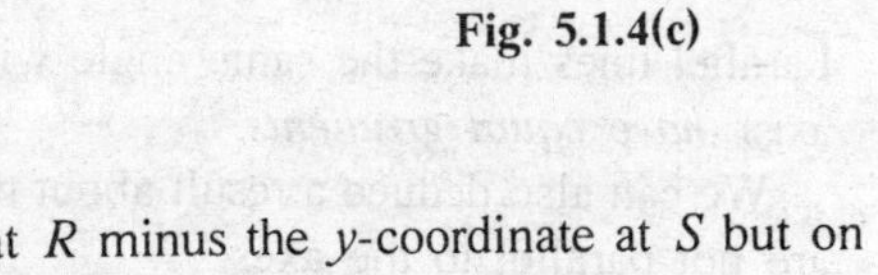

Fig. 5.1.4(d)

We have to accept that for any vertical line such as AB it is not possible to give a numerical value to its gradient. We simply state that the line AB is parallel to the y-axis with equation $x = 3$ in this case.

5.1.5 Parallel and Perpendicular Lines

The gradient of any straight line (which is not parallel to the y-axis) is found by calculating the tangent of the angle formed by rotating anticlockwise from the direction of the positive x-axis to the direction of the line.

See Fig. 5.1.5(a) and (b)

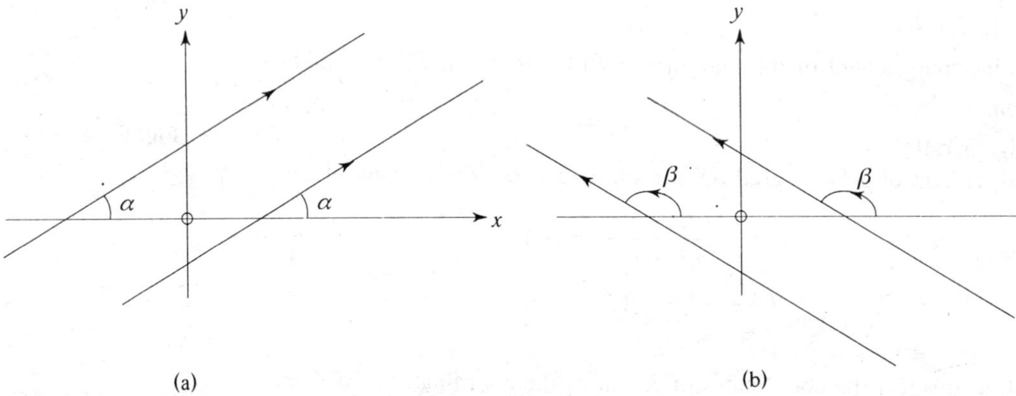

Fig. 5.1.5(a) and (b)

Parallel lines make the same angle with the x-axis so *parallel lines* (except those parallel to the y-axis) *have equal gradients.*

We can also deduce a result about the gradients for perpendicular lines in any case where the lines are not parallel to the axes.

Theorem for gradients of two perpendicular lines

If two lines, which are not parallel to the axes, meet at right angles, then the product of their gradients is -1, and conversely.

Proof (Direct)

Let the two straight lines be GE and HE, which meet at E, so that angle $GEH = 90°$, as in Fig. 5.1.5(c)

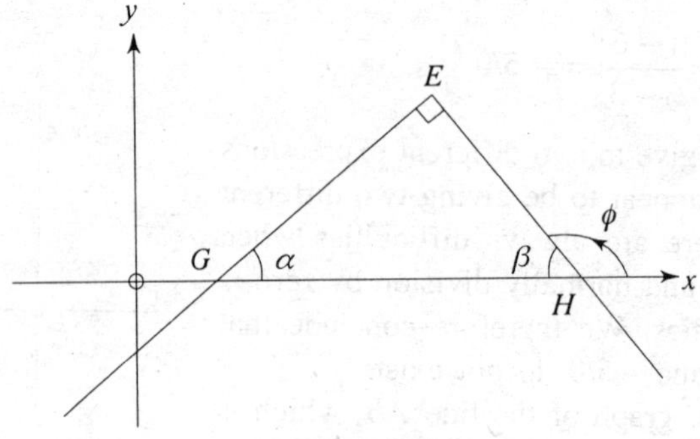

Fig. 5.1.5(c)

Let GE have gradient

$$m_1 = \tan \alpha$$

and HE have gradient

$$m_2 = \tan \phi$$

In the triangle GEH

$$m_1 = \tan \alpha = \frac{HE}{DE}$$

But also $m_2 = \tan \phi = -\tan \beta$

$$(\text{since } \beta + \phi = 180°)$$

But in triangle GEH

$$\tan \beta = \frac{GE}{HE} = \frac{1}{\tan \alpha}$$

So

$$m_2 = -\tan \beta = \frac{-1}{\tan \alpha} = \frac{-1}{m_1}$$

Hence the product of the gradients $m_1 \, m_2 = -1$.

Proof (Converse)

If the line HE is rotated slightly either anticlockwise or clockwise its new gradient will no longer be $-\tan \beta$ and so the product of the gradients of GE and HE will not be -1 and the lines GE and HE will not be perpendicular.

So the lines GE, HE will be perpendicular if and only if the products of their gradients is -1.

In general it is true that the statements 'two lines (not parallel to the axes) are perpendicular' and 'the product of the gradients of the two lines equals -1' are equivalent.

If a line has gradient m then the line perpendicular to it has gradient $-1/m$.

5.1.6 The Meaning of an Equation

If we make a statement such as 'P is the point (x, y)' then the point P can be anywhere in the Cartesian plane. Normally if we are required to locate P we are given a pair of numerical values for x and y which enable us to determine the position of P.

Suppose however the information we are given takes the form of an equation such as $y = x^2 - 4x$. This equation does not restrict us to only one pair of values for x and y, but it gives us as many pairs of values as we would like to find. However the point P is *not* free to take up any position on the plane, since for any chosen value of x there will be only one corresponding value of y. P must be restricted to taking up only those positions whose coordinates (x, y) satisfy the given equation $y = x^2 - 4x$.

As in subsection 3.3.1(d) of chapter III on Algebra it would normally be necessary to make up a table of values, such as the one shown below.

Selecting a domain $-2 \le x \le 6$ for x then:

x	-2	-1.5	-1	-0.5	0	0.5	1	1.5	2
x^2	4	2.25	1	0.25	0	0.25	1	2.25	4
$-4x$	8	6	4	2	0	-2	-4	-6	-8
$y = x^2 - 4x$	12	8.25	5	2.25	0	-1.75	-3	-3.75	-4

x	2.5	3	3.5	4	4.5	5	5.5	6
x^2	6.25	9	12.25	16	20.25	25	30.25	36
$-4x$	-10	-12	-14	-16	-18	-20	-22	-24
$y = x^2 - 4x$	-3.75	-3	-1.75	0	2.25	5	8.25	12

The graph for the domain $-2 \le x \le 6$ is shown in Fig. 5.1.6(a).

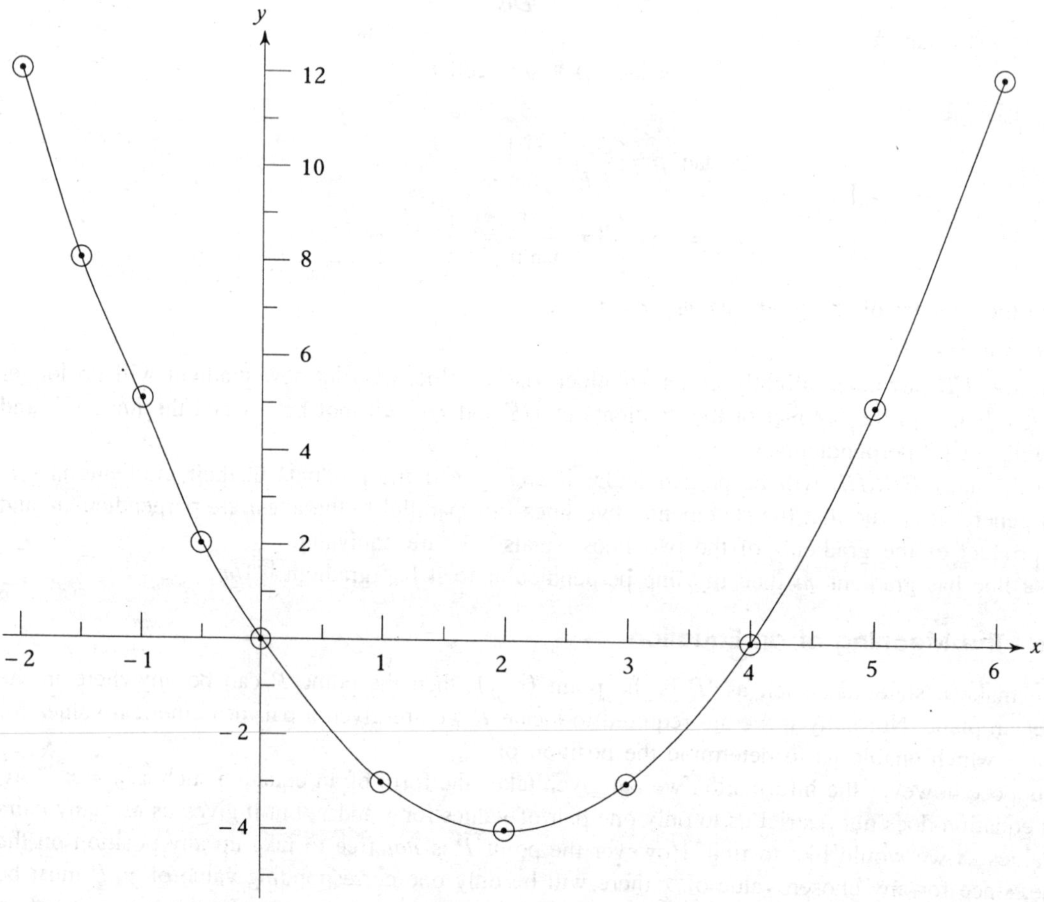

Fig. 5.1.6(a)

From the table it is clear that the points $(-2, 12)$, $(-1.5, 8.25)$, $(-1, 5)$, etc. have coordinates which satisfy the equation $y = x^2 - 4x$, and so by plotting these points and drawing a smooth curve through them as has already been done in chapter 3, sub-section 3.3.1(d), we are able to show all the possible positions of P corresponding to values of x on the domain $-2 \le x \le 6$.

In the same way as coordinates may be used to name a point so we may use an equation to name a curve, and we can refer to the curve '$y = x^2 - 4x$'. It is important to note that the equation is the condition that the point (x, y) should lie on the curve.

Thus only if $b = a^2 - 4a$ does the point (a, b) lie on the curve $y = x^2 - 4x$, and in such a case we may say that coordinates of the point satisfy the equation.

If $q \neq p^2 - 4p$ then the point (p, q) does *not* lie on the curve $y = x^2 - 4x$.

Example 5.1.6 (1)

Do the points (a) (6, –2), (b) (16, –14) lie on the line whose equation is $2x + 3y = 6$?

Solution:

(a) Test (6, –2) on the left hand side of the equation

$$2x + 3y = 6$$

$$\text{LHS} = 2 \times 6 + 3 \times -2 = 6 = \text{RHS}$$

so (6, – 2) does lie on the line whose equation is $2x + 3y = 6$.

(b) Test (16, –14) in a similar way

$$\text{LHS} = 2 \times 16 + 3 \times -14 = -10 \neq \text{RHS}$$

so (16, –14) does *not* lie on the line whose equation is $2x + 3x = 6$.

Example 5.1.6 (2)

Do the points (a) $\left(3\sqrt{3}, 4\right)$, (b) (2, 6) lie on the curve whose equation is

$$\frac{x^2}{36} + \frac{y^2}{64} = 1$$

Solution:

(a) Testing $\left(3\sqrt{3}, 4\right)$ on the left hand side of the equation.

$$\text{LHS} = \frac{9 \times 3}{36} + \frac{16}{64}$$

$$= \frac{27}{36} + \frac{16}{64} = \frac{3}{4} + \frac{1}{4} = 1 = \text{RHS}$$

So $\left(3\sqrt{3}, 4\right)$ does line on the curve whose equation is

$$\frac{x^2}{36} + \frac{y^2}{64} = 1$$

(b) Testing (2, 6) on the left hand side of the equation

$$\text{LHS} = \frac{4}{36} + \frac{36}{64}$$

$$= \frac{1}{9} + \frac{9}{16}$$

$$= \frac{97}{144} \neq \text{RHS}$$

The coordinates of the point (2, 6) *do not* satisfy the equation so (2, 6) *does not* lie on the curve whose equation is

$$\frac{x^2}{36} + \frac{y^2}{64} = 1.$$

The relationship between a curve and its equation may give rise to two main types of problems.

Firstly, these are the problems in which we are given the equation, and from it we are expected to find the curve. This is a problem with which you are already familiar, especially when you have been asked to solve quadratic and other equations graphically.

Secondly, there are problems in which we are given pureiy geometrical information about the curve, and we have to use this information to determine the equation of the curve. It is the second type of problem which arises more often in our study of coordinate geometry. First we shall examine a few more linear equations.

5.1.7 The Equation $y = mx + c$

We shall look at several examples which will lead us to the equation whose form is $y = mx + c$ where m and c are constants.

Example 5.1.7 (1)

Graph Fig. 5.1.7 (a) Describe the curve or line whose equation is $y = 2x$.

Solution:

This equation is satisfied by the coordinates of points (0, 0), (1, 2), (2, 4), (3, 6), etc. and is seen to represent a straight line through the origin of gradient 2. See Fig. 5.1.7 (a).

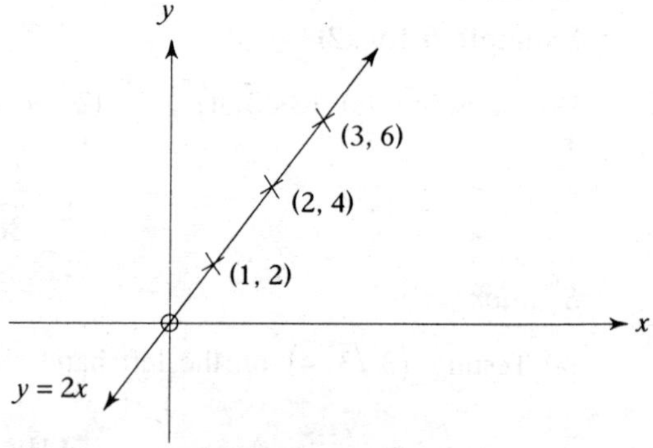

Fig. 5.1.7(a)

Example 5.1.7 (2)

Describe the curve or line whose equation is $x = 5$.

Solution:

This equation $x = 5$, means that the y-coordinate can take any value provided the x-coordinate is 5 and so for example the points (5, 0), (5, 1), (5, 2), (5, 3), etc. lie on the graph, which will be a straight line parallel to the y-axis, 5 units from it, on the side on which x is positive, See Fig. 5.1.7(b).

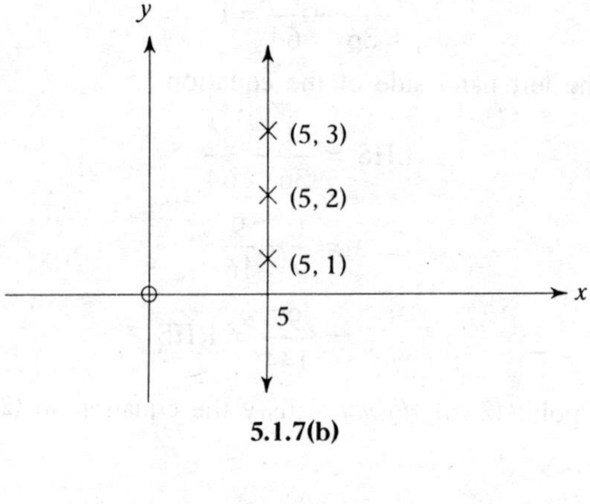

5.1.7(b)

Example 5.1.7 (3)

Find the equation of the line of gradient 3 which passes through the origin.

Solution:

From Fig. 5.1.7(c), the $P(x, y)$ to be a point on the line other than 0. Then the gradient of the line OP may be written as $\frac{y}{x}$.

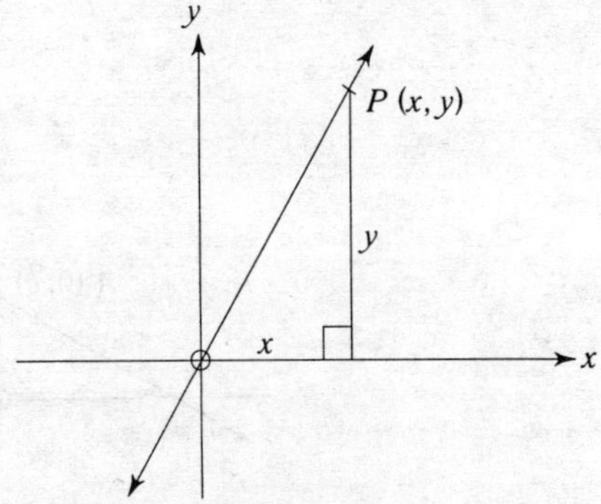

Fig. 5.1.7(c)

But from the information given the gradient is 3.

i.e. $\frac{y}{x} = 3$

so $y = 3x$ is the required equation.

Note that once we have cleared fractions then it is clear that the coordinates, of the origin (0, 0) also satisfy the equation $y = 3x$. However, when looking for an expression for the gradient it is essential to choose point $P(x, y)$ which is different from the given point (in this case the origin) which lies on the line.

Example 5.1.7 (4)

Determine the equation of the line of gradient -2 which passes through the point (0, 3).

Solution:

From Fig. 5.1.7 (d) take $P(x, y)$ to be a point on the line other than $A(0, 3)$. Then the gradient of the line AP may be written as

$$\frac{3-y}{0-x}$$

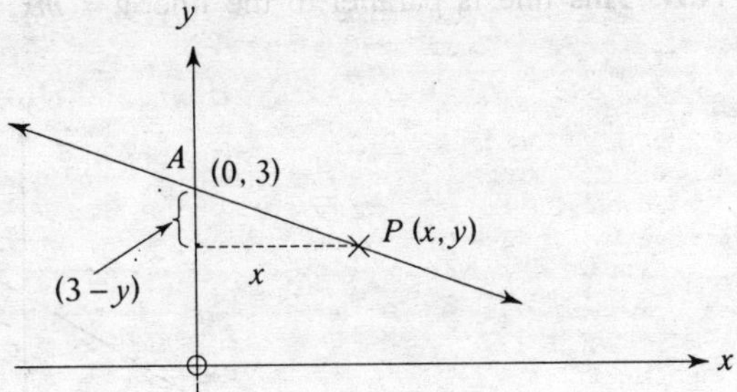

Fig. 5.1.7(d)

But the gradient of the line is given to be –2.

i.e. $\frac{3-y}{0-x} = -2$

i.e. $3 - y = 2x$

i.e. $y = -2x + 3$

on making y to be the subject, and this is the required equation.

If a straight line cuts the y-axis at the point $(0, c)$ the distance of this point from the origin is known as the intercept on the y-axis.

Now following the method of the last two examples we shall derive the equation of a line with given gradient m, through the given point (O, c) in the form $y = mx + c$.

Proof of Equation y = mx + c

Let $P(x, y)$ be any point on the line other than $A(0, c)$. Then the gradient of AP may be written

$$\frac{y-c}{x-0}$$

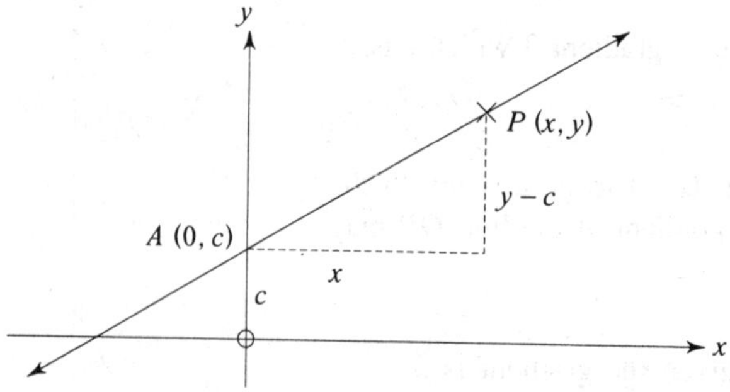

Fig. 5.1.7(e)

But the gradient of the line is given as m

i.e.
$$\frac{y - c}{x - 0} = m$$
i.e.
$$y - c = mx$$
i.e.
$$y = mx + c$$

N.B: This line is parallel to the line $y = mx$, which passes through the origin [See Fig. 5.1.7 (f)]

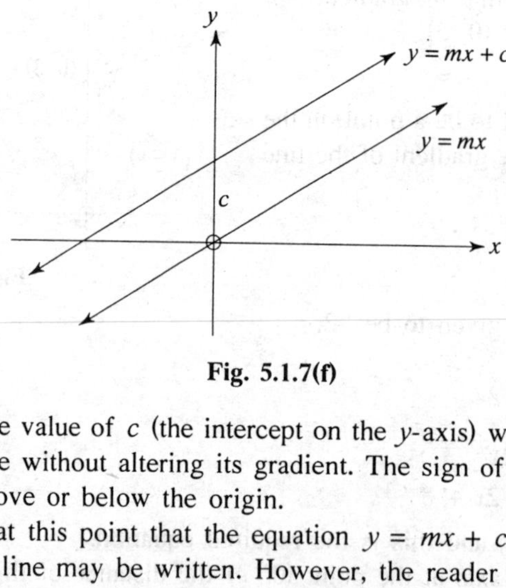

Fig. 5.1.7(f)

The effect of altering the value of c (the intercept on the y-axis) without altering the value of m, is to raise or lower the line without altering its gradient. The sign of c will determine whether the line will cut the y-axis above or below the origin.

It is tempting to think at this point that the equation $y = mx + c$ gives the form in which the equation of every straight line may be written. However, the reader should recall the problem we had when we tried to find the gradient of a line parallel to the y-axis. In such a case we cannot find a numerical value for m, and such an equation has to be written in the form $x = a$, where a is constant.

So far we can summarise the forms of an equation of a straight line:

$y = mx + c$ is a line of gradient m through $(0, c)$.

$y = mx$ is a line of gradient m through the origin

$y = c$ is a line of zero gradient (parallel to the x-axis)

$x = a$ is a line parallel to the y-axis.

Example 5.1.7 (5)

Determine the gradient of the straight line $3x + 5y = 30$, and also its intercepts on the axes.

Solution:

Re-write the equation with y as subject

from $\quad 3x + 5y = 30$

i.e. $\quad 5y = -3x + 30$

i.e. $\quad y = \frac{-3}{5}x + 6$

This is of the form $y = mx + c$, where $m = -\frac{3}{5}$ and $c = 6$

so we can see that its gradient is $-\frac{3}{5}$ and its intercept on the y-axis is 6.

To determine the intercepts on both axes it is usually easier to use the original equation.

$$3x + 5y = 30$$

on the y-axis $x = 0$

so $\quad 5y = 30$

i.e. $\quad y = 6$

The intercept on the y-axis is 6 as found earlier

To determine the intercept on the x-axis put $y = 0$

so $\quad 3x = 30$

i.e. $\quad x = 10$

The intercepts on the x-axis and the y-axis are 10 and 6 respectively.

5.1.8 Finding the Equation of a Straight Line

The method used in 5.1.7 will enable us to determine the equation of any straight line (not parallel to the y-axis), provided we know both the gradient and the coordinates of one point through which the line passes.

Example 5.1.8 (1)

Determine the equation of the straight line of gradient $-3/4$ which passes through the point $(-5, 2)$ and give your answer in the form $ax + by + c = 0$, where a, b, c are integers.

Solution:

If $P(x, y)$ is any point on the line other than $A(-5, 2)$, the gradient of the line may be written

$$\frac{y-2}{x-(-5)} \quad \text{or} \quad \frac{y-2}{x+5}$$

But this is given as $-3/4$

so $\quad \frac{y-2}{x-5} = \frac{-3}{4}$

i.e. $\quad 4(y - 2) = -3(x + 5)$

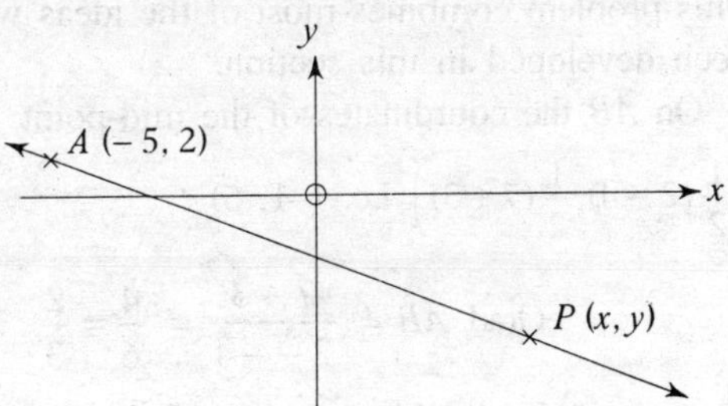

Fig. 5.1.8(a)

i.e. $4y - 8 = -3x - 15$

i.e. $3x + 4y + 7 = 0.$

Example 5.1.8 (2)

Determine the equation of the straight line joining the points (– 5, 2) and (4, – 4) and give your answer in the form $ax + by + c = 0$, where a, b, c are integers.

Solution:

See Fig. 5.1.8 (b)

The gradient of the line joining A (– 5, 2) and B (4, – 4) is

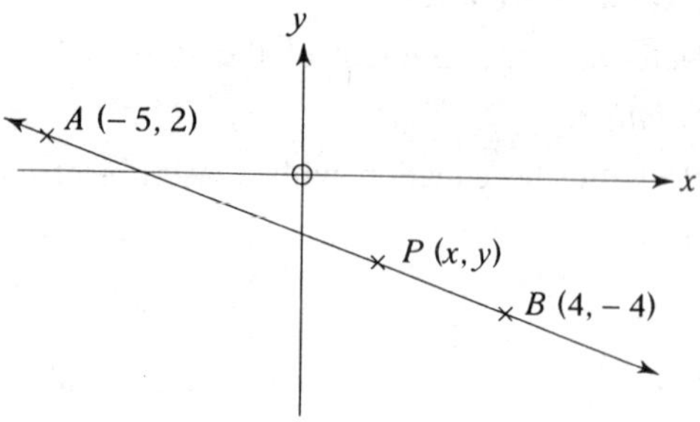

Fig. 5.1.8(b)

$$\frac{-4-2}{4-(-5)} = \frac{-6}{9} = \frac{-2}{3}$$

If $P(x, y)$ is any point on the line other than A (–5, 2) the gradient may be written $\dfrac{y-2}{x-(-5)} = \dfrac{y-2}{x+5}$

so $$\frac{y-2}{x+5} = \frac{-2}{3}$$

i.e. $3(y - 2) = -2(x + 5)$

i.e. $3y - 6 = -2x - 10$

i.e. $2x + 3y + 4 = 0$

The reader should check that the same equation would be obtained if instead of finding the gradient AP, the gradient of BP was found and then written equal to –2/3.

Example 5.1.8 (3)

The points A, B, C have coordinates (2, 7), (– 4, 3), and (4, 5) respectively. Determine in the form $ax + by = c$ where a, b, c are integers, the equations of the perpendicular bisectors of each of the line segments AB, BC, and hence determine the coordinates of the circumcentre of triangle ABC and the radius of the circumcircle of triangle ABC.

Solution:

See Fig. 5.1.8 (c)

This problem combines most of the ideas which have been developed in this section.

On AB the coordinates of the mid-point M will be

$$\left(\frac{1}{2}(2-4), \frac{1}{2}(7+3)\right) \text{ i.e. } (-1, 5)$$

$$\text{Grad } AB = \frac{7-3}{2--4} = \frac{4}{6} = \frac{2}{3}$$

The perpendicular bisector of AB has gradient –3/2 and passes through M (–1, 5)

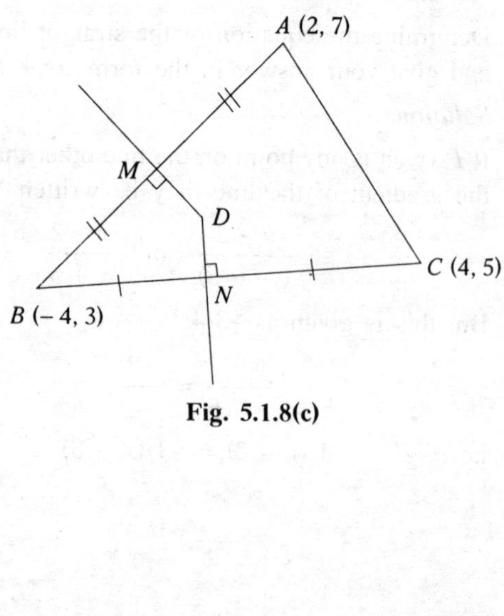

Fig. 5.1.8(c)

Its equation is

$$\frac{y-5}{x--1}=\frac{-3}{2}$$

i.e. $$2(y-5) = -3(x+1)$$

i.e. $$2y - 10 = -3x - 3$$

i.e. $$3x + 2y = 7 \qquad (1)$$

On *BC* the coordinates of the mid-point *N* will be ((1/2 (– 4 + 4), 1/2 (3 + 5)) i.e. (0, 4)

$$\text{Grad } BC = \frac{5-3}{4--4}=\frac{2}{8}=\frac{1}{4}$$

The perpendicular bisector of *BC* has gradient – 4 and passes through N (0, 4)

Its equation is

$$\frac{y-4}{x-0}=4$$

i.e. $$y - 4 = -4x$$

i.e. $$4x + y = 4 \qquad (2)$$

In order to determine the coordinates of the circumcentre *D* we have to solve the equations (1) and (2) for the two perpendicular bisectors simultaneously.

In (2) $$y = -4x + 4$$

and substituting in (1)

$$3x + 2(-4x + 4) = 7$$

i.e. $$3x - 8x + 8 = 7$$

i.e. $$1 = 5x$$

i.e. $$x = \frac{1}{5}$$

and substituting in (2)

$$y = -\frac{4}{5} + 4 = \frac{16}{5}$$

So the circumcentre *D* is at $\left(\frac{1}{5}, \frac{16}{5}\right)$.

To determine the radius of the circumcircle we can calculate the length *DA*

$$\mathbf{DA} = \mathbf{DO} + \mathbf{OA}$$

$$= \begin{pmatrix}-1/5\\-16/5\end{pmatrix} + \begin{pmatrix}2\\7\end{pmatrix} = \begin{pmatrix}9/5\\19/5\end{pmatrix}$$

so $$|\mathbf{DA}| = \sqrt{\{81/25 + 361/25\}} = \sqrt{\frac{442}{25}}$$

or $$\frac{1}{5}\sqrt{442} \quad \text{or} \quad 4.20 \text{ (two decimal places).}$$

The reader can check the last answer by calculating the lengths DB and DC.

5.2 INTRODUCTION TO COORDINATE GEOMETRY OF THE CIRCLE

Some of the results which have been used for straight line work may now be applied to work on the circle.

5.2.1 Equation of the Circle

In what follows we shall derive an equation satisfied by coordinates of all points lying on the circumference of a circle. So in this and the paragraphs which follow the phase 'equation of a circle will mean equation satisfied by the coordinates of all points on the circumference of a circle'.

What is the equation of a circle, centre O and radius r?

The radius r will be constant so if $P\ (x, y)$ is any point on the circumference. then, for all positions of P, OP is constant.

In Fig. 5.2.1(a)

as $$OP = r$$

$$OP^2 = r^2$$

i.e. $$x^2 + y^2 = r^2$$

So the equation of a circle, centre O, radius r, is $x^2 + y^2 = r^2$.

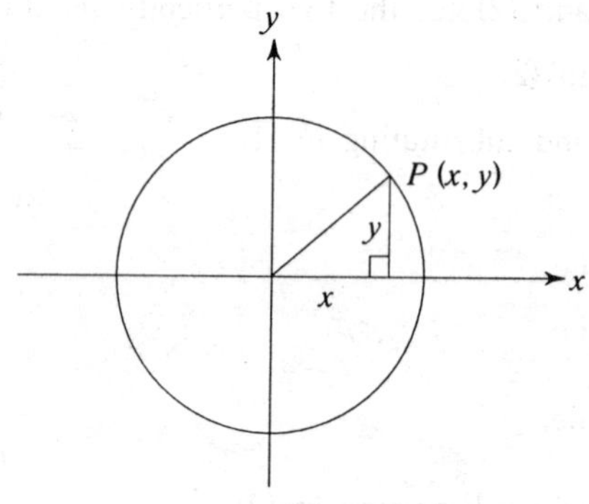

Fig. 5.2.1(a)

A similar argument may be used to determine the equation of a circle whose centre is (a, b). See Fig. 5.2.1(b).

If P is any point on the circumference, and $C\ (a, b)$ is the centre.

Draw CN parallel to the x-axis and NP parallel to the y-axis hence $\mathbf{CP} = \begin{pmatrix} x - a \\ y - b \end{pmatrix}$

But for all positions of $P\ (x, y)$, $|\mathbf{CP}| = r$

i.e. $$|\mathbf{CP}|^2 = r^2$$

i.e. $$(x - a)^2 + (y - b)^2 = r^2$$

so the equation of a circle, centre (a, b), radius r, is

$$(x - a)^2 + (y - b)^2 = r^2$$

This is the easiest form to use if the coordinates of the centre, and the radius of the circle, are known.

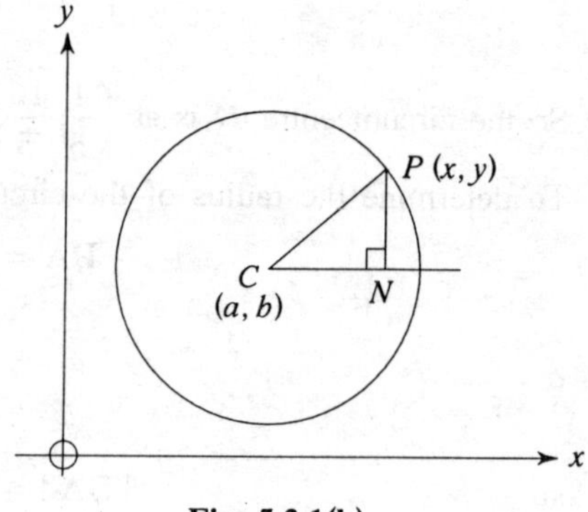

Fig. 5.2.1(b)

Example 5.2.1 (1)

Find the equation of the circle whose centre is (4, – 2) and whose radius is 3.

Solution:

The basic equation is $(x - 4)^2 + (y + 2)^2 = 3^2$

which may be expanded and simplified to give $x^2 - 8x + 16 + y^2 + 4y + 4 = 9$

i.e. $x^2 + y^2 - 8x + 4y + 11 = 0$

N.B: When the equation is written in expanded form, the coefficients of x^2 and y^2 are equal and the only other terms are linear (i.e. terms such as those which appear in the equation of a straight line).

In other problems we may be given an equation and asked to show that the equation is that of a circle and to identify the centre and radius.

Example 5.2.1 (2)

Show that the equation $2x^2 + 2y^2 - 8x + 5y + 10 = 0$

represents a circle and determine the coordinates of the centre and the length of the radius.

Solution:

In the given equation

$$2x^2 + 2y^2 - 8x + 5y + 10 = 0$$

the coefficients of x^2 and y^2 are equal, and the only other terms, $(- 8x + 5y + 10)$, are linear.

It is necessary to write the equation in the form

$$(x - a)^2 + (y - b)^2 = r^2$$

Firstly divide by 2

$$x^2 + y^2 - 4x + 5/2\ y + 5 = 0$$

Secondly, group together the terms in x^2 and x, and in y^2 and y

$$x^2 - 4x + y^2 + 5/2\ y = - 5$$

Thirdly complete the squares by adding the square of half the coefficient of x and the square of half the coefficient of y to both sides.

i.e. $$x^2 - 4x + 2^2 + y^2 + 5/2y + (5/4)^2$$

$$= 2^2 + (5/4)^2 - 5$$

i.e. $$(x - 2)^2 + (y + 5/4)^2 = 9/16$$

hence this is the equation of a circle, centre $(2, - 5/4)$ and radius $3/4$.

Example 5.2.1 (3)

Determine the equation of the circles through the points A (0, 2) and B (0, 18) which touch the x-axis.

Solution:

The sketch in Fig. 5.2.1 (c) suggests a method of solution.

For a start the centre must lie on the perpendicular bisector of AB, i.e. the line $y = 10$.

The circles both touch the x-axis and so radius $C_1 M_1$ = radius $C_2 M_2 = 10$ and so radius $C_1 A$ = radius $C_2 A = 10$.

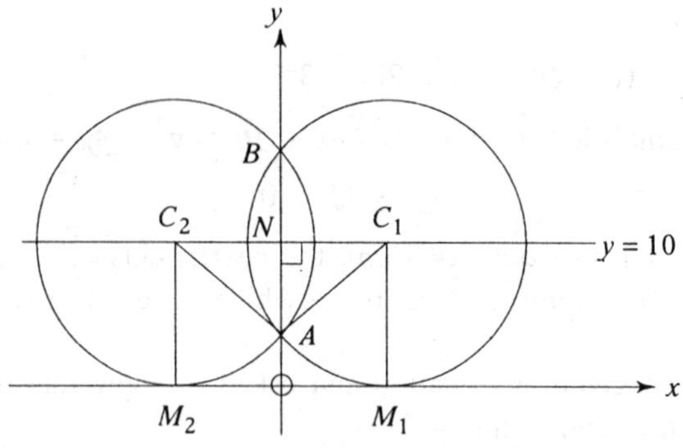

Fig. 5.2.1(c)

Since A is (0, 2) and N is (0, 10)

$$A\,N = 8$$

By Pythagoras

$$NC_1^2 = C_1A^2 - AN^2$$
$$= 10^2 - 8^2$$
$$= 36$$

so
$$NC_1 = 6$$

By a similar method we may show that $C_2N = 6$
hence C_1 has coordinates (6, 10) and C_2 has coordinates (– 6, 10)

The circle, centre C_1, has equation

$$(x - 6)^2 + (y - 10)^2 = 10^2$$

Which expanded and simplified leads to

$$x^2 + y^2 - 12x - 20y + 36 = 0$$

The circle, centre C_2, has equation

$$(x + 6)^2 + (y - 10)^2 = 10^2$$

which expanded and simplified leads to

$$x^2 + y^2 + 12x - 20y + 36 = 0.$$

5.2.2 Tangents to a Circle

A line which meets a circle at one and only one point is known as a tangent to a circle. A tangent to a circle and a radius drawn through the point of contact meet at right angles (See Fig. 5.2.2 (a))

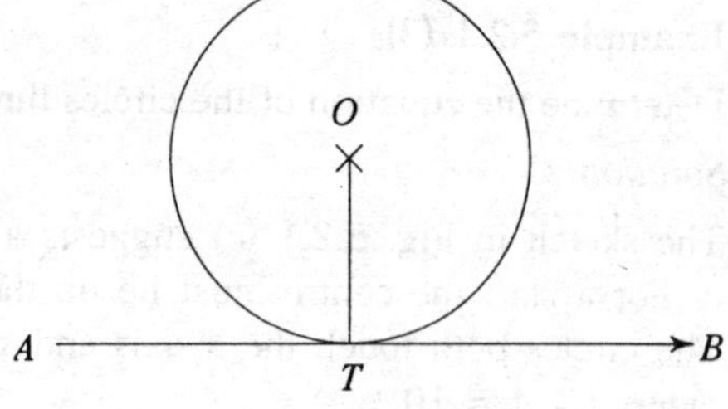

Fig. 5.2.2(a)

Example 5.2.2 (1)

Verify that the point (6, 2) lies on the circle whose equation is

$$x^2 - 14x + y^2 + 2y + 40 = 0$$

and find the equation of the tangent to the circle at that point.

Solution:

On substituting $x = 6$ and $y = 2$ in the left hand side of the given equation we have

$$6^2 - 14 \times 6 + 2^2 + 2 \times 2 + 40 = 36 - 84 + 4 + 4 + 40 = 0$$

so (6, 2) lies on the given circle.

The gradient of the tangent line can be found using the gradient of the radius through the point of contact, since these two lines are perpendicular.

The circle has equation

$$x^2 - 14x + y^2 + 2y = -40$$

i.e. $$x^2 - 14x + 7^2 + y^2 + 2y + 1^2 = 7^2 + 1^2 - 40$$

i.e. $$(x - 7)^2 + (y + 1)^2 = 10$$

so the centre is at (7, −1) and the gradient of the radius joining (7, −1) to (6, 2) is

$$\frac{2 - (-1)}{6 - 7} = \frac{3}{-1} = -3$$

The gradient of the tangent at (6, 2) will therefore be 1/3.

The equation of the tangent is

$$\frac{y - 2}{x - 6} = \frac{1}{3}$$

i.e. $$3y - 6 = x - 6$$

i.e. $$x = 3y$$

Example 5.2.2 (2)

Determine the lengths of the tangents from the point (10, 11) to the circle whose equation is

$$x^2 + y^2 - 12x - 14y + 81 = 0.$$

Solution:

Study Fig. 5.2.2 (b).

If we have the lengths r and d we can deduce the length t of each of the tangents from A using Pythagoras.

The equation of the given circle is

$$x^2 - 12x + y^2 - 14y = -81$$

i.e. $$x^2 - 12x + 6^2 + y^2 - 14y + 7^2 = 6^2 + 7^2 - 81$$

i.e. $$(x - 6)^2 + (y - 7)^2 = 4$$

so the circle has its centre at C (6, 7) and is of radius 2.

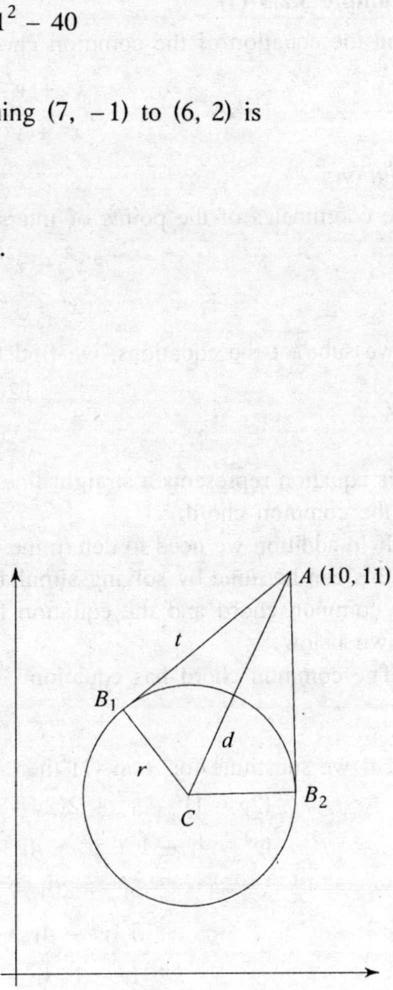

Fig. 5.2.2(b)

so $$d^2 = (10-6)^2 + (11-7)^2$$
$$= 4^2 + 4^2 = 32$$
and $$r^2 = 2^2 = 4$$
hence $$t^2 = d^2 - r^2$$
$$= 32 - 4$$
$$= 28$$

and the length of each tangent from A to the circle is $\sqrt{28}$. or $2\sqrt{7}$

5.2.3 Intersection of Two Circles

Example 5.2.3 (1)

Find the equation of the common chord of the circles

$$x^2 + y^2 + 2x - 20y + 16 = 0$$
$$x^2 + y^2 - 10x + 4y + 4 = 0.$$

Solution:

The coordinates of the points of intersection A and B of these circles satisfy the equations

$$x^2 + y^2 + 2x - 20y + 16 = 0 \quad (1)$$
$$x^2 + y^2 - 10x + 4y + 4 = 0 \quad (2)$$

If we subtract the equations, we find that the coordinates of A and B satisfy the equation

$$12x - 24y + 12 = 0$$
or $$x - 2y + 1 = 0$$

This equation represents a straight line and so it is the equation of the common chord.

If in addition we need to determine the coordinates of A and B, this can be done by solving simultaneously the equation for the common chord and the equation for one of the circles as shown below.

The common chord has equation

$$x = 2y - 1$$

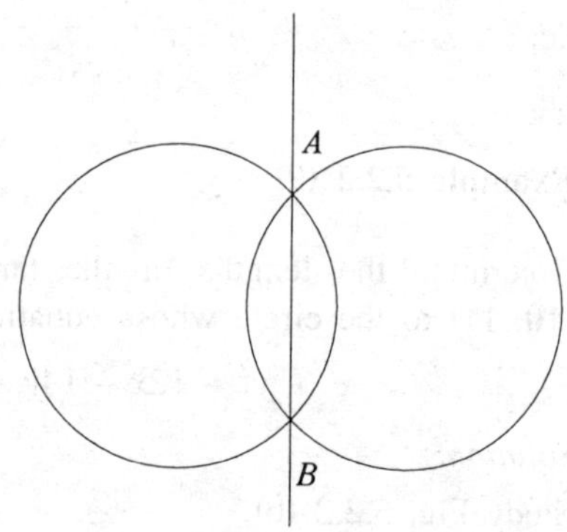

Fig. 5.2.3(a)

and if we substitute for x in (1) then

$$(2y-1)^2 + y^2 + 2(2y-1) - 20y + 16 = 0$$
i.e. $$4y^2 - 4y + 1 + y^2 + 4y - 2 - 20y + 16 = 0$$
i.e. $$5y^2 - 20y + 15 = 0$$
i.e. $$5\,(y^2 - 4y + 3) = 0$$
i.e. $$(y-1)\,(y-3) = 0$$
hence $$y = 1, \text{ or } y = 3$$
and on substituting $$x = 1 \text{ or } x = 5$$

hence the points of intersection are (1, 1) and (5, 3).

Two circles may not intersect but still by subtracting their equations we may obtain the equation of a straight line. What does such a line represent? Study the next example.

Example 5.2.3 (2)

What are the squares of the lengths of the tangents from the point $P\ (x_p, y_p)$ to each of the circles whose equations are

$$x^2 + y^2 - 1 = 0 \text{ and}$$
$$x^2 + y^2 + 12x - 16y + 75 = 0,$$

respectively?

Deduce the locus of the paint P such that the lengths of the tangents from P to both circles are equal.

Solution:

In Fig. 5.2.3 (b)

$$d^2 = OP^2 = x_p^2 + y_p^2$$

and the required squared length $AP^2 = t^2 = d^2 - 1$

i.e. $\quad x_p^2 + y_p^2 - 1$

In Fig. 5.2.3 (c) if we can determine the coordinates of the centre of the circle C and its radius r_1, we can then apply Pythagoras to find $BP^2 = t_1^2$.

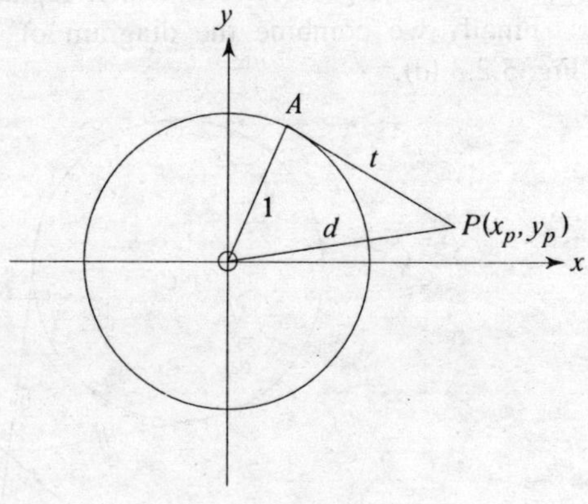

Fig. 5.2.3(b)

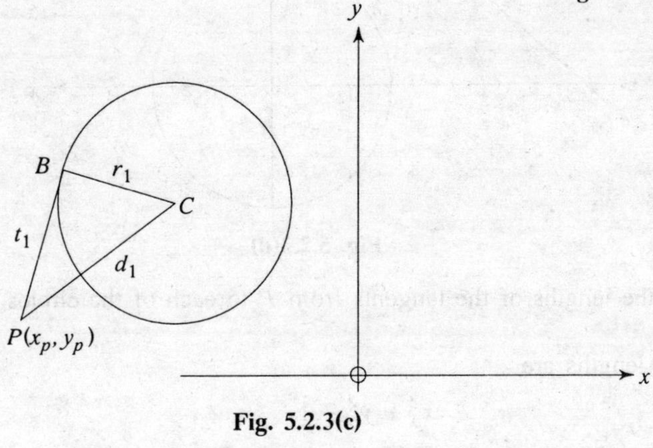

Fig. 5.2.3(c)

From the given equation of the second circle

$$x^2 + y^2 + 12x - 16y + 75 = 0$$

We may proceed to determine the centre and radius by

$$x^2 + 12x + y^2 - 16y = -75$$

i.e. $$x^2 + 12x + 36 + y^2 - 16y + 64 = 36 + 64 - 75$$

i.e. $$(x + 6)^2 + (y - 8)^2 = 25$$

i.e. the centre is at C (−6, 8) and the radius is 5 and so $t_1^2 = CP^2 - r_1^2$

$$= d_1^2 - r_1^2$$

$$= (x_p + 6)^2 + (y_p - 8)^2 - 25$$

expanding this expression

$$t_1^2 = x_p^2 + 12\ x_p + 36 + y_p^2 - 16y_p + 64 - 25$$

$$= x_p^2 + y_p^2 + 12\ x_p - 16y_p + 75$$

The discerning reader will have observed that this expression is that for the left hand side of the equation of the given circle with x replaced by x_p and y replaced by y_p.

Finally we combine the diagram of Fig. 5.2.3 (b) and (c) to deduce the required locus. In Fig. 5.2.3 (d).

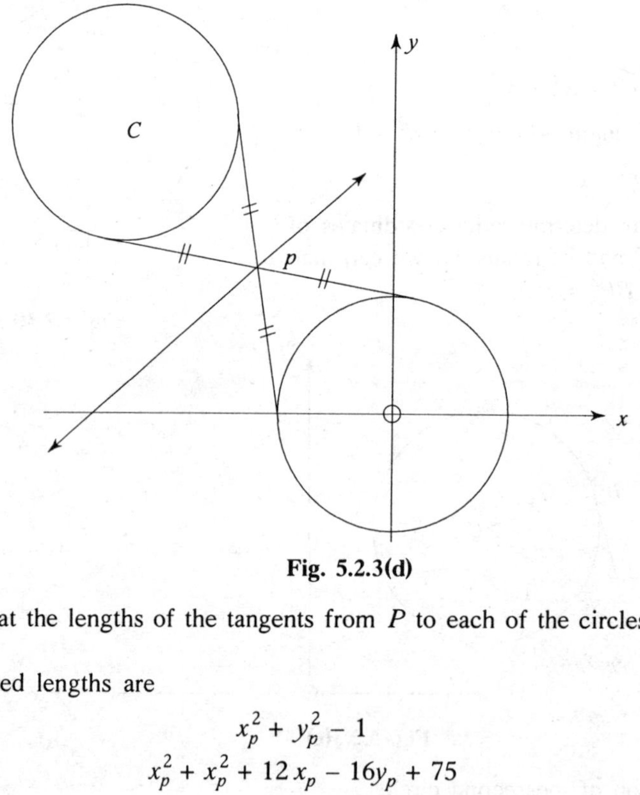

Fig. 5.2.3(d)

P is chosen so that the lengths of the tangents from P to each of the circles, centres O and C, are equal

Now the squared lengths are

$$x_p^2 + y_p^2 - 1$$

and $$x_p^2 + x_p^2 + 12\,x_p - 16y_p + 75$$

and these are equal,

so $$x_p^2 + y_p^2 + 12x_p - 16y_p + 75 = x_p^2 + y_p^2 - 1$$

i.e. $$12x_p - 16y_p + 76 = 0$$

Since this equation is true for all points P (x_p, y_p) for which the condition holds (i.e. the lengths of the tangents from P to each circle are equal) then the locus of P satisfies the equation $12x - 16y + 76 = 0$.

This equation is the equation which would have been obtained on subtracting the equation $x^2 + y^2 - 1 = 0$ from the equation

$$x^2 + y^2 + 12x - 16y + 75 = 0$$

In this case the straight line does *not* represent the common chord, since the circles do not intersect. It *does* represent the locus of points from which it is possible to draw tangents of equal length to both circles.

5.2.4 Orthogonal Circles

If the tangents to two circles at their points of intersection are perpendicular the circles are said to be *orthogonal.* Since the radius through a point of contact is perpendicular to the tangent, it follows that the tangent to one circle is a radius of the other at a point of contact. Thus if the centres of two orthogonal circles (radius r_1 and r_1) are a distance d apart it follows by Pythagoras that

$$d^2 = r_1^2 + r_2^2$$

Conversely if we find that two circles (radius r_1, r_2) are a distance d apart and $d^2 = r_1^2 + r_2^2$, then the circles are orthogonal. See Fig. 5.2.4 (a)

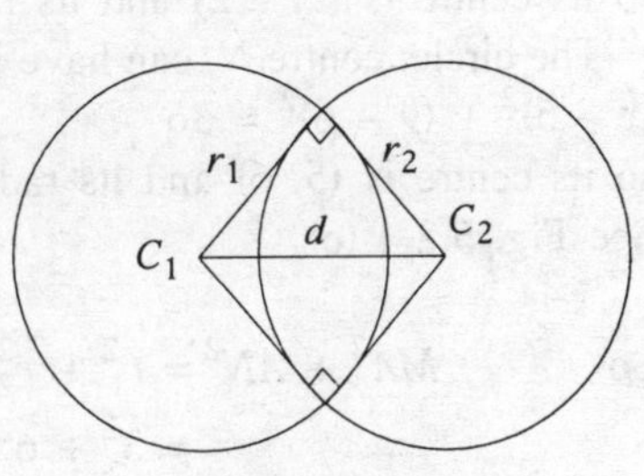

Fig. 5.2.4(a)

Example 5.2.4 (1)

Show that the circles with equations

$$x^2 + y^2 - 12x - 2y - 2 = 0$$

and
$$x^2 + y^2 + 2x - 4y - 6 = 0$$

are orthogonal.

Solution:

Proceeding as before to determine the coordinates of the centre and the lengths of the radius in the first circle

$$x^2 + y^2 - 12x - 2y = 2$$

$$x^2 - 12x + 36 + y^2 - 2y + 1 = 36 + 1 + 2$$

i.e.
$$(x - 6)^2 + (y - 1)^2 = 39$$

so the circle has centre (6, 1) and radius $\sqrt{39}$.

Similarly the second circle can be rewritten in the form

$$(x + 1)^2 + (y - 2)^2 = 11,$$

and it has centre (– 1, 2) and radius $\sqrt{11}$.

From Fig. 5.2.4 (b)

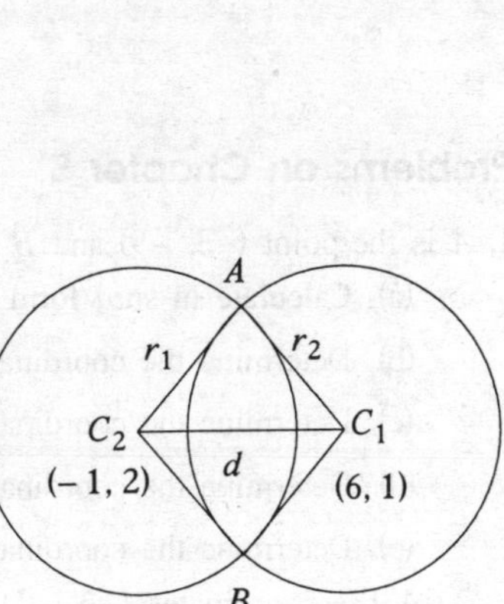

Fig. 5.2.4(b)

$$r_1^2 + r_2^2 = 11 + 39 = 50$$

and
$$d^2 = (6 - -1)^2 + (1 - 2)^2$$

$$= 49 + 1 = 50$$

since $\qquad d^2 = r_1^2 + r_2^2$

$\qquad$ (or $C_2 C_1^2 = C_2^2 A^2 + A\ C_1^2$)

then the circles are orthogonal.

Example 5.2.4 (2)

The circles $x^2 + y^2 - 4x + 4y - 1 = 0$, $x^2 + y^2 - 10x - 12y + 25 = 0$, have centres M, N respectively, and intersect at A and B.

Show that the angle MAN is an obtuse angle.

Solution:

Here we may use a method similar to that for the previous example.

The circle, centre M, can have its equation re-written in the form $(x - 2)^2 + (y + 2)^2 = 9$

so its centre is $(2, - 2)$ and its radius is 3.

The circle, centre N, can have its equation re-written in the form $(x - 5)^2 + (y - 6)^2 = 36$

so its centre is (5, 6) and its radius 6.

See Fig. 5.2.4 (c)

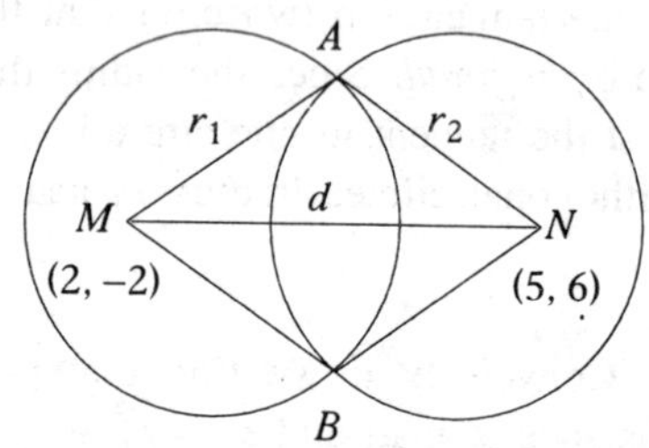

Fig. 5.2.4(c)

So $\qquad MA^2 + AN^2 = r_1^2 + r_2^2$

$$= 3^2 + 6^2 = 45$$

and $\qquad MN^2 = d^2 = (5 - 2)^2 + (6 - - 2)^2$

$$= 3^2 + 8^2 = 73$$

since $\qquad d^2 > r_1^2 + r_2^2$

the angle MAN is obtuse.

EXERCISE 5

Problems on Chapter 5

1. A is the point $(-3, -9)$ and B is the point $(5, 7)$.
   (a) Calculate in surd form the distance $|AB|$
   (b) Determine the coordinates of the mid-point M of AB.
   (c) Determine the coordinates of the point P which lies internally on AB so that $AP : PB = 3 : 1$
   (d) Determine the coordinates of the point Q which divides AB externally in the ratio $3 : 1$.
   (e) Determine the coordinates of the point R which divides AB externally in the ratio $1 : 3$.
2. A has coordinates $(-5, -7)$, B has coordinates $(-1, 3)$ and C has coordinates $(9 - 5)$. Determine the coordinates of the centroid G of triangle ABC.

3. Determine the gradient and the intercept on the y-axis, of each of the lines whose equations are given below.
   (a) $y = -3x - 5$
   (b) $y = \sqrt{3}\ x + 2\sqrt{3}$
   (c) $5x - 6y = 30$
   (d) $\sqrt{7}x + 9y = 63\sqrt{7}$
   (e) $x = 10$
4. Determine in the form $ax + by = c$, where a, b, c are integers the equation of the following straight lines:
   (a) the line through $\left(0, \frac{1}{2}\right)$ with gradient $\frac{1}{4}$
   (b) the line through (3, 7) with gradient $\frac{3}{4}$
   (c) the line through the points (–2, –6) and (4, –3).
   (d) the line through (5, –2) which is parallel to the line whose equation is $5x + 15y = 8$
   (e) the line through (7, –4) which is perpendicular to the line whose equation is $2x + 5y = 10$.
5. $ABCD$ is a parallelogram with A (5, 1) and C (8, 14). The equations of sides AB and AD are $x + 5y = 10$ and $3x - 2y = 13$ respectively. Determine
   (a) the gradient of AB,
   (b) the equation of the side CD,
   (c) the coordinates of the points B and D.
6. $ABCD$ is a rectangle and C has coordinates (14, 11). AD and AB have equations $x - y = 1$ and $x + y = 11$ respectively.
   (a) Determine the coordinates of A, B and D.
7. $ABCD$ is a quadrilateral
   AB has equation $7x - y = 23$
   BC has equation $x - y = -7$
   CD has equation $7x - y = -7$
   DA has equation $x - y = -1$
   (a) Determine the coordinates of A, B, C and D.
   (b) Show that $ABCD$ is a rhombous.
   (c) Determine the equation of the diagonals AC and BD, and also the coordinates of their point of intersection.
8. A (5, 7), B (11, 19) and C (53, –9) are the vertices of triangle ABC. Determine the equation of the altitude from C to AB and the coordinates of the orthocentre of triangle ABC.
9. $ABCD$ is a rectangle with A (5, 6) C(10, 16), and the diagonals meet at M. AD has equation $4x + 3y = 38$. Find,
   (a) the equation of AB,
   (b) the coordinates of B, D, M
   (c) the equations of side CD and diagonal AC,

(d) the area of the rectangle by calculating the lengths of the sides $\overline{AB}$ and $\overline{BC}$.

(e) Show that the ratio **AC** : **BC** = $\sqrt{5}$: 1

10. In triangle *ABC*, *A* has coordinates (12, 20), *B* has coordinates (3, 8), and *C* has coordinates (15, 16). *P* and *Q* divide $\overline{BC}$ internally and externally respectively in the ratio 3 : 1.

(a) Show that the ratio $\overline{BA} : \overline{AC}$ = 3 : 1

(b) Show that angle *PAQ* is a right angle.

11. *A*, *B*, *C*, *D* have coordinates (– 2, 15), (– 2, 5), (2, 3) and (4, 7) respectively.

(a) Show that *ABCD* is a kite

(b) Determine the equations of all the sides of *ABCD*, and also of the diagonals *BD* and *AC*.

(c) Show that the area of the kite is 40 unit^2.

(d) Show that $\overline{AC} : \overline{BD}$ = 2 and that $\overline{BD} : \overline{BC} = \sqrt{2}$

12. In triangle *ABC*, *M* is the mid-point of *BC*, *N* is the mid-point of *CA*, *P* is the mid-point of *AB*. The coordinates of *M*, *N*, *P* are (1, 7), (9, 14), (5, 12) respectively. Determine

(a) the gradient of *MP*

(b) the equation of *AC* (hint *MP* and *AC* are parallel)

(c) the coordinates of *C*

(d) the coordinates of *A* and *B*

(e) the coordinates of Centroid *G*

13. Determine in the form $ax + by = c$, where a, b, c are integers, the equations of the lines through (3, 5), which are parallel and perpendicular to the line whose equation is $2x - 3y = 12$

14. *A* (8, 9), *B* (0, 5) and *C* (2, 1) are three vertices of a rectangle *ABCD*. Determine

(a) the coordinates of the vertex *D* and the centre *Q* of the rectangle.

(b) the coordinates of the point *P* which lies on *AD* produced such that $\overline{AP} = 3\overline{AD}$

(c) the equation of the line *AP*.

*15. *P* is the point of intersection of the lines whose equations are $y = x\sqrt{3}$ and $x = 3 - y\sqrt{3}$. *Q* is the point where the line $x = 3 - y\sqrt{3}$ intersects the x-axis.
Determine the coordinates of *P* and of *Q*, the area of triangle *OPQ*, and the sizes of the angles of triangle *OPQ*.

16. (a) Calculate without using tables or calculator the size of the acute angle between the lines whose equations are

$$5x + 5y = 8, \text{ and } 3x + 3\sqrt{3}y = 6.$$

(b) Show that the acute angle between the lines whose equations are $x - y = 10$ and $3x - 4y = 12$, is $\tan^{-1} 1/7$.

(c) Show that the acute angle between the lines whose equations are $2x - y = 0$, and $7x - y = 21$, is $\tan^{-1} 1/3$.

(d) Determine without using tables or calculator the equation of the two lines, having the same intercept on the y-axis as the line whose equation is $x - 3y = 15$, and which are inclined at an angle $\tan^{-1} 1/7$ to that line.

*17. The points A, B, C have coordinates $(-3, -2)$, $(1, 2)$ and $(2, 3)$ respectively. Determine the coordinates of the point D which satisfy both of the conditions.

(a) the line segment DB is perpendicular to the line segment AC

(b) the vector **DC** is perpendicular to the vector

$$\bar{u} = \mathbf{DA} - \mathbf{AC}$$

18, Determine the coordinates of the centre, and the length of the radius, of each of the following circles whose equations are:

(a) $x^2 + y^2 = 9$

(b) $x^2 - 6x + y^2 - 4y - 12 = 0$

19. (a) Determine the coordinates of the centre and the length of the radius of the circle whose equation is :

$$x^2 + y^2 + 8x - 10y - 9 = 0$$

(b) Determine in the form $ax + by = c$, where a, b, c are integers, the equation of the tangent to the circle at the point $(1, 10)$.

*20. The circle whose equation is $x^2 + y^2 + 2gx + 2fy + c = 0$ passes through the origin and also through the points $(4, 0)$ and $(0, 6)$.

(a) Determine the values of g, f and c, and hence the coordinates of the centre and the length of the radius of the circle.

(b) Determine the length of each of the tangents from the point $(10, 10)$ to the circle.

21. Determine the coordinates of the points of intersection of the circle whose equation is $x^2 + y^2 = 100$, and the straight line whose equation is $2y = x + 10$

*22. (a) Show that the equation of the circle on AB as diameter, where A and B have coordinates (x_1, y_1) and (x_2, y_2) respectively, is given by $(x - x_1)(x - x_2) + (y - y_1)(y - y_2) = 0$

(b) Hence determine the equation of the circle on the line segment AB as diameter, when A and B have coordinates $(2, 4)$ and $(6, 10)$ respectively.

*23. Show that the circles whose equations are $x^2 + y^2 - 4x - 6y - 212 = 0$ and $x^2 + y^2 - 44x - 36y + 408 = 0$, are orthogonal. Find the equations of the tangents to each circle at their point of intersection which are parallel to the axes.

6

Calculus

INTRODUCTION

Calculus is the study of functions and their graphs, and of the methods of finding the gradient of a function at any point on its graph.

6.1 THE GRADIENT OF A FUNCTION

6.1.1 The Gradient of the Functions $y = x^n$ and $y = x^{-n}$ where n is a positive integer using the Definition of a Derivative

The reader has probably found the gradient to a given curve at a particular point by drawing a tangent line to the curve at that point 'by eye'. We now want to find the gradient to a given curve at any point on the curve for a wide variety of curves with different equations. Some understanding of the function notation is necessary, and the reader is referred to section 3.6.2 of Chapter III for a fuller explanation of the notation.

In particular the reader should understand that if the formula for any given function is known, i.e. $f(x)$ is known to be x^2, then it is possible to find a corresponding formula for f (x+h). If $f(x) = x^2$ then

$$f(x + h) = (x + h)^2$$

If $$f(x) = x^3 - 3x^2 + 2x - 1$$

then $$f(x + h) = (x + h)^3 - 3(x + h)^2 + 2\,(x + h) - 1$$

if $$f(x) = 1/x \text{ then } f(x + h) = \frac{1}{x + h}$$

if $$f(x) = \sin x \text{ then } f(x + h) = \sin(x + h).$$

if $$f(x) = 3^x \text{ then } f(x + h) = 3^{x + h}$$

In the beginning we shall not assume any known rules for finding the gradient of a curve but we shall establish these rules in many cases by using what is described as 'differentiation by first principles' or 'differentiation using the definition of a derivative'.

Geometrically this may be illustrated in the Figs. 6.1.1 (a) and 6.1.1 (b) which follow.

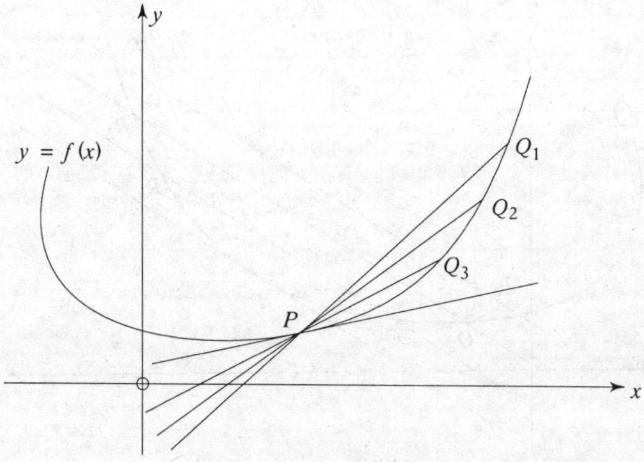

Fig. 6.1.1(a)

In Fig. 6.1.1 (a) for a function $y = f(x)$ we wish to find the gradient of the graph at point P, which is given by the gradient of the tangent line to the graph at P. We let Q be a neighbouring point on the graph. As Q takes up the positions Q_1, Q_2, Q_3, ..etc. Q becomes closer to P, and the chords PQ_1, PQ_2, PQ_3, etc become closer to the tangent line at P.

If P has coordinates $(x, f(x))$ and Q has coordinates $(x + h, f(x + h))$ we can find the gradient of the chord PQ as $\dfrac{f(x+h) - f(x)}{x + h - x}$.

As Q takes up successive positions Q_1, Q_2, Q_3, etc the value of h gets smaller, and we say that h becomes as close to zero as we would like (imagine a set of very small numbers which are non-zero such as 10^{-2}, 10^{-10}, 10^{-100}, $10^{-1\,000\,000}$ or even some negative numbers such as -10^{-1000}, $-10^{-10\,000}$) hence the gradient of the curve at P which is the gradient of the tangent line to the curve at P will become the $\lim\limits_{h \to 0}$ of the gradient of the chord PQ i.e.

$$\text{the } \lim_{h \to 0} \frac{f(x+h) - f(x)}{h}$$

It is important to note that we cannot allow h to be equal to zero, since division by zero is not permissable in the operations on real numbers, hence the use of the notation $h \to 0$ which means 'h becomes as close to zero as we would like'. This idea may be easier to understand after the reader has tried some examples on differentiation by first principles.

Another approach to finding the gradient, at the point P, to the curve whose equation is $y = f(x)$, is to select two points Q and R on either side of P and to estimate the gradient of the tangent line to the curve at P by using the gradient of the chord QR. As the points Q and R take up the positions Q_1, R_1; Q_2, R_2; Q_3, R_3; respectively then the gradient of the chord QR becomes closer to the gradient of the tangent line to the curve at P.

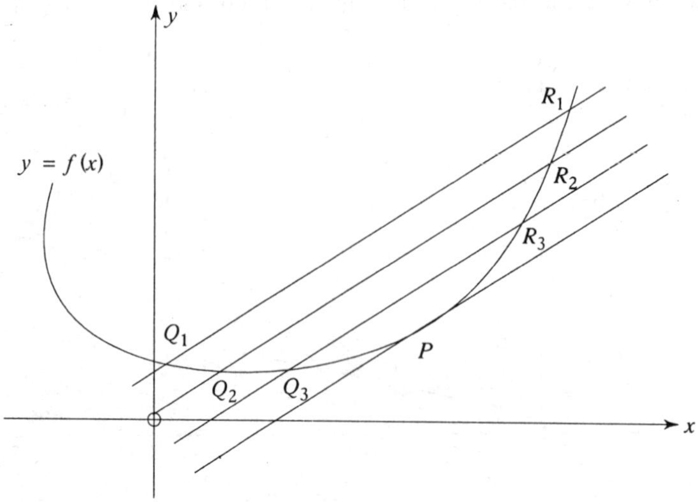

Fig. 6.1.1(b)

Normally we choose Q to have coordinates $((x - h), f(x - h))$ and R to have coordinates $(x + h)$, $f(x + h)$ so that the gradient of chord QR will be

$$\frac{f(x+h) - f(x-h)}{(x+h) - (x-h)}$$

or

$$\frac{f(x+h) - f(x-h)}{2h}$$

As in the previous case we may let h become as close to zero as we like, and obtain the gradient of the curve at P (the gradient of the tangent line to the curve at P) as the $\lim_{h \to 0}$ of the gradient of Chord QR i.e.

$$\text{the } \lim_{h \to 0} \frac{f(x+h) - f(x-h)}{2h}$$

Example 6.1.1 (1)

Find the gradient of the curve whose equation is $y = x^4$ using the definition of a derivative.

Solution: See Fig. 6.1.1 (c)

Let $P\ (x, x^4)$ and $Q\ (x + h, (x + h)^4)$ be neighbouring points on the curve.

$$\text{Gradient } PQ = \frac{(x+h)^4 - x^4}{x + h - x}$$

$$= \frac{x^4 + 4x^3h + 6x^2h^2 + 4xh^3 + h^4 - x^4}{h}$$

$$= \frac{4x^3h + 6x^2h^2 + 4xh^3 + h^4}{h}$$

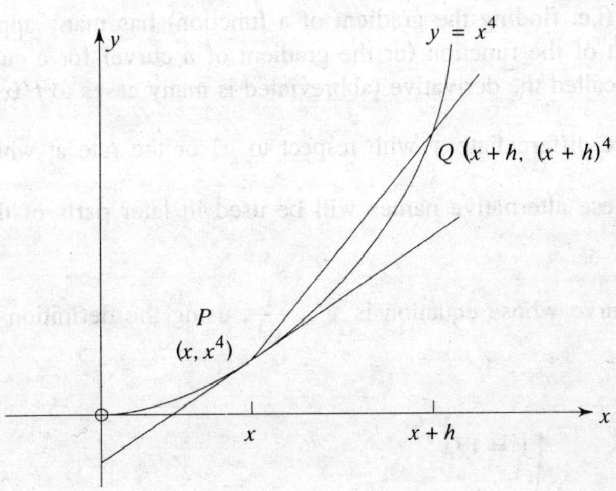

Fig. 6.1.1 (c)

$$\text{Gradient } PQ = \frac{h(4x^3 + 6x^2h + 4xh^2 + h^3)}{h}$$
$$= 4x^3 + 6x^2h + 4xh^2 + h^3$$

Before we take the limit as h becomes close to zero perhaps a numerical illustration of the gradient where $x = 3$ may help to convince the reader.

We take P as $(3, 3^4)$ and Q as $(3 + h, (3 + h)^4)$ and evaluate the gradient $PQ = 4x^3 + 6x^2h + 4xh^2 + h^3$

for successively smaller values of h

x	h	gradient PQ
3	0.1	113.521
3	0.01	108.541
3	0.001	108.054
3	10^{-4}	108.005
3	10^{-5}	108.001
3	10^{-6}	108.000

Taking the value of the gradient PQ correct to three decimal places it should be clear that as h gets smaller the value gets as close as we like to 108 which is the value of the first term ($4x^3$ evaluated at $x = 3$) in the expression for the gradient of the chord PQ.

If the reader is still not convinced, some further different numerical substitutions for x and h may be tried. This is left to the discretion of the reader.

However, it should be clear that, as h gets smaller, the expression for the gradient of PQ, for any value of x, can be made as close as we wish to $4x^3$ and we write that the gradient to the curve at

$$P = \lim_{h \to 0} \text{gradient } PQ$$

$$= \lim_{h \to 0} (4x^3 + 6x^2h + 4xh^2 + h^3) = 4x^3.$$

Because differentiation (i.e. finding the gradient of a function) has many applications the reader is advised that the gradient of the function (or the gradient of a curve) for a curve whose equation is given by $y = f(x)$ is also called the derivative (abbreviated is many cases to $f'(x)$) or the differential coefficient or $\frac{dy}{dx}$ (i.e. we differentiate y with respect to x) or the rate at which y changes with respect to x. Several of these alternative names will be used in later parts of this chapter.

Example 6.1.1 (2)

Find the gradient of the curve whose equation is $y = \frac{1}{x^3}$, using the definition of a derivative.

Solution:

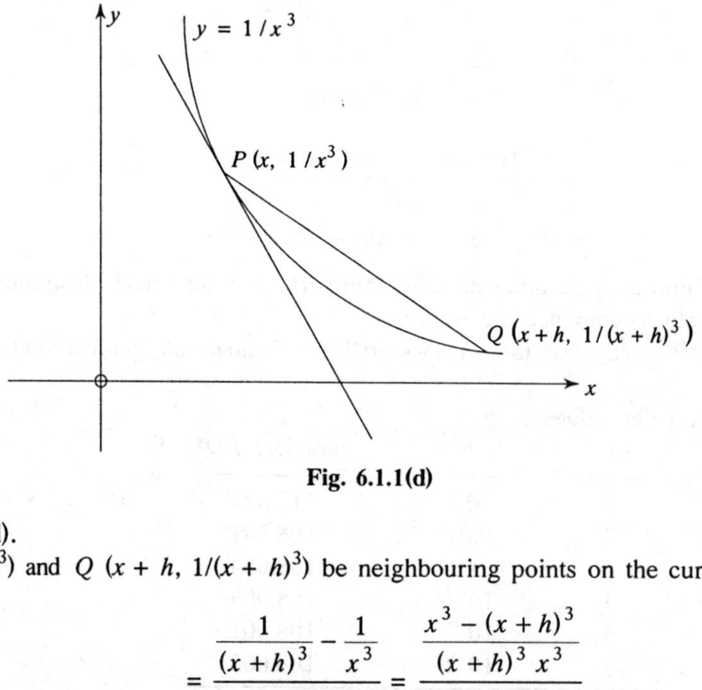

Fig. 6.1.1(d)

See Fig. 6.1.1(d).

Let P $(x, 1/x^3)$ and Q $(x + h, 1/(x + h)^3)$ be neighbouring points on the curve

$$\text{Gradient } PQ = \frac{\frac{1}{(x+h)^3} - \frac{1}{x^3}}{x + h - x} = \frac{\frac{x^3 - (x+h)^3}{(x+h)^3 x^3}}{h}$$

$$= \frac{x^3 - (x^3 + 3x^2 h + 3xh^2 + h^3)}{(x+h)^3 x^3} \times \frac{1}{h}$$

$$\text{Gradient } PQ = \frac{x^3 - x^3 - 3x^2 h - 3xh^2 - h^3}{(x+h)^3 x^3} \times \frac{1}{h}$$

$$= \frac{-3x^3 h - 3xh^2 - h^3}{(x+h)^3 x^3 h}$$

$$= \frac{h(-3x^2 - 3xh - h^2)}{(x+h)^3\, x^3 h}$$

$$= \frac{-3x^2 - 3xh - h^2}{(x+h)^3\, x^3}$$

The gradient to the curve at P

or $$\frac{dy}{dx} = \lim_{h \to 0} \frac{-3x^2 - 3xh - h^2}{(x+h)^3\, x^3}$$

$$= \frac{-3x^2}{x^3 \cdot x^3}$$

$$= -3/x^4 \text{ or } -3x^{-4}$$

N.B. As h becomes close to zero (i.e. $h \to 0$) $-3x^2 - 3xh - h^2$ becomes close to $-3x^2$ and $(x + h)^3$ becomes close to x^3.

A discerning reader may have observed that for the curve $y = x^4$ the gradient was $4x^3$ and for the curve $y = 1/x^3$ or x^{-3} the gradient was $= -3x^{-4}$, and may be asking whether we may establish a rule that the gradient of the curve $y = x^m$ will be mx^{m-1}. The next two examples illustrate a general method of attack, using the definition of a derivative, in the cases where n is a positive or negative integer. Establishing the result for fractional n will be examined later.

Example 6.1.1(3)

Find the gradient of the curve whose equation is $y = x^n$ (where n is a positive integer), using the definition of a derivative.

Solution: See Fig. 6.1.1(e).

Let P (x, x^n) and Q $(x + h, (x + h)^n)$ be neighbouring points on the curve

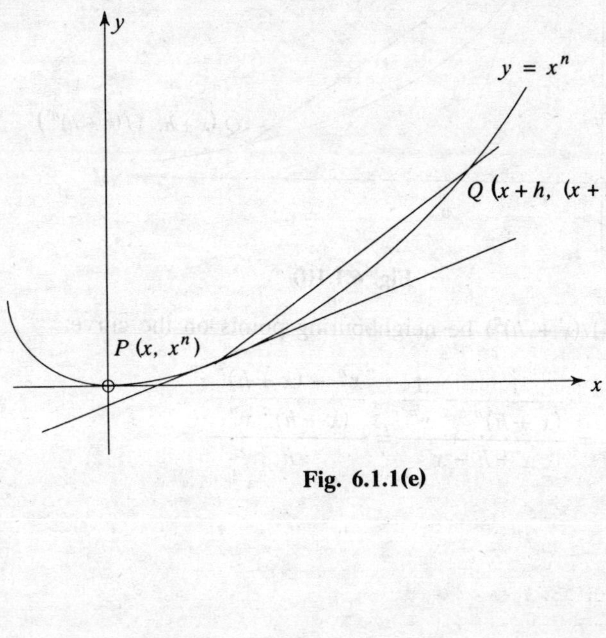

Fig. 6.1.1(e)

Gradient PQ $= \dfrac{(x+h)^n - x^n}{x+h-x}$

$$= \frac{x^n + nx^{n-1}h + {}_nC_2\, x^{n-2}h^2 + \ldots + {}_nC_r\, x^{n-r}h^r + \ldots + h^n - x^n}{h}$$

$$= \frac{nx^{n-1}h + {}_nC_2\, x^{n-2}h^2 + \ldots + {}_nC_r\, x^{n-r}h^r + \ldots + h^n}{h}$$

$$= \frac{h(nx^{n-1} + {}_nC_2\, x^{n-2}h + \ldots + {}_nC_r\, x^{n-r}h^{r-1} + \ldots + h^{n-1})}{h}$$

$$= n\,x^{n-1} + {}_nC_2\, x^{n-2}\,h + \ldots + {}_nC_r\, x^{n-r}\,h^{r-1} + \ldots + h^{n-1}$$

The gradient to the curve at P

or $\dfrac{dy}{dx} = \lim\limits_{h \to 0}$ gradient PQ

$$= \lim_{h\to 0}\,(nx^{n-1} + {}_nC_2\, x^{n-2}h + \ldots + {}_nC_r\, x^{n-r}\, h^{r-1} + \ldots + h^{n-1})$$

$$= nx^{n-1}$$

Example 6.1.1 (4)

Find the gradient of the curve whose equation is $y = x^{-n}$ (where n is a positive integer), using the definition of a derivative.

Solution: See Fig. 6.1.1(f)

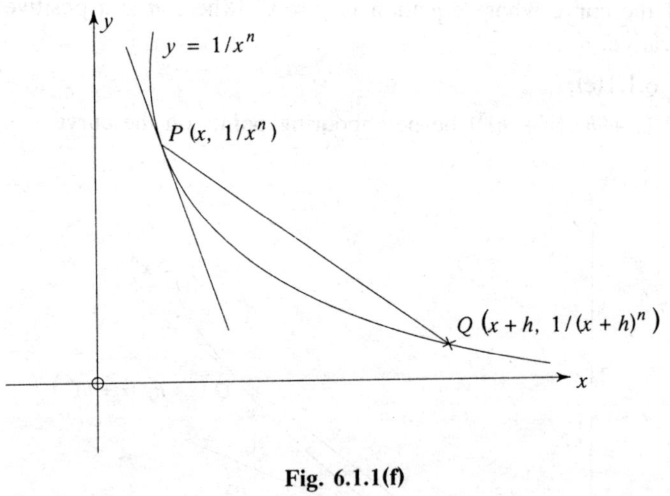

Fig. 6.1.1(f)

Let $P\,(x,\ 1/x^n)$, $Q\,(x + h,\ 1/(x + h)^n)$ be neighbouring points on the curve.

Gradient PQ $= \dfrac{\dfrac{1}{(x+h)^n} - \dfrac{1}{x^n}}{x+h-x} = \dfrac{\dfrac{x^n - (x+h)^n}{(x+h)^n\, x^n}}{h}$

$$= \frac{x^n - (x^n + nx^{n-1}h + {}_nC_2\, x^{n-2}h^2 + \ldots + {}_nC_r\, x^{n-r}h^r + \ldots + h^n)}{(x+h)^n\, x^n} \times \frac{1}{h}$$

$$= \frac{x^n - x^n - nx^{n-1}h - {}_nC_2\, x^{n-2}h^2 - \ldots - {}_nC_r\, x^{n-r}h^r - \ldots - h^n)}{(x+h)^n\, x^n\, h}$$

$$= \frac{-nx^{n-1}h - {}_nC_2\, x^{n-2}h^2 - \ldots - {}_nC_r\, x^{n-r}h^r - \ldots - h^n)}{(x+h)^n\, x^n h}$$

$$= \frac{h(-nx^{n-1} - {}_nC_2\, x^{n-2}h - \ldots - {}_nC_r\, x^{n-r}h^{r-1} - \ldots - h^{n-1})}{(x+h)^n\, x^n h}$$

$$= \frac{-nx^{n-1} - {}_nC_2\, x^{n-2}h - \ldots - {}_nC_r\, x^{n-r}h^{r-1} - \ldots - h^{n-1}}{(x+h)^n\, x^n}$$

The gradient to the curve at P

or
$$\frac{dy}{dx} = \underset{h \to 0}{\text{limit}} \text{ gradient } PQ$$

$$= \frac{-nx^{n-1}}{x^n\, x^n}$$

$$= -\, nx^{n-1-n-n}$$

$$= -nx^{-n-1} \text{ or } -n/x^{n+1}$$

The use of the $f'(x)$ notation (derivative) for the gradient will be useful in establishing further results in the next two examples.

Example 6.1.1 (5)

If for a given domain the functions $f(x)$ and $g(x)$ have derivatives $f'(x)$ and $g'(x)$ respectively then the function $f(x) + g(x)$ has derivative $f'(x) + g'(x)$.

Solution: From the definition of a derivative

$$f'(x) = \lim_{h \to 0} \frac{f(x+h) - f(x)}{h}$$

or more briefly
$$\lim_{h \to 0} \frac{f(x+h) - f(x)}{h}$$

and
$$g'(x) = \lim_{h \to 0} \frac{g(x+h) - gf(x)}{h}$$

The derivative of $f(x) + g(x)$ will be

$$\lim_{h \to 0} \frac{f(x+h) + g(x+h) - (f(x) + g(x))}{h}$$

$$= \lim_{h \to 0} \frac{f(x+h) + g(x+h) - f(x) - g(x)}{h}$$

$$= \lim_{h \to 0} \frac{f(x+h) - f(x) + g(x+h) - g(x)}{h}$$

$$= \lim_{h \to 0} \frac{f(x+h) - f(x)}{h} + \lim_{h \to 0} \frac{g(x+h) - g(x)}{h}$$

$$= f'(x) + g'(x)$$

Example 6.1.1 (6)

If $f(x)$ has a derivative $f'(x)$ over a given domain, and a is a constant, then the derivative of a $f(x)$ over the given domain is a $f'(x)$.

Solution: From the definition of a derivative

$$f'(x) = \lim_{h \to 0} \frac{f(x+h) - f(x)}{h}$$

The derivative of a $f(x)$ will be

$$\lim_{h \to 0} \frac{af(x+h) - af(x)}{h}$$

$$= \lim_{h \to 0} \frac{a(f(x+h) - f(x))}{h}$$

$$= a \lim_{h \to 0} \frac{f(x+h) - f(x))}{h} \qquad \text{as } a \text{ is a constant independent of } h$$

$$= a\, f'(x)$$

Example 6.1.1 (7)

Using the rules for differentiation determine the gradient function dy/dx in each of the following cases. (a) $y = 6x^7$ (b) $y = 5x^3 + 3/x^4 + 5$

Solution: As well as the rules established in examples 6.1.1 (3) and (4) we may also quote the rules established in examples 6.1.1 (5) for the sum of functions and in 6.1.1 (6) for a constant times a function.

hence in (a) $y = 6x^7$

$$\frac{dy}{dx} = 6 \times 7x^6 = 42\, x^6$$

(b)

$$y = 5x^3 + \frac{3}{x^4} + 5$$

$$= 5x^3 + 3x^{-4} + 5$$

$$\frac{dy}{dx} = 5 \times 3x^2 + 3\,(-4)\, x^{-5} + 0$$

$$= 15\, x^2 - 12/x^5$$

N.B. $5 = 5x^{\circ}$ so its gradient is $5 \times 0x^{-1} = 0$

Example 6.1.1 (8)

Find the equation of the tangent and the normal to the curve $y = x^2$ at the point A where $x = 1$. If the normal meets the curve again at a point B find the coordinates of B.

Solution:

At A $x = 1$ so $y = (1)^2 = 1$

A is the point (1, 1)

To determine the equation of the tangent we must determine its gradient.

At any point on the curve $y = x^2$ the gradient is $2x$.

Hence the gradient of the tangent at A (1, 1) is 2.

The equation of the tangent is therefore

$$\frac{y-1}{x-1} = 2$$

which (the reader can check) simplifies to

$$2x - y - 1 = 0$$

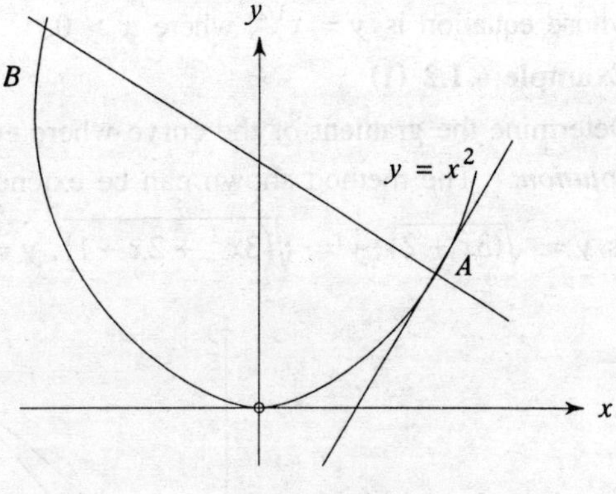

Fig. 6.1.1(g)

N.B. A common mistake made by beginners in this type of problem is to forget to substitute the value of the x coordinate when writing the gradient in the right hand side of the initial equation of the tangent, in this case we must write

$$\frac{y-1}{x-1} = 2 \text{ and not } \frac{y-1}{x-1} = 2x$$

The normal to a curve at any point is defined to be the line which is perpendicular to the tangent at that point.

Since at A (1, 1) we know that the gradient of the tangent is 2 then the gradient of the normal is –1/2.

The equation of the normal at A

is $$\frac{y-1}{x-1} = -1/2$$

which (the reader can check) simplifies to $x + 2y - 3 = 0$

Look back at Fig. 6.1.1 (g). The coordinates of the point B, where this line meets the curve again, must be satisfied by the simultaneous equations of the line and the curve,

i.e. $$y = x^2 \text{ and } x + 2y - 3 = 0$$

On substituting $y = x^2$ in the second equation we have $x + 2x^2 - 3 = 0$

i.e. $$2x^2 + x - 3 = 0$$

i.e. $$(2x + 3)(x - 1) = 0$$

i.e. $$x = -3/2 \text{ or } 1$$

Since we already knew that the line and the curve meet at A, the roots $x = 1$, $y = 1$, give the coordinates of A.

The value of y when $x = -3/2$

is given by $$y = \frac{1}{2}(3 - x) = \frac{1}{2}\left(3 - -\frac{3}{2}\right) = 9/4$$

hence B has coordinates $(-3/2, 9/4)$

6.1.2 The gradient of the function $y = x^{1/n}$ where n is an Integer

We shall first examine two methods in which we can establish the gradient function for the curve whose equation is $y = x^{1/2}$, where $x > 0$.

Example 6.1.2 (1)

Determine the gradient of the curve where equation is $y = x^{1/2}$, using the definition of a derivative.

Solution: The method shown can be extended to determining the gradient of other functions such as $y = \sqrt{(3x+2)}$, $y = \sqrt{(3x^2+2x+1)}$, $y = \sqrt{\cos x}$, when the reader encounters such problems.

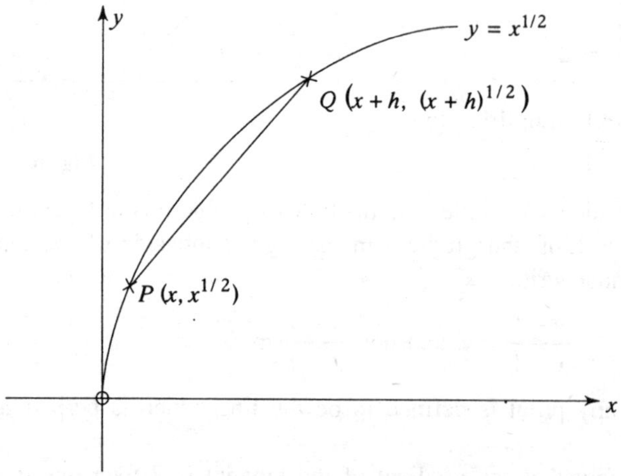

Fig. 6.1.2(a)

As shown in Fig. 6.1.2 (a) let $P\ (x, x^{1/2})$ and $Q\ (x + h, (x + h)^{1/2})$ be neighbouring points on the curve.

$$\text{Gradient } PQ = \frac{(x+h)^{1/2} - x^{1/2}}{x+h-x}$$

$$= \frac{(x+h)^{1/2} - x^{1/2}}{h}$$

Unfortunately there is no obvious factor in the numerator which can be cancelled with h, so we now try the method of ratioalising the numerator.

$$\text{Gradient } PQ = \frac{\left((x+h)^{1/2} - x^{1/2}\right)\left((x+h)^{1/2} + x^{1/2}\right)}{h\left((x+h)^{1/2} + x^{1/2}\right)}$$

$$= \frac{(x+h) - x}{h\left((x+h)^{1/2} + x^{1/2}\right)}$$

$$= \frac{x+h-x}{h\left((x+h)^{1/2} + x^{1/2}\right)}$$

i.e. Gradient $PQ \qquad = \dfrac{h}{h\left((x+h)^{1/2}+x^{1/2}\right)} = \dfrac{1}{\left((x+h)^{1/2}+x^{1/2}\right)}$

The gradient of the curve at P

$$= \lim_{h\to 0} \frac{1}{(x+h)^{1/2}+x^{1/2}}$$

$$= \frac{1}{x^{1/2}+x^{1/2}} = \frac{1}{2x^{1/2}} \quad \text{or} \quad \frac{1}{2}x^{-1/2}$$

The discerning reader might ask why we do not use the binomial theorem to expand $(x + h)^{1/2}$ after writing down the gradient of PQ. There are two reasons for this:

(i) The establishing of the series for the binomial theorem for a negative or fractional index actually depends on the knowledge of the gradient function for $y = x^{1/2}$, so we would be introducing a circular argument;

(ii) many beginners have difficulty in writing down the correct expression of $(x + h)^{1/2}$.

Example 6.1.2 (2)

Find the gradient of the curve $y = x^{1/2}$, using the knowledge of the gradient of the curve $y = x^2$

Solution: The gradient at the point $P(x, y)$ of the curve $y = x^{1/2}$ is given by the gradient of the tangent line at P and this is $\tan \alpha$ (see point R). At the point Q where the tangent line meets the y-axis at an angle β it is clear that $\alpha + \beta = 90°$.

If we interchange the axes and the curve (by reflecting in the line $y = x$) we have Fig. 6.1.2 (c).

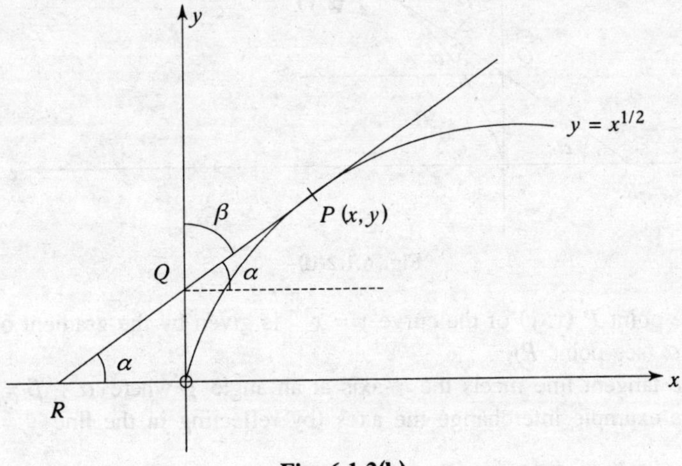

Fig. 6.1.2(b)

Naming the image points P_1, Q_1, R_1, respectively, we see that the gradient of the curve $x = y^2$ at P_1 is given by $\tan \beta$ and this gradient is already known to be $2y$.

So $\tan \beta = 2y$ As $\alpha + \beta = 90°$, $\tan \alpha = \cot \beta$

$$= \frac{1}{\tan \beta} = \frac{1}{2y}$$

So the gradient of the curve $y = x^{1/2}$ at P, given by $\tan \alpha$, is $\dfrac{1}{2y}$.

Since $y = x^{1/2}$
then the required gradient is

$$\frac{1}{2x^{1/2}} \text{ or } \frac{1}{2}x^{-1/2}$$

The next example repeats the process for the gradient of $y = x^{1/n}$ where n is any integer.

Example 6.1.2 (3)

Find the gradient of the curve $y = x^{1/n}$, n being any integer, using the knowledge of the gradient of the curve $y = x^n$.

Solution: See Fig. 6.1.2 (d).

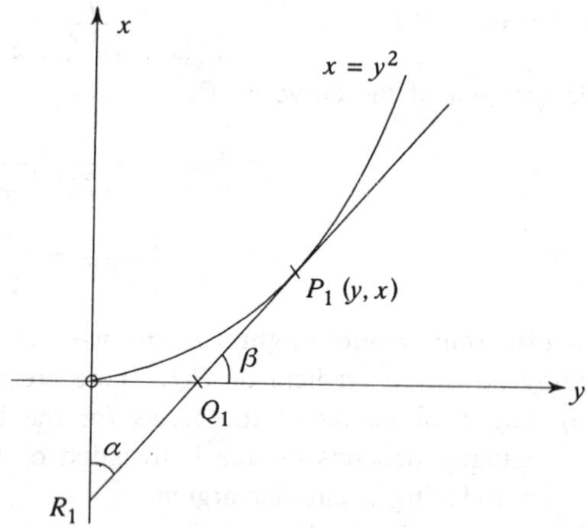

Fig. 6.1.2(c)

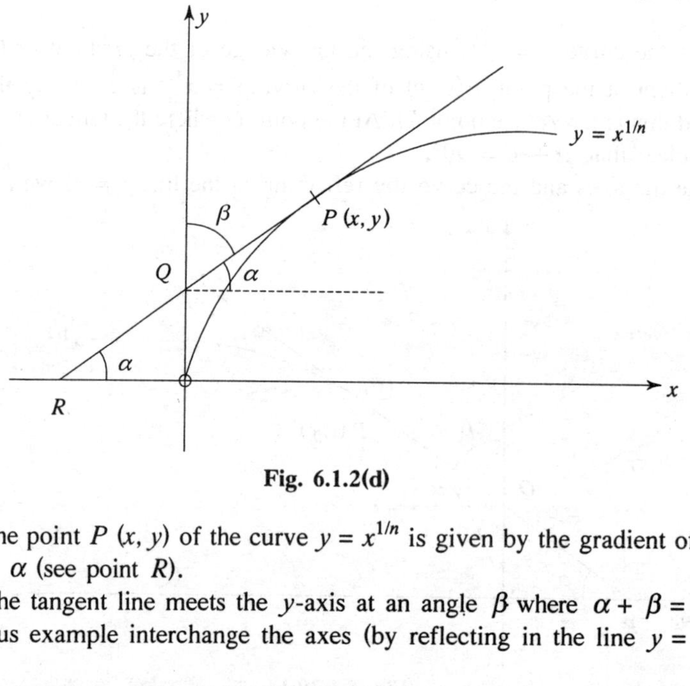

Fig. 6.1.2(d)

The gradient at the point P (x, y) of the curve $y = x^{1/n}$ is given by the gradient of the tangent line at P and this is $\tan\alpha$ (see point R).

At the point Q the tangent line meets the y-axis at an angle β where $\alpha + \beta = 90°$.

As in the previous example interchange the axes (by reflecting in the line $y = x$) and we have Fig. 6.1.2 (e)

The image points have been named P_1, Q_1, R_1, respectively.

The gradient of the curve $x = y^n$ at P_1 is given by $\tan\beta$ and we know that this gradient is ny^{n-1}.

So $\quad \tan\beta = ny^{n-1}$. As $\alpha + \beta = 90°$

$$\tan\alpha = \cot\beta = \frac{1}{\tan\beta} = \frac{1}{ny^{n-1}}$$

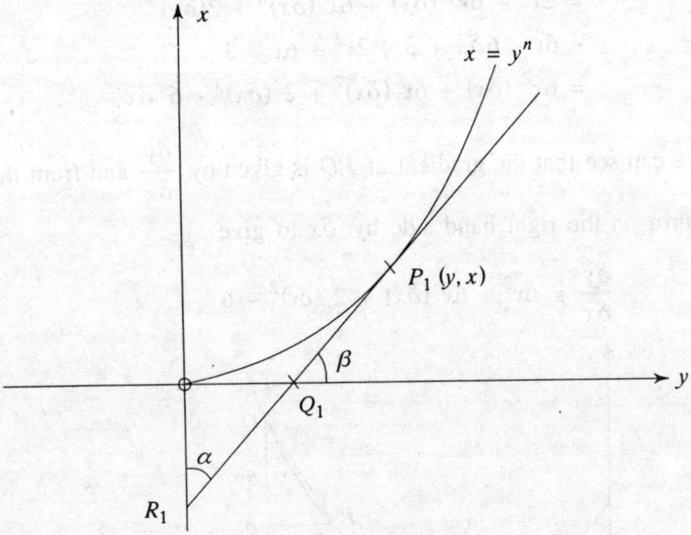

Fig. 6.1.2(e)

Since $y = x^{1/n}$

The required gradient is $\dfrac{1}{n\left(x^{1/n}\right)^{n-1}}$

$$= \frac{1}{n\,x^{1-1/n}} = \frac{1}{\dfrac{n\,x^{1}}{x^{1/n}}} = \frac{x^{1/n}}{n\,x^{1}} \quad \text{or} \quad \frac{1}{n}x^{1/n-1}$$

6.1.3 The Delta Notation

An alternative approach to the problem of finding the gradient of a curve using the definition of a derivative (or using first principles) is the use of the delta notation. If y is a function of x we let a small change in x be denoted by δx and the corresponding small change in y be denoted by δy so that the points $P\ (x, y)$ and $Q\ (x + \delta x, y + \delta y)$ will be neighbouring points on the curve.

The example below illustrates the use of the delta notation.

Example 6.1.3 (1)

Find the gradient of the curve whose equation is $y = 2x^3 - 6x + 3$, using first principles.

Solution: Let δx be an increment in x and δy the corresponding increment in y so that $P\ (x, y)$ and $Q\ (x + \delta x, y + \delta y)$ are neighbouring points on the curve so if

$$y = 2x^3 - 6x + 3$$

$$y + \delta y = 2\,(x + \delta x)^3 - 6(x + \delta x) + 3$$

by subtraction $\quad \delta y = 2\,(x + \delta x)^3 - 6(x + \delta x) + 3 - (2x^3 - 6x + 3)$

i.e. $\quad \delta y = 2\,(x^3 + 3x^2\,\delta x + 3x\,(\delta x)^2 + \delta x)^3\,)$

$$- 6x - 6\delta x + 3 - 2x^3 + 6x - 3$$

$$= 2x^3 + 6x^2 (\delta x) + 6x (\delta x)^2 + 2(\delta x)^3$$
$$- 6x - 6\delta x + 3 - 2x^3 + 6x - 3$$
$$= 6x^2 (\delta x) + 6x (\delta x)^2 + 2 (\delta x)^3 - 6 \delta x$$

From Fig. 6.1.3 (a) we can see that the gradient of PQ is given by $\dfrac{\delta y}{\delta x}$ and from the above working we can divide each term on the right hand side by δx to give

$$\frac{\delta y}{\delta x} = 6x^2 + 6x (\delta x) + 2 (\delta x)^2 - 6$$

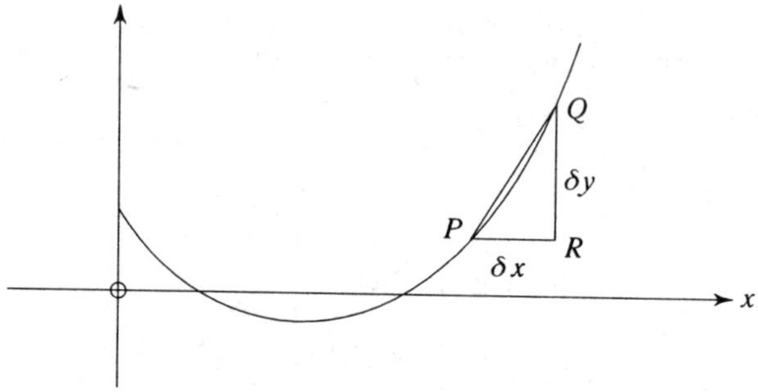

Fig. 6.1.3(a)

The gradient at P, denoted by

$$\frac{dy}{dx} = \lim_{\delta x \to 0} \frac{\delta y}{\delta x} = 6x^2 - 6$$

N.B. (i) the use of the 'd' appears only when we take the limit as δx approaches zero i.e. $\dfrac{dy}{dx}$ is the notation (used as shorthand) for the $\lim\limits_{\delta x \to 0} \dfrac{\delta y}{\delta x}$, otherwise the delta notation must appear in all of the earlier working,

(ii) δx is a single quantity, representing a small increment in x and so it is advisable to write the second and higher power of δx as $(\delta x)^2$, $(\delta x)^3$ etc. but never as $\delta^2 x^2$, $\delta^3 x^3$ etc. which would be meaningless.

6.1.4 The Chain Rule

Applying the rules of differentiation so far it would appear that if we are required to find the gradient function of say $y = (x + 5)^2$, $y = (x + 5)^8$, we would have to expand the brackets and then differentiate term by term.

For $$y = (x + 5)^2 = x^2 + 10x + 25$$

then $$\frac{dy}{dx} = 2x + 10 = 2 (x + 5)$$

This suggests that for $y = (x + 5)^8$

then $$\frac{dy}{dx} = 8\,(x + 5)^7$$

The result is correct. However if $y = (5x + 2)^6$, $\frac{dy}{dx}$ is not equal to $6\,(5x + 2)^5$

and if $y = (x^2 + 5x + 3)^7$, $\frac{dy}{dx}$ is not equal to $7\,(x^2 + 5x + 3)^6$

To solve such problems in differentiation it is necessary to establish a result known as the chain rule.

Proof of the Chain Rule

Let y be a function of t where t itself is a function of x.

Let δx, δt, δy be corresponding increments in x, t and y respectively such that neither δt nor δx is zero in any neighbourhood of the point (x, y) at which we are finding the gradient.

Then $$\frac{\delta y}{\delta x} = \frac{\delta y}{\delta t} \cdot \frac{\delta t}{\delta x}$$

taking the limit as $\delta x \to 0$

we have $$\frac{dy}{dx} = \frac{dy}{dt} \cdot \frac{dt}{dx}$$

This is the Chain Rule.

Example 6.1.4 (1)

Differentiate $y = (5x + 2)^6$ with respect to x.

Solution: We can write $y = t^6$ where $t = 5x + 2$ (i.e. use t as a function of x)

Then $$\frac{dy}{dx} = \frac{dy}{dt} \cdot \frac{dt}{dx} = 6t^5 \times 5 = 30\,t^5 = 30\,(5x + 2)^5$$

Example 6.1.4 (2)

Differentiate $y = (x^2 + 5x + 3)^7$ with respect to x.

Solution: We can write $y = t^7$ where $t = x^2 + 5x + 3$

then $$\frac{dy}{dx} = \frac{dy}{dt} \cdot \frac{dt}{dx} = 7t^6\,(2x + 5) \quad = 7\,(x^2 + 5x + 3)^6\,(2x + 5)$$

Example 6.1.4 (3)

Differentiate $y = \dfrac{1}{4 + \sqrt{x}}$ with respect to x.

Solution: Write $y = (\,4 + \sqrt{x}\,)^{-1} = t^{-1}$ where $t = 4 + x^{1/2}$

then $$\frac{dy}{dx} = \frac{dy}{dt} \cdot \frac{dt}{dx} = -1 \cdot t^{-2} \cdot \frac{1}{2}\,x^{-1/2}$$

i.e.
$$\frac{dy}{dx} = \frac{-1}{t^2 \cdot 2x^{1/2}} = \frac{-1}{2x^{1/2}\, t^2}$$
$$= \frac{-1}{2x^{1/2}(4 + x^{1/2})^2}$$

Example 6.1.4 (4)

Differentiate $y = (1 + x^2)^{1/4}$ with respect to x.

Solution: Write $y = t^{1/4}$ where $t = 1 + x^2$

then
$$\frac{dy}{dx} = \frac{dy}{dt}\cdot\frac{dt}{dx} = \frac{1}{4}t^{-3/4}\cdot 2x$$
$$= \frac{2x}{4t^{3/4}} = \frac{x}{2t^{3/4}} = \frac{x}{2(1+x^2)^{3/4}}$$

After practice the reader should be able to write down, and then simplify the answer, without the use of the intermediate variable t. For instance in the last example we had to differentiate some function raised to the power 1/4, which gives 1/4 times the function raised to the power –3/4 times the gradient of the function,

i.e. if
$$y = (1 + x^2)^{1/4} \text{ then } \frac{dy}{dx} = \frac{1}{4}(1 + x^2)^{-3/4}\cdot 2x$$

leading to the earlier answer.

Example 6.1.4 (5)

Differentiate (a) $y = x^{3/4}$

(b) $y = x^{p/q}$ (where p, q are integers) with respect to x.

Solution: In sub-section 6.1.2 of this section we showed that the gradient of the curve, whose equation is

$$y = x^{1/n}, \text{ is } \frac{1}{n}x^{1/n-1}$$

We are going to combine this result with the chain rule in each case.

(a) Let $y = x^{3/4} = (x^3)^{1/4} = t^{1/4}$ where $t = x^3$

then
$$\frac{dy}{dx} = \frac{dy}{dt}\cdot\frac{dt}{dx} = \frac{1}{4}t^{-3/4}\cdot 3x^2 = \frac{1}{4}(x^3)^{-3/4}\cdot 3x^2$$
$$= \frac{3}{4}x^{-9/4}.\,x^2 = \frac{3}{4}x^{-1/4}$$

(b) let $y = x^{p/q} = (x^p)^{1/q} = t^{1/q}$ where $t = x^p$

then
$$\frac{dy}{dx} = \frac{dy}{dt}\cdot\frac{dt}{dx} = \frac{1}{q}t^{1/q-1}\cdot px^{p-1}$$
$$= \frac{1}{q}(x^p)^{1/q-1}\cdot px^{p-1} = \frac{p}{q}x^{p/q-p}\cdot x^{p-1} = \frac{p}{q}x^{p/q-1}$$

We have now shown that, for any rational index m, the gradient of the curve, whose equation is $y = x^m$, is mx^{m-1}.

6.1.5 Products and Quotients

If we were asked to differentiate say $y = (x + 2)^7 (3x - 5)^4$ with respect to x, it would be possible, but also time consuming, to expand both brackets, multiply out term by term to get a final product, and then differentiate each of the individual terms in the final product.

If we were asked to differentiate $y = (x + 2)^{1/2} (x + 3)^{1/3}$ the above method would not even be possible.

In such cases it is necessary to learn rules for differentiating products and quotients.

Beware: many beginners believe that if it is a problem as differentiating a product it is only necesary to differentiate each factor in term and multiply the differentiated factors. This is wrong Take the gradient of the curve whose equation is $y = x^7$.

It is known that the gradient is $7x^6$. However if y is expressed as a product like $y = x^4 \cdot x^3$, it is clearly wrong to say that the gradient is $4x^3 \cdot 3x^2$.

Similarly many beginners believe that if it is a problem on differentiating a quotient it is only ncecessary to differentiate the numerator, differentiate the denominator, and then divide one result by the other.

As earlier for the case where $y = x^7 = \dfrac{x^{14}}{x^7}$ say, it is wrong to say that the gradient will be $\dfrac{14x^{13}}{7x^6}$

Having seen a popular wrong method of differentiating a product or a quotient it is time to explore the correct method.

Proof of Product Rule

Let $y = uv$ whre u and v are functions of x.

Let δx, δu, δv, δy be corresponding increments in x, u, v and y, with $\delta x \neq 0$.

then
$$y + \delta y = (u + \delta u)(v + \delta v)$$
$$= uv + v\,\delta u + u\,\delta v + \delta u\,\delta v$$

and
$$y = uv$$

so
$$\delta y = uv + v\,\delta u + u\delta v + \delta u\,\delta v - uv$$
$$= v\,\delta u + u\,\delta v + \delta u\,\delta v$$

divide by δx

then
$$\frac{\delta y}{\delta x} = v\frac{\delta u}{\delta x} + u\frac{\delta v}{\delta x} + \frac{\delta u}{\delta x}\cdot\frac{\delta v}{\delta x}\cdot\delta x$$

$$\frac{\delta y}{\delta x} = \lim_{\delta x \to 0}\frac{\delta y}{\delta x} = \lim_{\delta x \to 0}\left(v\frac{\delta u}{\delta x} + u\frac{\delta v}{\delta x} + \frac{\delta u}{\delta x}\cdot\frac{\delta v}{\delta x}\cdot\delta x\right)$$

$$= v\frac{du}{dx} + u\frac{dv}{dx} + 0$$

so if
$$y = uv$$

then $$\frac{dy}{dx} = v\frac{du}{dx} + u\frac{dv}{dx}$$

Example 6.1.5 (1)

Differentiate $y = (x^2 - 5)(x + 2)^2$ with respect to x, and simplify the result.

Solution:

If $$y = (x^2 - 5)(x + 2)^2$$

then $$u = x^2 - 5,\ v = (x + 2)^2$$

$$\frac{du}{dx} = 2x,\ \frac{dv}{dx} = 2(x + 2)$$

so
$$\begin{aligned}\frac{dy}{dx} &= v\frac{du}{dx} + u\frac{dv}{dx}\\ &= (x + 2)^2 \cdot 2x + (x^2 - 5) \cdot 2(x + 2)\\ &= 2(x + 2)[(x + 2)x + (x^2 - 5)]\\ &= 2(x + 2)[x^2 + 2x + x^2 - 5]\\ &= 2(x + 2)[2x^2 + 2x - 5]\end{aligned}$$

Example 6.1.5 (2)

Differentiate $y = (x^2 + 1)^3 (x^3 + 2)^2$ with respect to x, and simplify the result.

Solution:

If $$y = (x^2 + 1)^3 (x^3 + 2)^2$$

then $$u = (x^2 + 1)^3,\ v = (x^3 + 2)^2$$

$$\frac{du}{dx} = 3(x^2 + 1)^2 \cdot 2x,\ \frac{dv}{dx} = 2(x^3 + 2) \cdot 3x^2$$

$$\begin{aligned}\frac{dy}{dx} &= v\frac{du}{dx} + u\frac{dv}{dx}\\ &= (x^3 + 2)^2 .\ 3(x^2 + 1)^2 \cdot 2x + (x^2 + 1)^3 \cdot 2(x^3 + 2) \cdot 3x^2\\ &= 6x(x^2 + 1)^2 (x^3 + 2)[(x^3 + 2) + x(x^2 + 1)]\\ &= 6x(x^2 + 1)^2 (x^3 + 2)[x^3 + 2 + x^3 + x]\\ &= 6x(x^2 + 1)^2 (x^3 + 2)[2x^3 + x + 2]\end{aligned}$$

Example 6.1.5 (3)

Differentiate $y = (2x + 3)^{1/2} (2x + 5)^{3/2}$ with respect to x, and simplify the result.

Solution:

If $$y = (2x + 3)^{1/2} (2x + 5)^{3/2}$$

then $$u = (2x + 3)^{1/2},\ v = (2x + 5)^{3/2}$$

$$\frac{du}{dx} = \frac{1}{2}(2x + 3)^{-1/2} \cdot 2$$

$$\frac{dv}{dx} = \frac{3}{2}(2x + 5)^{1/2} \cdot 2$$

and

$$\frac{dy}{dx} = v\frac{du}{dx} + u\frac{dv}{dx}$$

$$= (2x + 5)^{3/2} \cdot \frac{1}{2}(2x + 3)^{-1/2} \cdot 2$$

$$+ (2x + 3)^{1/2} \cdot \frac{3}{2}(2x + 5)^{1/2} \cdot 2$$

$$\frac{(2x + 5)^{3/2}}{(2x + 3)^{1/2}} + 3(2x + 3)^{1/2}\,(2x + 5)^{1/2}$$

$$= \frac{(2x + 5)^{3/2} + 3(2x + 3)^{1/2}\,(2x + 3)^{1/2}\,(2x + 5)^{1/2}}{(2x + 3)^{1/2}}$$

$$= \frac{(2x + 5)^{1/2}\,[(2x + 5) + 3(2x + 3)]}{(2x + 3)^{1/2}}$$

i.e.

$$\frac{dy}{dx} = \frac{(2x + 5)^{1/2}\,[2x + 5 + 6x + 9]}{(2x + 3)^{1/2}}$$

$$= \frac{(2x + 5)^{1/2}\,(8x + 14)}{(2x + 3)^{1/2}}$$

$$= \frac{2(2x + 5)^{1/2}\,(4x + 7)}{(2x + 3)^{1/2}}$$

Example 6.1.5 (4)

Show that the point (2, 0) is a stationary point on the curve whose equation is

$$y = (x^2 - 1)\,(x - 2)^3.$$

Solution:

A stationary point on a curve is a point where the gradient is zero.

If

$$y = (x^2 - 1)\,(x - 2)^3$$

$$u = (x^2 - 1),\; v = (x - 2)^3$$

$$\frac{du}{dx} = 2x,\; \frac{dv}{dx} = 3\,(x - 2)^2$$

$$\frac{dy}{dx} = (x - 2)^3 \cdot 2x + (x^2 - 1) \cdot 3\,(x - 2)^2$$

$$= (x-2)^2\ [(x-2)\cdot 2x + 3\ (x^2-1)]$$
$$= (x-2)^2\ [2x^2 - 4x + 3x^2 - 3]$$
$$= (x-2)^2\ [\ 5x^2 - 4x - 3]$$

$\dfrac{dy}{dx} = 0$ for stationary points and $(x-2)^2$ is a factor of the gradient.

so $x = 2$ is a value of x for which $\dfrac{dy}{dx} = 0$

Substituting $x = 2$ in $y = (x^2-1)\ (x-2)^3$ we see that $y = 0$ when $x = 2$. So (2, 0) is a stationary point on the curve.

Proof of Quotient Rule

If $y = \dfrac{u}{v}$ where u, v are functions of x and $v \neq 0$,

let δx, δu, δv, δy be corresponding increments in x, u, v, y respectively, and $\delta x \neq 0$.

then
$$y + \delta y = \frac{u+\delta u}{v+\delta v} \quad \text{and} \quad y = \frac{u}{v}$$

so
$$\delta y = \frac{u+\delta u}{v+\delta v} - \frac{u}{v} = \frac{v\,(u+\delta u) - u\,(v+\delta v)}{(v+\delta v)\cdot v}$$

$$= \frac{uv + v\delta u - uv - u\delta v}{(v+\delta v)\cdot v} = \frac{v\delta u - u\delta v}{(v+\delta v)\cdot v}$$

Dividing by δx

$$\frac{\delta y}{\delta x} = \frac{v\delta u - u\delta v}{(v+\delta v)\cdot v}\cdot\frac{1}{\delta x}$$

and we divide each of the terms on the numerator by δx (the same as in working out $\dfrac{24-15}{7}\times\dfrac{1}{3}$ by first finding $\dfrac{24/3 - 15/3}{7}$ and finally arriving at $\dfrac{3}{7}$)

so
$$\frac{\delta y}{\delta x} = \frac{v\ \delta u/\delta x - u\ \delta v/\delta x}{(v+\delta v)\cdot v}$$

In the limit as $\delta x \to 0$, so $\delta v \to 0$, and

$$\frac{dy}{dx} = \lim_{\delta x\to 0}\frac{v\ \delta u/\delta x - u\ \delta v/\delta x}{(v+\delta v)\cdot v}$$

$$= \frac{v\ du/dx - u\ dv/dx}{v\cdot v}$$

So if
$$y = \frac{u}{v}$$

$$\frac{dy}{dx} = \frac{v\dfrac{du}{dx} - u\dfrac{dv}{dx}}{v^2}$$

Example 6.1.5 (5)

If $y = \dfrac{(x-4)^2}{(x+4)^2}$ and $x \neq -4$, find $\dfrac{dy}{dx}$ and simplify your answer.

Solution:

If $$y = \frac{(x-4)^2}{(x+4)^2}$$

$$u = (x-4)^2,\ v = (x+4)^2$$

$$\frac{du}{dx} = 2\ (x-4),\quad \frac{dv}{dx} = 2\ (x+4)$$

$$\frac{dy}{dx} = \frac{v\dfrac{du}{dx} - u\dfrac{dv}{dx}}{v^2}$$

$$= \frac{(x+4)^2 \cdot 2(x-4) - (x-4)^2 \cdot 2(x+4)}{(x+4)^4}$$

$$\frac{dy}{dx} = \frac{2(x-4)\ (x+4)\ [(x+4) - (x-4)]}{(x+4)^4}$$

$$= \frac{2(x-4)\ [x+4-x+4]}{(x+4)^3}$$

$$= \frac{16(x-4)}{(x+4)^3}$$

Example 6.1.5 (6)

If $y = \sqrt{\left(\dfrac{x^2-4}{x^2+4}\right)}$, find $\dfrac{dy}{dx}$ and simplify your answer.

Solution:

Set $$y = \frac{(x^2-4)^{1/2}}{(x^2+4)^{1/2}}$$

so $$u = (x^2-4)^{1/2},\ v = (x^2+4)^{1/2}$$

$$\frac{du}{dx} = \frac{1}{2}\ (x^2-4)^{-1/2} \cdot 2x\ ,\quad \frac{dv}{dx} = \frac{1}{2}\ (x^2+4)^{-1/2} \cdot 2x$$

i.e. $$\frac{du}{dx} = x(x^2-4)^{-1/2},\quad \frac{dv}{dx} = x(x^2+4)^{-1/2}$$

so $$\frac{dy}{dx} = \frac{(x^2+4)^{1/2}\, x(x^2-4)^{-1/2} - (x^2-4)^{1/2}\, x(x^2+4)^{-1/2}}{(x^2+4)}$$

$$= \frac{\dfrac{(x^2+4)^{1/2}x}{(x^2-4)^{1/2}} - \dfrac{(x^2-4)^{1/2}x}{(x^2+4)^{1/2}}}{(x^2+4)}$$

$$= \frac{\dfrac{x(x^2+4)^{1/2}(x^2+4)^{1/2} - x(x^2-4)^{1/2}(x^2-4)^{1/2}}{(x^2-4)^{1/2}(x^2+4)^{1/2}}}{(x^2+4)}$$

$$= \frac{x(x^2+4) - x(x^2-4)}{(x^2-4)^{1/2}(x^2+4)^{1/2}} \times \frac{1}{(x^2+4)}$$

$$= \frac{x^3+4x-x^3+4x}{(x^2-4)^{1/2}(x^2+4)^{3/2}} = \frac{8x}{(x^2-4)^{1/2}(x^2+4)^{3/2}}$$

6.1.6 Differentiation of Trigonometric Functions

We shall study how to find the gradient of $y = \sin x$ and of $y = \cos x$, using the definition of a derivative, and from these results we shall deduce the gradients of the remaining trigonomertic functions $y = \tan x$, $y = \cot x$, $y = \sec x$, $y = \operatorname{cosec} x$. However as a preliminary piece of work we shall explore the value of $\sin \theta/\theta$ where θ is small and positive, and θ is measured in radious.

Copy and complete the following table:

θ	$\sin\theta$	$\cos\theta$	$\tan\theta$	$\sin\theta/\theta$
0.05				
0.04				
0.03				
0.02				
0.01				
0.005				

You should have arrived at the answer shown below, depending on the number of significant figures used in your working.

θ	$\sin\theta$	$\cos\theta$	$\tan\theta$	$\sin\theta/\theta$
0.05	0.049979	0.998750	0.050042	0.99958
0.04	0.039989	0.999200	0.040021	0.99973
0.03	0.029996	0.99955	0.030009	0.99985
0.02	0.019999	0.99980	0.020003	0.99993
0.01	0.0099998	0.99995	0.010000	0.99998
0.005	0.00499998	0.999988	0.0050000	0.999996

It should be possible to see from these readings that:

$$\lim_{\theta\to 0} \sin\theta = 0,\quad \lim_{\theta\to 0} \cos\theta = 1$$

$$\lim_{\theta\to 0} \tan\theta = 0,\quad \lim_{\theta\to 0} \sin\theta/\theta = 1$$

and if we were to investigate it (but this should be clear from the table), $\lim_{\theta\to 0} \tan\theta/\theta = 1$.

Indeed if we make $\theta = 0.001$,
$\sin\theta = 0.00099999983$, $\cos\theta = 0.9999995$, $\tan\theta = 0.0010000003$, $\sin\theta/\theta = 0.9999998$, $\tan\theta/\theta = 1.0000003$.

Indeed if θ is made as well as 0.0001, the readings for $\sin\theta$ and $\tan\theta$ are also 0.0001 within the range of the average electronic calculator, while the reading for $\cos\theta$ is 0.999 999 9, thereby confirming what we suspected earlier.

There is of course no difficulty in deducing the values of sin 0, cos 0, tan 0, as a glance at the angle θ in the first quadrant when θ is very small shows $x \simeq r$, and $y \simeq 0$ (see Fig. 6.1.6 (a))

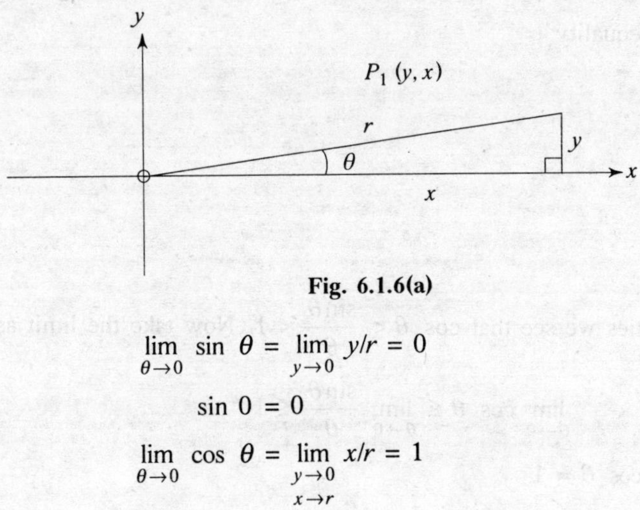

Fig. 6.1.6(a)

Hence $$\lim_{\theta\to 0} \sin\theta = \lim_{y\to 0} y/r = 0$$

hence $$\sin 0 = 0$$

$$\lim_{\theta\to 0} \cos\theta = \lim_{\substack{y\to 0\\ x\to r}} x/r = 1$$

$$\lim_{\theta\to 0} \tan\theta = \lim_{\substack{y\to 0\\ x\to r}} y/x = 0/r = 0$$

However it is necessary to examine the $\lim_{\theta\to 0} \sin\theta/\theta$ using another approach, where we have to make use of our knowledge of areas of plane figures and of the geometry of the circle. From the calculation carried out earlier it would appear that $\lim_{\theta\to 0} \sin\theta/\theta = 1$

***Proof of* $\lim_{\theta\to 0}$ *sin* $\theta/\theta = 1$**

In Fig. 6.1.6 (b) draw an arc *AB* of a circle, centre *O*, radius *r*, with angle $AOB = \theta > 0$ where θ is small and measured in radians.

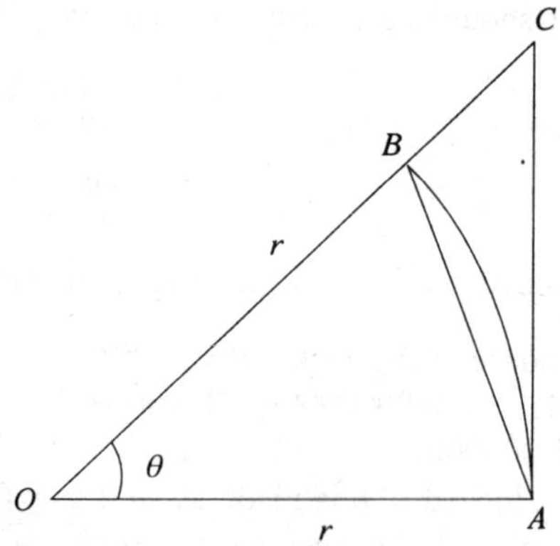

Fig. 6.1.6(b)

At A draw the tangent line to meet OB produced in C.

Then angle $OAC = 90°$ as the tangent line to a circle is perpendicular to the radius at the point of contact.

so $$AC = r \tan \theta.$$

Now in size,

area of $\Delta\ OAB$ < area of sector OAB < area of $\Delta\ OAC$

i.e. $$\frac{1}{2} r^2 \sin \theta < \frac{1}{2} r^2 \theta < \frac{1}{2} r^2 \tan \theta$$

dividing throughout by $\frac{1}{2} r^2$

$$\sin \theta < \theta < \tan \theta$$

(this inequality was true in the table prepared earlier for the values of $\sin \theta$, θ and $\tan \theta$). Taking the left hand inequality $\sin \theta < \theta$ and dividing by θ, $\frac{\sin\theta}{\theta} < 1$

as $\theta > 0$

Taking the right hand inequality

$$\theta < \tan \theta$$

i.e. $$\theta < \frac{\sin\theta}{\cos\theta}$$

i.e. $$\theta \cos \theta < \sin \theta$$

i.e. $$\cos \theta < \frac{\sin\theta}{\theta}$$

Combining both inequalities we see that $\cos \theta < \frac{\sin\theta}{\theta} < 1$. Now take the limit as $\theta \to 0$

* i.e. $$\lim_{\theta\to 0} \cos \theta \le \lim_{\theta\to 0} \frac{\sin\theta}{\theta} \le 1$$

however we know $\lim_{\theta\to 0} \cos \theta = 1$

so $$1 \le \lim_{\theta\to 0} \frac{\sin\theta}{\theta} \le 1$$

i.e. $$\lim_{\theta\to 0} \frac{\sin\theta}{\theta} = 1$$

Using the result that $\lim_{\theta\to 0} \sin \theta/\theta = 1$. we may now proceed to the original quest, of finding the derivatives of $y = \sin x$, $y = \cos x$.

Note that the limit is still 1 even if θ is negative, as if $\theta < 0$, say $\theta = -\alpha$ where $\alpha > 0$, then

$$\lim_{\theta\to 0} \sin \theta/\theta = \lim_{\alpha\to 0} \sin (-\alpha)/(-\alpha)$$

$$\lim_{\alpha \to 0} \frac{-\sin \alpha}{-\alpha} = \lim_{\alpha \to 0} \frac{\sin \alpha}{\alpha} = 1$$

Example 6.1.6 (1)

Differentiate $y = \sin x$, using the definition of a derivative.

Solution: See Fig. 6.1.6 (c)

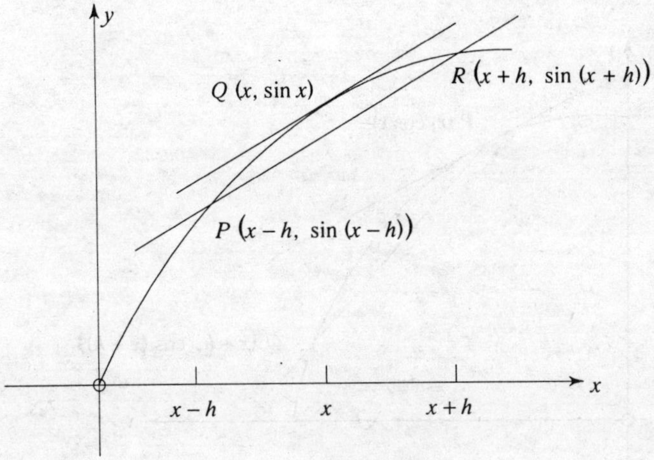

Fig. 6.1.6(c)

Here we let Q $(x, \sin x)$ be the point at which we wish to determine the gradient of the tangent line to the curve $y = \sin x$, and we take P $(x - h, \sin (x - h))$ and R $(x + h, \sin (x + h))$ as two neighbouring points to Q on either side of Q, where P and R also lie on the curve. Note that x is measured in radians.

$$\text{Gradient } PR = \frac{\sin (x+h) - \sin (x-h)}{(x+h) - (x-h)}$$

$$= \frac{\sin x \cos h + \cos x \sin h - (\sin x \cos h - \cos x \sin h)}{x + h - x + h}$$

$$= \frac{\sin x \cos h + \cos x \sin h - \sin x \cos h + \cos x \sin h}{2h}$$

$$= \frac{2 \cos x \sin h}{2h} = \frac{\cos x \sin h}{h}$$

$$\text{Gradient at } Q = \lim_{h \to 0} \text{Gradient } PR$$

i.e.

$$\text{gradient at } Q = \lim_{h \to 0} \cos x \frac{\sin h}{h} = \cos x \cdot 1 = \cos x$$

Hence if $y = \sin x$ the gradient of the curve $\dfrac{dy}{dx} = \cos x$.

Example 6.1.6 (2)

Differentiate $y = \cos x$ using the definition of a derivative.

Solution:

An approach similar to that used in the previous example may be used and this is left as an exercise to the reader.

For interest we shall use an alternative approach which makes use of just two neighbouring points on the curve. See Fig. 6.1.6 (d)

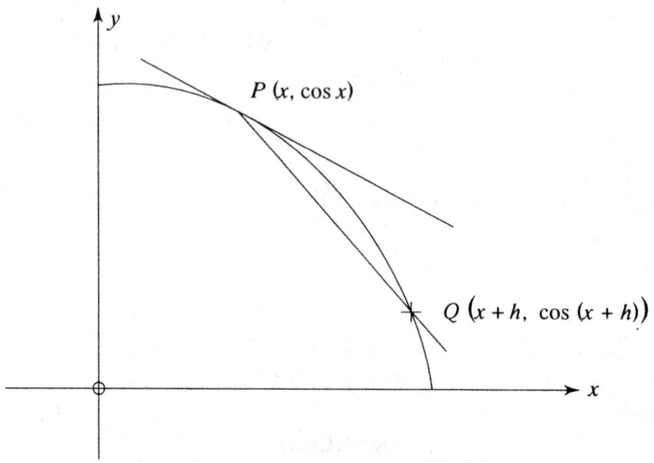

Fig. 6.1.6(d)

Let $P\ (x, \cos x)$, $Q\ (x + h, \cos (x + h))$ be neighbouring points on the curve,

gradient PQ $= \dfrac{\cos(x+h) - \cos x}{x + h - x}$

$$= \frac{\cos(x+h) - \cos x}{h}$$

$$= \frac{-2\sin\frac{1}{2}(2x+h)\sin\frac{1}{2}h}{h} \quad \text{(using the factor formulae)}$$

$$= \frac{-\sin\frac{1}{2}(2x+h)\sin\frac{1}{2}h}{\frac{1}{2}h}$$

(on dividing numerator and denomina tor by 2 so as to obtain 1/2 h on the denominator)

Gradient at P

$$= \lim_{h \to 0} \text{gradient } PQ$$

$$= \lim_{h \to 0} - \sin \frac{1}{2}(2x + h) \cdot \frac{\sin \frac{1}{2}h}{\frac{1}{2}h} = \left(-\sin \frac{1}{2} \cdot 2x\right) \cdot 1 = -\sin x$$

hence if $y = \cos x$, the gradient of the curve $\frac{dy}{dx} = -\sin x$

Example 6.1.6 (3)

Differentiate with respect to x,

(a) $y = \sin 3x$

(b) $y = \cos (4x + 5)$

(c) $y = \sin (5x + 3\pi/4)$

(d) $y = \cos^3 x$

(e) $y = \cos (x^3)$

(f) $y = \cos x°$

Solution:

(a) The answer may look simple, but it is important to write $y = \sin 3x = \sin t$ where $t = 3x$ in order to see the application of the chain rule in this and subsequent problems in this example.

$$\frac{dy}{dx} = \frac{dy}{dt} \cdot \frac{dt}{dx} = (\cos t)3$$

$$= 3 \cos 3x$$

(b) $y = \cos (4x + 5) = \cos t$ where $t = 4x + 5$

so
$$\frac{dy}{dx} = \frac{dy}{dt} \cdot \frac{dt}{dx} = (-\sin t)4$$

$$= -4 \sin (4x + 5)$$

(c) we shall try this and the problem (*f*) in the example by direct chain rule where appropriate.

$$y = \sin (5x + 3\pi/4)$$

so
$$\frac{dy}{dx} = (\cos (5x + 3\pi/4))\ 5$$

$$= 5 \cos (5x + 3\pi/4)$$

(d)
$$y = \cos^3 x = (\cos x)^3$$

$$= t^3 \text{ where } t = \cos x$$

(**N.B.** we have first found $\cos x$ and then cubed the result so the 'outside function' is t^3)

$$\frac{dy}{dx} = \frac{dy}{dt} \cdot \frac{dt}{dx} = 3t^2 (-\sin x)$$

$$= -3\cos^2 x \sin x$$

(e) $y = \cos(x^3) = \cos t$ where $t = x^3$

(**N.B.** unlike problem (d) here we have first found the cube of x, then taken the cosine, so the 'outside function' is $\cos t$)

$$\frac{dy}{dx} = \frac{dy}{dt} \cdot \frac{dt}{dx} = (-\sin t)\, 3x^2$$

$$= -3x^2 \sin(x^3)$$

(f) $y = \cos x^\circ = \cos \pi x/180$

so $$\frac{dy}{dx} = (-\sin \pi x/180)\; \pi/180$$

$$= (-\pi/180) \sin \pi x/180$$

i.e. $$\frac{dy}{dx} = \frac{-\pi \sin x^\circ}{180}$$

Example 6.1.6 (4)

Using the known results for the derivatives of $y = \sin x$ and $y = \cos x$ determine the derivatives of (a) $y = \tan x$ (b) $y = \operatorname{cosec} x$.

Solution:

It is possible to find these using the definition of a derivative, and there will be an opportunity for the reader to try this approach as an exercise.

However here we shall use the two known results along with appropriate application of the chain rule and the quotient rule.

(a) if $$y = \tan x = \frac{\sin x}{\cos x} = \frac{u}{v} \text{ say}$$

$$\frac{dy}{dx} = \frac{\cos x(\cos x) - \sin x(-\sin x)}{\cos^2 x}$$

using $$du/dx = \cos x,\ \frac{dv}{dx} = -\sin x$$

i.e. $$\frac{dy}{dx} = \frac{\cos^2 x + \sin^2 x}{\cos^2 x} = \frac{1}{\cos^2 x} = \sec^2 x$$

hence if $$y = \tan x,\ \frac{dy}{dx} = \sec^2 x$$

(b) if $y = \operatorname{cosec} x = (\sin x)^{-1}$

then $$\frac{dy}{dx} = -1\,(\sin x)^{-2} \cdot \cos x$$

$$= \frac{-\cos x}{\sin^2 x} = \frac{-1}{\sin x} \cdot \frac{\cos x}{\sin x}$$

$$= -\operatorname{cosec} x \cot x$$

hence if $y = \operatorname{cosec} x$

$$\frac{dy}{dx} = -\operatorname{cosec} x \cot x$$

In a similar way we may show that if $y = \cot x$, $\frac{dy}{dx} = -\operatorname{cosec}^2 x$ and if $y = \sec x$, $\frac{dy}{dx} = \sec x \tan x$.

This is left as an exercise to the reader. So in summary, for all six trigonometrical functions,

if $y = \sin x$, $\frac{dy}{dx} = \cos x$

if $y = \cos x$, $\frac{dy}{dx} = -\sin x$

if $y = \tan x$, $\frac{dy}{dx} = \sec^2 x$

if $y = \cot x$, $\frac{dy}{dx} = -\operatorname{cosec}^2 x$

if $y = \sec x$, $\frac{dy}{dx} = \sec x \tan x$

if $y = \operatorname{cosec} x$, $\frac{dy}{dx} = -\operatorname{cosec} x \cot x$.

Example 6.1.6 (5)

Determine $\frac{dy}{dx}$ if (a) $y = \tan 5x$ (b) $y = \sqrt{\tan x}$ (c) $y = x^4 \sec^2 x$ (d) $y = \frac{\cot x}{(x+2)^3}$

Solution:

(a) if $y = \tan 5x$

$$\frac{dy}{dx} = (\sec^2 5x)5 = 5 \sec^2 5x$$

(b) if $y = \sqrt{\tan x} = (\tan x)^{1/2}$

$$\frac{dy}{dx} = \frac{1}{2} (\tan x)^{-1/2} \cdot \sec^2 x$$

$$= \frac{\sec^2 x}{2(\tan x)^{1/2}} = \frac{\sec^2 x}{2\sqrt{(\tan x)}}$$

(c) if $y = x^4 \sec^2 x$

$$\frac{dy}{dx} = 4x^3 \sec^2 x + x^4 (2 \sec x) \cdot \sec x \tan x$$

$$= 2x^3 \sec^2 x (2 + x \tan x)$$

(N.B. for $v = \sec^2 x = (\sec x)^2$

then $\frac{dv}{dx} = 2 (\sec x)^1 (\sec x \cdot \tan x)$

(d) if $y = \frac{\cot x}{(x+2)^3}$

$$\frac{dy}{dx} = \frac{(x+2)^3(-\operatorname{cosec}^2 x) - (\cot x)(3(x+2)^2)}{(x+2)^6}$$

$$= \frac{(x+2)^2\left[-(x+2)\operatorname{cosec}^2 x - 3\cot x\right]}{(x+2)^6}$$

$$= \frac{-(x+2)\operatorname{cosec}^2 x - 3\cot x}{(x+2)^4}$$

6.1.7 Differentiation of Exponential and Logarithmic Functions

Figs. 6.1.7 (a), 6.1.7 (b) show the graph of the equation $y = a^x$ (for $a > 1$ in (a) and for $0 < a < 1$ in (b))

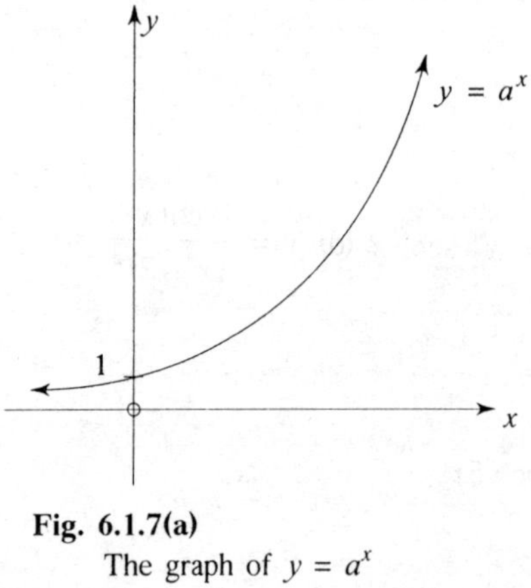

Fig. 6.1.7(a)
The graph of $y = a^x$ for $a > 1$

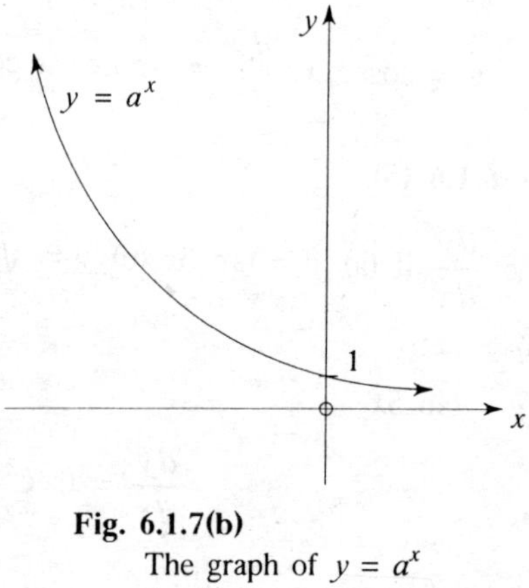

Fig. 6.1.7(b)
The graph of $y = a^x$ for $0 < a < 1$

We can approach the problem of determining the gradient of this curve in a way similar to the approach for the other types of functions we looked at earlier.

In particular if we look at Fig. 6.1.7 (c)

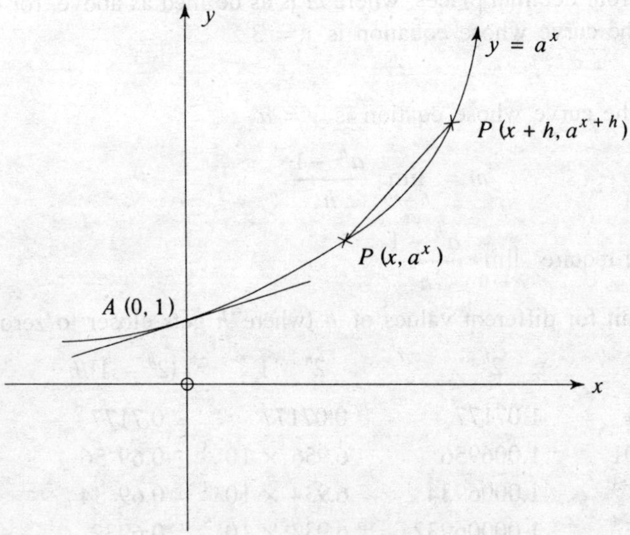

Fig. 6.1.7(c)

Let $P\ (x,\ a^x)$, $Q\ (x + h,\ a^{x+h})$ be neighbouring points on the curve whose equation is $y = a^x$

$$\text{gradient } PQ = \frac{a^{x+h} - a^x}{x + h - x} = \frac{a^x\,(a^h - 1)}{h}$$

So that $\frac{dy}{dx}$ is given by the gradient of the tangent line to the curve at P.

i.e.
$$\frac{dy}{dx} = a^x \lim_{h\to 0} \frac{a^h - 1}{h}$$

Now consider the gradient (say its value is m) to the curve at the point A (0, 1)

This will be
$$\lim_{h\to 0} \frac{a^{0+h} - a^0}{h - 0} = \lim_{h\to 0} \frac{a^h - 1}{h}$$

i.e.
$$m = \lim_{h\to 0} \frac{a^h - 1}{h}$$

and at any point $P\ (x,\ a^x)$ on the curve the gradient

$$\frac{dy}{dx} = a^x \lim_{h\to 0} \frac{a^h - 1}{h} = a^x \cdot m = m\,a^x$$

We now explore some possible values of m, having found that for the curve where equation $y = a^x$ the gradient is $m\ a^x$ where m is the value of gradient at the point (0, 1).

Example 6.1.7 (1)

Determine m, correct to four decimal places, where m is as defined as above, for (a) the curve whose equation is $y = 2^x$ (b) the curve whose equation is $y = 3^x$.

Solution:

We have seen that for the curve whose eqation is $y = a^x$

then $$m = \lim_{h \to 0} \frac{a^h - 1}{h}$$

in (a) we put $a = 2$ and require $\lim_{h \to 0} \frac{a^h - 1}{h}$

An evaluation of the limit for different values of h (where h gets closer to zero) is shown below).

h	2^h	$2^h - 1$	$(2^h - 1)/h$
0.1	1.07177	0.07177	0.7177
0.01	1.006956	6.956×10^{-3}	0.69 56
10^{-3}	1.0006934	6.934×10^{-4}	0.69 34
10^{-4}	1.00006932	6.932×10^{-5}	0.6932
10^{-5}	1.000006931	6.931×10^{-6}	0.69 31
10^{-6}	1.000 0006931	6.931×10^{-7}	0.69 31

Correct to four decimal places $m = 0.6931$

so for $y = 2^x$ the gradient at any point on the curve, $\frac{dy}{dx}$, is given by $(0.6931)\ 2^x$ approximately.

In (b) we put $a = 3$ and require $\lim_{h \to 0} \frac{a^h - 1}{h}$

An evaluation of the limit for different values of h (where h gets closer to zero) is shown below.

h	3^h	$3^h - 1$	$(3^h - 1)/h$
0.1	1.11612	0.11612	1.1612
0.01	1.011047	0.011047	1.1047
10^{-3}	1.0010992	1.0992×10^{-3}	1.0992
10^{-4}	1.00010987	1.0987×10^{-4}	1.0987
10^{-5}	1.000010986	1.0986×10^{-5}	1.0986
10^{-6}	1.00000 10986	1.0986×10^{-6}	1.0986

Correct to four decimal places $m = 1.0986$

so for $y = 3^x$ the gradient at any point on the curve, $\frac{dy}{dx}$, is given by $(1.0986)3^x$ approximately.

What we have found in this example is that for the curve whose equation is $y = 2^x$, $m \simeq 0.6931$ and the gradient at any point is given by $(0.6931)\ 2^x$ approximately, and for the curve whose equation is $y = 3^x$, $m \simeq 1.0986$ and the gradient at any point is given by $(1.0986)\ 3^x$ approximately.

This suggests that there is a number e, where $2 < e < 3$, for which m is one. It can be shown (but this is beyond the size of this book) that $e \simeq 2.7183$ to four decimal places.

Definition

We define the number e to be the number such that the gradient of the curve, whose equation is $y = e^x$, at (0, 1), is equal to 1. The function e^x is known as the exponential function. So if $y = e^x$, then $\frac{dy}{dx} = 1 \cdot e^x = e^x$.

N.B. Since x may be any real number the domain of the exponential may be taken as the real number line (i.e. all points on the x-axis). However $e^x > 0$ for all real x, so the range of the exponential function will be all positive real numbers (i.e. all points on the y-axis for which $y > 0$).

Much theoretical work in Science, Engineering and Agriculture is based on mathematical models which use the exponential function and its associated functions. In particular it is often useful to perform theoretical calculations using logarithms to base e where the number e is as defined above.

The idea of logarithms was discovered independently in the early seventeenth century by John Napier (from Edinburgh, Scotland, U.K.) and Burgi (from Switzerland). Later Napier discussed with a friend Henry Briggs how logarithms to different bases could be used to solve problems which occurred at that time, especially with respect to distances to be measured in the study of navigation and in the study of astronomy.

Some texts denote the function 'the logarithm, to base e, of x' as $\log_e x$, but many texts also denote this function as ln x. The latter notation is in honour of John Napier, and logarithms to base e are sometimes described as Napierian logarithms or Natural logarithms.

We now investigate how to determine the gradient $\frac{dy}{dx}$ for the curve whose equation is given by

$$y = \ln x$$

Example 6.1.7 (2)

Determine the gradient of the curve whose equation is $y = \ln x$.

Solution:

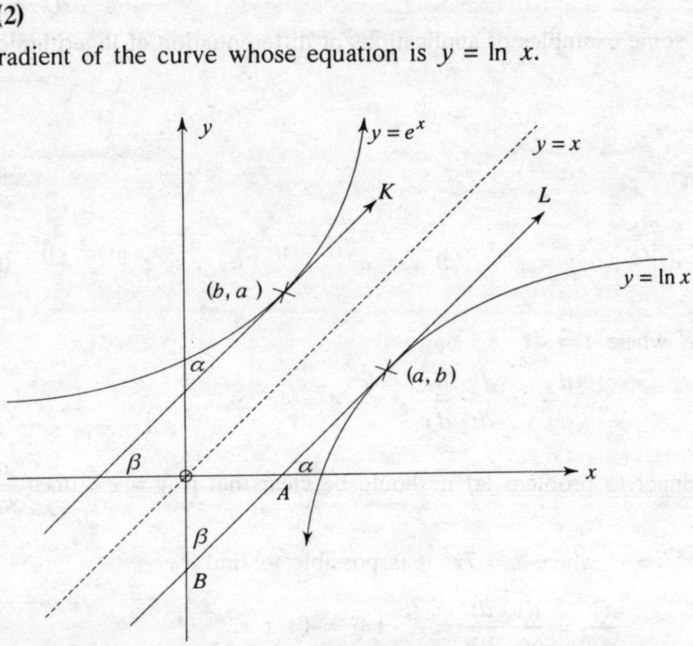

Fig. 6.1.7(d)

In Fig. 6.1.7 (d) it is seen that the graph for the curve whose equation $y = \ln x$ can be obtained by reflecting the graph for the curve of the inverse function $y = e^x$ in the line of reflection $y = x$.

Under any such reflection any point (a, b) on the graph of $y = \ln x$ will have an image point (b, a) which lies on the graph of $y = e^x$.

In particular the tangent line L to the curve whose equation is $y = \ln x$, making an angle α with the x-axis and an angle β with the y-axis, will have an image line which will be the tangent line K to the curve whose equation is $y = e^x$, making an angle α with the y-axis and an angle β with the x-axis.

Now the gradient to the curve whose equation is $y = \ln x$ will be given by $\tan \alpha$. By examining triangle AOB in Fig. 6.1.7 (d) it is clear that $\alpha + \beta = 90°$.

Also since (a, b) lies on the curve whose equation is given by $y = \ln x$, then $b = \ln a$. However we can re-write the relation as $a = e^b$. It is already known that the point (b, a) or (b, e^b) lies on the image curve whose equation is $y = e^x$ and that the gradient of the curve at the point (b, a) is given by $\tan \beta$.

Hence $\tan \beta = e^b$ (as the gradient of the curve $y = e^x$ is e^x) and

$$\tan \alpha = \tan (90° - \beta) = \cot \beta = \frac{1}{\tan \beta} = \frac{1}{e^b}$$

Hence the gradient, at (a, b), of the curve whose equation is $\ln x$ is

$$1/e^b \text{ or } 1/a \text{ (since } a = e^b).$$

However, we took (a, b) to represent any point, say $(x, \ln x)$ on the curve, so we can conclude that for the equation $y = \ln x$ (known as the logarithmic function) the gradient $\frac{dy}{dx} = \frac{1}{x}$.

There now follow some examples of applications of differentiation of logarithmic and exponential functions.

Example 6.1.7 (3)

Determine $\frac{dy}{dx}$ when

(a) $y = e^{4x}$ (b) $y = e^{-5x}$ (c) $y = e^{7x^2}$ (d) $y = e^{\sqrt{(x^2 + 4)}}$ (e) $y = e^{\sin (3x^2 + 1)}$ (f) $y = \frac{e^{-x^2}}{\cos x}$

Solution:

(a) Let $y = e^{4x} = e^t$ where $t = 4x$

$$\frac{dy}{dx} = \frac{dy}{dt} \cdot \frac{dt}{dx} = e^t \cdot 4 = 4e^{4x}$$

(b) In a similar manner to problem (a) it should be clear that if $y = e^{-5x}$ then $\frac{dy}{dx} = -5e^{-5x}$

(c) Writing $y = e^{7x^2} = e^t$ where $t = 7x^2$ it is possible to find

$$\frac{dy}{dx} = \frac{dy}{dt} \cdot \frac{dt}{dx} = e^t \cdot 14x = 14\, x\, e^{7x^2}$$

(d) $y = e^{(x^2+4)^{1/2}} = e^t$ whre $t = (x^2 + 4)^{1/2}$

$$\frac{dy}{dx} = \frac{dy}{dt}\cdot\frac{dt}{dx} = e^t \cdot \frac{1}{2}(x^2 + 4)^{-1/2} \cdot 2x$$

$$\frac{e^{(x^2+4)^{1/2}}(2x)}{2(x^2+4)^{1/2}} = \frac{x\,e^{(x^2+4)^{1/2}}}{(x^2+4)^{1/2}}$$

If the reader still has difficulty with $\frac{dt}{dx}$ then introduce another variable $u = x^2 + 4$ so that $t = u^{1/2}$

and $\frac{dt}{dx} = \frac{dt}{du}\frac{du}{dx}$

(e) $y = e^{\sin(3x^2+1)}$

There is a similar approach to that for problem (d)

and so $$\frac{dy}{dx} = e^{\sin(3x^2+1)}\,[\cos(3x^2 + 1)]6x$$

$$= 6x\cos(3x^2 + 1)\,e^{\sin(3x^2+1)}$$

(f) for $y = \frac{e^{-x^2}}{\cos x}$

use the quotient rule to find

$$\frac{dy}{dx} = \frac{\cos x(-2x e^{-x^2}) - e^{-x^2}(-\sin x)}{\cos^2 x}$$

$$= \frac{e^{-x^2}(\sin x - 2x\cos x)}{\cos^2 x}$$

Example 6.1.7 (4)

Determine $\frac{dy}{dx}$ when

(a) $y = \ln 5x$

(b) $y = \ln x^4$

(c) $y = \ln(x^2 + 2x + 5)$

(d) $y = \ln \frac{x}{(x^3+1)^{1/2}}$

(e) $y = \ln\cos x$

(f) $y = \ln(\text{cosec } x + \cot x)$

N.B. since the domain of any logarithmic function is the set of positive real numbers (see Fig. 6.1.7 (d)) it is assumed that the logarithm of any function can only be defined for positive values of that function.

Thus for (a) and (d) we can only define the logarithm for a domain of $x > 0$ and for (e) we can only define x for say a domain $-\pi/2 < x < \pi/2$.

Solution:

(a) This can be done in two ways

(i) $y = \ln 5x = \ln t$ where $t = 5x$

$$\frac{dy}{dx} = \frac{dy}{dt}\cdot\frac{dt}{dx} = \frac{1}{t}\cdot 5 = \frac{5}{5x} = \frac{1}{x}$$

(ii) $y = \ln 5x$

$$= \ln 5 + \ln x \text{ so } \frac{dy}{dx} = \frac{1}{x}$$

as the derivative of ln 5 (a constant) is zero.

(b) $y = \ln x^4 = 4 \ln x$

$$\frac{dy}{dx} = \frac{4}{x}$$

(the reader can try this using $y = \ln t$ where $t = x^4$ as an exercise)

(c) $y = \ln (x^2 + 2x + 5) = \ln t$ where $t = x^2 + 2x + 5$

$$\frac{dy}{dx} = \frac{dy}{dt}\cdot\frac{dt}{dx} = \frac{1}{t}(2x + 2) = \frac{2(x+1)}{x^2 + 2x + 5}$$

(d) $y = \ln\left(\dfrac{x}{(x^3+1)^{1/3}}\right)$

$$= \ln x - \frac{1}{3}\ln (x^3 + 1)$$

$$\frac{dy}{dx} = \frac{1}{x} - \frac{1}{3}\cdot\frac{3x^2}{(x^3+1)}$$

(the reader can check this by writing $t = (x^3 + 1)$)

i.e.

$$\frac{dy}{dx} = \frac{1}{x} - \frac{x^2}{x^3+1} = \frac{x^3 + 1 - x^3}{x(x^3+1)} = \frac{1}{x(x^3+1)}$$

(e) $y = \ln \cos x$

$$\frac{dy}{dx} = \frac{1}{\cos x} \times -\sin x = \frac{-\sin x}{\cos x} = -\tan x$$

(f) $y = \ln (\text{cosec } x + \cot x)$

$$\frac{dy}{dx} = \frac{-\text{cosec } x \cot x - \text{cosec}^2 x}{\text{cosec } x + \cot x}$$

$$= \frac{-\text{cosec } x (\cot x + \text{cosec } x)}{(\text{cosec } x + \cot x)} = -\text{cosec } x$$

6.2 APPLICATION OF DIFFERENTIATION TO VELOCITY AND ACCELERATION AND STATIONARY VALUES

6.2.1 Displacement and Velocity

When an object is moving in a straight line it is helpful to consider its position at any time t (normally measured in seconds) by giving its displacement x from a given fixed point (usually the origin). For x to be fully defined both the distance and the direction must be given, and in this case the direction can be indicated using a positive or negative sign. The displacement is normally measured in metres.

Now consider the displacement time graph and two neighbouring points $P\ (t, x)$, $Q\ (t + \delta t, x + \delta x)$ on the graph, as shown in Fig. 6.2.1 (a).

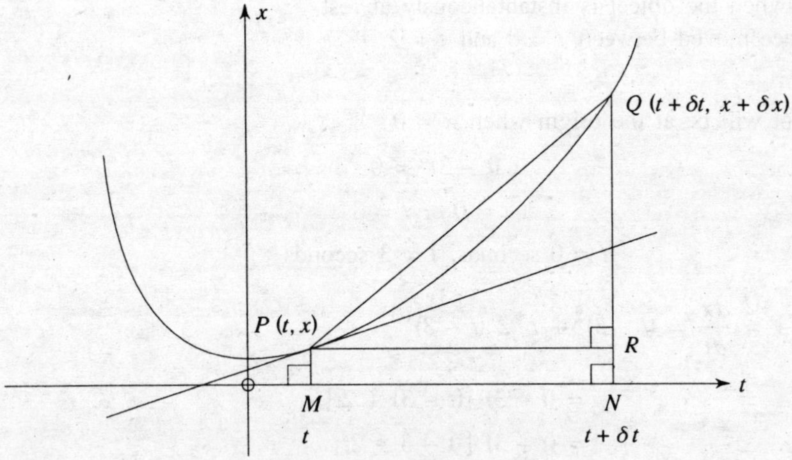

Fig. 6.2.1(a)

Displacement at time $t = MP = x$, displacement at time $t + \delta t = NQ = x + \delta x$

Change in displacement $= NQ - MP$

$= NQ - NR$

$= RQ$

$= \delta x$

Change in time $= MN = PR = \delta t$

We define 'Average Velocity'

as $$\frac{\text{Change in Displacement}}{\text{Change in Time}}$$

i.e. $$\frac{RQ}{PR} = \frac{\delta x}{\delta t}$$

If x is a known function of t we can see that if δt becomes close to zero then the average velocity will become close to the velocity at the instant t.

In this case the chord PQ will become close to the tangent to the curve at P, and the gradient of this tangent will give the measure of the velocity V at the instant t.

In this case V, the velocity at the instant t, will be $\lim\limits_{\delta t \to 0} \dfrac{\delta x}{\delta t} = \dfrac{dx}{dt}$

or $$V = \frac{dx}{dt}$$

Example 6.2.1 (1)

An object moving in a straight line has its displacement x metres from an origin O at time t seconds given by $x = t\,(t - 3)^2$.

Determine

(a) the time when the object is at the origin,
(b) the time when the object is instantaneously at rest,
(c) the distance moved between $t = 0$ and $t = 2$.

Solution:

(a) The object will be at the origin when $x = 0$

i.e. $t\,(t - 3)^2 = 0$

i.e. at $t = 0$ or $t = 3$

i.e. $t = 0$ seconds, $t = 3$ seconds

(b) Velocity $V = \dfrac{dx}{dt} = (t - 3)^2 + t \cdot 2\,(t - 3)^1$

$$= (t - 3)\,[(t - 3) + 2t]$$
$$= (t - 3)\,[\,t - 3 + 2t]$$
$$= (t - 3)\,(3t - 3)$$
$$= 3\,(t - 3)\,(t - 1)$$

i.e. $V = 0$, when $t = 3$ or 1

so the object is instantaneously at rest at $t = 1$ second, $t = 3$ seconds.

(c) A glance at Fig. 6.2.1 (b) shows that when $t = 1$ $x = 4$

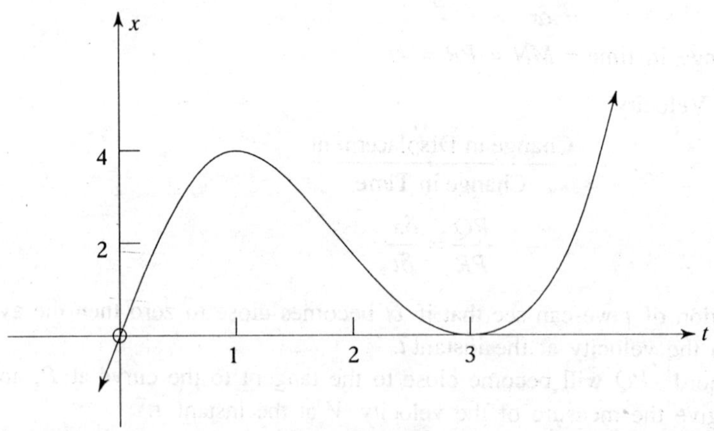

Fig. 6.2.1(b)

Now between $t = 0$ and $t = 1$ the velocity is positive and the object moves from position $x = 0$ to position $x = 1\,(1 - 3)^2 = 4$

Between $t = 1$, and $t = 3$ the velocity is negative and the object moves from position $x = 4$ to position $x = 0$

However the distance moved by the object between $t = 0$ and $t = 2$ will be given by 4 + the positive difference between the values of x at times $t = 1$, $t = 2$ respectively.
i.e. between $t = 1$ and $t = 2$ the object moves from $x = 4$ to $x = 2\ (2 - 3)^2 = 2$

So the total distance covered will be (4 + 2) metres = 6 metres.

6.2.2 The Velocity-Time Graph

Velocity and Acceleration

Consider the velocity-time graph shown in Fig. 6.2.1 (c), where an object moving in a straight line has a velocity v at time t, and a velocity $v + \delta v$ at time $t + \delta t$.

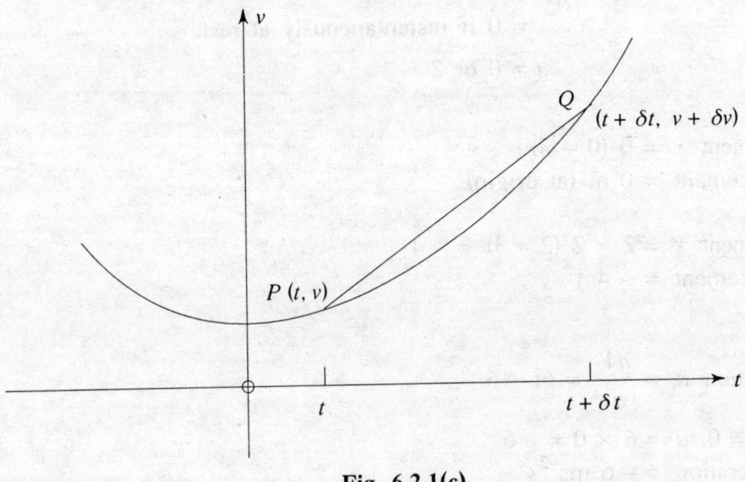

Fig. 6.2.1(c)

Using an argument similar to that at the beginning of sub-section 6.2.1 we define Average Acceleration

$$= \frac{\text{Change in Velocity}}{\text{Change in Time}}$$

In the graph this is given by the gradient of the chord PQ, taken over the time interval $[t, t + \delta t]$

i.e. $\dfrac{\delta v}{\delta t}$

The gradient of the tangent at the point P will give the acceleration at instant t. And in this case if the velocity v is a known function of t then the instantaneous acceleration a at time t will be the

$$\lim_{\delta t \to 0} \frac{\delta v}{\delta t} \quad \text{i.e.} \quad a = \frac{dv}{dt}$$

Hence we can determine the acceleration at an instant t by differentiating v with respect to t.

Normally we measure displacement (and distance) in metres, and time in seconds, so the units of velocity are then given in metres per second (often written m/s or ms^{-1}) and the units of acceleration are given in metres per second per second (which can be written m/s/s or more often m/s^2 or ms^{-2}.

Example 6.2.2 (1)

An object moving in a straight line has its displacement x metres at time t seconds given by $x = t^2 (t - 3)$.

Detrermine the displacement and the acceleration at the times when the object is instantaneously at rest.

Solution:

$$x = t^2 (t - 3) = t^3 - 3t^2$$

$$\text{velocity } V = \frac{dx}{dt} = 3t^2 - 6t = 3t (t - 2)$$

$$= 0 \text{ if instantaneously at rest.}$$

i.e. $t = 0$ or 2

when $t = 0$

the displacement $x = 0 (0 - 3)$

i.e. displacement $= 0$ m (at origin)

When $t = 2$

the displacement $x = 2 \times 2 (2 - 3) = -4$

i.e. displacement $= -4$m

From $V = 3t^2 - 6t$

$$\text{the acceleration } a = \frac{dV}{dt} = 6t - 6$$

When $t = 0$, $a = 6 \times 0 = -6$

i.e. acceleration $= -6 \text{ ms}^{-2}$

When $t = 2$, $a = 6 \times 2 - 6 = 6$

i.e. acceleration $= 6 \text{ ms}^{-2}$

Example 6.2.2 (2)

A particle is moving along the x-axis and its displacement x metres from the origin after t seconds is given by $x = e^{-\sqrt{3}t} \sin t$.

Determine its displacement, velocity and acceleration when $t = \pi/2$ and also the smallest positive value of t for which the particle is instantaneously at rest.

Solution:

From $x = e^{-\sqrt{3}t} \sin t$, differentiate to get

velocity $V = -\sqrt{3}\, e^{-\sqrt{3}t} \sin t + e^{-\sqrt{3}t} \cos t$ and differentiate again to get acceleration

$$a = 3e^{-\sqrt{3}t} \sin t - \sqrt{3}\, e^{-\sqrt{3}t} \cos t$$

$$- \sqrt{3}\, e^{-\sqrt{3}t} \cos t - e^{-\sqrt{3}t} \sin t$$

$$= 2e^{-\sqrt{3}t} \sin t - 2\sqrt{3}\, e^{-\sqrt{3}t} \cos t$$

substituting $t = \pi/2$

then $$x = e^{-\sqrt{3}\pi/2} \cdot 1 = e^{-\sqrt{3}\pi/2}$$

$$V = -\sqrt{3}\ e^{-\sqrt{3}\pi/2} \cdot 1 + e^{-\sqrt{3}\pi/2} \cdot 0 = -\sqrt{3}\ e^{-\sqrt{3}\pi/2}$$

$$a = 2e^{-\sqrt{3}\pi/2} - 2\sqrt{3}\ e^{-\sqrt{3}\pi/2} \cdot 0 = 2e^{-\sqrt{3}\pi/2}$$

so at $t = \pi/2$

The displacement $= e^{-\sqrt{3}\pi/2}$ metres

the velocity $= -\sqrt{3}\ e^{-\sqrt{3}\pi/2}\ \text{ms}^{-1}$

and the acceleration $= 2\ e^{-\sqrt{3}\pi/2}\ \text{ms}^{-2}$

When the particle is instantaneously at rest $V = 0$

i.e. $$-\sqrt{3}\ e^{-\sqrt{3}t} \sin t + e^{-\sqrt{3}t} \cos t$$

$$= e^{-\sqrt{3}t}(-\sqrt{3} \sin t + \cos t) = 0$$

i.e. $-\sqrt{3} \sin t + \cos t = 0$ as $e^{-\sqrt{3}t} > 0$ for all t

i.e. $$\cos t = \sqrt{3} \sin t$$

i.e. $$\frac{1}{\sqrt{3}} = \frac{\sin t}{\cos t}$$

i.e. $$\frac{1}{\sqrt{3}} = \tan t$$

the general solution is $t = \dfrac{\pi}{6} + n\pi,\ n \in Z$ and the smallest positive value of t for which the particle is instantaneously at rest is $t = \pi/6$.

6.2.3 Velocity and Acceleration as Functions of Displacement

In some problems we may be given the velocity or the acceleration as a function of the displacement x.

We know that in terms of t the velocity $v = \dfrac{dx}{dt}$ and the acceleration $a = \dfrac{dv}{dt}$ for an object moving in a straight line with displacement x metres at time t seconds.

Using the chain rule

we may write $a = \dfrac{dv}{dt}$

$$= \frac{dv}{dx} \cdot \frac{dx}{dt} = \frac{dv}{dx} \cdot v$$

i.e. $$a = v\frac{dv}{dx}$$

also since $$\frac{d}{dx}\left(\frac{1}{2}v^2\right) = \frac{1}{2}\,2v\,\frac{dv}{dx} = v\,\frac{dv}{dx}$$

We may use either of the forms $a = v\dfrac{dv}{dx}$ or $a = \dfrac{d}{dx}\left(\dfrac{1}{2}v^2\right)$

in cases where the velocity is given as a function of the displacement x.

Example 6.2.3 (1)

An object moving in a straight line has its velocity v m s^{-1} at any time t seconds given by $v^2 = x^4 - 32x + 16$, where x metres is its displacement at time t seconds. Find where the object is when the acceleration is zero.

Solution:

Using $$a = \frac{d}{dx}\left(\frac{1}{2}v^2\right)$$

$$a = \frac{d}{dx}\left(\frac{1}{2}(x^4 - 32x + 16)\right) = \frac{1}{2}\,(4x^3 - 32)$$

$$= \frac{1}{2}\cdot 4\,(x^3 - 8) = 2\,(x^3 - 8) = 0 \text{ when } x = 2$$

so the object is at displacement +2 metres from the origin when the acceleration is zero.

Example 6.2.3 (2)

A particle moving in a straight line has velocity ms^{-1} and displacement x metres at time t seconds where $$v = (x^2 - 4)^{3/2} \text{ and } x \geq 2.$$

Show that the acceleration is zero when the v particle is instantaneously at rest.

Solution:

Using $$a = v\frac{dv}{dx}$$

$$a = (x^2 - 4)^{3/2}\cdot 3/2\,(x^2 - 4)^{1/2}\cdot 2x.$$

$$= 3x\,(x^2 - 4)^2 = 0 \text{ when } x = 0 \text{ or } x = \pm 2$$

However we are given $x \geq 2$

hence the acceleration is zero, when $x = 2$ metres

then $$v = (x^2 - 4)^{3/2} = 0$$

i.e. the particle is instantaneously at rest at the same time as the acceleration is zero.

6.2.4 Maximum, Minimum and Points of Inflexion

A stationary value of any function $y = f(x)$ is a value of $f(x)$ for which its rate of change with respect to x is zero, i.e. stationary values of $y = f(x)$ will occur when $\dfrac{dy}{dx}$ (or $f'(x)$) $= 0$

In the graph of $y = f(x)$ a stationary value will occur when the gradient of the curve is zero, i.e. the tangent to the curve wil be parallel to the x-axis (see Fig. 6.2.4(a))

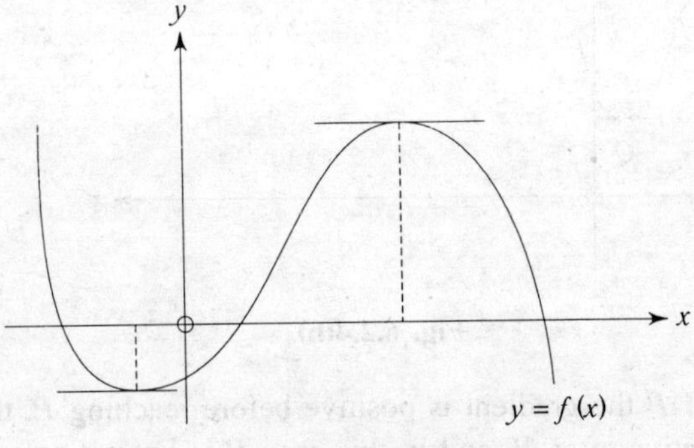

Fig. 6.2.4(a)

From Fig. 6.2.4 (a) it can abe seen that the stationary values of $y = f(x)$ are given by the y-coordinates of the points on the curve whre the tangent is parallel to the x-axis.

Example 6.2.4 (1)

Find the stationary values of $y = x^3 - 6\ x^2 + 1$

Solution: If $y = x^3 - 6x^2 + 1$

then $$\frac{dy}{dx} = 3x^2 - 12x$$

At stationary values of y

$$\frac{dy}{dx} = 0$$

i.e. $3x^2 - 12x = 0$

i.e. $3x\ (x - 4) = 0$

i.e. the stationary values occur when $x = 0$ and when $x = 4$.

The values are given by $y = 1$ and $y = 4^3 - 6 \times 4^2 + 1$ respectively,

i.e. by $y = 1,\ y = -31$.

N.B. for the curve $y = x^3 - 6x^2 + 1$

$$\frac{dy}{dx} = 3x^2 - 12x = 0 \text{ when } x = 0 \text{ or } x = 4$$

so the gradient of the curve is zero at the points (0, 1), and (4, −31).

It is possible that the gradient of a curve is equal to zero at several points on the curve. The shape of the curve in the immediate neighbourhood of such points belongs to one of the three categories shown in Fig. 6.2.4 (b).

Moving along the curve from left to right, i.e. in the positive direction of the x-axis,

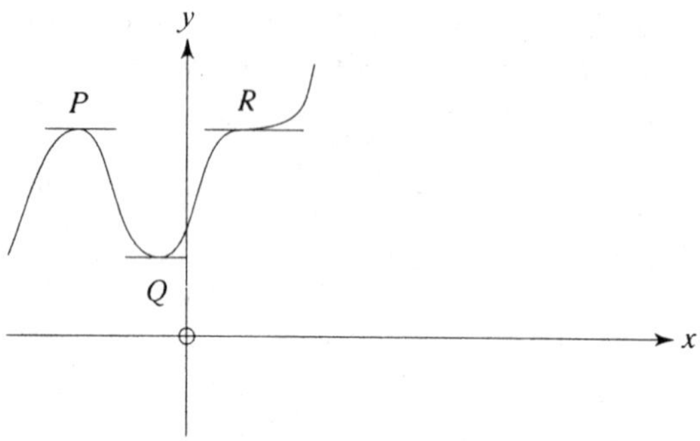

Fig. 6.2.4(b)

(i) in the neighbourhood of P the gradient is positive before reaching P, the gradient is zero at P, and the gradient is negative after P, and in this case P is known as a maximum turning point and the y-coordinate of P is called the maximum value of y;

(ii) in the neighbourhood of Q the gradient is negative before reaching Q, the gradient is zero at Q, and the gradient is positive after Q, and in this case Q is known as a minimum turning point and the y-coordinate of Q is called the minimum value of y;

(iii) at R the curve does not turn, although the gradient is zero at R; in this case the gradient does not change sign as the curve moves through R, but the sense in which the curve is turning does change (from clockwise to anticlockwise);

R is an example of a point of inflexion.

Any point on a curve where the sense of turning changes is called a point of inflexion.

In Fig. 6.2.4 (b) apart from R there are two other points of inflexion on the curve, one between P and Q, and one between Q and R. Note that the gradient at a point of inflexion will not necessarily be equal to zero.

Note also that the gradient at a maximum or a minimum turning point must be zero but that the terms maximum and minimum may not necessarily mean the greatest and least values of the function respectively.

The terms 'maximum' and 'minimum' will apply to the behaviour of the function only in the immediate neighbourhood of the function. It is normally a straight forward exercise to find the coordinates of points on a curve with equation $y = f(x)$ at which y will have stationary values. However it is necessary to examine the behaviour of the function $f(x)$ and its gradient in the neighbourhood of the stationary values in order to distinguish whether these values are maxima, minima or points of inflexion. In particular it is important to find out what is happening just before, and just after, a stationary point, as we move along the curve from left to right.

Consider in Fig. 6.2.4 (c)

what is happening at the points P_1 to the left of P, P_2 to the right of P, and close to P; Q_1 to the left of Q, Q_2 to the right of Q, and close to Q; R_1 to the left of R, R_2 to the right of R, and close to R.

If we examine the values of y; at P (a maximum value)

$$y \text{ at } P_1 < y \text{ at } P$$

and

$$y \text{ at } P_2 < y \text{ at P}$$

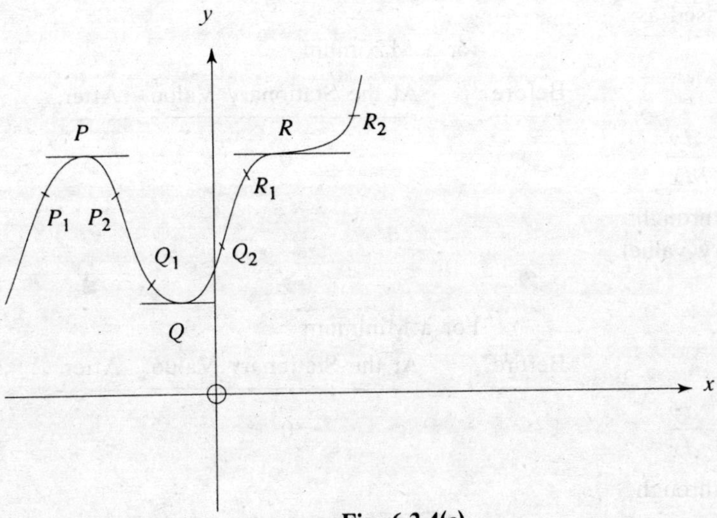

Fig. 6.2.4(c)

at Q (a minimum value)

$$y \text{ at } Q_1 > y \text{ at } Q$$

and

$$y \text{ at } Q_2 > y \text{ at } Q$$

At R (a point of inflection)

$$Y \text{ at } R_1 < y \text{ at } R \text{ and } y \text{ at } R_2 > y \text{ at } R$$

Summarising for each type of stationary point:

	Maximum	Minimum	Inflection
Values of y on either side of stationary point	Both less	Both greater	One less and the other greater.

Consider also the behaviour of the gradient dy/dx in the neighbourhod of each stationary point.

In the neighbourhood of P,

at P_1 $dy/dx > 0$

at P $dy/dx = 0$

at P_2 $dy/dx < 0$

In the neighbourhood of Q,

at Q_1 $dy/dx < 0$

at Q $dy/dx = 0$

at Q_2 $dy/dx > 0$

In the neighbourhood of R,

at R_1 $dy/dx > 0$

at R $dy/dx = 0$

at R_2 $dy/dx > 0$

This can be summarised as:

for a Maximum

	Before	At the Stationary Value	After
Sign of $\frac{dy}{dx}$ (moving through a stationary value)	+	0	–
y	↗	→	↘

For a Minimum

	Before	At the Stationary Value	After
Sign of $\frac{dy}{dx}$ (moving through a stationary point)	–	0	+
y	↘	→	↗

for a Point of Inflexion

Case (1)

	Before	At the Stationary Value	After
Sign of $\frac{dy}{dx}$ (moving through a stationary point)	+	0	+
y	↗	→	↗

Case (2)

	Before	At the Stationary Value	After
Sign of $\frac{dy}{dx}$ (moving through a stationary point)	–	0	–
y	↘	→	↘

When passing through P, $\frac{dy}{dx}$ has changed from positive to negative and so has decreased as x has increased, meaning that the rate of increase of $\frac{dy}{dx}$ with respect to x is negative, or $\frac{d}{dx}\left(\frac{dy}{dx}\right)$ is negative.

The cumbersome notation for the rate of increase of $\frac{dy}{dx}$ with respect to x can be shortened to $\frac{d^2y}{dx^2}$.

Similarly when passing through Q, we observe that $\frac{dy}{dx}$ has changed from negative to positive,

i.e. $\frac{dy}{dx}$ has increased as x has increased, or $\frac{d^2y}{dx^2} > 0$.

Using the sign of $\frac{d^2y}{dx^2}$ can only point with certainity to the type of stationary value when $\frac{d^2y}{dx^2} > 0$ (which implies that the stationary value is a minimum) or when $\frac{d^2y}{dx^2} < 0$ (which implies that the stationary value is a maximum).

It is true that $\frac{d^2y}{dx^2} = 0$ at a point of inflexion but $\frac{d^2y}{dx^2}$ can also be zero at maxima and minima.

When testing to determine the type of stationary point it is always possible to do this by testing the sign of the gradient $\frac{dy}{dx}$ before and after the stationary point.

The tests should become clearer after studying the examples which follow.

Example 6.2.4 (2)

Determine the coordinates of the stationary points on the curve whose equation is

$$y = 2x^3 - 21\,x^2 + 60\,x + 12$$

and determine their nature.

Solution:

If $\quad y = 2x^3 - 21\,x^2 + 60x + 12$

$$\frac{dy}{dx} = 6x^2 - 42x + 60 = 0 \text{ for stationary points}$$

i.e. $\quad 6\,(x - 2)\,(x - 5) = 0$

i.e $\quad x = 2 \quad \text{or} \quad x = 5$

when $\quad x = 2,\ y = 2 \times 2^3 - 21 \times 2^2 + 60 \times 2 + 12 = 64$

when $\quad x = 5,\ y = 2 \times 5^3 - 21 \times 5^2 + 60 \times 5 + 12 = 37$

Using the sign test on $\frac{dy}{dx}$, we examine the sign of each of the factors of $\frac{dy}{dx}$ point before the stationary point, at the stationary point, and just after the stationary point.

For instance just before $x = 2$ (indicated by 2 – in the table) the value of x could be 1 for which 6 is positive, but $x - 2 = 1 - 2 = -1$ is negative, and $x - 5 = 1 - 5 = -4$ is negative.

The product of a positive number and two negative numbers is positive, thus the sign of $\frac{dy}{dx}$ will be positive before $x = 2$.

The reader could try substituting other values of x just before $x = 2$, such as $x = 1.5$, or $x = 1.9$, and should be convinced that the signs of $x - 2$ and $x - 5$ would still both be negative.

The symbol 2 + in the table means a value of x just more than $x = 2$, and corresponding meanings apply to the symbols 5– and 5+.

x	2–	2	2+	5–	5	5+
6	+	+	+	+	+	+
$x - 2$	–	0	+	+	+	+
$x - 5$	–	–	–	–	0	+
$\frac{dy}{dx}$	+	0	–	–	0	+
y	↗	→	↘	↘	→	↗

From the sign test it can be seen that (2, 64) is a maximum point and (5, 37) is a minimum point.

In this case it is also easy to find $\frac{d^2y}{dx^2} = 12x - 42$

At $$x = 2, \frac{d^2y}{dx^2} = 12 \times 2 - 42$$

$$= -18 < 0$$

meaning (2, 64) is a maximum point

At $$x = 5, \frac{d^2y}{dx^2} = 12 \times 5 - 42$$

$$= 18 > 0$$

meaning (5, 37) is a minimum point.

Example 6.2.4 (3)

Determine the coordinates and nature of the turning points on the curve whose equation is

$$y = 4x^3 - 5x^2 + 2x + 1$$

Solution:

If $$y = 4x^3 - 5x^2 + 2x + 1$$

$$\frac{dy}{dx} = 12\,x^2 - 10x + 2 = 0 \text{ for turning points}$$

i.e. $$2\,(3x - 1)\,(2x - 1) = 0$$

i.e. $$x = \frac{1}{3} \quad \text{or} \quad x = \frac{1}{2}$$

When $$x = \frac{1}{3},$$

$$y = 4 \times \frac{1}{27} - 5 \times \frac{1}{9} + 2 \times \frac{1}{3} + 1 = \frac{34}{27}$$

When $$x = \frac{1}{2}, y = 4 \times \frac{1}{8} - 5 \times \frac{1}{4} + 2 \times \frac{1}{2} + 1 = \frac{5}{4}$$

Using the sign test we need to take care with the values of x that we consider when x is $\frac{1}{3}+$ and when x is $\frac{1}{2}-$.

When x is $\frac{1}{3}+$ it might be tempting to substitute $x = 1$ to determine the signs of the factor to the right of $x = \frac{1}{3}$. However $x = 1$ *cannot* be in the neighbourhood of $x = \frac{1}{3}$ in this example, since we also have another turning point when $x = \frac{1}{2}$, as shown in Fig. 6.2.4 (d).

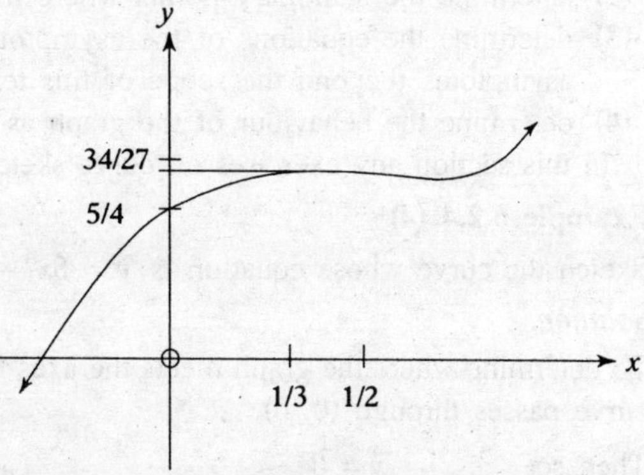

Fig. 6.2.4(d)

When considering $x = \frac{1}{3}+$ it is necessary to take a value of x which is more than $\frac{1}{3}$ but less than $\frac{1}{2}$, say $x = \frac{5}{12}$, for which the factor 2 is positive, the factory $3x - 1$ is positive, and the factory $2x - 1$ is negative, hence $\frac{dy}{dx}$, being the product of two positive factors and a negative factor, will be negative.

Similarly for $x = \frac{1}{2}-$ it would be *wrong* to consider a value such as $x = 0$, since we have ignored the previous stationary point where $x = \frac{1}{3}$. A value such as $x = 9/20$ would be appropriate.

The table for the sign test is shown below:

x	$\frac{1}{3}-$	$\frac{1}{3}$	$\frac{1}{3}+$	$\frac{1}{2}-$	$\frac{1}{2}$	$\frac{1}{2}+$
2	+	+	+	+	+	+
$3x - 1$	–	0	+	+	+	+
$2x - 1$	–	–	–	–	0	+
$\frac{dy}{dx}$	+	0	–	–	0	+
y	↗	→	↘	↘	→	↗

From the table it can be seen that $\left(\frac{1}{3}, \frac{34}{27}\right)$ is a maximum point and $\left(\frac{1}{2}, \frac{5}{4}\right)$ is a minimum point.

This example can be solved by finding $\frac{d^2y}{dx^2}$ and substituting the values $x = \frac{1}{3}$, $x = \frac{1}{2}$ in that expression, and this is left as an exercise for the reader.

When sketching graphs it is necessary only to show the main features, rather than plotting a number of points laboriously which is a requirement for accurately drawing a graph.

In sketching there are four main items to look for, although one of these is beyond the scope of this text. The items are:

(1) determine the points where the graph meets the axes if these can be easily found.
(2) determine the stationary points where these can be easily found,
(3) determine the equations of the asymptotes and the behaviour of the curve at points near the asymptotes (beyond the scope of this text).
(4) determine the behaviour of the graph as $x \to \infty$ and as $x \to -\infty$.

In this section any exercises on curve sketching will not involve curves which have asymptotes.

Example 6.2.4 (4)

Sketch the curve whose equation is $y = 5x^4 - x^5$

Solution:

To determine where the graph meets the axes first set $x = 0$ in the given equation to give $y = 0$. The curve passes through (0, 0).

Then set $\quad y = 0$

to give $\quad 0 = 5x^4 - x^5$

i.e. $\quad 0 = x^4 (5 - x)$

i.e. $\quad x = 0$ or $x = 5$

For the stationary points

$$\frac{dy}{dx} = 20x^3 - 5x^4$$

$$= 5x^3 (4 - x) = 0$$

i.e. $\quad x = 0$ or $x = 4$

and the stationary points are (0, 0) and (4, 256). Using the value of $\frac{d^2y}{dx^2}$ to test for the type of stationary point

$$\frac{d^2y}{dx^2} = 60x^2 - 20x^3$$

if $\quad x = 4, \; \frac{d^2y}{dx^2} = 60 \times 4^2 - 20 \times 4^3 = -320 < 0$

so $\quad$ (4, 256) is a maximum point.

If $\quad x = 0, \; \frac{d^2y}{dx^2} = 0$

and this does *not* imply that (0, 0) is a point of imflexion.

At (0, 0) both $\dfrac{dy}{dx}$ and $\dfrac{d^2y}{dx^2}$ are zero, so we need to test the sign of $\dfrac{dy}{dx}$ before and after $x = 0$

x	0−	0	0+
5	+	+	+
x^3	−	0	+
$4 - x$	+	+	+
dy/dx	−	0	+
y	↘	→	↗

From the test we see that (0, 0) is a minimum point,

The turning points should have given guidance as to how the curve behaves as x becomes large and positive, and as x becomes large and negative.

For $x \to -\infty$ consider the equation of the curve.

$$y = 5x^4 - x^5$$

If x is large and negative say -10^6

$$y = 5 \times 10^{24} - (-10^{30})$$

or $\quad y \simeq 10^{30}$ (the value of the dominating term $-x^5$; x^5 is of higher degree than any other term)

So as $\quad x \to -\infty, y \to +\infty$

If x is large and positive say 10^6

$$y = 5 \times 10^{24} \quad 10^{30}$$

or $\quad y \simeq -10^{30}$ (based on the dominating term)

So as $\quad x \to +\infty, y \to -\infty.$

The sketch can now be drawn in Fig. 6.2.4 (e), showing the main features.

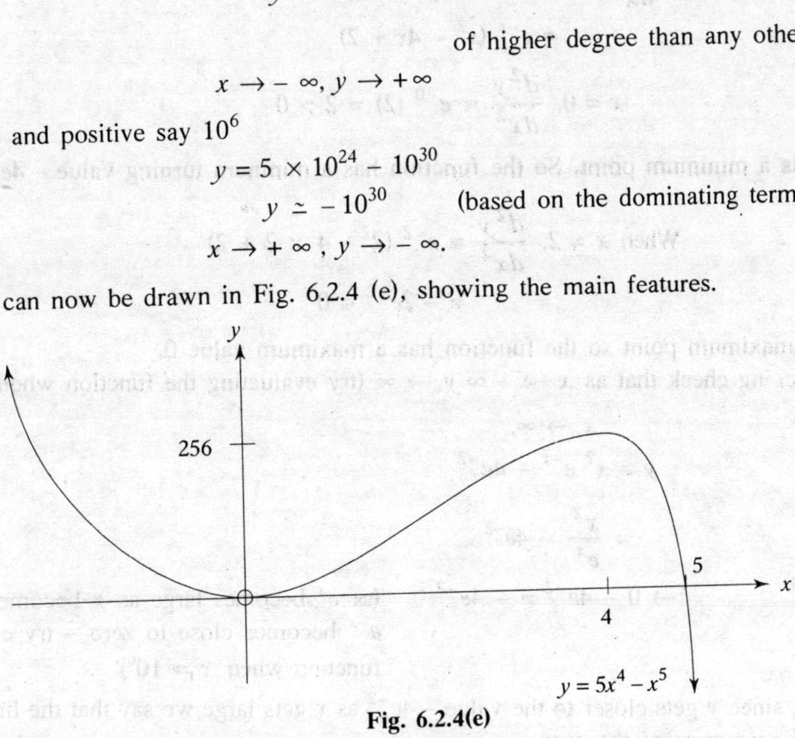

Fig. 6.2.4(e)

Example 6.2.4 (5)

Find the first and second derivatives of the function $y = x^2 e^{-x} - 4e^{-2}$. Determine the maximum and minimum turning values of the function and hence sketch the curve representing the functions.

Solution:

$$y = x^2 e^{-x} - 4e^{-2}$$

$$\frac{dy}{dx} = -x^2 e^{-x} + 2xe^{-x}$$

$$= xe^{-x}(2 - x)$$

for stationary points

$$\frac{dy}{dx} = 0$$

i.e. $\quad xe^{-x}(2 - x) = 0$

i.e. $\quad x = 0 \text{ or } x = 2$

When $\quad x = 0, y = 0 - 4e^{-2}$

when $\quad x = 2, y = 4e^{-2} - 4e^{-2} = 0$

so the stationary points are $(0, -4e^{-2})$, $(2, 0)$

$$\frac{d^2y}{dx^2} = x^2 e^{-x} - 2xe^{-x} - 2xe^{-x} + 2e^{-x}$$

$$= e^{-x}(x^2 - 4x + 2)$$

when $\quad x = 0, \dfrac{d^2y}{dx^2} = e^{-0}(2) = 2 > 0$

so $(0, -4e^{-2})$ is a minimum point. So the function has a mininum turning value $-4e^{-2}$

$$\text{When } x = 2, \frac{d^2y}{dx^2} = e^{-2}(2^2 - 4 \times 2 + 2)$$

$$= -2e^{-2} < 0$$

so (2, 0) is a maximum point so the function has a maximum value 0.

Before sketci ing check that as $x \to -\infty$ $y \to \infty$ (try evaluating the function when $x = -10^6$)

As $\quad x \to \infty,$

$$y = x^2 e^{-x} - 4e^{-2}$$

$$= \frac{x^2}{e^x} - 4e^{-2}$$

$$\to 0 - 4e^{-2} = -4e^{-2}$$

(as e^x becomes large as x becomes large, so e^{-x} becomes close to zero – try evaluating the function when $x = 10^6$)

In this case, since y gets closer to the value $-4e^{-2}$ as x gets large we say that the line $y = -4e^{-2}$ is a horizontal asymptate of the curve.

Note also that we have shown that (2, 0) is a maximum turning point and so the graph touches the x-axis at (2, 0).

It is not easy to see at first when the curve crosses the negative x-axis but by inspection it may be found to be at (–2, 0).

The sketch has been drawn in Fig. 6.2.4 (f)

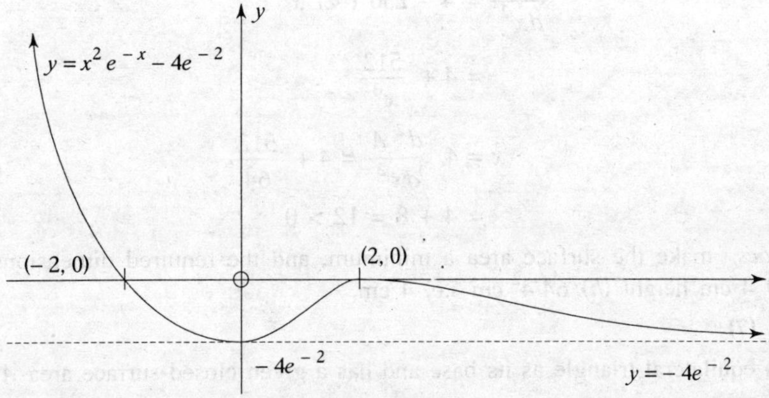

Fig. 6.2.4(f)

Our knowledge of how to determine stationary points and stationary values of a function can be applied to a variety of problems, especially those that involve area or volume.

Example 6.2.4 (6)

A closed rectangular container has a square base and is required to have a volume of 64 cm^3. If the container is made of thin metal find the dimensions which will minimise the surface area.

Solution:

Take the dimensions as given in Fig. 6.2.4 (g). The total surface area A is given by $A = 2x^2 + 4xh$.

However A is given as a function of two variables x and h. It is necessary to use the given information that the volume of the container is 64 cm^3 in order to express one of the variables in terms of the other.

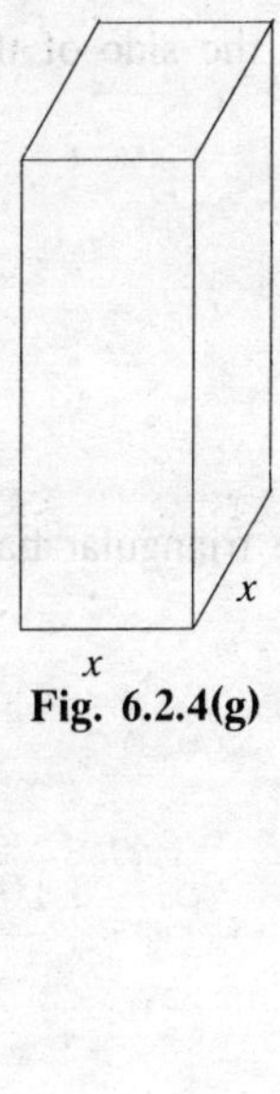

Fig. 6.2.4(g)

$$x^2 h = 64$$

so
$$h = \frac{64}{x^2}$$

Substitute for h in the expression for A to give $A = 2x^2 + 4x \cdot \dfrac{64}{x^2}$

$$= 2x^2 + 256\ x^{-1}$$

$$\frac{dA}{dx} = 4x - 256x^{-2} = 0 \text{ for}$$

stationary values. i.e.
$$4x^3 - 256 = 0$$

which leads to $x = 4$.

However we need to test whether $x = 4$, will give a minimum value of A.

Find $\frac{d^2 A}{dx^2}$ using the expression as found for $\frac{dA}{dx}$ and *not* the simplified expression $4x^3 - 256$.

Using
$$\frac{dA}{dx} = 4x - 256x^{-2}$$

$$\frac{d^2 A}{dx^2} = 4 - 256\,(-2)\,x^{-3}$$

$$= 4 + \frac{512}{x^3}$$

When
$$x = 4, \frac{d^2 A}{dx^2} = 4 + \frac{512}{64}$$

$$= 4 + 8 = 12 > 0$$

hence $x = 4$ *does* make the surface area a minimum, and the required dimensions are length of square base (x) 4 cm height (h) $64/4^2$ cm i.e. 4 cm.

Example 6.2.4 (7)

A prism has an equilateral triangle as its base and has a given closed surface area A. Show that its maximum volume is $\frac{A^{3/2}}{3^{9/4}\,2^{1/2}}$.

Solution:

Let the side of the triangular base be $2x$ and its length h as in Fig. 6.2.4 (h)

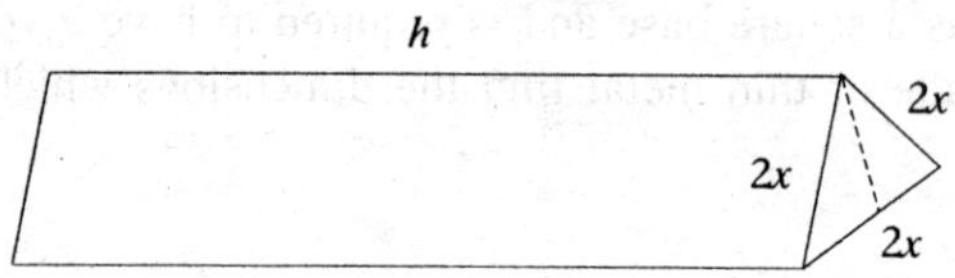

Fig. 6.2.4(h)

The triangular base will have altitude $x\sqrt{3}$ and area $\frac{1}{2} \cdot 2x \cdot x\sqrt{3}$

i.e.
$$x^2\sqrt{3}$$

so
$$V = x^2\sqrt{3}\,h$$

and
$$A = 2x^2\sqrt{3} + 6xh$$

So
$$h = \frac{A - 2x^2\sqrt{3}}{6x}$$

and
$$V = \frac{x^2\sqrt{3}\left(A - 2x^2\sqrt{3}\right)}{6x}$$

$$= \frac{x\sqrt{3}\,(A - 2x^2\sqrt{3})}{6}$$

$$= Ax\ \frac{\sqrt{3}}{6} - \frac{2x^3\sqrt{3}\sqrt{3}}{6}$$

$$= Ax\ \frac{\sqrt{3}}{6} - x^3$$

$$\frac{dV}{dx} = A\ \frac{\sqrt{3}}{6} - 3x^2 = 0 \text{ for stationary values.}$$

i.e. $$x^2 = \frac{A\sqrt{3}}{18}$$

i.e. $$x = \pm\ \frac{A^{1/2}\ 3^{1/4}}{3 \cdot 2^{1/2}}$$

$$= \pm\ \frac{A^{1/2}}{3^{3/4} \cdot 2^{1/2}}$$

as x is measuring length take $x > 0$ and so $x = \dfrac{A^{1/2}}{3^{3/4} \cdot 2^{1/2}}$

also to test whether the value of x gives a maximum find

$$\frac{d^2V}{dx^2} = -6x$$

If $$x = \frac{A^{1/2}}{3^{3/4} \cdot 2^{1/2}}$$

$$\frac{d^2V}{dx^2} = \frac{-6A^{1/2}}{3^{3/4} \cdot 2^{1/2}} < 0$$

So this value of x does give maximum volume.

Substituting $x = \dfrac{A^{1/2}}{3^{3/4} \cdot 2^{1/2}}$ or $x^2 = \dfrac{A\sqrt{3}}{18}$ in the expression for V.

Where $$V = \frac{x\sqrt{3}\left(A - 2x^2\sqrt{3}\right)}{6}$$

$$= \frac{\dfrac{3^{1/2}\ A^{1/2}}{3^{3/4}\ 2^{1/2}}\left(A - \dfrac{2A\sqrt{3}\sqrt{3}}{18}\right)}{6}$$

$$= \frac{3^{1/2}\ A^{1/2}}{3^{3/4}\ 2^{1/2}}\left(A - \frac{6A}{18}\right) \times \frac{1}{6}$$

$$= \frac{3^{1/2}\ A^{1/2}}{3^{3/4}\ 2^{1/2}} \cdot \frac{2A}{3} \cdot \frac{1}{6}$$

which simplifies to $A^{3/2}/(3^{\ 9/4}\ 2^{1/2})$

6.3 INTEGRATION AS THE REVERSE OF DIFFERENTIATION

In many problems we may know the gradient function and the question we have to ask is what is the original function of which the given function is the gradient?

The process of finding a function $f(x)$ from its derived function $f'(x)$ is called integration, and is the reverse of differentiation.

6.3.1 The Constant of Integration

If we know that $y = x^2 + 9$ then $\dfrac{dy}{dx} = 2x$

But if $y = x^2 + 6$, $\dfrac{dy}{dx}$ is also $2x$ and if $y = x^2 - 5$, $\dfrac{dy}{dx}$ is also $2x$. In fact $2x$ is the derivative, not only of x^2, but of x^2 + any constant. It could be the derivative of say $x^2 - 1$, x^2, $x^2 + 1$, $x^2 + 2$ etc. as shown in Fig. 6.3.1 (a).

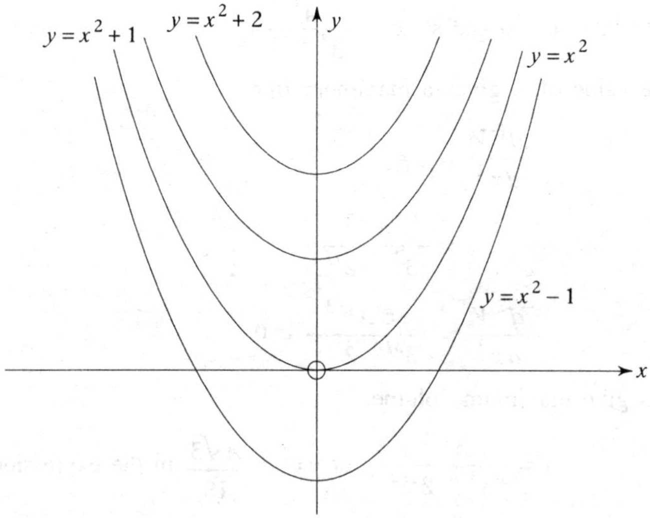

Fig. 6.3.1(a)

This means that the result of integrating $2x$, which we call the *integral* of $2x$, is not a unique function but is of the form $x^2 + C$ where C is an arbitrary constant, called the constant of integration.

This is written as $\int 2x\, dx = x^2 + C$

where $\int ...\, dx$ means the integral of ... with respect to x.

As integration is the reverse of differentiation, for a function $f(x)$ we have

$$\int f'(x)\, dx = f(x) + C$$

If we have information about the coordinates of any point which lies on the curve we can determine the arbitrary constant C.

Thus if we are told that $\dfrac{dy}{dx} = 2x$

then $$y = \int 2x\, dx$$

i.e. $$y = x^2 + C$$

and if we are told that the point (3, 5) lies on the curve then clearly we can substitute $x = 3$, $y = 5$ in the given equation to get $5 = 9 + C$

i.e. $$C = -4$$

6.3.2 The Integration of x^n

for $y = x^5$, we know $\dfrac{dy}{dx} = 5x^4$

and so $$\int 5x^4\, dx = x^5 + C$$

If $$y = \frac{1}{5}x^5$$

$$\frac{dy}{dx} = \frac{1}{5} \cdot 5x^4 = x^4$$

and so $$\int x^4\, dx = \frac{x^5}{5} + C$$

In general if $y = x^{n+1}$

then $$\frac{dy}{dx} = (n + 1)\, x^n$$

and if $$y = \frac{1}{n+1}\, x^{n+1}$$

then $$\frac{dy}{dx} = \frac{(n+1)}{(n+1)}\, x^n = x^n$$

we may conclude that

$$\int x^n\, dx = \frac{x^{n+1}}{n+1} + C$$

provided $n \neq -1$

This means that to integrate a power of x, increase the index by 1 and divide by the new index, apart from the case where the original power is -1.

In the case of $\int x^{-1}\, dx$ we know that if $y = \ln x$

then $$\frac{dy}{dx} = \frac{1}{x} = x^{-1}$$

as we established in example 6.1.7 (2). It follows that $\int x^{-1}\,dx = \ln x + C$

Example 6.3.2 (1)

(a) If $\dfrac{dy}{dx} = x^{10}$ determine y.

(b) If $\dfrac{dy}{dx} = x^{-5}$ determine y.

(c) If $\dfrac{dy}{dx} = x^{-3/4}$ determine y.

Solution:

(a) If $\dfrac{dy}{dx} = x^{10}$

$$y = \int x^{10}\,dx = \frac{x^{11}}{11} + C$$

(b) If $\dfrac{dy}{dx} = x^{-5}$

$$y = \int x^{-5}\,dx = \frac{x^{-4}}{-4} + C$$

$$= -\frac{x^{-4}}{4} + C \quad \text{or} \quad -\frac{1}{4}x^{-4} + C$$

(c) If $\dfrac{dy}{dx} = x^{-3/4}$

$$y = \int x^{-3/4}\,dx$$

$$= \frac{x^{1/4}}{1/4} + C = 4x^{1/4} + C$$

6.3.3 Integrating a Constant k

We know that the result of differentiating Kx is K so $\int K\,dx = Kx + C$.

6.3.4 Integration of $K\,x^n$ and of a Sum of Functions

We know that if $y = Kx^{n+1}$

then
$$\frac{dy}{dx} = K\,(n+1)\,x^n$$

and if
$$y = \frac{K}{(n+1)}\,x^{n+1}$$

then $$\frac{dy}{dx} = \frac{K(n+1)}{(n+1)} x^n = K x^n$$

hence $$\int K x^n \, dx = \frac{K x^{n+1}}{(n+1)} + C$$

where K is a constant.

We also showed in example 6.1.1 (5) that if we differentiate a sum of two functions this can be performed by differentiating the functions separately and then adding. The principle can be applied in differentiating a sum of any number of functions.

Since integration is the reverse operation, whenever we are required to integrate a sum of functions we can integrate each function and then add.

Example 6.3.4 (1)

If $\frac{dy}{dx} = 20\, x^3 - 12\, x^2 + 5$ for a particular curve and if $y = 40$ when $x = 2$, determine y in terms of x.

Solution:

As $$\frac{dy}{dx} = 20\, x^3 - 12\, x^2 + 5$$

$$y = \int (20x^3 - 12\, x^2 + 5)\, dx$$

$$= \frac{20x^4}{4} - \frac{12x^3}{3} + 5x + C$$

i.e. $$y = 5x^4 - 4x^3 + 5x + C$$

But $$y = 40 \text{ when } x = 2$$

so $$40 = 5 \times 2^4 - 4 \times 2^3 + 5 \times 2 + C$$

i.e. $$40 = 58 + C$$

or $$-18 = C$$

hence the equation of the curve is

$$y = 5x^4 - 4x^3 + 5x - 18$$

Example 6.3.4 (2)

Find $$\int (5x^9 - 35x^6 + \frac{1}{2} x^3 - x^{-2} + x^{1/3})\, dx$$

Solution:

$$\int (5x^9 - 35x^6 + \frac{1}{2} x^3 - x^{-2} + x^{1/3})\, dx$$

$$= \frac{5x^{10}}{10} - \frac{35x^7}{7} + \frac{1}{2}\cdot\frac{x^4}{4} - \frac{x^{-1}}{(-1)} + \frac{x^{4/3}}{4/3} + C$$

$$= \frac{1}{2} x^{10} - 5x^7 + \frac{1}{8} x^4 + \frac{1}{x} + \frac{3}{4} x^{4/3} + C$$

6.3.5 Integration of $(ax + b)^n$

Consider the curve whose equation is $y = (5x + 3)^4$

We can find $\frac{dy}{dx}$ using the chain rule as in 6.1.4. so that $\frac{dy}{dx} = 4\ (5x + 3)^3 \cdot 5$

$$= 20\ (5x + 3)^3$$

so $$\int 20\ (5x + 3)^3\ dx = (5x + 3)^4 + C$$

In the same way if

$$y = \frac{(5x + 3)^4}{20}$$

then $$\frac{dy}{dx} = \frac{4(5x + 3)^3 \cdot 5}{20} = (5x + 3)^3$$

so $$\int (5x + 3)^3\ dx = \frac{(5x + 3)^4}{4.5} + C$$

$$= \frac{(5x + 3)^4}{20} + C$$

In the general case when $y = (ax + b)^{n+1}$ where a, b are constants.

then $$\frac{dy}{dx} = (n + 1)\ (ax + b)^n \cdot a$$

$$= a\ (n + 1)\ (ax + b)^n$$

so $$\int a\ (n + 1)\ (ax + b)^n\ dx = (ax + b)^{n+1} + C$$

Now if $$y = \frac{(ax + b)^{n+1}}{(n + 1)a} \quad \text{and} \quad n \neq -1$$

then $$\frac{dy}{dx} = \frac{(n + 1)(ax + b)^n \cdot a}{(n + 1) \cdot a} = (ax + b)^n$$

so that $$\int (ax + b)^n\ dx = \frac{(ax + b)^{n+1}}{(n + 1) \cdot a} + C \text{ if} \quad n \neq -1$$

While this is a useful result to remember note that

(1) the reader will not need to refer to the result after doing adequate practice in this type of integration.

(2) this particular use of reverse chain rule can only be applied for a linear function taken to a power; it is not true that

$$\int (3x^2 + 1)^{1/2}\ dx \text{ is } \frac{(3x^2 + 1)^{3/2}}{3/2 \cdot 6x} + C$$

as $(3x^2 + 1)$ is *not* a linear function of x. The reader should try differentiating $\frac{(3x^2+1)^{3/2}}{3/2 \cdot 6x}$ and thereby be convinced that the answer is *not* $(3x^2 + 1)^{1/2}$;

(3) In case of any doubt about an answer to any integration the reader should always check by differentiating the answer and finding out whether that will be the same as the given function which had to be integrated.

Note (3) is a general rule which applies for many difficult integrations.

Example 6.3.5 (1)

Find (a) $\int (4x - 3)^5 \, dx$

(b) $\int (2x - 3)^{-5} \, dx$

(c) $\int (4x - 3)^{1/4} \, dx$ if $x \geq 3/4$

Solution:

(a) $$\int (4x - 3)^5 \, dx = \frac{(4x-3)^6}{6 \times 4} + C = \frac{(4x-3)^6}{24} + C$$

(b) $$\int (2x - 3)^{-5} \, dx = \frac{(2x-3)^{-4}}{-4 \times 2} + C = \frac{-1}{8(2x-3)^4} + C$$

(c) $$\int (4x - 3)^{1/4} \, dx = \frac{(4x-3)^{5/4}}{5/4 \times 4} + C = \frac{(4x-3)^{5/4}}{5} + C$$

6.3.6 Integration of Trigonometric Functions

Whenever a given function to be integrated is recognized as the derivative $f'(x)$ of a function $f(x)$ then clearly if $y = f(x)$

then $$\frac{dy}{dx} = f'(x)$$

so that $$\int f'(x) \, dx = f(x) + C$$

From the results of 6.1.6

We know:

$f(x)$	$f'(x)$
$\sin x$	$\cos x$
$\cos x$	$-\sin x$
$\tan x$	$\sec^2 x$
$\cot x$	$-\text{cosec}^2 x$
$\sec x$	$\sec x \tan x$
$\text{cosec } x$	$-\text{cosec } x \cot x$

So that

$$\int \cos x \, dx = \sin x + C$$

$$\int \sin x \, dx = -\cos x + C$$

$$\int \sec^2 x \, dx = \tan x + C$$

$$\int \text{cosec}^2 x \, dx = -\cot x + C$$

$$\int \sec x \tan x \, dx = \sec x + C$$

$$\int \text{cosec } x \cot x \, dx = -\text{cosec } x + C$$

For trigonometric functions of linear functions of x we may proceed in a manner similar to 6.3.5 should we be required to integrate such functions.

We can also extend the idea of 6.3.4 when integrating a constant times a trigonometrical function. Study the examples which follow

Example 6.3.6 (1)

Integrate with respect to x (a) $5 \sec^2 x$ (b) $\cos 4x$ (c) $\sin \frac{3x}{2}$ (d) $\sin (5x + 2)$ (e) $\sec^2 (3x + 5\pi 4)$ (f) $\text{cosec } (\pi/3 - 4x) \cot (\pi/3 - 4x)$ (g) $\tan^2 x + x$

Solution:

(a) $$\int 5 \sec^2 x \, dx = 5 \tan x + C$$

(b) $$\int \cos 4x \, dx = \frac{1}{4} \sin 4x + C$$

(in case the reader is not sure let $y = \sin 4x$ then $\frac{dy}{dx} = 4 \cos 4x$ then let $y = \frac{1}{4} \sin 4x$ so that $\frac{dy}{dx} = \frac{1}{4} \cdot 4 \cos 4x = \cos 4x$)

(c) $$\int \sin \frac{3x}{2} \, dx = \frac{-\cos \frac{3x}{2}}{\frac{3}{2}} + C = -\frac{2}{3} \cos \frac{3x}{2} + C$$

(d) $$\int \sin (5x + 2) \, dx = -\frac{1}{5} \cos (5x + 2) + C$$

(e) $$\int \sec^2 (3x + 5\pi/4) \, dx = \frac{1}{3} \tan (3x + 5\pi/4) + C$$

(f) $$\int \text{cosec } (\pi/3 - 4x) \cot (\pi/3 - 4x) \, dx = -\left(\frac{1}{-4} \text{cosec } (\pi/3 - 4x)\right) + C$$

$$= \frac{1}{4} \text{cosec } (\pi/3 - 4x) + C$$

(g) This presents a problem at first sight as $\tan^2 x$ cannot be recognized as a derivative of a known function, but we can write $\tan^2 x$ as $\sec^2 x - 1$ and proceed as follows:

$$\int(\tan^2 x + x)\, dx = \int(\sec^2 x - 1 + x)\, dx = \tan x - x + \frac{1}{2}x^2 + C$$

In case the reader has any doubts about any of the answers try differentiating the answer.

6.3.7 Integration of Exponential Functions

We know from 6.1.6 that if $y = e^x$ then $\dfrac{dy}{dx} = e^x$

so $$\int e^x\, dx = e^x + C$$

Also if $y = K\, e^x$, where K is a constant,

then $$\frac{dy}{dx} = K\, e^x$$

so $$\int K\, e^x\, dx = K\, e^x + C$$

for e raised to a linear function of x, say $y = e^{ax+b}$ where a, b are constants.

then $$\frac{dy}{dx} = e^{ax+b} \cdot a = a\, e^{ax+b}$$

so $$\int a\, e^{ax+b}\, dx = e^{ax+b} + C$$

If $$y = \frac{1}{a}\, e^{ax+b}$$

then $$\frac{dy}{dx} = \frac{1}{a} \cdot a\, e^{ax+b} = e^{ax+b}$$

and so $$\int e^{ax+b}\, dx = \frac{1}{a}\, e^{ax+b} + C$$

Note again that this process of reverse chain rule only applies for e raised to a power which is a linear function of x

$$\int e^{x^2}\, dx \neq \frac{e^{x^2}}{2x} + C$$

and the reader can check this by differentiating $\dfrac{e^{x^2}}{2x}$.

In fact so far no mathematician has been able to determine a known function of x whose derivative is equal to e^{x^2}.

Example 6.3.7 (2)

Find (a) $\int e^{4x}\,dx$ (b) $\int e^{-x}\,dx$

(c) $\int 3e^{-6x}\,dx$ (d) $\int e^{2-3x}\,dx$

Solution:

(a) $$\int e^{4x}\,dx = \frac{1}{4}\,e^{4x} + C$$

(b) $$\int e^{-x}\,dx = \frac{e^{-x}}{-1} + C = -\,e^{-x} + C$$

(c) $$\int 3\,e^{-6x}\,dx = \frac{3e^{-6x}}{-6} + C = -\,\frac{1}{2}e^{-6x} + C$$

(d) $$\int e^{2-3x}\,dx = -\,\frac{1}{3}\,e^{2-3x} + C$$

Example 6.3.7 (2)

Find $\int a^x\,dx$ using two methods (i) by first expressing a^x as a power of e, (ii) by first finding the derivative of a^x. The value of a is constant.

Solution:

(i) Let $a^x = e^u$

then taking logarithms to base e, $x \ln a = u$

so that $$a^x = e^u = e^{x \ln a}$$

hence $$\int a^x\,dx = \int e^{x \ln a}\,dx = \frac{1}{\ln a}\,e^{x \ln a} + C$$

or $$\int a^x\,dx = \frac{1}{\ln a}\cdot a^x + C$$

(ii) If $$y = a^x$$

taking logarithms to base e

$$\ln y = x \ln a$$

or $$x = \frac{1}{\ln a}\cdot \ln y$$

The gradient of the inverse function is

$$\frac{1}{\ln a}\cdot\frac{1}{y} \quad \text{or} \quad \frac{1}{y \ln a}$$

and by an argument similar to that of examples 6.1.2 (2) and (3) and also of example 6.1.7(2) we may show that the gradient of $y = a^x$ is the reciprocal of the gradient of the inverse function.

i.e. if $$y = a^x$$

then $$\frac{dy}{dx} = y \ln a \quad \text{or} \quad a^x (\ln a)$$

hence $$\int a^x (\ln a)\, dx = a^x + C$$

so if $$y = \frac{1}{\ln a} \cdot a^x$$

$$\frac{dy}{dx} = \frac{1}{\ln a} \cdot a^x \ (\ln a) = a^x$$

and $$\int a^x\, dx = \frac{1}{\ln a} \cdot a^x + C$$

One difficulty which occurs when handling $\int \frac{1}{x}\, dx = \ln x + C$ is the question of negative values of x.

For $x < 0$ stating $\int \frac{1}{x}\, dx = \ln x + C$ is not valid since the logarithmic function does not exist for negative values of x.

However we can write if $x < 0$

$$\int \frac{1}{x}\, dx = \int \frac{-1}{-x}\, dx = \ln (-x) + C.$$

so that for $\quad x > 0, \int \frac{1}{x}\, dx = \ln x + C$

for $\quad x < 0, \int \frac{1}{x}\, dx = \ln (-x) + C$

i.e. for all values of $\quad x \neq 0$

we can state that $$\int \frac{1}{x}\, dx = \ln |x| + C$$

6.3.8 Integration of a Fraction Where the Numerator is the Derivative of the Denominator

Integration of expressions which include fractions is a very difficult topic and we shall not attempt to cover it in this text. However, if we have to integrate an expression which can be identified as the form $\frac{f'(x)}{f(x)}$ the answer is readily available.

First consider $y = \ln (x^3 + 3x^2 - 4x + 2)$ and note that

$$\frac{dy}{dx} = \frac{1}{x^3 + 3x^2 - 4x + 2} \times (3x^2 + 6x - 4)$$

$$= \frac{3x^2 + 6x - 4}{x^3 + 3x^2 - 4x + 2}$$

Observe that, in the answer, the numerator is the derivative of the denominator and also that

$$\int \frac{(3x^2 + 6x - 4)\,dx}{(x^3 + 3x^2 - 4x + 2)}$$

$$= \ln (x^3 + 3x^2 - 4x + 2) + C$$

or strictly $\ln |x^3 + 3x^2 - 4x + 2| + C$

In general if $y = \ln | f(x)|$

then $\frac{dy}{dx} = \frac{1}{f(x)} \times f'(x) = \frac{f'(x)}{f(x)}$

and so $\int \frac{f'(x)}{f(x)}\,dx = \ln | f(x)| + C$

In such cases where we notice that, in the fraction we are required to integrate, the numerator is the exact derivative of the denominator, then the answer will be the logarithm, to e, of the modulus of the denominator.

Example 6.3.8 (1)

Find

(a) $\int \frac{\cos x}{2 + \sin x}\,dx$ (b) $\int \frac{\sin 2x}{4 + \sin^2 x}\,dx$ (c) $\int \frac{1}{3x - 2}\,dx$

(d) $\int \frac{x^2}{x^3 - 6}\,dx$ (e) $\int \frac{e^{4x}}{e^{4x} + 3}\,dx$ (f) $\int \frac{e^{2x} + 1}{e^{2x} - 1}\,dx$

(g) $\int \cot x\,dx$

Solution:

(a) $\int \frac{\cos x}{2 + \sin x}\,dx = \ln | 2 + \sin x | + C$ (the derivative of $2 + \sin x$ is $\cos x$)

(b) $\int \frac{\sin 2x}{4 + \sin^2 x}\,dx = \int \frac{2 \sin x \cos x}{4 + \sin^2 x}\,dx$

$= \ln (4 + \sin^2 x) + C$ (the derivative of $4 + \sin^2 x$ is $2 \sin x \cos x$, and $4 + \sin^2 x > 0$ for all x so $(4 + \sin^2 x) = |4 + \sin^2 x|$)

(c) $\int \frac{1}{3x - 2}\,dx$

Here the numerator is not the derivative of the denominator, but we can make 3 appear on the numerator if we multiply both numerator and denominator by 3.

thus $$\frac{1}{3}\int \frac{3}{3x-2}\,dx = \frac{1}{3}\ln|3x-2| + C$$

(d) $$\int \frac{x^2}{x^3-6}\,dx$$

In a similar manner to that of (c) we write the integral as

$$\frac{1}{3}\int \frac{3x^2}{x^3-6}\,dx = \frac{1}{3}\ln|x^3-6| + C$$

(e) $$\int \frac{e^{4x}}{e^{4x}+3}\,dx = \frac{1}{4}\int \frac{4e^{4x}}{e^{4x}+3}\,dx = \frac{1}{4}\ln(e^{4x}+3) + C$$

Note that $e^{4x} + 3 > 0$ for all x

(f) $$\int \frac{e^{2x}+1}{e^{2x}-1}\,dx$$

This expression can be re-written as

$$\int \frac{e^{x}+e^{-x}}{e^{x}-e^{-x}}\,dx$$

and now the numerator is the derivative of the denominator so

$$\int \frac{e^{x}+e^{-x}}{e^{x}-e^{-x}}\,dx = \ln|e^{x}-e^{-x}| + C$$

(g) $\int \cot x\,dx$ can be written $\int \frac{\cos x}{\sin x}\,dx = \ln|\sin x| + C$

N.B: Any factor which is included in both numerator and denominator in order to make the integral of the form $\int \frac{f'(x)}{f(x)}\,dx$ should be a constant. For example $\int \frac{dx}{x^2+4}$ cannot be found by multiplying numerator and denominator by a factor $2x$ and then writing the answer as

$$\frac{1}{2x}\int \frac{2x\,dx}{x^2+4} = \frac{\ln(x^2+4)}{2x} + C$$

The reader who thinks this answer is correct should differentiate it in order to convince himself/herself that the answer is wrong.

6.4 INTEGRATION APPLIED TO AREA

Integration has many applications in Biology, Chemistry, Engineering and Physics. One very important applications is that in many cases it can be used as a tool to calculate the exact area enclosed by curves.

6.4.1 The Area Under a Curve

Consider the curve whose equation is given by $y = f(x)$, as drawn in Fig. 6.4.1 (a), and the problem of calculating the area enclosed by the curve, the x-axis, and the lines $x = a$, $x = b$.

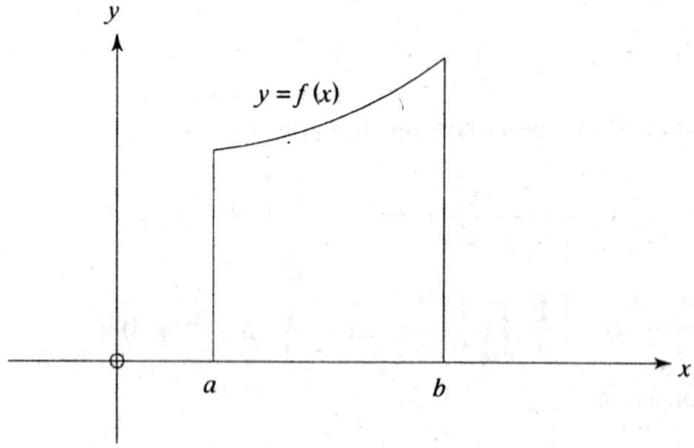

Fig. 6.4.1(a)

Now consider two neighbouring points P, Q, on the curve, where $y = f(x)$ is increasing (a similar argument can be followed if $y = f(x)$ is decreasing).

Let P be the point with coordinates $(x, f(x))$ and Q the point with coordinates $(x + \delta x, f(x + \delta x))$, as in Fig. 6.4.1(b).

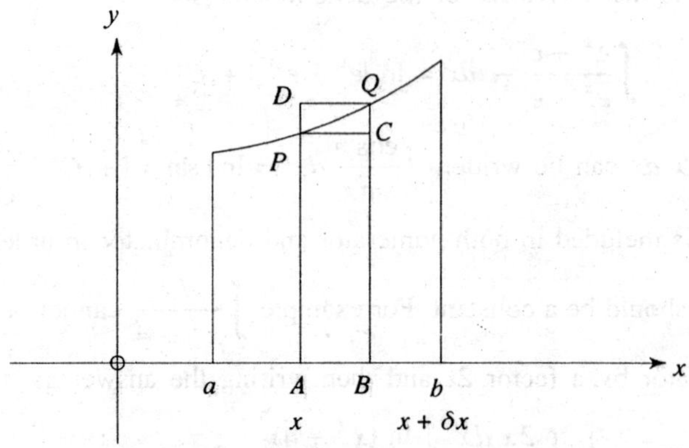

Fig. 6.4.1(b)

Part of the area we wish to calculate is enclosed by the part of the curve from P to Q, the line segments AP, BQ, and the x-axis. Let the area of $PA\ BQ$ be δA.

Now in terms of size, area of rectangle $ABCP$ < area of $PABQ$ < area of rectangle $ABQD$

or
$$f(x)\ \delta x < \delta A < f(x + \delta x) \cdot \delta x$$

since AP will be of length $f(x)$, the y-coordinate at P, and BQ will be of length $f(x + \delta x)$, the y-coordinate at Q.

Taking $\delta x > 0$ we have $f(x) < \dfrac{\delta A}{\delta x} < f(x + \delta x)$ and taking the limit as $\delta x \to 0$

$$f(x) \le \lim_{\delta x \to 0} \frac{\delta A}{\delta x} \le \lim_{\delta x \to 0} f(x + \delta x)$$

i.e.
$$f(x) \le \frac{dA}{dx} \le f(x).$$

So we can now define an area function A whose derivative is $f(x)$.

i.e.
$$\frac{dA}{dx} = f(x)$$

so
$$A = \int f(x)\, dx = F(x) + C$$

where
$$F'(x) = f(x).$$

However if we are required to calculate the area under the curve from $x = a$ to $x = b$ then clearly the value of the area A at $x = a$ will be zero since there will be no area at that point,

so if
$$A = F(x) + C$$

and
$$A = 0 \text{ when } x = a$$

$$0 = F(a) + C$$

i.e.
$$C = -F(a)$$

i.e.
$$A = F(x) - F(a)$$

Now the required area is that enclosed by the curve, the x-axis, and the lines $x = a$, $x = b$ and the value of the area A at $x = b$ will be given by $F(b) - F(a)$.

Example 6.4.1 (1)

Find the area enclosed by the curve $y = x^5$, the x-axis and the lines $x = 1$, $x = 3$.

Solution:

If the area function is A

$$\frac{dA}{dx} = x^5$$

so
$$A = \frac{1}{6}x^6 + C$$

At
$$x = 1,\ A = 0$$

so
$$0 = \frac{1}{6} + C$$

i.e.
$$C = -\frac{1}{6}$$

so
$$A = \frac{1}{6}x^6 - \frac{1}{6}$$

At
$$x = 3,\ A = \frac{1}{6} \times 3^6 - \frac{1}{6} = \frac{728}{6} = 121\frac{1}{3}$$

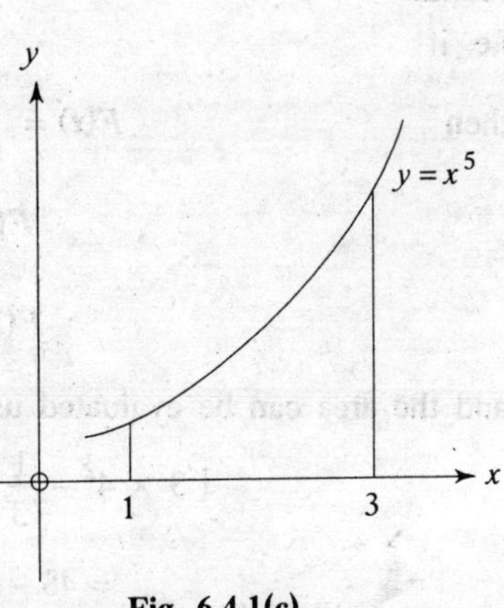

Fig. 6.4.1(c)

Example 6.4.1 (2)

Find the area enclosed by the curve $y = 6x - x^2$, the x-axis and the lines $x = 2$, $x = 4$.

Solution:

A quick check on the equation shows that the graph passes through (0, 0) and (6, 0) and also the graph, being a quadratic with a negative coefficient of x^2, will be a mountain peak shape. Proceeding as in the last example, if the area function is A,

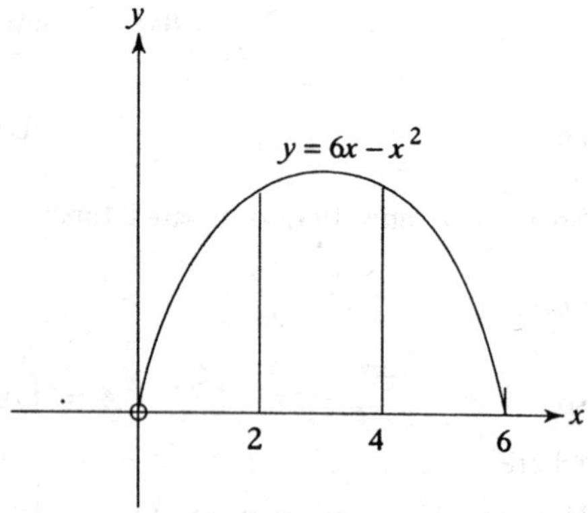

Fig. 6.4.1(d)

$$\frac{dA}{dx} = 6x - x^2$$

so $$A = 3x^2 - \frac{1}{3}x^3 + C$$

At $$x = 2,\ A = 0$$

so $$0 = 3 \times 4 - \frac{1}{3} \times 8 + C$$

i.e. $$C = \frac{-28}{3}$$

So $$A = 3x^2 - \frac{1}{3}x^3 - \frac{28}{3}$$

We require the area from $x = 2$ to $x = 4$

When $x = 4$,

$$A = 3 \times 4^2 - \frac{1}{3} \times 4^3 - \frac{28}{3} = 48 - \frac{64}{3} - \frac{28}{3}$$

Required area $$= 17\frac{1}{3}$$

The reader may have observed that in order to calculate the area it is possible to find the function $F(x)$ whose derivative is $f(x)$, then evaluate $F(x)$ at $x = 4$, and at $x = 2$, then subtract the two values found.

i.e. if $$f(x) = 6x - x^2$$

then $$F(x) = \int (6x - x^2)\,dx = 3x^2 - \frac{1}{3}x^3 + C$$

$$F(4) = 3 \times 4^2 - \frac{1}{3} \times 4^3 + C$$

$$F(2) = 3 \times 2^2 = -\frac{1}{3} \times 2^3 + C$$

and the area can be evaluated using

$$[\,3 \times 4^2 - \frac{1}{3} \times 4^3 + C] - [\,3 \times 2^2 - \frac{1}{3} \times 2^3 + C\,]$$

$$= 48 - \frac{64}{3} + C - 12 + \frac{8}{3} - C = 17\frac{1}{3}$$

We can shorten the argument by writing the required area as $\int_2^4 (6x - x^2)\,dx$ where the limits $x = 2$, $x = 4$ for the area are indicated at the lower and upper ends of the integral sign. $\int$.

On finding the function $F(x)$ of which $6x - x^2$ is the derivative,

i.e. $$3x^2 - \frac{1}{3}x^3 + C$$

we can write the instruction to evaluate $F(x)$ at $x = 4$, to evaluate $F(x)$ at $x = 2$, then subtract using the notation.

$$[\, 3x^2 - \frac{1}{3}x^3 + C\,]_2^4$$

The integral $3x^2 - \frac{1}{3}x^3 + C$, involving the arbitrary constant of integration, is called an *indefinite integral*.

However, when limits are given, and the integral can be evaluated, as in this case $[\, 3x^2 - \frac{1}{3}x^3 + C\,]_2^4$ it is called a definite integral.

The reader will have noticed that, in evaluating a definite integral, the constant of integration is cancelled out, and so it is not necessary to write it in a square bracket when we are evaluating the definite integral.

Example 6.4.1 (3)

Find the area between the curve whose equation is $y = 2x + \sin x + e^{-x}$, the x-axis, and the lines $x = 0$, $x = \pi/2$.

Solution:

A glance at the equation and the limits for the values of x should convince the reader that the curve is above the x-axis between $x = 0$, $x = \pi/2$ (see Fig. 6.4.1 (e))

The required area

$$= \int_0^{\pi/2} (2x + \sin x + e^{-x})\,dx$$

$$= [\, x^2 - \cos x - e^{-x}\,]_0^{\pi/2}$$

$$= [\, \frac{\pi^2}{4} - \cos \pi/2 - e^{-\pi/2}\,] - [\, 0 - \cos 0 - e^{-0}\,]$$

$$= \frac{\pi^2}{4} - 0 - e^{-\pi/2} - 0 + 1 + 1 = 2 + \frac{\pi^2}{4} - e^{-\pi/2}$$

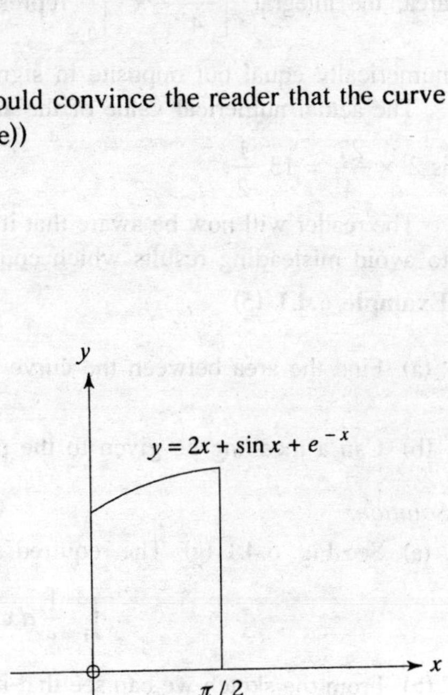

Fig. 6.4.1(e)

Example 6.4.1 (4)

Find the area between the curve whose equation is $y = x^2 (x - 3)$, the x-axis, and (a) the lines $x = 0$, $x = 3$ (b) the lines $x = 3$, $x = 4$.

Solution:

(a) By considering the sign of the term in x with the highest power (here x^3), and the points of inter-section with the x-axis, the sketch of the graph can be produced as in Fig. 6.4.1 (f).

The required area

$$= \int_0^3 (x^3 - 3x^2)\, dx = \left[\frac{x^4}{4} - x^3\right]_0^3$$

$$= \frac{3^4}{4} - 3^3 = \frac{-27}{4}$$

(b) The required area

$$= \int_3^4 (x^3 - 3x^2)\, dx = \left[\frac{x^4}{4} - x^3\right]_3^4$$

$$= \left[\frac{4^4}{4} - 4^3\right] - \left[\frac{3^4}{4} - 3^3\right]$$

$$= 0 - \left(\frac{-27}{4}\right) = \frac{27}{4}$$

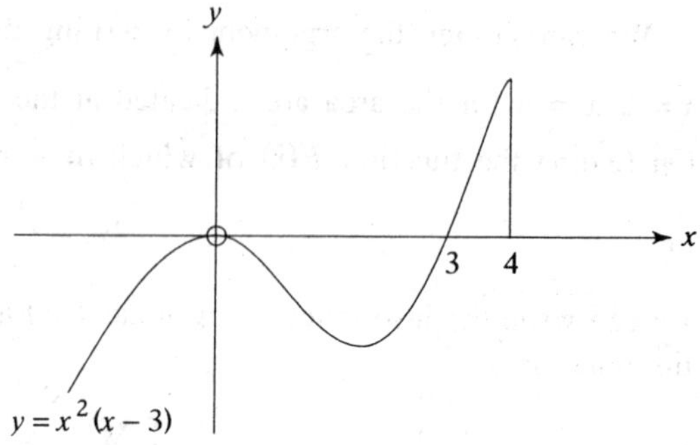

Fig. 6.4.1(f)

The answer for part (a) illustrates that the area under a curve is negative below the x-axis. It can be verified that $\left[\frac{x^4}{4} - x^3\right]_0^4$ is zero. Since we are now aware of a convention about the sign of an area, the integral $\left[\frac{x^4}{4} - x^3\right]_0^4$ represents the sum of two areas we have evaluated, which are numerically equal but opposite in sign.

The actual numerical value of the area formed between the curve, the x-axis, and the line $x = 4$, is $2 \times \frac{27}{4} = 13\ \frac{1}{2}$.

The reader will now be aware that it is essential to make a sketch of the curve under consideration to avoid misleading results which could come from a perfectly correct calculation.

Example 6.4.1 (5)

(a) Find the area between the curve $y = \frac{1}{x}$, the x-axis and the lines $x = 1$, $x = 3$

(b) Can a meaning be given to the phrase 'the area under $y = \frac{1}{x}$ from $x = -1$ to $x = 3$?

Solution:

(a) See Fig. 6.4.1 (g) The required area

$$= \int_1^3 \frac{1}{x}\, dx = [\ \ln x\ ]_1^3 = \ln 3 - \ln 1 = \ln 3$$

(b) From the sketch we can see that if we try to find the area under the curve from $x = -1$ to $x = 3$, between these limits is the value $x = 0$ for which y has no value (it can be seen that $y \to \infty$ as $x \to 0+$, and $y \to -\infty$ as $x \to 0-$)

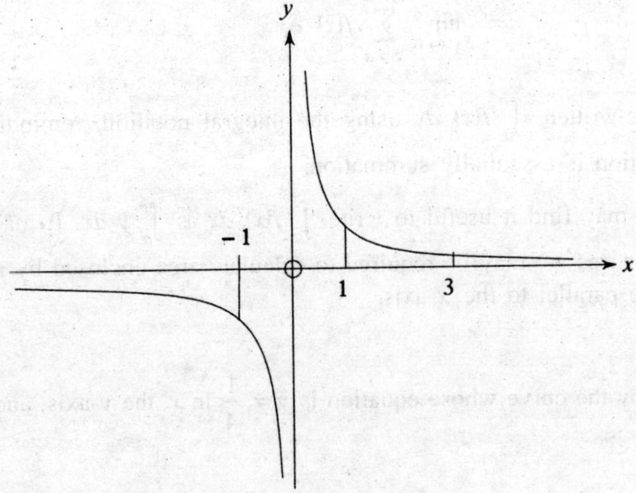

Fig. 6.4.1(g)

The curve comprised two separate branches. It is possible to evaluate

$$[\ln x]_{-1}^{3} = \ln^{3} - \ln^{|-1|} = \ln^{3}$$

but the result is meaningless.

If we break the area into two parts and integrate from −1 to 0, then from 0 to 3, in each case we have a meaningless term ln 0 (remember ln x is only defined for $x > 0$).

This second part of example 6.4.1 (5) illustrates that in order to calculate the area under a curve, the curve must have no breaks between the limits of x involved or in other words, the curve must be continuous throughout the domain of integration (i.e. for all values of x between the limits).

The reader will also have observed from Fig. 6.4.1 (b), reproduced below as Fig. 6.4.1 (h) that if we make the width $AB = \delta x$ sufficiently small, we can make the element of area δA as close as we like to $f(x)\ \delta x$. The required area enclosed by the curve $y = f(x)$, the lines $x = a$, $x = b$, and the x-axis could now be considered as a sum of such small areas δA, and taking the limit as $\delta x \to 0$.

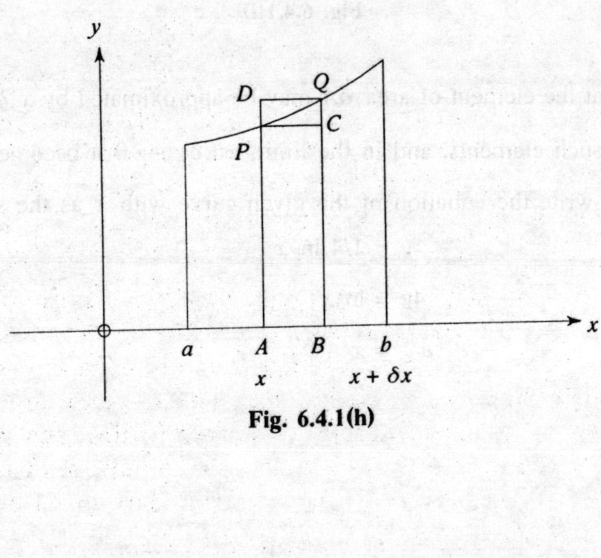

Fig. 6.4.1(h)

i.e. the required area $= \lim_{\delta x \to 0} \sum_{x=a}^{x=b} f(x)\ \delta x$

and this limit can be re-written $\int_a^b f(x)\ dx$ using the integral notation, reminding us that, in this context of area, integration is essentially summation.

In same contexts we may find it useful to write $\int_a^b f(x)\ dx$ as $\int_a^b y\ dx$. It may have occurred to the reader that in some cases we may be required to calculate area enclosed by a curve, the y-axis, and two lines which are parallel to the x-axis.

Example 6.4.1 (6)

Find the area enclosed by the curve whose equation is $y = \frac{1}{4} \ln x$, the y-axis, and the lines $y = \ln 2$, $y = \ln 4$.

Solution:

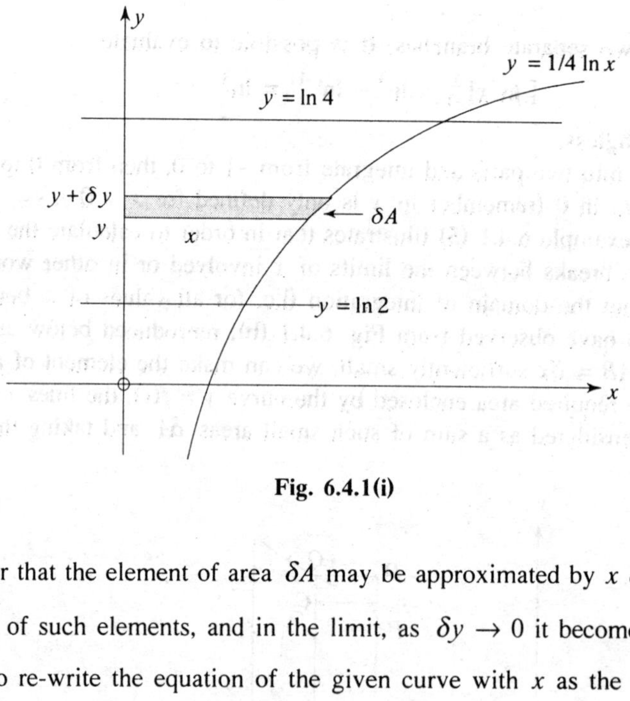

Fig. 6.4.1(i)

See Fig. 6.4.1 (i)

It should be clear that the element of area δA may be approximated by $x\ \delta y$. and so the required area will be a sum of such elements, and in the limit, as $\delta y \to 0$ it becomes $\int_{\ln 2}^{\ln 4} x\ dy$.

It is necessary to re-write the equation of the given curve with x as the subject so as

$$y = 1/4 \ln x$$

$$4y = \ln x$$

i.e. $$e^{4y} = x$$

Hence the required area is

$$\int_{\ln 2}^{\ln 4} e^{4y}\ dy = \left[\frac{1}{4}\ e^{4y}\right]_{\ln 2}^{\ln 4} = \frac{1}{4}e^{4\ln 4} - \frac{1}{4}\ e^{4\ln 2}$$

$$= \frac{1}{4}\ e^{\ln 4^4} - \frac{1}{4}\ e^{\ln 2^4} = \frac{1}{4} \times 4^4 - \frac{1}{4} \times 2^4 = 60$$

(the reader who is unable to spot the change from $e^{\ln 4^4} = 4^4$ should write $u = e^{\ln 4^4}$ and then $\ln u = \ln 4^4$ hence $u = 4^4$)

The use of elements of area is also helpful in cases where we may be required to calculate the area between two curves, or between a line and a curve.

Example 6.4.1 (7)

Find the finite area enclosed by the curves whose equations are $y = 16 - x^2$ and $y = x^2 - 4x$

Solution:

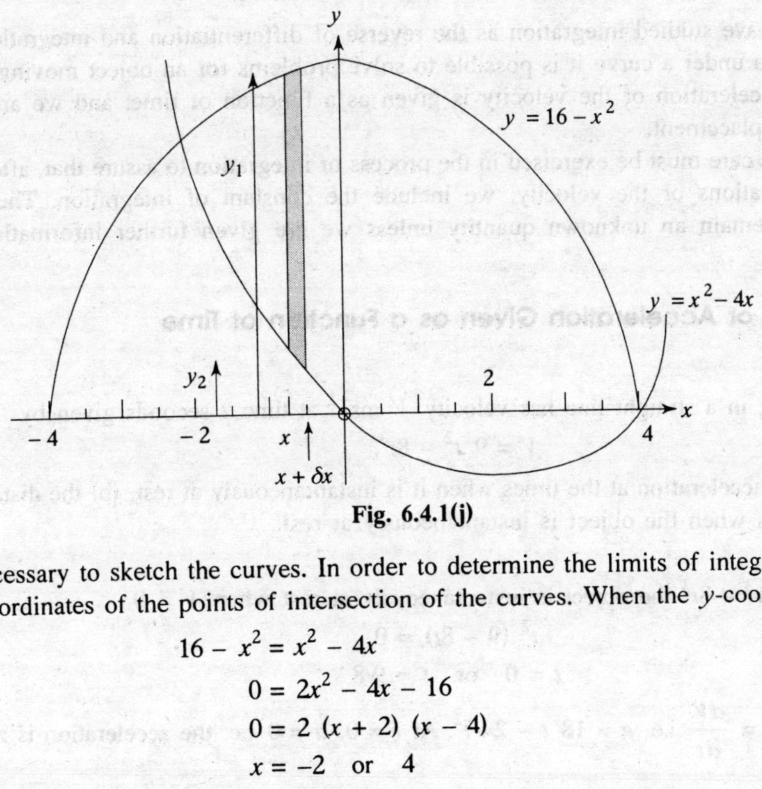

Fig. 6.4.1(j)

First it is necessary to sketch the curves. In order to determine the limits of integration we have to find the x-coordinates of the points of intersection of the curves. When the y-coordinates are equal

then $\quad 16 - x^2 = x^2 - 4x$

i.e. $\quad 0 = 2x^2 - 4x - 16$

i.e. $\quad 0 = 2\ (x + 2)\ (x - 4)$

so $\quad x = -2$ or 4

The sketch is completed in Fig. 6.4.1 (j)

The element of area δA is shaded in the Figure, and

$$\delta A = (y_1 - y_2)\ \delta x$$

where $\quad y_1 = 16 - x^2,\ y_2 = x^2 - 4x$

the required area

$$= \int_{-2}^{4} (y_1 - y_2)\,dx = \int_{-2}^{4} \left[(16 - x^2) - (x^2 - 4x)\right] dx$$

$$= \int_{-2}^{4} \left[16 + 4x - 2x^2\right] dx \;= \left[16x + 2x^2 - \frac{2}{3}x^3\right]_{-2}^{4}$$

$$= \left[16 \times 4 + 2 \times 4^2 - \frac{2}{3} \times 4^3\right] - \left[16 \times -2 + 2 \times (-2)^2 - \frac{2}{3} \times (-2)^3\right] = 72$$

6.5 INTEGRATION APPLIED TO DISPLACEMENT, VELOCITY AND ACCELERATION

In sub-sections 6.2.1 and 6.2.2 we showed how, for an object moving in a straight line with displacement x after time t, the velocity V and acceleration a could be found by using the formulae.

$$V = \frac{dx}{dt},\; a = \frac{dV}{dt} \quad \text{or} \quad V\frac{dV}{dx} \quad \text{or} \quad \frac{d}{dx}\left(\frac{1}{2}V^2\right)$$

Now that we have studied integration as the reverse of differentiation and integration applied to calculation of area under a curve it is possible to solve problems for an object moving in a straight line where the acceleration or the velocity is given as a function of time, and we are required to determine the displacement.

However, great care must be exercised in the process of integration to assure that, after integrating either the accelerations or the velocity, we include the constant of integration. The constant of integration will remain an unknown quantity unless we are given further information about the motion.

6.5.1 Velocity or Acceleration Given as a Function of Time

Example 6.5.1(1)

An object moving in a straight line has velocity V ms^{-1} at time t seconds given by

$$V = 9\,t^2 - 8t^3.$$

Determine (a) its acceleration at the times when it is instantaneously at rest, (b) the distance covered between the times when the object is instantaneously at rest.

Solution:

(a) From $V = 9t^2 - 8t^3$ the object is instantaneously at rest when $V = 0$

i.e. $$t^2\,(9 - 8t) = 0$$

i.e. $$t = 0 \quad \text{or} \quad t = 9/8$$

the acceleration $a = \dfrac{dV}{dt}$ i.e. $a = 18\,t - 24\,t^2$. At $t = 0$, $a = 0$ i.e. the acceleration is zero at $t = 0$ seconds.

at $$t = \frac{9}{8},\; a = 18 \times \frac{9}{8} - 24\left(\frac{9}{8}\right)^2 = -\frac{81}{8}$$

i.e. acceleration is $-\dfrac{81}{8}$ at $t = \dfrac{9}{8}$ seconds or there is a retardation of $\dfrac{81}{8}$ ms^{-2} at $t = \dfrac{9}{8}$ seconds.

(b) From Fig. 6.5.1 (a) we can see that the required distance will be given by the area between the graph of $V = 9t^2 - 8t^3$ and the t-axis, between the limits $t = 0$, $t = \dfrac{9}{8}$ and this is

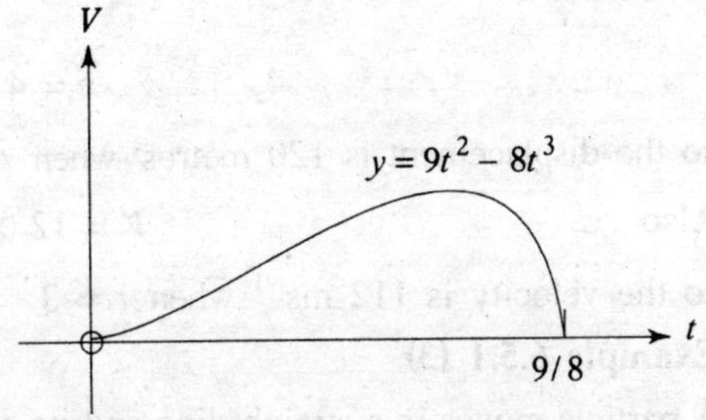

Fig. 6.5.1(a)

$$\int_0^{9/8} (9t^2 - 8t^3)dt = \left[(3t^3 - 2t^4\right]_0^{9/8}$$

$$= 3 \times \left(\frac{9}{8}\right)^3 - 2\left(\frac{9}{8}\right)^4 = \frac{2187}{2048} \text{ m}$$

i.e. the distance covered is $\dfrac{2187}{2048}$ m or 1.07 m (to 3 significant figures).

Example 6.5.1 (2)

A particle is moving in a straight line and its acceleration after t seconds is given by 24 t ms^{-2}.

(a) Find an expression for the displacement x metres after t seconds, in terms of t.

(b) Given that at $t = 0$, $x = 0$ and that at $t = 2$, $x = 40$, find the displacement and the velocity when $t = 3$.

Solution:

(a) if the velocity after t seconds is V ms^{-1}

then
$$\frac{dV}{dt} = 24\ t$$

and on integrating
$$V = \int 24\,t\,dt = 12t^2 + C_1$$

from
$$\frac{dx}{dt} = 12\ t^2 + C_1$$

$$x = \int (12\ t^2 + C_1)\ dt = 4\ t^3 + C_1\ t + C_2$$

(b) At present we have two unknown constants included in the expressions for V and x.

Now at $\quad t = 0,\ x = 0$

so $\quad 0 = 4\ x\ 0^3 + C_1 \times 0 + C_2$

i.e. $\quad C_2 = 0$

At $\quad t = 2,\ x = 40$

so $\quad 40 = 4 \times 2^3 + 2C_1$

i.e. $\quad C_1 = 4$

so after t seconds the displacement and velocity are given by

$$x = 4t^3 + 4t$$

$$V = 12\ t^2 + 4 \qquad \text{respectively.}$$

When $t = 3$

$$x = 4 \times 3^3 + 4 \times 3 = 120$$

so the displacement is 120 metres when $t = 3$

Also $$V = 12 \times 3^2 + 4 = 112$$

so the velocity is 112 ms^{-1} when $t = 3$

Example 6.5.1 (3)

A particle moves in a straight line and its acceleration after t seconds is $\cos^2 t$. Determine expressions for its velocity V ms^{-1} and its displacement x metres after t seconds, given that at $t = 0$, the velocity is 2 ms^{-1} and the displacement is – 4 metres. Determine also the velocity and the displacement after $\pi/3$ seconds.

Solution:

The velocity V can be determined from $\frac{dV}{dt} = \cos^2 t$

$$\text{so } V = \int \cos^2 t \, dt$$

So far we have only learnt how to integrate sines and cosines of multiple angles and *not* products of these functions so we have to write

$$\cos^2 t = \frac{1}{2}[1 + \cos 2t]$$

so $$V = \int \frac{1}{2}[1 + \cos 2t]dt$$

i.e. $$V = \frac{1}{2}t + \frac{1}{4}\sin 2t + C_1$$

At $$t = 0, \; V = 2,$$

so $$2 = \frac{1}{2} \times 0 + \frac{1}{4} \times 0 + C_1$$

i.e. $$C_1 = 2$$

so $$V = \frac{dx}{dt} = \frac{1}{2}t + \frac{1}{4}\sin 2\,t + 2$$

i.e. velocity after t seconds is $\left(\frac{1}{2}t + \frac{1}{4}\sin 2t + 2\right) ms^{-1}$

also $$x = \int\left(\frac{1}{2}t + \frac{1}{4}\sin 2t + 2\right)dt$$

$$= \frac{1}{4}t^2 - \frac{1}{8}\cos 2t + 2\,t + C_2$$

at $$t = 0, \; x = -4$$

so
$$-4 = \frac{1}{4} \times 0 - \frac{1}{8} \times 1 + 2 \times 0 + C_2$$

i.e.
$$C_2 = -\frac{31}{8}$$

so
$$x = \frac{1}{4} t^2 - \frac{1}{8} \cos 2t + 2t - \frac{31}{8}$$

and displacement after t seconds is

$$\left(\frac{1}{4} t^2 - \frac{1}{8} \cos 2t + 2t - \frac{31}{8}\right) \text{ metres.}$$

When $t = \pi/3$

$$x = \frac{1}{4} \times \frac{\pi^2}{9} - \frac{1}{8} \cos \frac{2\pi}{3} + \frac{2\pi}{3} - \frac{31}{8}$$

$$= \frac{\pi^2}{36} - \frac{1}{8} \times -\frac{1}{2} + \frac{2\pi}{3} - \frac{31}{8}$$

$$= \frac{\pi^2}{36} + \frac{1}{16} + \frac{2\pi}{3} - \frac{31}{8}$$

$$= \frac{4\pi^2 + 96\pi - 549}{144}$$

i.e. the displacement is

$$\frac{4\pi^2 + 96\pi - 549}{144} \text{ metres after } \pi/3 \text{ seconds.}$$

Also when $t = \pi/3$

$$V = \frac{1}{2} \times \frac{\pi}{3} + \frac{1}{4} \sin \frac{2\pi}{3} + 2$$

$$= \frac{\pi}{6} + \frac{1}{4} \times \frac{\sqrt{3}}{2} + 2$$

$$= \frac{4\pi + 3\sqrt{3} + 48}{24}$$

i.e. the velocity is $\dfrac{4\pi + 3\sqrt{3} + 48}{24}$ ms^{-1} after $\pi/3$ seconds.

6.5.2 Acceleration Given as a Function of Displacement

In problems where the acceleration is given as a function of displacement x, it is possible to integrate the equation $a = V \dfrac{dV}{dx} = \dfrac{d}{dx}\left(\dfrac{1}{2} V^2\right)$ with respect to x in order to obtain an expression for the velocity V as a function of the displacement x.

Example 6.5.2 (1)

A particle moves along the positive x-axis and its acceleration is 16 x ms^{-2} where x metres is its displacement from the origin 0. Its initial velocity is 3 ms^{-1} when $x = 0$, find a general formula for its velocity in terms of x and find also its velocity when it is 1 metre from the origin.

Solution:

Using $$\frac{d}{dx}\left(\frac{1}{2}V^2\right) = 16x$$

then $$\frac{1}{2}V^2 = \int 16x\,dx$$

i.e. $$\frac{1}{2}V^2 = 8x^2 + C$$

but $$V = 3 \text{ when } x = 0$$

so $$\frac{1}{2} \times 9 = C$$

i.e. $$\frac{1}{2}V^2 = 8x^2 + \frac{9}{2}$$

or $$V^2 = 16x^2 + 9$$

so $$V = \pm\sqrt{(16x^2 + 9)}$$

Since the initial velocity and the acceleration for $x > 0$ are both positive then the subsequent velocity is positive so we take the positive square root.

i.e. velocity V is $$\sqrt{(16x^2 + 9)}\ \text{ms}^{-1}$$

when the particle is 1m from the origin, $x = 1$ and the velocity is $\sqrt{(16 \times 1 + 9)} = 5$ ms^{-1}

Example 6.5.2 (2)

A particle moves along a straight line and when its displacement from a fixed point O is x metres its retardation is $(2 + x)$ ms^{-2}.

Given that its velocity is $2\sqrt{3}$ ms^{-1}, when $x = 0$, express its velocity V in term of x and find also the value of x when the particle first comes to rest.

Solution:

Since the retardation is $(2 + x)$

we write $$\frac{d}{dx}\left(\frac{1}{2}V^2\right) = -2 - x$$

so $$\frac{1}{2}V^2 = \int(-2 - x)\,dx$$

i.e. $$\frac{1}{2}V^2 = -2x - \frac{1}{2}x^2 + C_1$$

But when $x = 0,\ V = 2\sqrt{3}$

so $\frac{1}{2} \times 4 \times 3 = C_1$

or $C_1 = 6$

hence $\frac{1}{2} V^2 = -2x - \frac{1}{2} x^2 + 6$

or $V^2 = 12 - 4x - x^2$

hence $V = \pm \sqrt{(12 - 4x - x^2)}$

Due to the retardation the velocity will be negative after first coming to rest so in this case we have to keep both solutions for V. The particle comes to rest when

$$12 - 4x - x^2 = 0$$

i.e. $x^2 + 4x - 12 = 0$

i.e. $(x + 6)(x - 2) = 0$

i.e. $x = -6$ or 2

However the particle is initially moving in the positive direction so it will first come to rest when $x > 0$, i.e. when $x = 2$.

Example 6.5.2 (3)

A particle moves along a straight line and its retardation is $10\ e^{2x}$ ms^{-2} when its displacement is x metres from a fixed point O.

Given that its initial velocity (when $x = 0$) is 10 ms^{-1} find correct to three significant figures the value of x when the particle come to rest.

Solution:

Since the retardation is $10\ e^{2x}$ ms^{-2}, if the velocity is V ms^{-1}

then $\frac{d}{dx}\left(\frac{1}{2} V^2\right) = -10\ e^{2x}$

so $\frac{1}{2} V^2 = \int -10\ e^{2x}\, dx$

i.e. $\frac{1}{2} V^2 = -5e^{2x} + C$

When $x = 0,\ V = 10$

i.e. $\frac{1}{2} \times 100 = -5 + C$

i.e. $C = 55$

so $\frac{1}{2} V^2 = -5e^{2x} + 55$

or $V^2 = -10e^{2x} + 110$

The particle comes to rest when $V = 0$

i.e. $0 = -10e^{2x} + 110$

i.e. $10e^{2x} = 110$

i.e. $e^{2x} = 11$

i.e. $2x = \ln 11$

or $x = \frac{1}{2} \ln 11 = 1.20$ (to three significant figures.)

Example 6.5.2 (4)

A particle moves along a straight line and its retardation is 4 V ms^{-2} where it has velocity V ms^{-1} when its displacement is x metres from a fixed point O.

Given that its initial velocity (when it is at the origin) is 80 ms^{-1} find the value of x when the particle comes to rest.

Solution:

It is easier in this case to used the formula $V \frac{dV}{dx}$ for acceleration.

So $V \frac{dV}{dx} = -4V$ (as we are given retardation)

i.e. $\frac{dV}{dx} = -4$

so $V = \int -4\, dx = -4x + C$

At $x = 0,\ V = 80$

So $C = 80$

hence $V = 80 - 4x$

the particle comes to rest when $V = 0$.

i.e. $80 - 4x = 0$

or $x = 20$

6.6 INTEGRATION BY SUBSTITUTION

In sub-section 6.3.5 we have discussed how to find $\int (ax + b)^n\, dx$ and found that this integral can be easily determined so long as the function in the bracket was of the form $ax + b$, where a and b were constants, i.e. a linear function of x.

We shall now consider some harder problems where it is necessary to apply reverse chain rule in order to perform an integration for example if $y = e^u$ where u is a function of x, we know by the chain rule that $\frac{dy}{dx} = e^u \frac{du}{dx}$

So that is may be possible to integrate a function of the form $e^u \dfrac{du}{dx}$ by inspection.

6.6.1 Reverse Chain Rule by Inspection

In this part the reader is encouraged to try to spot an integration, where the given function is a derivative of another function as a result of applying the Chain Rule.

Example 6.6.1 (1)

Determine (a) $\int 4x^3 e^{x^4} dx$

(b) $\int \sin x \, e^{\cos x} dx$ (c) $\int x^5 e^{x^6} dx$

Solution:

(a) If $y = e^{x^4}$ i.e. $y = e^u$

where $u = x^4$

$$\frac{dy}{dx} = 4x^3 e^{x^4}$$

so $$\int 4x^3 e^{x^4} dx = e^{x^4} + C$$

(b) On a similar argument

$$\int \sin x \, e^{\cos x} dx = -e^{\cos x} + C \quad \text{(start with } y = e^u \text{ where } u = \cos x \text{ and find } \frac{dy}{dx}\text{)}$$

(c) Starting with $y = e^{x^6}$

$$\frac{dy}{dx} = 6x^5 e^{x^6}$$

so if $y = \dfrac{1}{6} e^{x^6}$

$$\frac{dy}{dx} = \frac{1}{6} \cdot 6x^5 e^{x^6} = x^5 e^{x^6}$$

so $$\int x^5 e^{x^6} dx = \frac{1}{6} \int 6x^5 e^{x^6} dx$$

$$= \frac{1}{6} e^{x^6} + C$$

Example 6.6.1 (2)

Determine (a) $\int x (3x^2 + 5)^4 dx$ (b) $\int \sin^2 5x \cos 5x \, dx$

Solution:

(a) if
$$y = (3x^2 + 5)^5$$
$$\frac{dy}{dx} = 5\ (3x^2 + 5)^4\ 6x = 30x\ (3x^2 + 5)^4$$

This is 30 times the function we are required to integrate.

If
$$y = \frac{1}{30}\ (3x^2 + 5)^5$$
$$\frac{dy}{dx} = \frac{1}{30} \cdot 5\ (3x^2 + 5)^4 \cdot 6x = x\ (3x^2 + 5)^4$$

So
$$\int x\ (3x^2 + 5)^4\ dx = \frac{1}{30}\ (3x^2 + 5)^5 + C$$

or
$$\frac{1}{5.6}\ (3x^2 + 5)^5 + C$$

with the factor 5 coming on the denominator as a result of raising $(3x^2 + 5)$ to the power 5 and the factor 6 being the coefficient of x when we differentiate $3x^2$ with respect to x.

(b) In $\int \sin^2 5x \cos 5x\ dx$ the answer appears to involve $\sin^3 5x$

since if
$$y = \sin^3 5x$$
$$\frac{dy}{dx} = (3 \sin^2 5x)\ (5 \cos 5x)$$

thus suggesting
$$\int \sin^2 5x \cos 5x\ dx = \frac{\sin^3 5x}{3 \times 5} + C$$

or
$$\frac{1}{15} \sin^3 5x + C$$

The checking of the answer, by differentiation, is left as an exercise to the reader.

6.6.2 Integration by Substitution

Suppose we have a function of x to be integrated in which we wish to make a substitution to a new variable u where u itself is a known function of x.
Now if $y = f(u)$ (as we had in the case of the Chain Rule)

We know that
$$\frac{dy}{dx} = f'(u)\ \frac{du}{dx}$$

where $f'(u)$ denotes $\frac{d}{du}\ [f(u)]$

So we see that any product of the form $f'(u)\ \frac{du}{dx}$ can be integrated using $$\int \left(f'(u) . \frac{du}{dx} \right) dx = f(u) + C \qquad \text{(i)}$$

but also $$f'(u) = \frac{d}{du} (f(u))$$

so $$\int f'(u)\, du = f(u) + C \qquad \text{(ii)}$$

If we compare (i) and (ii) it can be seen that

$$\int \left(f'(u) . \frac{du}{dx} \right) dx = \int f'(u)\, du$$

or $$\int \dots \frac{du}{dx} \cdot dx = \int \dots du$$

So that $\frac{du}{dx} \cdot dx$ and du are equivalent operations in the sense that they will give the same results when applied to the same function.

With practice the reader will learn how to obtain the equivalent operation.

Example 6.6.1 (2)

Determine (a) $\int x\,(3x^2 + 5)^4\, dx$ (b) $\int \sin^2 5x \cos 5x\, dx$

Solution:

(a) Substitute $$u = 3x^2 + 5$$

so $$\frac{du}{dx} = 6x, \quad \dots\; \frac{du}{dx} \cdot dx = \dots du$$

i.e. $$\dots\; 6x \cdot dx = \dots\; du$$

The pair of equivalent operators can be obtained directly from $\frac{du}{dx} = 6x$

by separating the variables i.e. treating $\frac{du}{dx}$ as if it were a fraction giving $\dots du = \dots\; 6x\, dx$

N.B: This is not an equation or identity, it is an equivalence of operator. Some functions can be integrated in the way by a suitable substitution (i.e. change of variable). A function, expressed as a product in which one factor can be identified as the derivative of the function (normally inside a bracket) in the other factor, can normally be integrated using this method.

From $$\dots\; du = \dots\; 6x\, dx$$

then $$\int x\,(3x^2 + 5)^4\, dx = \frac{1}{6} \int 6x\,(3x^2 + 5)^4\, dx$$

$$= \frac{1}{6} \int (3x^2 + 5)^4\, 6x\, dx = \frac{1}{6} \int u^4\, du \qquad \text{(since we substitute } u = 3x^2 + 5)$$

$$= \frac{1}{6} \cdot \frac{u^5}{5} + C$$

so $$\int x\,(3x^2 + 5)^4\,dx = \frac{1}{30}\,(3x^2 + 4)^5 + C$$

as in example 6.6.1 (2) (a)

(b) Writing $\int \sin^2 5x \cos 5x\,dx$ in the form $\int (\sin 5x)^2 \cos 5x\,dx$ a suitable substitution could be $u = \sin 5x$

so $$\text{..}\ du = \text{...}\ 5 \cos 5x\,dx$$

so $$\int \sin^2 5x \cos 5x\,dx$$

$$= \frac{1}{5} \int \sin^2 5x\,(5 \cos 5x) dx = \int \frac{1}{5}\,u^2\,du = \frac{1}{5} \cdot \frac{u^3}{3} + C$$

$$= \frac{1}{15}\,\sin^3 5x + C \quad \text{as in example 6.6.1 (b)}$$

The reader may wish to ask how we can know the substitution that should be used. There is no unique answer. The knowledge of the best substitution can only come from practice and experience. Changing the variable from x to u must include both the correct substitution and the correct identification of the differential operators (... dx = ... du).

It is like boarding a bus at a bus station where several companies are operating the route on which you wish to travel. First you have to travel and secondly you must ensure that you have the right ticket for that bus.

Example 6.6.2 (2)

Determine

(a) $\int x^2\,(2x^3 + 5)^{3/2}\,dx$

(b) $\int \frac{(\ln x)^4}{x}\,dx$

(c) $\int \sec^2 3x\, e^{2 \tan 3x}\,dx.$

Solution:

(a) For $\int x^2\,(2x^3 + 5)^{3/2}\,dx$ it would seem a substitution

$u = 2\,x^3 + 5$ might work.

$$u = 2x^3 + 5, \text{ .... } du = \text{ ... } 6x^2\,dx$$

so $$\int x^2\,(2x^3 + 5)^{3/2}\,dx = \frac{1}{6} \int (2x^3 + 5)^{3/2}.\ 6x^2\,dx$$

$$= \frac{1}{6} \int u^{3/2}\,du = \frac{1}{6} \cdot \frac{u^{5/2}}{5/2} + C$$

$$= \frac{1}{6} \cdot \frac{(2x^3 + 5)^{5/2}}{(5/2)} + C$$

$$= \frac{1}{15} (2x^3 + 5)^{5/2} + C$$

(b) for $\int \frac{(\ln x)^4}{x} dx$

since $$\frac{d}{dx} (\ln x) = \frac{1}{x}$$

a substitution $u = \ln x$ might work.

$$u = \ln x, \quad ....du = ... \frac{1}{x} dx$$

so $$\int \frac{(\ln x)^4}{x} dx = \int (\ln x)^4 \cdot \frac{1}{x} dx$$

$$= \int u^4 \, du = \frac{1}{5} u^5 + C = \frac{1}{5} (\ln x)^5 + C$$

N.B: $(\ln x)^5$ is not the same as $\ln x^5$.

(c) for $\int \sec^2 3x \, e^{2 \tan 3x} dx$ since $\frac{d}{dx} (\tan 3x) = 3 \sec^2 3x$ a substitution $u = \tan 3x$ might work.

$$u = \tan 3x, \; ... \; du = ... 3 \sec^2 3x \, dx$$

so $$\int \sec^2 3x \, e^{2 \tan 3x} dx$$

$$= \frac{1}{3} \int e^{2 \tan 3x} \cdot 3 \sec^2 3x \, dx = \frac{1}{3} \int e^{2u} \cdot du$$

$$= \frac{1}{3} \cdot \frac{1}{2} e^{2u} + C = \frac{1}{6} e^{2 \tan 3x} + C$$

N.B: In many cases of substitution a correct matching of factors is necessary for success. For example:

$\int x \sqrt{(x^2 + 1)} \, dx$ can now be solved easily using the substitution $u = x^2 + 1$.

but $\int \sqrt{(x^2 + 1)} \, dx$ cannot be solved by such a substitution.

6.6.3 Substitution for Integrals Which Contain $\sqrt{(a^2 - x^2)}$

Many integrals contain a term in $\sqrt{(a^2 - x^2)}$ on powers of $\sqrt{(a^2 - x^2)}$ where a is a constant. In such cases it is well worth trying a substitution of the form $x = a \sin \theta$, as in the example shown below.

Example 6.6.3 (1)

Determine $$\int \sqrt{(9 - x^2)}\ dx$$

Solution:

If we try $x = 3 \sin \theta$ ... $dx = ... 3 \cos \theta\, d\theta$

so
$$\int \sqrt{(9 - x^2)}\ dx = \int \sqrt{(9 - 9 \sin^2 \theta)}\ 3 \cos \theta\, d\theta$$
$$= \int \sqrt{(9 \cos^2 \theta)}\ 3 \cos \theta\, d\theta$$
$$= \int 3 \cos \theta \cdot 3 \cos \theta\, d\theta$$
$$= \int 9 \cos^2 \theta\, d\theta$$
$$= \frac{9}{2} \int (1 + \cos 2\theta)\ d\theta \quad \text{(as in example 6.5.1 (3))}$$
$$= \frac{9}{2} [\theta + \frac{1}{2} \sin 2\theta] + C$$

Here we have a difficulty in re-writing the answer in term of the given variables x. However noting that $x = 3 \sin \theta$

then $$\sin \theta = \frac{x}{3}$$

so that $$\theta = \sin^{-1} \frac{x}{3}$$

and $$\cos \theta = \sqrt{\left(1 - \frac{x^2}{9}\right)}$$

as $$\cos^2 \theta = 1 - \sin^2 \theta$$

In the answer we may express $\sin 2\theta$ as $2 \sin \theta \cos \theta$.

Hence we can write $\int \sqrt{(9 - x^2)}\ dx$
$$= \frac{9}{2} [\sin^{-1} \frac{x}{3} + \frac{1}{2} \cdot 2 \sin \theta \cos \theta] + C$$
$$= \frac{9}{2} \left[\sin^{-1} \frac{x}{3} + \frac{x}{3} \sqrt{\left(1 - \frac{x^2}{9}\right)}\right] + C$$
$$= \frac{9}{2} \sin^{-1} \frac{x}{3} + \frac{1}{2} x \sqrt{(9 - x^2)} + C \quad \text{(the final step is left as an exercise to the reader).}$$

6.6.4 Substitution in Definite Integrals: Change of Limits

We have already compared the method of substitution with that of boarding a bus at a bus station where several companies are operating the route on which you wish to travel.

You have to board the bus on which you wish to travel (i.e. correct substitution) AND have the right ticket (correct identification of the differential operators).

And on a long journey have your luggage stored in the bus on which you wish to travel (correct limits for the new variable).

The last step should be clear if you follow the next few examples.

Example 6.6.4 (1)

Evaluate $\int_0^2 x^2 \sqrt{(x^3+8)}\ dx$

Solution:

Let $u = x^3 + 8$

then $\ldots du = \ldots 3x^2\ dx$

also if $x = 0$ (the lower limit)

$u = 8$

if $x = 2$ (the upper limit)

$u = 2^3 + 8 = 16$

Hence $\int_0^2 x^2 \sqrt{(x^3+8)}\ dx$

$$= \frac{1}{3}\int_0^2 \sqrt{(x^3+8)} \cdot 3x^2\ dx = \frac{1}{3}\int_8^{16} \sqrt{u}.du$$

$$= \frac{1}{3}\left[\frac{u^{3/2}}{3/2}\right]_8^{16} = \frac{1}{3}\left[\frac{2}{3}u^{3/2}\right]_8^{16}$$

$$= \frac{1}{3}\left[\frac{2}{3}\times 16^{3/2} - 2/3\times 8^{3/2}\right]$$

$$= \frac{1}{3}\left[\frac{2}{3}\times 64 - \frac{2}{3}\times 16\sqrt{2}\right]$$

So $\int_0^2 x^2 \sqrt{(x^3+8)}\ dx = \frac{32}{9}\left[4-\sqrt{2}\right]$

Example 6.6.4(2)

Evaluate $\int_0^2 \frac{dx}{(16-x^2)^{3/2}}$

Solution:

The denominator is the third power of $\sqrt{(16-x^2)}$ so we try the substitution $x = 4\sin\theta$

then $\ldots dx = \ldots 4\cos\theta\, d\theta$

when $x = 0$, then $4\sin\theta = 0$

i.e. $\theta = 0$

when $x = 2$ then $4 \sin \theta = 2$

i.e. $\sin \theta = \frac{2}{4} = \frac{1}{2}$

so $\theta = \pi/6$

Hence $\int_0^2 \frac{1}{(16 - x^2)^{3/2}} . dx$

$$= \int_0^{\pi/6} \frac{1}{(16 - 16\sin^2 \theta)^{3/2}} \cdot 4 \cos \theta \, d\theta$$

$$= \int_0^{\pi/6} \frac{1}{(16\cos^2 \theta)^{3/2}} \cdot 4 \cos \theta \, d\theta$$

$$= \int_0^{\pi/6} \frac{1}{64\cos^3 \theta} \cdot 4 \cos \theta \, d\theta$$

$$= \int_0^{\pi/6} \frac{1}{16} \sec^2 \theta \, d\theta$$

$$= \frac{1}{16} [\tan \theta]_0^{\pi/6}$$

$$= \frac{1}{16} \left[\frac{1}{\sqrt{3}} - 0\right] = \frac{\sqrt{3}}{48}$$

Example 6.6.4 (3)

Evaluate $\int_{-1/2}^{2} x (2x + 5)^{1/2} \, dx$

Solution:

We shall solve this problem by substituting $u = 2x + 5$ but the reader may try solving the same problem using the substitution $t^2 = 2x + 5$

if $u = 2x + 5$ ... du = ... $2 \, dx$

If $x = -\frac{1}{2}$, $u = 4$

If $x = 2$, $u = 9$

Hence $\int_{-1/2}^{2} x (2x + 5)^{1/2} \, dx$

$$= \frac{1}{2} \int_{-1/2}^{2} x (2x + 5)^{1/2} \, 2 \, dx$$

$$= \frac{1}{2}\int_4^9 \frac{1}{2}(u-5)\,u^{1/2}\,du$$ (remember that the original function must be expressed in terms of u)

$$= \frac{1}{4}\int_4^9 \left(u^{3/2} - 5u^{1/2}\right) du = \frac{1}{4}\left[\frac{u^{5/2}}{5/2} - \frac{5u^{3/2}}{3/2}\right]_4^9$$

$$= \frac{1}{4}\left[\frac{2}{5}9^{5/2} - \frac{10}{3}.9^{3/2}\right] - \frac{1}{4}\left[\frac{2}{5}.4^{5/2} - \frac{10}{3}.4^{3/2}\right]$$

$$= \frac{1}{4}\left[\frac{2}{5}\times 243 - \frac{10}{3}\times 27\right] - \frac{1}{4}\left[\frac{2}{5}\times 32 - \frac{10}{3}\times 8\right] = \frac{79}{15}$$

(the reader can check the final steps as an exercise)

6.6.5 Integration of Powers of Trigonometric Functions

If we have to integrate an even power of sin x or cos x the double angle formulae for $\sin^2 x$ and $\cos^2 x$ are useful.

Example 6.6.5 (1)

Find $\int \sin^4 x\, dx$

Solution:

Any reader who might be tempted to write an answer such as $(-\cos x)^4 + C$ or $\frac{\sin^5 x}{5\cos x} + C$ should differentiate the suggested answer. The reader should then be convinced that the suggested answer is wrong. As was done in example 6.5.1 (3)

We have to use the double angle formula and get $\int \sin^4 x\, dx$

$$= \int (\sin^2 x)^2\, dx$$

$$= \int [\tfrac{1}{2}(1-\cos 2x)]^2\, dx$$

$$= \int \frac{1}{4}(1 - 2\cos 2x + \cos^2 2x)\, dx$$

$$= \int \frac{1}{4}[1 - 2\cos 2x + \frac{1}{2}(1+\cos 4x)]dx$$ (applying the double angle formulae to get $\cos^2 2x$ in terms of $\cos 4x$)

$$= \frac{1}{4}\int [\frac{3}{2} - 2\cos 2x + \frac{1}{2}\cos 4x]dx$$

$$= \frac{1}{4}[\frac{3}{2}x - \sin 2x + \frac{1}{2}\frac{1}{4}\sin 4x] + C$$

$$= \frac{3}{8}x - \frac{1}{4}\sin 2x + \frac{1}{32}\sin 4x + C$$

If we have to integrate an odd power of sin x or cos x we may make use of the identity $\cos^2 x + \sin^2 x = 1$ as shown in the following example.

Example 6.6.5 (2)

Find $\int \sin^7 x\, dx$

Solution:

Use
$$\sin^7 x = \sin x \sin^6 x = \sin x\, (\sin^2 x)^3$$
$$= \sin x\, (1 - \cos^2 x)^3$$
$$= \sin x\, (1 - 3\cos^2 x + 3\cos^4 x - \cos^6 x)$$

$$\text{so } \int \sin^7 x\, dx$$

$$= \int \sin x\, (1 - 3\cos^2 x + 3\cos^4 x - \cos^6 x)\, dx$$

$$= \int (\sin x - 3\cos^2 x \sin x + 3\cos^4 x \sin x - \cos^6 x \sin x)\, dx$$

$$= -\cos x + \cos^3 x - \frac{3}{5}\cos^5 x + \frac{1}{7}\cos^7 x + C$$

on using reverse chain rule or a substitution $u = \cos x$ (an exercise for the reader).

If we have to integrate powers of tan x it is helpful to use the identity $\tan^2 x + 1 = \sec^2 x$

Example 6.6.5 (3)

Find $\int \tan^4 x\, dx$

Solution:

$$\int \tan^4 x\, dx = \int \tan^2 x \tan^2 x\, dx$$
$$= \int \tan^2 x\, (\sec^2 x - 1)\, dx$$
$$= \int \tan^2 x \sec^2 x\, dx - \int \tan^2 x\, dx$$
$$= \frac{1}{3}\tan^3 x - \int (\sec^2 x - 1)\, dx$$
$$= \frac{1}{3}\tan^3 x - \tan x + x + C$$

The integration of the first expression can be found by inspection (reverse chain rule) or by substituting $u = \tan x$.

6.7 INTEGRATION BY PARTS

In many of the problems we met in 6.6 we could find an integral involving a function with two factors by using reverse chain rule or substitution, examples being

$$\int x\, e^{x^2}\, dx = \frac{1}{2}\, e^{x^2} + C$$

or
$$\int x \cos (x^2 + 3)\, dx = \frac{1}{2} \sin (x^2 + 3) + C$$

However integration such as $\int xe^x\, dx$, $\int x \cos x\, dx$ cannot be found by substitution, but can be found using integration by parts.

6.7.1 Reverse Product Rule

Based on the product rule we know that

if
$$y = uv$$
$$\frac{dy}{dx} = v\frac{du}{dx} + u\frac{dv}{dx}$$

or
$$\frac{d}{dx}(uv) = v\frac{du}{dx} + u\frac{dv}{dx}$$

so that
$$uv = \int v\frac{du}{dx} \cdot dx + \int u\frac{dv}{dx} \cdot dx$$

if we make the latter expression on the right hand side to be the subject then

$$\int u\frac{dv}{dx}\, dx = uv - \int V\frac{du}{dx} \cdot dx$$

Example 6.7.1 (1)

Find $\int x \sin x\, dx$

Solution:

Choose
$$u = x, \quad \text{and} \quad \sin x = \frac{dv}{dx}$$

then
$$\frac{du}{dx} = 1, \quad \text{and} \quad v = -\cos x$$

then $\int x \sin x\, dx$

$$= x(-\cos x) - \int (-\cos x)\, 1\, dx$$

$$= -x\cos x + \int \cos x\, dx$$

$$= -x\cos x + \sin x + C$$

As with the technique of substitution, mastery of the technique of integration by parts will come only with practice and experience.

However, there are two guiding principles:

(i) the choice of which factor should be taken as u and which factor as $\frac{dv}{dx}$ should be made bearing in mind that it must be possible to integrate $\frac{dv}{dx}$ easily.

(ii) check whether $\int v \frac{du}{dx} \cdot dx$ is easier than the original integral, and in this case if the factor includes a power of x, the aim will be to reduce the power of x.

Example 6.7.2

Find $\int x^3 \sin x \, dx$

Solution:

If we select $x^3 = \frac{dv}{dx}$

then $v = \frac{x^4}{4}$, making the power of x become worse in the second integral.

So take $u = x^3$, $\frac{dv}{dx} = \sin x$

hence $\frac{du}{dx} = 3x^2$ $v = -\cos x$

and $\int x^3 \sin x \, dx = x^3 (-\cos x) - \int (-\cos x) \, 3x^2 \, dx$

The second integral will now require a further application of parts.

$\int 3x^2 \cos x \, dx$ can be made simpler by taking

$$u = 3x^2, \ \frac{dv}{dx} = \cos x$$

so $\frac{du}{dx} = 6x$, $v = \sin x$

Hence $\int 3x^2 \cos x \, dx = 3x^2 \sin x - \int \sin x \, (6x) \, dx$ and the integral on the right hand side will give $6x (\cos x) - \int 6 \cos x$ on further integration by parts.

so $\int x^3 \sin x \, dx = -x^3 \cos x + 3x^2 \sin x + 6x \cos x - 6 \sin x + C.$

Example 6.7.3

Evaluate $\int x^3 \ln x \, dx$

Solution:

At first sight this particular integration looks difficult. We cannot identify $\frac{dv}{dx}$ as $\ln x$ since there is no known function v for which $\ln x$ is the derivative.

We are forced back to taking $\frac{dv}{dx} = x^3$ and $v = \frac{1}{4}x^4$, making the power of x become worse,

However if we take $\frac{dv}{dx} = x^3$

then $$u = \ln x$$

and $$v = \frac{1}{4}x^4, \quad \frac{du}{dx} = \frac{1}{x}$$

so $$\int_1^2 x^3 \ln x \, dx = [\frac{1}{4}x^4 \ln x\,]_1^2 - \int_1^2 3x^2 \cdot \frac{1}{x}\, dx$$

$$= \frac{1}{4} \times 16 \ln 2 - \int_1^2 3x \, dx$$

$$= 4 \ln 2 - \left[\frac{3}{2}x^2\right]_1^2 = 4 \ln 2 - \frac{9}{2}$$

The reader might wish to try solving this problem by making a substitution $x = e^u$. This is left as an exercise to the reader.

Example 6.7.4

Show that $\int_0^2 x^2 e^{2x} \, dx = \frac{1}{4}[\,5\, e^4 - 1\,]$

Solution:

Taking $$u = x^2, \qquad \frac{dv}{dx} = e^{2x}$$

then $$\frac{du}{dx} = 2x, \qquad v = \frac{1}{2}e^{2x}$$

and $$\int_0^2 x^2 e^{2x} \, dx = [\,x^2 \cdot \frac{1}{2}e^{2x}\,]_0^2 - \int_0^2 \frac{1}{2}e^{2x} \cdot 2x \, dx$$

$$= 4 \times \frac{1}{2}e^4 - 0 - \int_0^2 x\, e^{2x} \, dx$$

and the second integral can be found using

$$u = x, \qquad \frac{dv}{dx} = e^{2x}$$

then $$\frac{du}{dx} = 1, \qquad v = \frac{1}{2}e^{2x}$$

i.e. $$\int_0^2 xe^{2x} \, dx = [\frac{1}{2}xe^{2x}\,]_0^2 - \int_0^2 \frac{1}{2}e^{2x} \, dx$$

$$= \frac{1}{2} \times 2\, e^4 - 0 - [\,\frac{1}{4}e^{2x}\,]_0^2$$

$$= e^4 - \frac{1}{4} e^4 + \frac{1}{4} e^0 = \frac{3}{4} e^4 + \frac{1}{4}$$

so
$$\int_0^2 x^2 e^{2x}\, dx = 2\, e^4 - [\frac{3}{4} e^4 + \frac{1}{4}]$$

$$= 2\, e^4 - \frac{3}{4} e^4 - \frac{1}{4} = \frac{5}{4} e^4 - \frac{1}{4} = \frac{1}{4} (5e^4 - 1)$$

EXERCISE 6

Problems on Chapter 6

1. Find the gradient of the curve whose equation is given below in (a) to (h) using the definition of a derivative in the form $\lim_{h \to 0} \frac{f(x+h) - f(x)}{h}$ in each case.

   (a) $y = x^3$ (b) $y = x^2$ (c) $y = x^6$ (d) $y = x^{10}$

   (e) $y = \frac{1}{x}$ (f) $y = \frac{1}{x^2}$ (g) $y = \frac{1}{x^6}$ (h) $y = \frac{1}{x^{20}}$

2. Find the gradient of the curve whose equation is given below using the definition of a derivative in the form

$$\lim_{h \to 0} \frac{f(x+h) - f(x-h)}{2h}$$

   in each case.

   (a) $y = x^5$ (b) $y = x^7$ (c) $y = \frac{1}{x^4}$ (d) $y = \frac{1}{x^5}$

3. Determine the equation of the tangent to the curve whose equation is $y = x^3$ at the point A where $x = 1$. If the tangent meets the curve again in point B determine the coordinates of B and the equation of the tangent to the curve at B.

4. Determine the equation of the tangent and the normal to the curve whose equation is $y = 4 - x^2$ at the point C where $x = 1$. If the normal at C meets the curve again at D, determine the coordinates of D, and also the coordinates of the point of intersection E of the tangents to the curve at the points C and D.

5. Determine the gradient of the curve whose equation is given in (a) and (b) below using the definition of a derivative.

   (a) $y = \frac{1}{x^{1/2}}$ (b) $y = (4x + 3)^{1/2}$

6. Determine the gradient of the curve whose equations are given in (a) to (d) below using the definition of a derivative and the delta notation.

   (a) $y = x^2 - 5x + 3$ (b) $y = 3x^4 - 6x + 2$ (c) $y = \frac{3}{x}$ (d) $y = \frac{-5}{x^2}$

7. Write down the derivatives of each of the following functions in (a) to (j) by rule. Simplify where appropriate.

(a) $y = x^{40}$

(b) $y = 5x^{20} - 80\,x^{10} + 50x^5 - 7$

(c) $y = 7\,x^2 - \frac{1}{x} + \frac{2}{x^3}$ if $x \neq 0$

(d) $y = 6x^{3/2}$

(e) $y = \frac{-5}{7x^{2/5}}$ if $x \neq 0$

(f) $y = 2x^{3/2} - 4x^{1/2} + \frac{2}{x^{1/2}} + 1$ if $x \neq 0$

(g) $y = (x^3 - 3)^2$

(h) $y = \frac{(x-5)^2}{x}$ if $x \neq 0$

(i) $y = \frac{(x^{1/2} - 2)^2}{x^{1/2}}$ if $x \neq 0$

(j) $y = \frac{(x^{1/3} - 2)^3}{x^{2/3}}$ if $x \neq 0$

8. Find the derivatives of each of the following functions in (a) to (h) and simplify your answer.

(a) $y = (6x - 5)^{1/3}$

(b) $y = (6x - 2)^{1/2}$

(c) $y = (4x - 1)^{-1/2}$ if $x > 1/4$

(d) $y = \frac{1}{(x^{1/2} + 2)^3}$ if $x > 0$

(e) $y = (3x^2 + 7)^{1/3}$

(f) $y = (6x^2 + 4x + 3)^{1/2}$

(g) $y = 1/(10x^2 + 2x + 5)^{1/2}$

(h) $y = \frac{1}{(2x + 2x^{1/2} + 5)^2}$ if $x > 0$

9. Find the derivatives of each of the following functions in (a) to (h) and simplify your answer.

(a) $y = (x + 3)^7\,(x + 5)^{10}$

(b) $y = (3x + 5)^{1/2}\,(5x + 2)^{3/2}$ if $x > -2/5$

(c) $y = (x^2 + 1)^{1/2}\,(x^2 + 3)^{1/3}$

(d) $y = (x^3 + 4)^{1/3}\,(x^3 + 6)^{2/3}$

(e) $y = \frac{x^2 - 4}{x^2 + 4}$

(f) $y = \frac{(2x + 3)^{1/2}}{(5x + 2)^{3/2}}$

(g) $y = \frac{x^2 - 9}{x^2 + 9}$

(h) $y = \frac{(1 + x^{1/2})}{(1 - x^{1/2})}$ if $0 \le x < 1$

10. Differentiate each of the following functions of x in (a) to (h) and simplify your answer where appropriate.

(a) $y = \sin 4x$

(b) $y = \cos \frac{3}{2}\,x$

(c) $y = \sin^4 5x$

(d) $y = \sqrt{\cos 3x}$ if $0 \le x < \pi/6$

(e) $y = \sin^5 x + \cos^5 x$

(f) $y = \sin^2 x + \cos^2 x$

(g) $y = \sin^3 6x - \cos^2 6x$

(h) $y = \cos^4 8x - \sin^4 4x$

11. Differentiate each of the following functions of x in (a) to (h) and simplify your answer where appropriate.

(a) $y = \tan 8\,x$ (b) $y = \dfrac{1}{8}\sec 8x$

(c) $y = \sec^5 x$ (d) $y = \text{cosec}^2\, 5x + 3\cot^2 5x$

(e) $y = \sqrt{\cot x}$ if $0 < x < \pi/2$ (f) $y = \sec^2 x - \tan^2 x$

(g) $y = \sqrt{(\sec^2 + 2))}$ (h) $y = (\sec^2 5x + 2\tan^2 5x + 3)^{1/3}$

12. If $y = \dfrac{\sin x - \cos x}{\sin x + \cos x}$, $x \neq \dfrac{3\pi}{4} + 2\pi n,\ n \in Z$, show that $\dfrac{dy}{dx} = \dfrac{2}{1 + \sin 2x}$

13. If $y = \dfrac{\sec x - \tan x}{\sec x + \tan x}$ and $0 < x < \pi/2$ show that $\dfrac{dy}{dx} = \dfrac{-2\sec x}{(\sec x + \tan x)^2}$

14. Differentiate y with respect to x when

(a) $y = x^2 \cos x$ (b) $y = \sqrt{(1 + \sin x)}$ for $0 \leq x < \dfrac{3\pi}{2}$

(c) $y = \sin^3 x$ (d) $y = \sin(x^3)$

15. If $y = \dfrac{\cos 6x}{\cos 3x}$ show that $\dfrac{dy}{dx} = -3\tan 3x \sec 3\,x - 6\sin 3x$

16. Determine $\dfrac{dy}{dx}$ in each of (a) to (j) below, simplifying your answer where appropriate.

(a) $y = e^{6x}$ (b) $y = e^{-10x}$ (c) $y = e^{4x^3}$ (d) $y = e^{\sqrt{(9x^2+16)}}$

(e) $y = e^{\cos(8x^3+4)}$ (f) $y = \dfrac{e^{-x^3}}{x^2}$ (g) $y = e^{-x}\sin 3x$ (h) $y = e^{4x}\sec 4x$

(i) $y = e^{x^2}\sin x$ (j) $y = x^4 e^{-2x}$

17. If $y = \dfrac{e^{4x} - e^{-4x}}{e^{4x} + e^{-4x}}$ show that $\dfrac{dy}{dx} = \dfrac{16}{\left(e^{4x} + e^{-4x}\right)^2}$

18. Determine $\dfrac{dy}{dx}$ when

(a) $y = \ln 6x$ (b) $y = \ln x^6$

(c) $y = \ln(x^3 + 3x^2 + 4)$ (d) $y = \ln\left\{\dfrac{(x+2)}{(x^4+16)^{1/4}}\right\}$

(e) $y = \ln(\cos^2 x)$ (f) $y = \ln(\sin(x^3))$

(g) $y = \ln\left(\sqrt{(1 + \sin x)}\right)$ (h) $y = \ln(\sin^3 x + \cos^3 x)^{1/3}$

*19. Given that $y = x - \ln(\sec x + \tan x)$ on the domain $0 \le x < \frac{\pi}{2}$,

(a) determine $\frac{dy}{dx}$ in a form which involves only one trigonometric function.

(b) deduce that $y < 0$ for $0 < x < \frac{\pi}{2}$.

20. The velocity V ms^{-1} of a particle moving in a straight line at time t seconds is given by $V = t(t - 6)$. Find the acceleration at the times when the velocity is zero.

*21. The displacement x metres of a particle moving in a straight line at time t seconds is given by $x = \frac{1}{3t - 2}$

(a) show that the velocity is always negative.

(b) Find the values of t for which the velocity is $-\frac{1}{3}$ ms^{-1} and find the acceleration at these times.

*22. A particle moves along the x-axis so that at time t seconds its displacement x metres from origin O is given by $x = t^3 - 3t^2 + 9t$.

(a) show that the velocity V ms^{-1} is always positive, and find its least value.

(b) sketch the graphs of V and x indicating clearly the point corresponding to the least velocity.

23. A particle moving in a straight line has displacement x metres from an origin O when its velocity is V ms^{-1}, and V is given by the equation.

$$V = 4(30 - x).$$

Initially $\qquad x = 10.$

Find the acceleration of the particle when the particle has a velocity which is half its initial velocity.

*24. A particle moves along a straight line so that its distance X from a fixed point O at time t is given by the equation $X = A \cos nt$ where A and n are constants.

(a) Find its acceleration in terms of X.

(b) Show that its speed at time t is $n\sqrt{(A^2 - X^2)}$

(c) When the particle passes through the points P, Q, R in that order, its velocities are 0 ms^{-1}, 2ms^{-1} and $-$ 1 ms^{-1} respectively.

Given that PQ = 2m, and PR = 8m determine (i) the amplitude of the motion, (ii) the period of the motion.

25. If a curve has its equation given by $y = (3x - 2)^2 (2x + 5)^2$ determine the coordinates of all the points on the curve for which $\frac{dy}{dx} = 0$

26. Determine the coordinates and nature of the turning points of the curve $y = 2x^3 - 3x^2 - 36x$ and sketch the curve.

27. Determine the coordinates and nature of the turning points on the curve $y = x^4 - 2x^3$ and sketch the curve.

28. A curve has equation $y = (x - 2)e^{-x}$.

(a) Determine the coordinates of the points where the curve meets the axes.

(b) Determine the coordinates and nature of the turning point.

(c) Sketch the curve, given that $y \to -\infty$ as $x \to -\infty$, and $y \to 0$ as $x \to \infty$.

29. A closed rectangular box has its length exactly twice its width and is to be designed to hold 9 m^3 of powder. Determine the dimensions which will make its surface area a minimum.

30. A closed cardboard box is designed to hold a volume of 288 cm^3. The length is $3x$ cm and the width x cm. Show that the total surface area A cm^2 is given by $A = 6x^2 + \dfrac{768}{x}$ and find the dimensions of the box which will make A a minimum.

*31. A hollow cylinder of radius r with a base and no lid has to be made using only A square units of metal. Show that the volume V cubic units is given by the formula

$$V = \frac{Ar}{2} - \frac{\pi r^3}{2}$$

hence find the value of r which will make the volume a maximum and show that this maximum volume is $\dfrac{A^{3/2}}{3\sqrt{(3\pi)}}$ cubic units.

*32. An open rectangular box is made from a thin piece of cardboard which is rectangular in shape, of length 12 cm and width 8 cm, by cutting squares of side x cm away from each corner and folding so that the cut edges are joined.

Show that the volume V cm^3 of the box is given by $V = 4x^3 - 44x^2 + 96x$.

Find the value of x which will make the volume a maximum, and also the maximum volume.

33. The curve whose equation is $y = 2x^3 + ax^2 + bx + 3$ has stationary values at $x = 1$ and $x = 2$. Find the values of the constants a and b. Find also the equation of the tangent to the curve at the point where $x = \dfrac{1}{2}$.

34. Find each of the indefinite integrals in (a) to (j) below:

(a) $\int x^{30}\, dx$ (b) $\int x^{-25}\, dx$

(c) $\int x^{8/3}\, dx$ (d) $\int\left(10x^4 - 6x^2 + \dfrac{3}{x^{5/2}} - \dfrac{14}{x^{9/2}}\right)dx$

(e) $\int (5x + 12)^{1/3}\, dx$ (f) $\int \sin\left(3x + \dfrac{\pi}{6}\right) dx$

(g) $\int 12\,\text{cosec}^2\left(4x + \dfrac{\pi}{4}\right) dx$ (h) $\int 25\, e^{-5x}\, dx$

(i) $\int \dfrac{dx}{3x + 4}$ (j) $\int \cot 3x\, dx$

35. Find each of the following indefinite integrals in (a) to (j) below.

(a) $\int \dfrac{4dx}{4x + 1}$ (b) $\int \dfrac{\cos x\, dx}{5 + 2\sin x}$

(c) $\int \frac{\sin x \sin 2x\, dx}{8+\sin^3 x}$ (d) $\int \frac{x^3\, dx}{x^4+81}$

(e) $\int \frac{(x^2+2x+3)\, dx}{x^3+3x^2+9x+6}$ (f) $\int \frac{e^{-3x}\, dx}{4-e^{-3x}}$

(g) $\int \frac{\sin 2x\, dx}{2\cos^2 x+3}$ (h) $\int \frac{e^{2x}-1}{e^{2x}+1} dx$

(i) $\int \frac{\tan^4 x \sec^2 x\, dx}{\tan^5 x+1}$ (j) $\int \frac{(\sec^6 x \tan x + e^{6x})\, dx}{\sec^6 x + e^{6x}}$

36. If $y = \ln (x^2 + 5x + 6)$ find $\frac{dy}{dx}$ and show that

$$\int_1^2 \frac{(4x+10)\, dx}{x^2+5x+6} = \ln\left(\frac{25}{9}\right)$$

37. If $y = x \sin^2 x$ find $\frac{dy}{dx}$ and hence evaluate $\int_0^{\pi/4} (x \sin 2x - \cos^2 x)\, dx$.

38. Find the area enclosed by the curve whose equation is $y = x^2$, the x-axis, and the lines $x = 1$, $x = 2$.

39. Evaluate $\int_0^3 (x^3 - 2)\, dx$ and explain with the aid of a sketch why the integral does not give you the total area formed by the curve whose equation is $y = x^3 - 2$, the x-axis, and the lines $x = 0$, $x = 3$.

40. Find the equation of the curve whose gradient function is given by $\frac{dy}{dx} = x - 4$ and which passes through (4, 1).

41. Find the function $f(x)$ for which $f'(x) = x^2 + 4x$ and $f(3) = 15$.

42. Evaluate each of the integrals (a) to (h).

(a) $\int_1^4 \sqrt{x}\,(x+3)\, dx$ (b) $\int_1^2 \frac{(2x+1)^2}{x^4} dx$

(c) $\int_{1/4}^1 \frac{(x-1)}{\sqrt{x}} dx$ (d) $\int_1^{27} x^{1/3}\, dx$

(e) $\int_0^{\pi/4} \sin 2x\, dx$ (f) $\int_0^2 \frac{e^{2x}}{e^{2x}+e^2} dx$

(g) $\int_0^2 \frac{x^3}{x^4+16} dx$ (h) $\int_{\pi/6}^{\pi/3} \cot x\, dx$

43. (a) Criticise the following procedure for determining the total area enclosed by the curve whose equation is $y = x^2 - 3x + 2$, the x-axis and the lines $y = 0$, $x = 0$ and $x = 3$.

$$\text{Area} = \int_0^3 (x^2 - 3x + 2)\, dx$$

$$= \left[\frac{1}{3}x^3 - \frac{3x^2}{2} + 2x\right]_0^3$$

$$= \frac{1}{3} \times 27 - \frac{3 \times 9}{2} + 2 \times 3 - 0 = 1\frac{1}{2}$$

(b) Determine the correct value of the area.

44. A curve has equation $y = 8x + 20 - x^2$.

(a) Determine the coordinates of the points where the curve meets the axes.

(b) Determine the coordinates and nature of the turning point.

(c) Sketch the curve.

(d) Find the area bounded by the axes and the curve in the first quadrant.

45. In each of the cases (a) to (d), calculate the coordinates of the points of intersection of the two given curves (or given curve and given line), and also the finite area enclosed by the two curves (or curve and line).

(a) Curve S which has equation $y = x^3$ and line L which has equation $y = 4x$.

(b) Curve S which has equation $y = x^{3/2}$ and line L which has equation $y = 2x$.

(c) Curve S which has equation $y = 2x^{1/2}$ and curve T which has equation $y = \frac{1}{4}x^2$.

(d) Curve S which has equation $y = x^2$ and curve T which has equation $y = 8 - x^2$.

*46. A curve has equation $y = 2\,e^{-4x} - e^{-2x}$.

(a) Show that $y = 0$, when $x = \frac{1}{2}\ln 2$ without tables or calculator.

(b) Find the area bounded by the x-axis, the y-axis, and the part of the curve joining the points where $x = 0$ and $x = \frac{1}{2}\ln 2$.

47. Calculate the area bounded by the parabola $y = 2 - x^2$ and the straight line $y = x$.

48. Determine the area of the region bounded by the lines whose equations are $y = x + 5$ and $y = 2(x + 1)$, and the curve whose equation is $y = (5 - x)(1 + x)$.

*49. The area of a triangle of sides a, b, c, is given by $A = \sqrt{\{s(s-a)(s-b)(s-c)\}}$ where A is the area and S in the semi-perimeter.

A triangle has two sides $a = 6$ and $b = x$. Its perimeter is 16. By considering an expression for A^2 or otherwise, determine the value of x for which the area of the triangle is a maximum.

50. The acceleration a ms^{-2} of a particle moving in a straight line at time t seconds is given by $a = 24t^2 - 6t$. If the particle starts from rest at $t = 0$, determine

(a) an expression for its velocity V ms^{-1} at any time t seconds.

(b) the other value of t for which the particle is at rest.

(c) the distance moved by the particle between the two times when the particle is at rest.

51. A particle moving in a straight line has velocity V ms^{-1} at time t seconds given by $V = t(t - 3)^2$.

(a) Determine the times when the acceleration is zero and the velocity at these times, distinguishing between the local maximum and minimum velocities.

(b) Sketch the graph of velocity against time.

(c) Calculate the distance covered between the times when the particle is momentarily at rest.

*52. A particle is moving in a straight line and its acceleration t seconds after passing the origin is given as $(4 \sin 2t + 36e^{-6t})$ ms^{-2}.

Given that the particle was at the origin when $t = 0$ and was initially at rest determine its velocity and displacement at time $\frac{\pi}{3}$ seconds.

53. A particle moving in a straight line has a displacement x metres from an origin O when its velocity is V ms^{-1}, and its acceleration is given as $(4 - 2x)$ ms^{-2}.

If the velocity is $2\sqrt{6}$ ms^{-1} when the particle is at the origin, obtain an expression for V^2 in terms of x, and the values of x for which the particle is at rest.

54. A particle moving in a straight line has a displacement x metres from an origin O when its velocity is V ms^{-1}, and its acceleration is given as $(3x^2 - 4)$ ms^{-2}.

If the particle is initially at rest at the origin, find its other displacement when it is at rest.

55. A particle moving in a straight line has a displacement x metres from an origin O when its velocity is V ms^{-1} and its acceleration is given as $(e^{4x} - e^{2x})$ ms^{-2}.

Given that its velocity at the origin is 1 ms^{-1} show that this is the least numerical value of the velocity.

56. Determine

(a) $\int x^2 e^{x^3}\, dx$ (b) $\int \sec^2 x\, e^{\tan x}\, dx$

(c) $\int (3 + x)\, e^{(3 + x)^2}\, dx$ (d) $\int x\, (2x^2 + 5)^7\, dx$

(e) $\int x\, (3x^2 + 5)^{1/6}\, dx$ (f) $\int \sec^2 x \tan^4 x\, dx$

(g) $\int \operatorname{cosec}^2 2x \cot 2x\, dx$ (h) $\int \sin^3 4x \cos 4x\, dx$

(i) $\int \cos^7 6x \sin 6x\, dx$ (j) $\int \sin^2 4x \cos 4x\, e^{\sin^3 4x}\, dx$

57. Determine

(a) $\int \frac{\ln (x^3)}{x}\, dx$ (b) $\int \frac{1}{x\,(2 + 3 \ln x)}\, dx$ if $x > e^{-2/3}$

(c) $\int \frac{\cos (\ln x)}{x}\, dx$ (d) $\int \frac{e^{4x}\, dx}{\sqrt{(16 + e^{4x})}}$

58. Show that

$$\int \frac{x\, dx}{(1 + 4x)^{1/3}} = \frac{1}{160}\left[(1 + 4x)^{2/3}\,(24x - 9)\right] + C$$

59. Show that $\int_{0}^{3/2} \frac{x\,dx}{\sqrt{(9-x^2)}} = \frac{3}{2}(2-\sqrt{3})$

*60. Show that $\int \frac{x\,dx}{(1-2x)^2} = \frac{1}{4}\left\{\frac{1}{(1-2x)} + \ln(1-2x)\right\} + C$ if $x < \frac{1}{2}$

61. Show that $\int_{0}^{9.5} x\,(27 - 2x)^{2/3}\,dx \simeq 263.5$

62. Evaluate $\int_{0}^{\sqrt{2}} \frac{dx}{\sqrt{(4-x^2)}}$

63. Evaluate $\int_{0}^{2} \frac{dx}{\sqrt{(16-x^2)}}$

64. Show that $\int_{0}^{2\sqrt{3}} \frac{x^2\,dx}{\sqrt{(16-x^2)}} = \frac{2}{3}\,(4\pi - 2\sqrt{3})$.

65. Evaluate each of the integrals:

(a) $\int_{0}^{\pi/3} (\cos^6 x)\,dx$ (b) $\int_{0}^{\pi/2} \sin^7 x\,dx$

(c) $\int_{0}^{\pi/4} \tan^5 x\,dx$ (d) $\int_{0}^{\pi/4} \sec^6 x\,dx$

66. Determine:

(a) $\int (x + 4)\cos x\,dx$ (b) $\int (x + 10)\,e^{10x}\,dx$

(c) $\int x^8 \ln x\,dx$ (d) $\int x\,\mathrm{cosec}^2\,x\,dx$

67. Show that $\int_{0}^{1} x^2 e^x\,dx = 2 - 5\,e^{-1}$.

68. Evaluate $\int_{0}^{\pi/2} (x + 1)\sin x\,dx$

69. Evalute each of the integals:

(a) $\int_{0}^{\pi} x \sin x\,dx$ (b) $\int_{0}^{\pi} x^2 \sin x\,dx$ (c) $\int_{0}^{\pi} x^3 \sin x\,dx$

*70. Given that $I_n = \int (\ln x)^n\,dx$ show that $I_n = x\,(\ln x)^n - n\,I_{n-1}$ and use this result to evaluate $\int_{1}^{e} (\ln x)^3\,dx$.

[Hint: in I_n take $u = (\ln x)^n$ and $\frac{dv}{dx} = 1$]

*71. (a) Simplify $\dfrac{1}{1-u}+\dfrac{1}{1+u}$ where $u \neq \pm 1$.

(b) By writing $\int \operatorname{cosec} x\, dx$ in the form $\int \dfrac{\sin x\, dx}{\sin^2 x}$ show that, using the substitution $u = \cos x$,

$\int \operatorname{cosec} x\, dx = -\int \dfrac{du}{1-u^2}$ and hence show that

$\int \operatorname{cosec} x\, dx = -\dfrac{1}{2} \ln \left(\dfrac{1+\cos x}{1-\cos x}\right) + C,$ using the result in (a).

(c) By writing $\cos x = \dfrac{1-t^2}{1+t^2}$ where $t = \tan \dfrac{1}{2} x$ and using your answer in (b), determine $\int \operatorname{cosec} x\, dx$

as a function of $t = \tan \dfrac{1}{2} x$.

*72. (a) Using the appropriate double angle formula show that $\tan \dfrac{\pi}{8} = \sqrt{2} - 1$.

(b) Using your answer in Question 71 (c) show that

$$\int_{\pi/4}^{\pi/2} (x + 2) \cot x \operatorname{cosec} x\, dx = \frac{(\sqrt{2}-1)(8-\pi\sqrt{2})}{4} + \ln (\sqrt{2}+1)$$

7

Answers to Exercises

Exercise I Answers

(1) Mean = 151.1
Median = 151
Mode = 148

(2)

Grade	Mark	No. of Students
A	88–99	10
B	68–87	20
C	40–67	42
D	24–39	16
E	0–23	12
	Mean = 53.7	

(3) See Fig. 7.1.1(a)

(4) (a) 31.72 (b) 5.14

(5) (a) Fig. 7.1.1 (b)
(a) Mode = 19.8 cm (b) Mean = 18.9 cm
Median = 19.1 cm

(6) (a) 5 cm (b) *lcb* 44.5 cm, *ucb* 49.5 cm
(c) 50–54 cm (d) 52.42 cm
(e) 5.65 cm (f) Median = 52.3 cm
Lower Quartile = 49.0 cm
Upper Quartile = 56.3 cm

(7) (a) 42 cm^3 (b) *ucb* 59.5 cm^3, *lcb* 54.5 cm^3
(c) 51.65 cm^3 (d) 4.60 cm^3
(e) Median = 51.94 cm^3
Lower Quartile = 48.75 cm^3
Upper Quartile = 54.90 cm^3

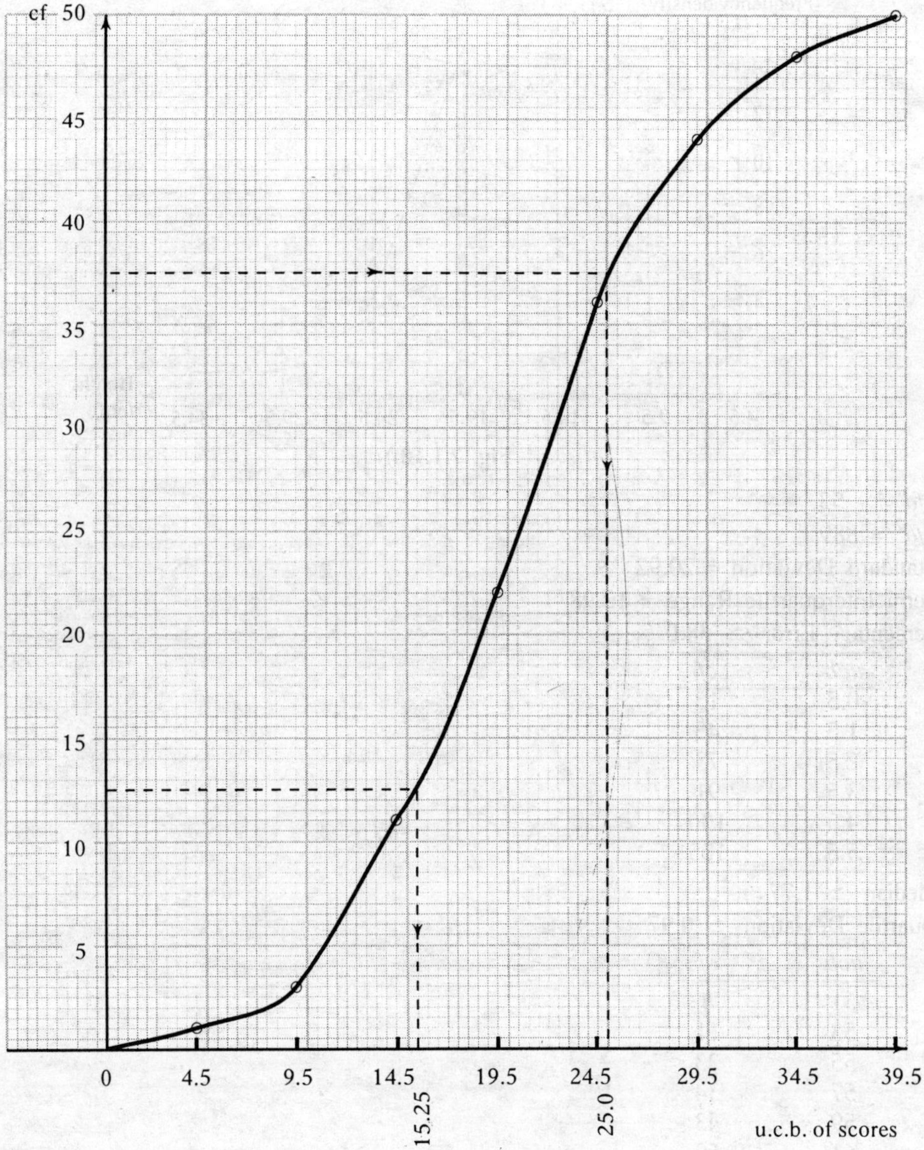

Fig. 7.1.1(a) Quartile Deviation $\simeq$ 5

(8) M = 5.17
S = 6.46

(9) (a) 5 kg. (b) *ucb* 67.5 kg, *lcb* 62.5 kg

(c) 73 – 77 kg (d) Mean = 72.2 kg
Standard Deviation = 7.98 kg

(e) 83.6 kg

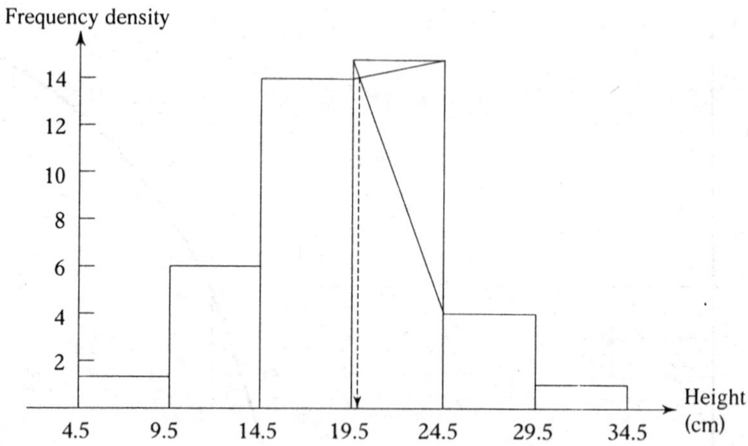

Fig. 7.1.1(b)

(10) $\Sigma fd = -53$
$\Sigma fd^2 = 889$
Standard Deviation = 20.92
Semi Interquartile Range = 12.98
Percentage Error $\simeq$ 38.0 %

(11)

ucb	*cf*
0.5	15
1.5	45
2.5	73
3.5	91
4.5	100
5.5	105

Median $\simeq$ 1.77
Quartile Deviation $\simeq$ 0.97

(12)

ucb	*cf*
51	4
53	11
55	23
57	34
59	43
61	50

Median $\simeq$ 55.4 kg
Quartile Deviation $\simeq$ 2.27 kg

(13)

x	f	z	fz	fz^2
50	2	−3	−6	18
55	12	−2	−24	48
60	12	-1	−12/−42	12
65	25	0	0	0

70	27	1	27	27
75	10	2	20	40
80	9	3	27	81
85	3	4	12/86	48
			$\Sigma fz = 44$	$\Sigma fz^2 = 274$

Mean = 67.2
Standard Deviation = 7.98

(14) Mean = 76
Standard Deviation = 8.17

(15) Mean = 314.9 ohms
Standard Deviation = 1.71 ohms

(16) Mean = 95 000 Buranian dollars
Standard Deviation = 37.75 thousand Buranian dollars.

(17) (a)

ucb	*cf*
9.5	4
19.5	10
29.5	25
39.5	52
49.5	91
59.5	136
69.5	161
79.5	181
89.5	195
99.5	200

Median $\simeq$ 51.5

(b) (i) 47 (ii) 74

(18) (a)

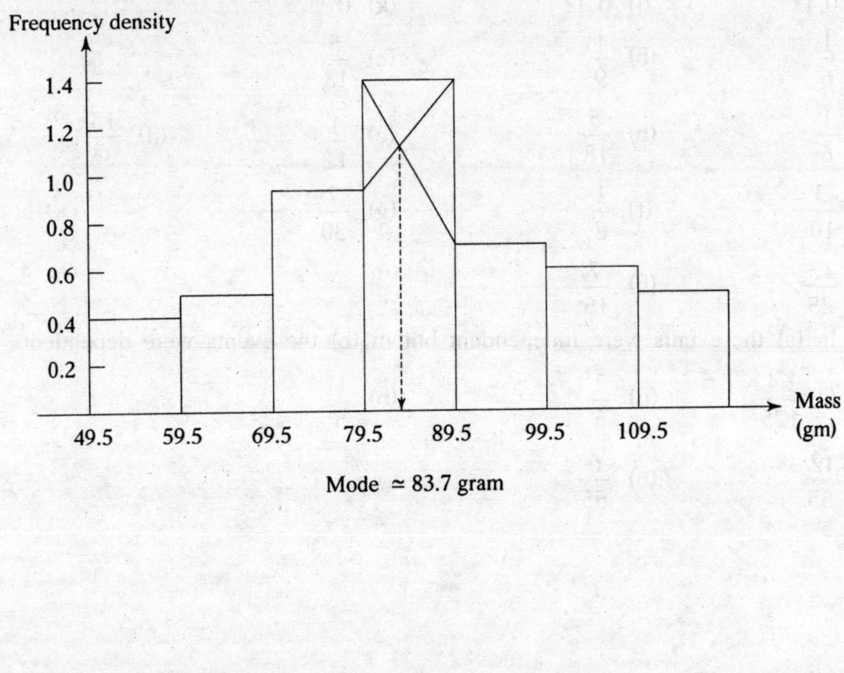

Mode $\simeq$ 83.7 gram

(b) Mean = 85.1 gram
Median = 84.5 gram

(19)

ucb (cm)	*cf*
3	6
4	13
5	21
7	31
10	40

Median $\simeq$ 4.9 cm

(20) (a) (i) 56 (ii) 20% i.e. 10 students (iii) 44
(b) (i) $Q = 55$
(ii)

Mark (x)	≤ 25	≤ 45	≤ 55	≤ 65	≤ 80	≤ 90	≤ 100
cf	0	5	20	40	60	75	80

(21) (a) 31.1 cm (b) 44.3 litres (c) 138.8 kg

(22) $a = 31,\ b = 89$

(23) $a = 21,\ b = 41,\ c = 15$

(24) $105 \leq a \leq 134,\ 75 \leq b \leq 104$, upper quartile is 159.5 when $a = 134$.

Exercise II Answers

(1) (a) 1/6 (b) 7/12

(2) (a) $\frac{3}{16}$ (b) $\frac{3}{16}$ (c) $\frac{3}{8}$ (d) $\frac{1}{2}$

(3) (a) 0.49 (b) 0.33 (c) 0.20 (d) 0.14
(e) 0.06 (f) 0.04 (g) 0.02 (h) 0.47
(i) 0.43 (j) 0.32 (k) 0

(4) (a) $\frac{1}{6}$ (b) $\frac{1}{9}$ (c) $\frac{5}{18}$

(5) (a) $\frac{1}{6}$ (b) $\frac{5}{18}$ (c) $\frac{1}{12}$ (d) $\frac{13}{36}$
(e) $\frac{3}{10}$ (f) $\frac{1}{2}$ (g) $\frac{7}{30}$

(6) (a) $\frac{13}{25}$ (b) $\frac{7}{15}$
(c) In (a) the events were independent but in (b) the events were dependent.

(7) (a) (i) $\frac{64}{125}$ (ii) $\frac{2}{5}$ (b) $\frac{1}{5}$

(8) (a) $\frac{12}{55}$ (b) $\frac{6}{55}$

(9) 0.347

(10) (a) $\frac{1}{11}$ (b) $\frac{10}{11}$

(11) (a) $\frac{1}{36}$ (b) $\frac{5}{18}$ (c) $\frac{2}{9}$

(12) (a) $\frac{1}{8}$ (b) $\frac{1}{6}$

(13) (a) $\frac{1}{27}$ (b) $\frac{7}{27}$

(14) $625 + 500x + 150x^2 + 20x^3 + x^4$, 635.06.

(15) $10^6 + 6 \times 10^5 x + 1.5 \times 10^5 x^2 + 2 \times 10^4 x^3$, 1 012 060

(16) $10^8 - 8 \times 10^7 x + 2.8 \times 10^7 x^2 - 5.6 \times 10^6 x^3$, 98 411 155

(17) (a) $\frac{15}{64}$ (b) $\frac{11}{32}$

(18) (a) $\frac{25}{216}$ (b) $\frac{19}{144}$

(19) (a) 0.91 (b) 0.83

(20) (a) 0.29 (b) 0.80

(21) (a) 0.11 (b) 0.578

(22) (a) 0.205 (b) 0.737

(23) (a) 0.273 (b) 0.377

(24) (a) $\frac{11}{32}$ (b) $\frac{7}{64}$

(25) (a) 0.311 (b) 0.544

Exercise III Answers

(1) $\sqrt{\frac{54}{125}}$

(2) $\frac{2}{5}$

(3) (a) 26 (b) $\frac{118 + 22\sqrt{15}}{17}$

(4) 5.999421

(5) 28.284272

(6) (a) 0 (b) 1

(7) (a) (i) $\frac{1}{2^{11}}$ (ii) $\frac{1}{3^2}$

(b) 1.54

(c) (i) 5 (ii) 3.32 (iii) 5 or 3.32

(8) 0.672

(9) (a) 2 (b) 3 (c) $\frac{5}{3}$ (d) $\frac{5}{3}$

(10) (a) –5 (b) 1.585 (c) –5 or 1.585

(11) (a) 0.36 (b) 3

(12) $\sqrt{5}$ or 25

(13) $\sqrt{5}$ or 5

(14) 2

(15) (a) (i) 5.4972 (ii) $\bar{8}.4916$

(b) 1

(16) (b) 81

(17) $\frac{1}{27}$

(18) (b) 5

(19) (a) –6 or –9 (b) 3.39 or –0.89

(20) See graph—Fig. 7.1.1 (c)

(21) (a) $-2 < x < 5$ (b) $x > \frac{4}{5}$ or $x < \frac{-3}{4}$ (c) $-3 \le x \le 4$ (d) $x \ge \frac{6}{5}$ or $x \le -1$

(22) –27

(23) 18

(24) (a) 2 or 6 (b) $K > 6$ or $K < 2$

(25) (a) 420 (b) 7

(26) 23

(27) 16

(28) 5216

(29) (a) 125 (b) 30

(30) (b) ${}_{n+2}C_9$

(31) (a) 2520 (b) 71

(32) (a) 15 (b) 20 (c) 15

(33) 27

(34) (a) 600 (b) 312

(35) (a) 3840 (b) 72000

(36) 9032

(37) (a) 4, 6 (b) 444 (c) 10

(38) 42, 240

(39) 729, $\frac{1}{3}$, 1093.5

(43) 150 500

(44) 16 731

(45) $x > 0$ or $x < -2$, $1 + x$

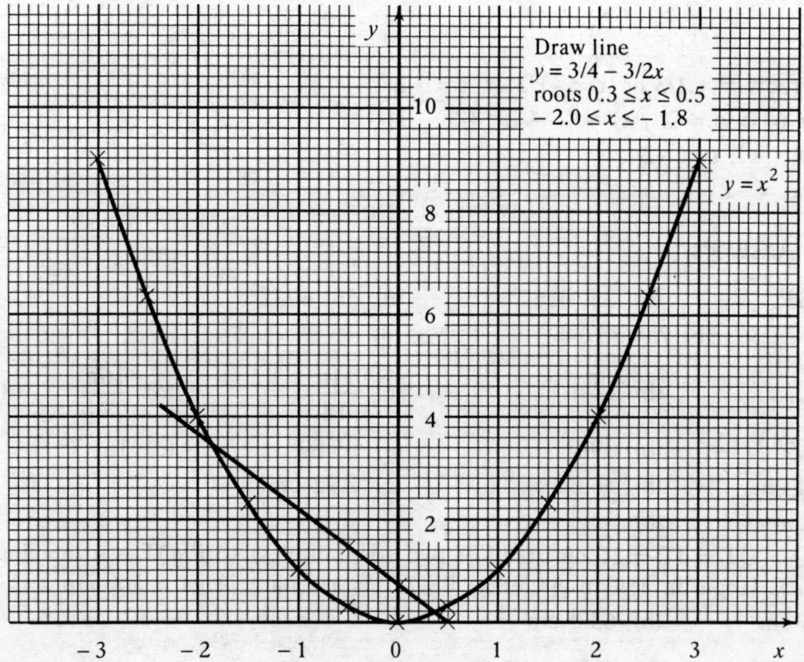

Fig. 7.1.1(c)

(46) 10 years
(47) 11 years
(48) 55 353 Buranian dollars
(49) (a) 1060 Buranian dollars
(b) 2183.6 Buranian dollars
(d) 13972 Buranian dollars

(51) $1 - \frac{1}{3}x - \frac{1}{9}x^2 - \frac{5}{81}x^3$, 9.9833

(52) $1 - \frac{1}{4}x^2 - \frac{3}{32}x^4 - \frac{7}{128}x^6$

(53) $1 + \frac{1}{2}x - \frac{1}{8}x^2 + \frac{1}{16}x^3$, 5.1962

(54) 1.220
(55) 3.315
(56) (b) 1.091
(58) $a = 2,\ b = -9$
$f(x) \equiv (x - 2)(2x + 1)(x - 3)$
(59) $a = 4,\ b = -1$

(60) $\frac{37x}{5} + \frac{9}{5}$

(61) $a = 13$, roots $x = 1/3$ or $x = -1/2$ or $x = -2$

(62) $a = 2$ or 6, if $a = 2$, $f(x) \equiv (x - 2)^2 (2x - 1)$

(63) $a = 6$, $b = -1$, $c = -2$

(64) $a = -7$, $b = -2$

Exercise IV Answers

(1) (a) 28.55° (b) 35. 6167° (c) 56. 6922°

(2) (a) 63° 21′36″ (b) 75° 25′ 45″

(3) (a) $\frac{3\pi}{4}$ (b) $\frac{3\pi}{8}$ (c) $\frac{11\pi}{18}$ (d) $\frac{16\pi}{9}$ (e) $\frac{17\pi}{3}$

(4) (a) 427.5° (b) 390° (c) 65.89° (d) 16.47°

(5) (a) 7.54 cm (b) 45.24 cm^2 (c) 2.92 cm^2

(6) (a) 90 cm (b) 3.58 cm^2

(7) PS = SQ = QT = TP = 10 cm (radii of two equal circles) so PSQT is a rhombus. Also angle PST = 90° (angle in a semi-circle) so PSQT is a square. Area = 100 cm^2.

(8) (a) 50.8 cm^2 (b) 63.0 cm

(9) 3

(10) (a) 1 (b) 0 (c) 1

(11) (a) $\frac{\pi}{6}$ or $\frac{5\pi}{6}$ (b) $\frac{2\pi}{3}$ or $\frac{4\pi}{3}$

(c) $\frac{\pi}{3}$ or $\frac{2\pi}{3}$ or $\frac{4\pi}{3}$ or $\frac{5\pi}{3}$ (d) $\frac{\pi}{4}$ or $\frac{3\pi}{4}$ or $\frac{5\pi}{4}$ or $\frac{7\pi}{4}$

(12) $\theta = \frac{\pi}{6}$ or $\frac{5\pi}{6}$ or $\frac{3\pi}{6}$ on $[0, 2\pi]$

$\theta = \frac{\pi}{6} + 2k\frac{\pi}{3} \quad k \in Z$

$\theta = \frac{7\pi}{2}$ on $(3\pi, 4\pi)$

(13) $C = 108°$, $b = 13.38$ cm, $c = 19.02$ cm, $R = 10.00$ cm

(14) $b = 5$ cm, $c = 7$ cm, $A = 21.79°$, B = 38.21°

(15) $DC = 138.9$ m, $AC = 1626$ m

(16) $A = 120°$, $B = 38.21°$, $C = 21.79°$, 25.98 cm^2

(17) 19.27 cm or 4.98 cm

(18) (i) two (ii) one (iii) none (iv) one

(19) $R = 8$ cm, $A = 48.59°$ or 131.41°, $B = 101.41°$ or 18.59°

(20) (a) $\cos A = \frac{5}{7}$, $\cos C = 1/5$

(b) $\cos \frac{A}{2} = \sqrt{\frac{6}{7}}$, $\sin \frac{A}{2} = \frac{1}{\sqrt{7}}$

(c) $35k/13$

(21) (a) $C = 135°$ (c) 4 unit^2

(22) (a) $\pm 30° + 360n°$, $n \varepsilon Z$ (b) $\pm 10° + 120n°$, $n \varepsilon Z$

(c) $60° + 180n°$, $n \varepsilon Z$ (d) $20° + 60n°$, $n \varepsilon Z$

(e) $\pm 60° + 180n°$, $n \varepsilon Z$ (f) $72n°$, $n \varepsilon Z$

(g) $15° + 60n°$, $n \varepsilon Z$

(23) (a) 70.5° or 120° or 240° or 289.5°

(b) 30° or 150° or 199.5° or 340.5°

(c) 63.43° or 71.57° or 243.43° or 251.57°

(d) 120° or 109.47° or 240° or 250.53°

(24) (a) 120° or 240° (b) 41.4° or 120° or 240° or 318.6°

(c) 180° (d) 90° or 194.48° or 345.52°

(27) $\sin (A - B) = \frac{16}{65}$

$\cos (2A - B) = \frac{204}{325}$

(28) $\pm 90° + 360n°$ or $\pm 41.6° + 360n°$, or $\pm 138.4° + 360n°$, $n \varepsilon Z$

(30) $2 r^2 \theta - r^2 \sin 2\theta$

(33) 1

(34) $\frac{\pi}{6} + \frac{n\pi}{3}$, $n \varepsilon Z$, $\theta = \frac{23\pi}{6}$

(36) (b) $\sin \frac{5\pi}{12}$

(37) (a) $5 \sin (x + 36.87°)$ (b) Minimum -5 when $x = 233.13°$

(c) 0° or 106.26°

(38) (a) $\frac{-5\pi}{6}$ or $\frac{5\pi}{6}$ or $\frac{7\pi}{6}$ (b) $-\frac{\pi}{4}$ or $\frac{7\pi}{4}$

(39) 6.87° or 246.87°

(40) $a = 24$, $b = 1.287$

(41) $K = 12$, $\alpha = 1.18$, $x = 0$ or 2.36 or 6.28

(42) 7

(43) $x = 0°$ or 120° or 240° or 360° or 82.5° or 202.5° or 322.5°

(44) (a) $K = 3$ (b) $R = 4$ (c) Period 120°, amplitude 4

(45) (i) $a = 2$ (ii) $b = 3$ (iii) $c = -90°$, $x = 40°$, 80°, 160°

(46) (a) $R = 2$ $A = 53.2°$ (b) Period = 180°, amplitude = 2,

Phase constant = 53.2° (first interpretation) or 26.6° (second interpretation).

47. (a) $f(x) \equiv R \cos (2x - \alpha) \equiv 17 \cos (2x - 28.1°)$

(b) Maximum $f(x) = 17$ at $x \simeq 14.05°$ or $194.05°$

Minimum $f(x) = -17$ at $x \simeq 104.05°$ or $284.05°$

(c) $x \simeq 59.05°$ or $149.05°$ or $239.05°$ or $329.05°$

(d)

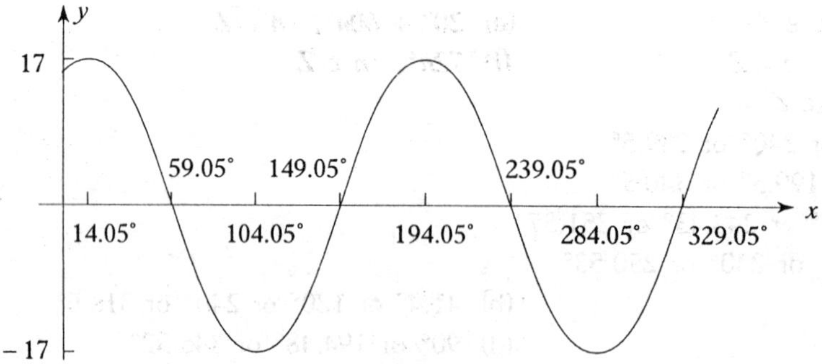

Fig. 7.4.1(a)

(e) Period 180°, amplitude 17, phase constant $\simeq 28.1°$ (first interpretation) or $14.05°$ (second interpretation)

(48) (a) $f(x) \equiv 1 \sin (2x - 30°)$

(b)

	$x = 10°$	$x = 45°$	$x = 90°$
$2x$	20°	90°	180°
$2x - A$	–10°	60°	150°
$f(x)$	–0.17	0.87	0.50
$3x$	30°	135°	270°
$3x + 60°$	90°	195°	330°
$\cos (3x + 60°)$	0	–0.97	0.87
$g(x)$	0	–1.94	1.74

(48) (c) See graph—Fig. 7.4.1(b)

(48) (d) $x \simeq 11.5°$ or $78.5°$

(49) $x = 30°$ or $90°$ or $150°$ or $210°$ or $270°$ or $330°$

(50) (b) 38.0 cm

(52) (b) (i) $\dfrac{\sin (\alpha - \beta)}{\cos \gamma}$

Exercise V Answers

(1) (a) $8\sqrt{5}$ (b) (1, –1) (c) (3, 3) (d) (9, 15) (e) (–7, –17)

(2) (1, –3)

(3) (a) –3, –5 (b) $\sqrt{3}$, $2\sqrt{3}$ (c) $\dfrac{5}{6}$, –5 (d) $-\dfrac{\sqrt{7}}{9}$, $7\sqrt{7}$

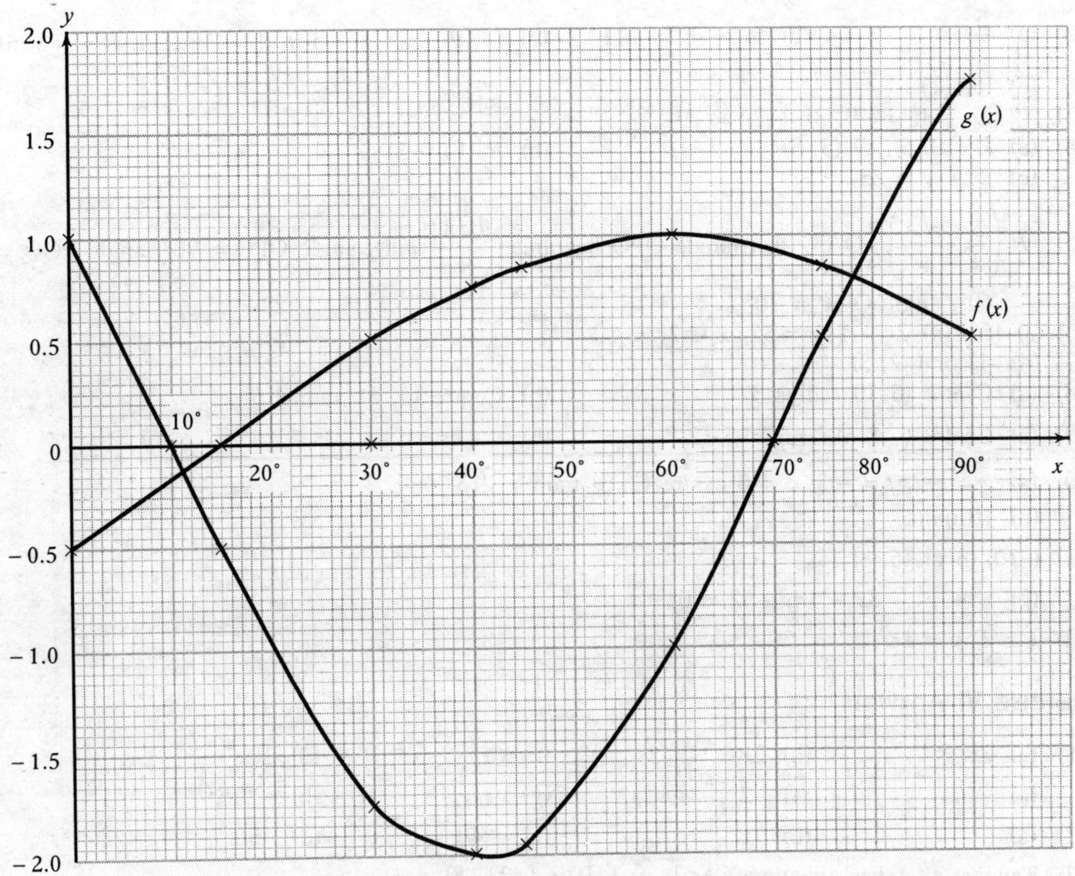

Fig. 7.4.1(b)

(e) no gradient can be defined, and the line does not meet the y-axis, as the line $x = 10$ is a vertical line parallel to the y-axis.

(4) (a) $x - 4y = -2$, (b) $3x - 4y = -19$ (c) $x - 2y = 10$ (d) $x + 3y = -1$
(e) $5x - 2y = 43$

(5) (a) $-\dfrac{1}{5}$, (b) $x + 5y = 78$ (c) B (0, 2), D (13, 13)

(6) A(6, 5), B (13, 12), D(7, 4)

(7) (a) A (4, 5), B (5, 12), C (0, 7) D(–1, 0).
(b) all four sides have length $5\sqrt{2}$
(c) AC: $x + 2y = 14$, BD : $2x - y = -2$ meet at (2, 6)

(8) Altitude CD has equation $x + 2y = 35$ orthocentre H is at (9, 13)

(9) (a) $3x - 4y = -9$, (b) B (13, 12), D(2, 10), M (7.5, 11)
(c) CD: $3x - 4y = -34$, AC: $2x - y = 4$, (d) 50 unit^2

(11) (a) $\overline{BC} = \overline{CD} = 2\ 2\sqrt{5}$ and $\overline{AB} = \overline{DA} = 10$ hence ABCD is a kite
(b) AB: $x = -2$, BC: $x + 2y = 8$, CD: $2x - y = 1$, DA: $4x + 3y = 37$

(12) (a) $\frac{5}{4}$ (b) $5x - 4y = -11$ (c) (5, 9) (d) $A = (13, 19)$, $B = (-3, 5)$

(e) (5, 11)

(13) $2x - 3y = -9$, $3x + 2y = 19$

(14) (a) D (10, 5), Q (5, 5) (b) P (14, –3)

(c) $2x + y = 25$

(15) $P\left(\frac{3}{4}, \frac{3\sqrt{3}}{4}\right)$, Q (3, 0), triangle OPQ has area $\frac{9\sqrt{3}}{8}$ unit^2, angle OPQ = 90°, angle POQ = 60°, angle OQP = 30°

(16) (a) 15°, (d) $2x - 11y = 55$, $x - 2y = 10$

(17) (4, –1) or (–2, 5)

(18) (a) Centre (0, 0) radius 3 (b) Centre (3, 2), radius 5

(19) (a) Centre (– 4, 5), radius $5\sqrt{2}$ (b) $x + y = 11$

(20) (a) $g = -2$, $f = -3$, $c = 0$ centre (2, 3) radius $\sqrt{13}$

(b) 10 units

(21) (–10, 0) or (6, 8)

(22) (b) $x^2 + y^2 - 8x - 14y + 52 = 0$

(23) $x = 2$, $y = 18$

Exercise VI Answers

(1) (a) $3x^2$ (b) $2x$ (c) $6x^5$ (d) $10x^9$

(e) $-1/x^2$ (f) $-2/x^3$ (g) $-6/x^7$ (h) $-20/x^{21}$

(2) (a) $5x^4$ (b) $7x^6$ (c) $-4/x^5$ (d) $-5/x^6$

(3) Tangent at A has equation $y = 3x - 2$. B is (–2, –8)

Tangent at B has equation $y = 12x + 16$

(4) Tangent at C: $y = -2x + 5$

Normal at C: $2y = x + 5$

D is $\left(\frac{-3}{2}, \frac{7}{4}\right)$

E is $\left(\frac{31}{20}, \frac{19}{10}\right)$

(5) (a) $-\frac{1}{2x^{3/2}}$ (b) $\frac{2}{(4x+3)^{1/2}}$

(6) (a) $2x - 5$ (b) $12x^3 - 6$ (c) $\frac{-3}{x^2}$ (d) $\frac{10}{x^3}$

(7) (a) $40\,x^{39}$ (b) $50\,x^4\,(2x^{15} - 16x^5 + 5)$

(c) $\frac{14x^5 + x^2 - 6}{x^4}$ (d) $9x^{1/2}$

(e) $\dfrac{2}{7x^{7/5}}$ (f) $\dfrac{(3x-1)(x+1)}{x^{3/2}}$

(g) $6x^2(x^3-3)$ (h) $\dfrac{(x-5)(x+5)}{x^2}$

(i) $\dfrac{x-4}{2x^{3/2}}$ (j) $\dfrac{x-12x^{1/3}+16}{3x^{5/3}}$

(8) (a) $\dfrac{2}{(6x-5)^{2/3}}$ (b) $\dfrac{3}{(6x-2)^{1/2}}$

(c) $\dfrac{-2}{(4x-1)^{3/2}}$ (d) $\dfrac{-3}{2x^{1/2}(x^{1/2}+2)^4}$

(e) $\dfrac{2x}{(3x^2+7)^{2/3}}$ (f) $\dfrac{2(3x+1)}{(6x^2+4x+3)^{1/2}}$

(g) $\dfrac{-10x-1}{(10x^2+2x+5)^{3/2}}$

(9) (a) $(x+3)^6(x+5)^9(17x+65)$ (b) $\dfrac{3(5x+2)^{1/2}(20x+27)}{2(3x+5)^{1/2}}$

(c) $\dfrac{x(5x^2+11)}{2(x^2+1)^{1/2}(x^2+3)^{2/3}}$ (d) $\dfrac{x^2(3x^3+14)}{(x^3+4)^{2/3}(x^3+6)^{1/3}}$

(e) $\dfrac{16x}{(x^2+4)^2}$ (f) $\dfrac{-20x-41}{2(2x+4)^{1/2}(5x+2)^{5/2}}$

(g) $\dfrac{36x}{(x^2+9)^2}$ (h) $\dfrac{1}{x^{1/2}(1-x^{1/2})^2}$

(10) (a) $4\cos 4x$ (b) $\dfrac{-3}{2}\sin\dfrac{3}{2}x$

(c) $20\sin^3 5x$ (d) $\dfrac{-3\sin 3x}{2\sqrt{\cos}\,3x}$

(e) $5\sin x\cos x(\sin^3 x-\cos^3 x)$ or $5\sin x\cos x(\sin x-\cos x)(1+\sin x\cos x)$

(f) 0

(g) $3\sin 12x(3\sin 6x+2)$ (h) $-16(2\cos^3 8x\sin 8x+\sin^3 4x\cos 4x)$

(11) (a) $8\sec^5 8x$ (b) $\sec 8x\tan 8x$

(c) $5\sec^5 x\tan x$ (d) $-40\operatorname{cosec}^2 5x\cot 5x$

(e) $\dfrac{\operatorname{cosec}^2 x}{2\cot^{1/2} x}$ (f) 0

(g) $\dfrac{\sec^2 x\tan x}{(\sec^2 x+2)^{1/2}}$ (h) $\dfrac{10\sec^2 5x\tan 5x}{(\sec^2 5x+2\tan^2 5x+3)^{2/3}}$

(14) (a) $x(2\cos x-x\sin x)$ (b) $\dfrac{\cos x}{2(1+\sin x)^{1/2}}$

(c) $3 \sin^2 x \cos x$ (d) $3x^2 \cos (x^3)$

(16) (a) $6e^{6x}$ (b) $-10\ e^{-10x}$

(c) $12\ x^2\ e^{4x^3}$ (d) $\dfrac{9x\,e^{\sqrt{(9x^2+16)}}}{\sqrt{(9x^2+16)}}$

(e) $-24\ x^2 \sin (8x^3 + 4)\ e^{\cos(8x^3+4)}$ (f) $\dfrac{-e^{-x^3}(3x^3+2)}{x^3}$

(g) $e^{-x} (3 \cos 3x - \sin 3x)$ (h) $4e^{4x} \sec 4x\ (1 + \tan 4x)$

(i) $e^{x^2} (\cos x + 2x \sin x)$ (j) $2x^3\ e^{-2x}\ (2 - x)$

(18) (a) $\dfrac{1}{x}$ (b) $\dfrac{6}{x}$

(c) $\dfrac{3x(x+2)}{x^3+3x^2+4}$ (d) $\dfrac{2(8-x^3)}{(x+2)(x^4+16)}$ or $\dfrac{2(2-x)(4+2x+x^2)}{(x+2)(x^4+16)}$

(e) $-2 \tan x$ (f) $3x^2 \cot (x^3)$

(g) $\dfrac{\cos x}{2(1+\sin x)}$ (h) $\dfrac{\sin x \cos x (\sin x - \cos x)}{\sin^3 x + \cos^3 x}$

(19) (a) $1 - \sec x$

(b) $\dfrac{dy}{dx} = 1 - \sec x < 0$ for $0 \le x < \dfrac{\pi}{2}$ so y is decreasing for $0 \le x < \dfrac{\pi}{2}$, but when $x = 0, y = 0$

hence $y < 0$ for $0 < x < \dfrac{\pi}{2}$

(20) $-6\ \text{ms}^{-2}$ at $t = 0$, $6\ \text{ms}^{-2}$ at $t = 6$

(21) (a) $\dfrac{dx}{dt} = \dfrac{-3}{(3t-2)^2} < 0$ for all t

(b) $t = \dfrac{5}{3}$ when acceleration is $\dfrac{2}{3}\ \text{ms}^{-2}$

$t = -\dfrac{1}{3}$ when acceleration is $-\dfrac{2}{3}\ \text{ms}^{-2}$

(22) (a) $v = 3\ (t - 1)^2 + 6 > 0$ for all t least $v = 6$

(b)

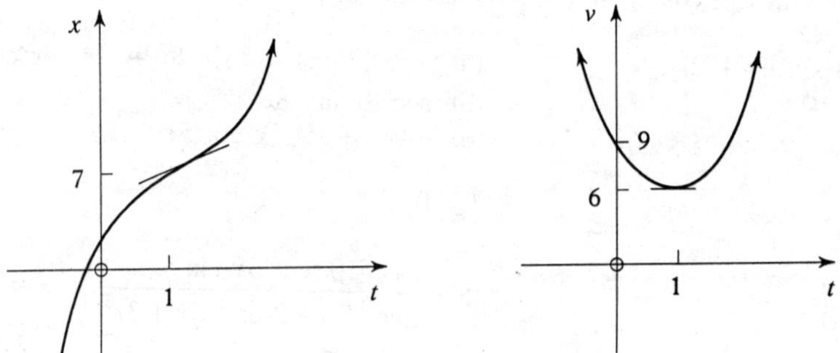

(a) (b)

Fig. 7.7.1

(23) -160 ms^{-2}

(24) (a) $-n^2X$ (c) (i) 4.2 m (ii) $\frac{8\pi}{5}\sqrt{5}$ or 11.2 seconds approx.

(25) $\left(\frac{2}{3}, 0\right), \left(\frac{-11}{12}, \frac{130320}{576}\right), \left(\frac{-5}{2}, 0\right)$

(26) Maximum turning point at (–2, 44)
Minimum turning point at (3, –81)

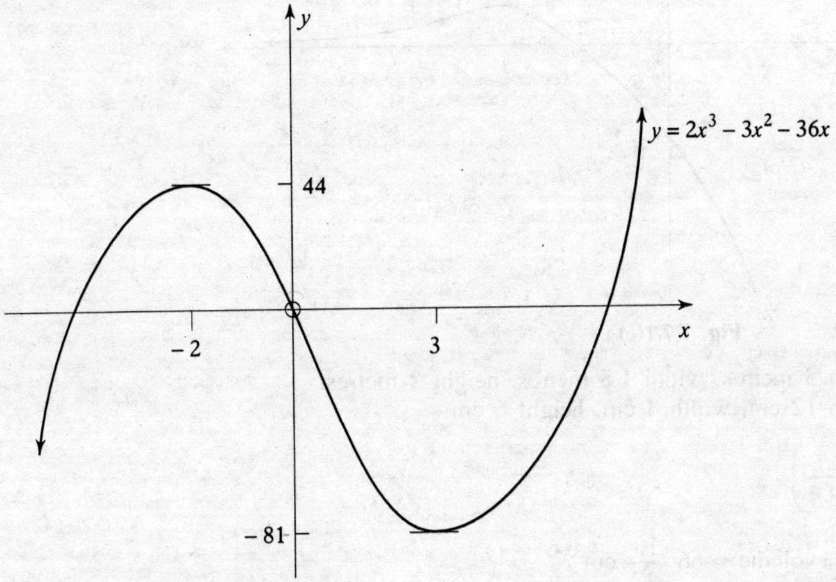

Fig. 7.7.1(c)

(27) Point of inflexion (0, 0)
Minimum turning point $\left(\frac{3}{2}, \frac{-27}{16}\right)$

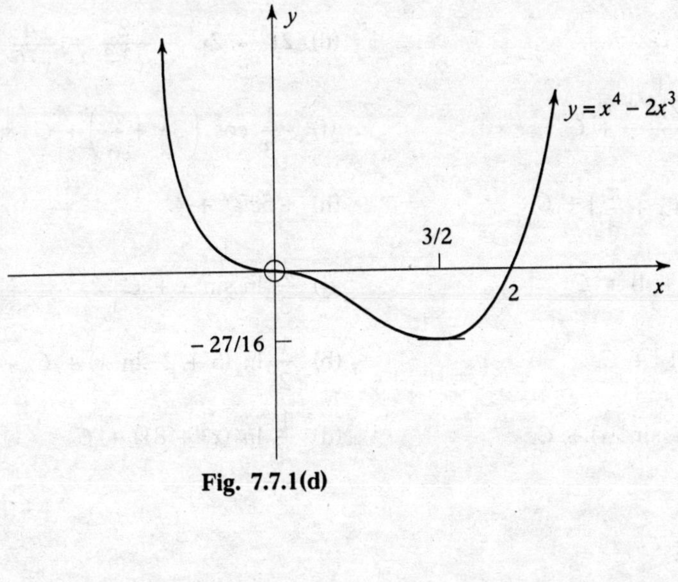

Fig. 7.7.1(d)

(28) (a) (0, –2), (2, 0)
(b) $(3, e^{-3})$ maximum turning point.
(c)

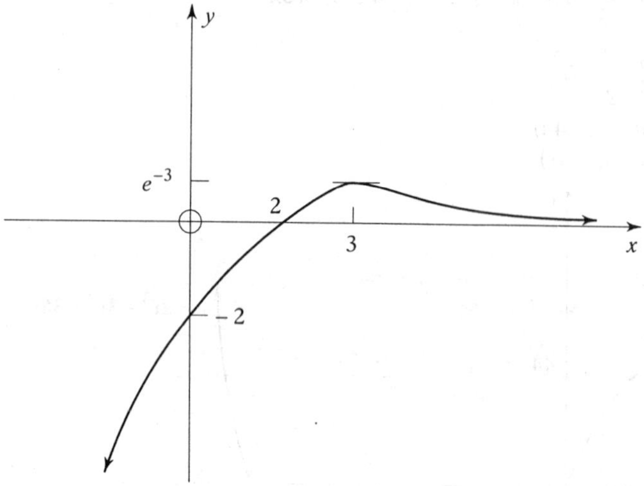

Fig. 7.7.1(e)

(29) Length 3 metres, width 1.5 metres, height 2 metres.
(30) Length 12 cm, width 4 cm, height 6 cm.

(31) $r = \left(\dfrac{A}{3\pi}\right)^{1/2}$

(32) $x = \dfrac{4}{3}$, volume $= 66\ \dfrac{10}{27}$ cm^3.

(33) $a = -9$, $b = 12$, equation of tangent: $18x - 4y + 19 = 0$

(34) (a) $\dfrac{1}{31}x^{31} + C$ (b) $-\dfrac{1}{24x^{24}} + C$

(c) $\dfrac{3x^{11/3}}{11} + C$ (d) $2x^5 - 2x^3 - \dfrac{2}{x^{3/2}} + \dfrac{4}{x^{7/2}} + C$

(e) $\dfrac{3(5x+12)^{4/3}}{4} + C$ (f) $-\dfrac{1}{3}\cos\left(3x + \dfrac{\pi}{6}\right) + C$

(g) $-3\cot\left(4x + \dfrac{\pi}{4}\right) + C$ (h) $-5e^{-5x} + C$

(i) $\dfrac{1}{3}\ln(3x + 4) + C$ (j) $\dfrac{1}{3}\ln\sin x + C$

(35) (a) $\ln|4x + 1| + C$ (b) $\dfrac{1}{2}\ln(5 + 2\sin x) + C$

(c) $\dfrac{2}{3}\ln(8 + \sin^3 x) + C$ (d) $\dfrac{1}{4}\ln(x^4 + 81) + C$

(e) $\frac{1}{3}\ \ln|x^3 + 3x^2 + 9x + 6| + C$ (f) $\frac{1}{3}\ \ln|4 - e^{-3x}| + C$

(g) $-\frac{1}{2}\ \ln|\cos 2x + 4| + C$ (h) $\ln(e^x + e^{-x}) + C$

(i) $\frac{1}{5}\ \ln|\tan^5 x + 1| + C$ (j) $\frac{1}{6}\ \ln(\sec^6 x + e^{6x}) + C$

(36) $\dfrac{2x+5}{x^2+5x+6}$

(37) $-\dfrac{\pi}{8}$

(38) $\dfrac{7}{3}$

(39) $\dfrac{57}{4}$

Part of the curve lies below the x-axis, so the integral represents area A – area B.

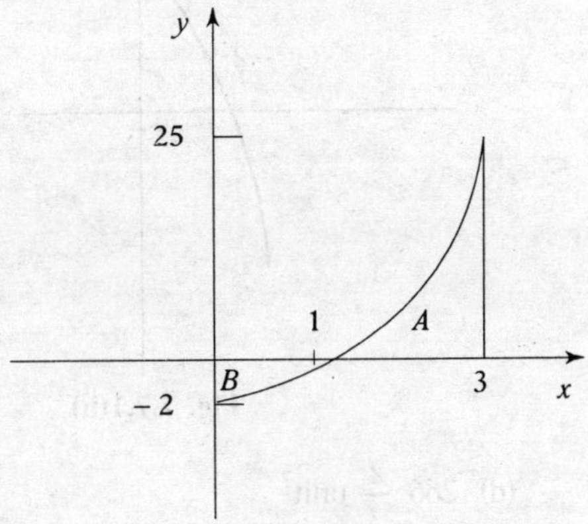

Fig. 7.7.1(f)

(40) $y = \frac{1}{2}x^2 - 4x + 9$

(41) $f(x) = \frac{1}{3}x^3 + 2x^2 - 12$

(42) (a) $\dfrac{132}{5}$ (b) $\dfrac{91}{24}$

(c) $\dfrac{-5}{12}$ (d) 60

(e) $\dfrac{1}{2}$ (f) 1 (g) $\dfrac{1}{4}\ln 2$ (h) $\dfrac{1}{2}\ln 3$

(43) (a)

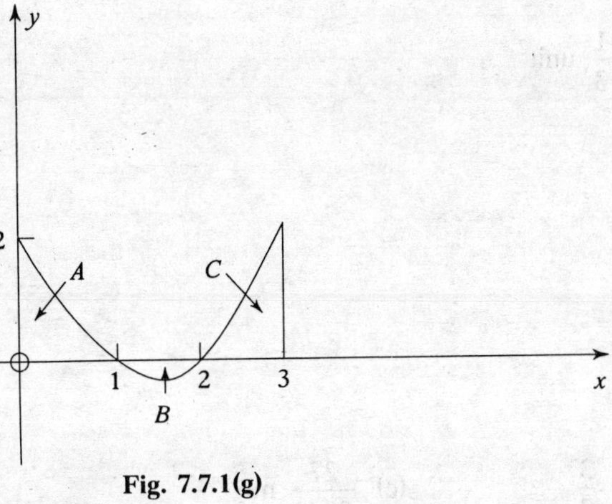

Fig. 7.7.1(g)

Part of the area (shown as B) lies below the x-axis so the given integral will not give the total numerical value of the area, as the integral give area A + area C – area B.

(b) $\frac{11}{6}$

(44) (a) (0, 20), (–2, 0), (10, 0) (b) (4, 36) maximum turning point

(c)

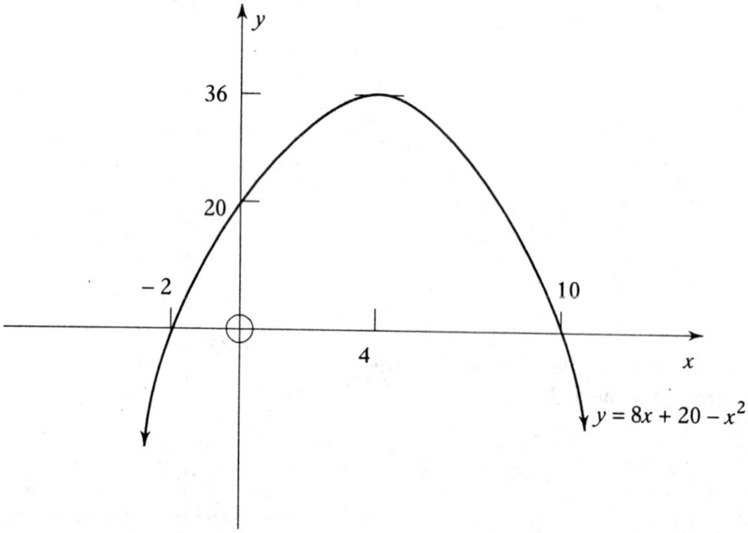

Fig. 7.7.1(h)

(d) $266\ \frac{2}{3}$ unit^2

(45) (a) (0, 0), (2, 8), (–2, –8) two areas of 4 unit^2 so 8 unit^2

(b) (0, 0), (4, 8) area 3.2 unit^2

(c) (0, 0), (4, 16) area $5\ \frac{1}{3}$ unit^2

(d) (2, 4), (–2, 4) area $21\ \frac{1}{3}$ unit^2

(46) (b) $\frac{1}{8}$

(47) $4\frac{1}{2}$

(48) $6\frac{1}{6}$

(49) $x = 5$

(50) (a) $V = 8\,t^3 - 3t^2$ (b) $t = \frac{3}{8}$ (c) $\frac{27}{2048}$ m

(51) (a) $t = 1$ sec, v has max $4\ ms^{-1}$

$t = 3$ sec, v has min $0\ ms^{-1}$

(b)

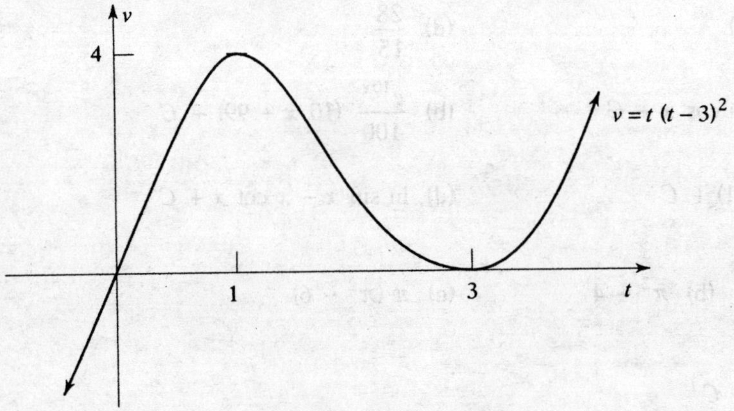

Fig. 7.7.1(i)

(c) $6\frac{3}{4}$ m

(52) $(9 - 6e^{-2\pi})\ ms^{-1}$, $\left[\frac{1}{6}\left(16\pi - 2\sqrt{3} - 4\right) + e^{-2\pi}\right]$m

(53) $V^2 = 24 + 8x - 2x^2$

$x = -2$ or 6 when the particle is at rest

(54) $x = 4$ metres

(56) (a) $\frac{1}{3} e^{x^3} + C$ (b) $e^{\tan x} + C$

(c) $\frac{1}{2} e^{(3+x)^2} + C$ (d) $\frac{1}{32} (2x^2 + 5)^8 + C$

(e) $\frac{1}{7} (3x^2 + 5)^{7/6} + C$ (f) $\frac{1}{5} \tan^5 x + C$

(g) $-\frac{1}{4} \cot^2 2x + C$ (h) $\frac{1}{16} \sin^4 4x + C$

(i) $-\frac{1}{48} \cos^8 6x + C$ (j) $\frac{1}{12} e^{\sin^3 4x} + C$

(57) (a) $\frac{3}{2} (\ln x)^2 + C$ (b) $\frac{1}{3} \ln (2 + 3 \ln x) + C$

(c) $\sin (\ln x) + C$ (d) $\frac{1}{2} \sqrt{(16 + e^{4x})} + C$

(62) $\frac{\pi}{4}$

(63) $\frac{\pi}{6}$

(65) (a) $\frac{1}{96}(10\pi + 9\sqrt{3})$ (b) $\frac{16}{35}$

(c) $\frac{1}{4}(2 \ln 2 - 1)$ (d) $\frac{28}{15}$

(66) (a) $(x + 4)\sin x + \cos x + C$ (b) $\frac{e^{10x}}{100}(10x + 99) + C$

(c) $\frac{x^9}{81}(9 \ln x - 1) + C$ (d) $\ln \sin x - x \cot x + C$

(68) 2

(69) (a) π (b) $\pi^2 - 4$ (c) $\pi(\pi^2 - 6)$

(70) $4e + 6$

(71) (c) $\ln\left(\tan \frac{1}{2}x\right) + C$